# Proteomics: Advanced Concepts and Perspectives

# Proteomics: Advanced Concepts and Perspectives

Edited by **Arthur Handley**

R CALLISTO REFERENCE

New York

Published by Callisto Reference,
106 Park Avenue, Suite 200,
New York, NY 10016, USA
www.callistoreference.com

**Proteomics: Advanced Concepts and Perspectives**
Edited by Arthur Handley

International Standard Book Number: 978-1-63239-526-9 (Hardback)

Printed in the United States of America.

# Contents

# Preface

Advances in the fields of biology and computational science during the past twenty years have opened new possibilities. One such discipline which has emerged is Omics fields. Due to a lack of correlation between mRNA and proteins, and growing significance of post transcriptional regulations and protein modifications in protein functions and human diseases, proteomics has become an increasingly important research field. Proteomics is the study of the proteome in the cell, which represents the complete set of proteins encoded by the genome. Proteomics is different and more complicated as compared to genomics. This is because while an organism's genome is more or less constant, the proteome differs across cells from time to time.

Protein research has expanded from its biochemical characterization of individual proteins, to the high throughput proteomics analysis of a cell, complex cell populations, and even an entire organism. This extremely important development highlights the potential of using proteomics methods in studies related to protein functions and human diseases. An important application of proteomics is in in the field of oncology, which is the discovery and validation of prognostic and predictive biomarkers, and plays a fundamental role in personalized therapy for cancer patients. Though Proteomics essentially focuses on proteins, it often finds its applications in areas such as protein purification and mass spectrometry.

I especially wish to acknowledge the contributing authors, without whom a work of this magnitude would clearly not have been realizable. I thank all the contributing authors for allocating much of their scarce time to this project. Not only do I appreciate their participation, but also their adherence as a group to the time parameters set for this publication.

I hope that this book proves to be a resourceful guide for both basic and advanced concepts in proteomics.

**Editor**

# High Mass Accuracy Phosphopeptide Identification Using Tandem Mass Spectra

**Rovshan G. Sadygov**

*Sealy Center for Molecular Medicine, Department of Biochemistry and Molecular Biology, The University of Texas Medical Branch, Galveston, TX 77555, USA*

Correspondence should be addressed to Rovshan G. Sadygov, rgsadygo@utmb.edu

Academic Editor: Qiangwei Xia

Phosphoproteomics is a powerful analytical platform for identification and quantification of phosphorylated peptides and assignment of phosphorylation sites. Bioinformatics tools to identify phosphorylated peptides from their tandem mass spectra and protein sequence databases are important part of phosphoproteomics. In this work, we discuss general informatics aspects of mass-spectrometry-based phosphoproteomics. Some of the specifics of phosphopeptide identifications stem from the labile nature of phosphor groups and expanded peptide search space. Allowing for modifications of Ser, Thr, and Tyr residues exponentially increases effective database size. High mass resolution and accuracy measurements of precursor mass-to-charge ratios help to restrict the search space of candidate peptide sequences. The higher-order fragmentations of neutral loss ions enhance the fragment ion mass spectra of phosphorylated peptides. We show an example of a phosphopeptide identification where accounting for fragmentation from neutral loss species improves the identification scores in a database search algorithm by 50%.

## 1. Introduction

The reversible phosphorylation of proteins regulates many aspects of cell life [1–3]. Phosphorylation and dephosphorylation, catalyzed by protein kinases and protein phosphatases, can change the function of a protein, for example, increase or decrease its biological activity, stabilize it or mark it for destruction, facilitate or inhibit movement between subcellular compartments, initiate or disrupt protein-protein interactions [1]. It is estimated that 30% of all cellular proteins are phosphorylated on at least one residue [4]. Abnormal phosphorylation is now recognized as a cause or consequence of many human diseases. Several natural toxins and tumor promoters produce their effects by targeting particular protein kinases [5, 6] and phosphatases. Protein kinases catalyze the transfer of the γ-phosphate from ATP to specific amino acids in proteins; in eukaryotes, these are usually Ser, Thr, and Tyr residues.

Mass-spectrometry-based proteomics has emerged as a powerful platform for the analysis of protein phosphorylations [7]. In particular, the shotgun proteomics [8], using liquid chromatography coupled with mass spectrometry (LC-MS), has been successfully employed for comprehensive analysis of global phosphoproteome [6, 9, 10]. The advances in the phosphoproteomics were driven by developments in mass spectrometry (high resolution and mass accuracy), peptide/protein separation, phosphopeptide/protein enrichment, peptide fragmentation [11, 12], quantification, and bioinformatics data processing, Figure 1. Currently, thousands of the phosphopeptides can be detected and quantified in just one experiment. Excellent recent reviews describe experimental procedures involved in phosphoproteomics [13, 14]. Bioinformatics processing is recognized as an integral part of phosphoproteome analysis. Several applications have been developed for phosphopeptide identifications [15, 16], phosphorylation site localization [17, 18], and quantification [19]. Tandem mass spectra are searched for phosphopeptides from protein sequences with potential modifications on Ser, Thr, and Tyr residues. The searches are not targeted. Every modifiable residues can be either modified or unmodified. The effective peptide search space increases exponentially leading to computational complexity

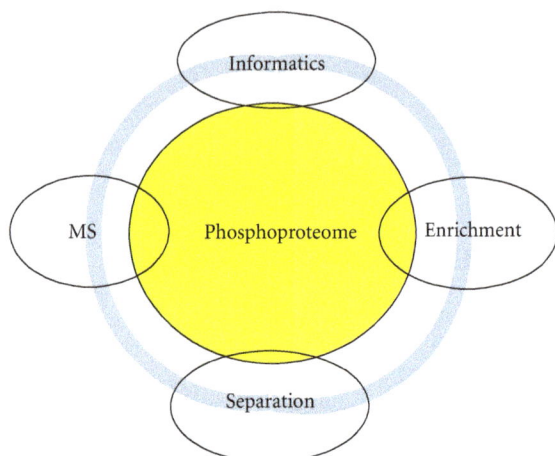

FIGURE 1: Phosphoproteomics and its constituent parts.

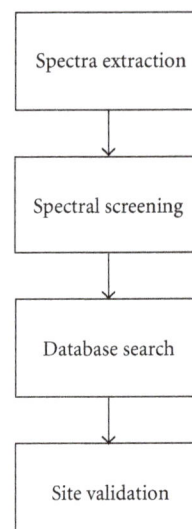

FIGURE 2: General informatics flowchart of a phosphoproteomics analysis.

as well as possible false identifications. High mass accuracy afforded by the modern mass spectrometers enables reducing the complexity of the search space by applying tighter bounds on peptide masses.

Lu and coworkers [20, 21] have developed models based on support vector machine (SVM) to screen for phosphopeptide spectra and validate their identifications. Their approach accurately explains spectra from phosphorylated peptides. However, SVM also acts like a black box, and it is difficult to gain insights into specifics of its decision making. Another development had used dynamic programming to relate spectra of modified and unmodified forms of a peptide [22]. This approach identifies modified peptides by comparing their tandem mass spectra with the annotated tandem mass spectra of unmodified peptides. The search space is restricted to peptides positively identified in unmodified form.

Here, we describe the informatics aspects of phosphopeptide identifications using protein sequence databases and mass spectral data from high mass accuracy and resolution instruments. Database identifications of phosphorylated peptides are done in a dynamic mode—assuming that in a peptide sequence Ser, Thr, and Tyr may or may not be are modified. For database searches, it effectively means exponential increase in the size of database. About 17% of amino acid residues (of which Ser 8.5%, Thr 5.7%, Tyr 3.0%) [23] in human proteome can potentially be phosphorylated. In general, if there are N amino acid residues which can potentially be phosphorylated, the effective database size could increase by as much as $2^N$ times.

## 2. Informatics Aspects of the Phosphoproteomics

### 2.1. Spectra Extraction. 
LTQ-Orbitrap mass spectrometer [24] stores the mass spectra in a proprietary "raw" file format (ThermoFisher Scientific, San Jose, CA). extract_msn algorithm extracts spectral information from the raw file and converts it into text file format for further data processing.

It uses a built-in module to evaluate isotopic envelope of mass species. From the isotope distribution, extract_msn determines the monoisotopic mass and charge state of a peptide. Both of these are critically important and used by database search algorithms.

Normally, the full MS scan is recorded in the Orbitrap mass analyzer which is a high resolution and mass accuracy mass analyzer. The routine mass accuracy of intact peptides is in the range of ±5–10 part-per-million (ppm). This is a very high mass accuracy and is very important for reducing false discovery rates of peptide identifications. The accuracy of the intact peptide's mass affects the number of candidate peptides from the database that will be considered in matching to the spectra. The candidate peptides are filtered based on the mass of the intact peptide and accuracy with which the mass has been measured. The higher the accuracy, the smaller the number of candidate peptides, and as a result the smaller the possibility of false positives. Fragment ion masses are recorded in ion trap mass analyzer. This is a very sensitive mass analyzer. However, the mass accuracy of measured ions is nominal, and normally in the range of ±0.5 Da. Figure 2 summarizes the informatics flowchart of a phosphoproteomics analysis.

### 2.2. Database Searching. 
Peptide identification using tandem mass spectra and protein databases is an integral part of proteomics. It is important that peptide assignments are determined with high accuracy and are verifiable. In high-throughput experiments, when thousands of tandem mass spectra are searched, it is not practical for an expert user to manually assign every spectrum and the assignments are made by software. The software uses a concept, either heuristic or probabilistic model, to measure similarity between experimental tandem mass spectrum and an amino acid sequence. For high quality spectra, when signal-to-noise ratio is high and spectra contains clearly defined ion

series, most of the programs and concepts perform very well. However, when the peptide fragmentation is poor and the spectrum contains very few distinguishable peaks, or if the peptide amino acid sequence is not in the database, a number of factors lead to a wrong peptide assignment—a false identification. Depending on the software, sometimes false identifications may yield high assignment scores. High mass accuracy may substantially reduce the false identification by restricting the search space of allowed candidate sequences.

Another important aspect of the database search algorithm is the modeling of the fragmentation pattern. This is especially true for cases when there is significant difference from the routine fragmentation pattern, caused, for example, by posttranslational modifications. Thus, it is known that, in CID, peptides containing phosphorylated amino acid residues, Ser or Thr, tend to lose the phosphor group(s) before they fragment along the peptide backbone. We show here that accounting for product ions in this pathway significantly improves the identification scores of phosphopeptides in SEQUEST database search algorithm.

*2.3. Assignment of Phosphorylation Sites.* It is often difficult to differentiate between possible phosphorylation sites in a peptide and to uniquely assign phosphorylated amino acid residues. It has to do with the presence of several Ser, Thr, and Tyr residues in a peptide and low overall intensity of peaks from product ions of phosphorylated peptides. To address this problem, Beausoleil and coworkers [17] have developed a probability-based approach to determine phosphorylation sites of peptides from the results of SEQUEST database search algorithm [16]. Their model divides the spectra into the mass intervals of equal width. In every interval, only 6 to 8 (dependent on the intensity) peaks are retained and the rest of the peaks are ignored. In matching to the experimental peaks, only the modified fragments are considered. The probability of phosphorylation site determination is estimated via a binomial probability. This model has been successful in many practical applications especially for high scoring peptides.

# 3. Discussions

There are several mass spectral characteristics of phosphorylated peptides. Thus, in collision-induced dissociation reactions, one of the most prevalent pathways of the molecular dissociation is the neutral loss of the labile phosphor group of Ser or Thr residues. The presence of the relevant ion often serves as a diagnostic feature for phosphorylated peptides [25]. Often the product ions in spectra include two ion series, one from the phosphorylated peptide and the other from the precursor peptide that has lost the phosphor group(s). We have previously modified SEQUEST database search algorithm to account for the two ion sequences when identifying phosphorylated peptides [26]. The development has helped to improve the sensitivity of the phosphor peptide identifications and location of phosphorylation sites. Here, we demonstrate, in the example of this algorithm, the

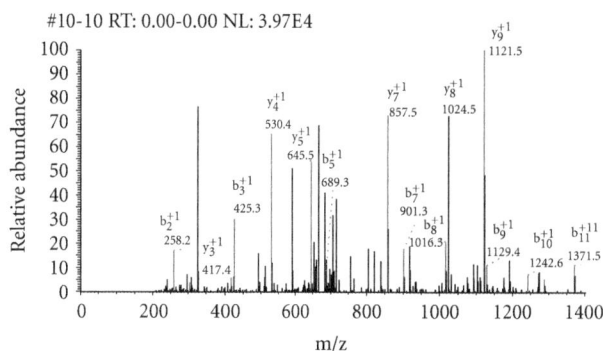

FIGURE 3: An example spectrum of phosphorylated peptide, R.TRS*PS*PDDILER.V. The cross-correlation scores before and after SEQUEST fragment ion modifications are 4.97 and 3.14, respectively.

advantages of using extensive fragmentation pathways for targeted analysis.

In general, fragmentation of peptides via CID produces strong b- and y-ion series [27]. Therefore, most database search algorithms generate b- and y-ions (and corresponding water and ammonia losses from them) for the theoretical spectrum of a candidate amino acid sequence. In the tandem mass spectra of phosphorylated peptides, additional fragmentation patterns are observed. In addition to the b- and y-ions of the original phosphorylated peptide, ions that originate from phosphor group losses, 98 Da (from Ser or Thr residues), are also present. We accounted for these fragments by augmenting the fragmentation pattern correspondingly to add neutral loss fragments from the phosphorylated amino acid residues. The fragmentation model [28] was used for both, preliminary and cross-correlation scores in SEQUEST. Cross-correlation scores of phosphorylated peptides generated from new fragmentation pattern were about 50% higher.

The interpretations of the tandem mass spectra of phosphorylated peptides may be complicated. The main reason for this is the low abundance of fragment ions due to the alternative fragmentations. One experimental approach used for enhancing phosphopeptide identifications is to do a higher-order mass spectrometry on the fragment ions of original precursor. In these experiments, neutral loss fragment ions generated during CID in $MS^2$ are further dissociated generating $MS^3$ spectra. In spectra collected with this approach, there are number of ions corresponding to phosphoric acid group losses from b- and y-ions. An example of such a tandem mass spectrum is shown in Figure 3. SEQUEST matched this spectrum to the phosphorylated peptide sequence, R.TRS*PS*PDDILER.V. The cross-correlation and preliminary scores in the model not including phosphoric acid loss fragmentations were 3.14 and 1029.6, respectively. When we included the neutral loss ions, the cross-correlation and preliminary scores were 4.97 and 2353.6, respectively. The total number of theoretical ions generated for this peptide was 44. 35 of these ions matched to product ions in the spectrum. In contrast, 16 of the 22 theoretical ions matched the tandem mass spectrum in the original model (ignoring neutral loss fragments). The results on this and other phosphorylated peptide spectra showed that a realistic model of product ions of phosphorylated peptides needs to account for the fragments resulting from neutral (phosphoric acid group) loss of the b- and y-ions. The procedure has been automated and is used in the case of dedicated $MS^n$ experiments to enhance fragmentation spectra of phosphopeptides.

## 4. Conclusion

Increased mass accuracy for precursor ions combined with enhanced fragmentation pathways helps bioinformatics methods to improve phosphopeptide identifications from tandem mass spectra and protein sequence databases. Normally identifications of phosphorylated peptides yield small cross-correlation scores. This has partially to do with the theoretical fragmentation models, which take into account only b- and y-ions generated from the peptide bond fragmentations of phosphorylated precursor peptides. We augmented the fragmentation pattern (in SEQUEST) [26] introducing theoretical peaks for b- and y-ions from neutral loss precursors and fragments. Cross-correlation scores of phosphorylated peptides increased by up to 50% using the enhanced fragmentation model.

## Abbreviations

Da: Dalton
ppm: Parts per million
Ser: Serine
Thr: Threonine
Tyr: Tyrosine
CID: Collision induced dissociation
LC: Liquid chromatography
MS: Mass spectrometry.

## Acknowledgments

This work was supported, in part, by UL1RR029876 UTMB CTSA (ARB), HHSN272200800048C NIAID Clinical Proteomics Center (ARB), and NIH-NLBIHHSN2682010000037C NHLBI Proteomics Center for Airway Inflammation (Alex Kurosky, UTMB).

## References

[1] P. Cohen, "The origins of protein phosphorylation," *Nature Cell Biology*, vol. 4, no. 5, pp. E127–E130, 2002.
[2] J. A. Ubersax and J. E. Ferrell Jr., "Mechanisms of specificity in protein phosphorylation," *Nature Reviews Molecular Cell Biology*, vol. 8, no. 7, pp. 530–541, 2007.
[3] G. Manning, D. B. Whyte, R. Martinez, T. Hunter, and S. Sudarsanam, "The protein kinase complement of the human genome," *Science*, vol. 298, no. 5600, pp. 1912–1934, 2002.
[4] P. Cohen, "The regulation of protein function by multisite phosphorylation—a 25 year update," *Trends in Biochemical Sciences*, vol. 25, no. 12, pp. 596–601, 2000.
[5] J. N. Andersen, S. Sathyanarayanan, A. Di Bacco et al., "Pathway-based identification of biomarkers for targeted therapeutics: personalized oncology with PI3K pathway inhibitors," *Science Translational Medicine*, vol. 2, no. 43, Article ID 43ra55, 2010.
[6] A. Moritz, Y. Li, A. Guo et al., "Akt-RSK-S6 kinase signaling networks activated by oncogenic receptor tyrosine kinases," *Science Signaling*, vol. 3, no. 136, article ra64, 2010.
[7] H. Zhou, J. D. Watts, and R. Aebersold, "A systematic approach to the analysis of protein phosphorylation," *Nature Biotechnology*, vol. 19, no. 4, pp. 375–378, 2001.
[8] A. J. Link, J. Eng, D. M. Schieltz et al., "Direct analysis of protein complexes using mass spectrometry," *Nature Biotechnology*, vol. 17, no. 7, pp. 676–682, 1999.
[9] E. L. Huttlin, M. P. Jedrychowski, J. E. Elias et al., "A tissue-specific atlas of mouse protein phosphorylation and expression," *Cell*, vol. 143, no. 7, pp. 1174–1189, 2010.
[10] B. Zhai, S. A. Beausoleil, J. Mintseris, and S. P. Gygi, "Phosphoproteome analysis of Drosophila melanogaster embryos," *Journal of Proteome Research*, vol. 7, no. 4, pp. 1675–1682, 2008.
[11] J. J. Coon, B. Ueberheide, J. E. P. Syka et al., "Protein identification using sequential ion/ion reactions and tandem mass spectrometry," *Proceedings of the National Academy of Sciences of the United States of America*, vol. 102, no. 27, pp. 9463–9468, 2005.
[12] M. P. Jedrychowski, E. L. Huttlin, W. Haas, M. E. Sowa, R. Rad, and S. P. Gygi, "Evaluation of HCD- and CID-type fragmentation within their respective detection platforms for murine phosphoproteomics," *Molecular & Cellular Proteomics*, vol. 10, no. 12, article M111, 2011.
[13] F. Wang, C. Song, K. Cheng, X. Jiang, M. Ye, and H. Zou, "Perspectives of comprehensive phosphoproteome analysis using shotgun strategy," *Analytical Chemistry*, vol. 83, no. 21, pp. 8078–8085, 2011.
[14] C. L. Nilsson, "Advances in quantitative phosphoproteomics," *Analytical Chemistry*, vol. 84, no. 2, pp. 735–746, 2012.
[15] B. E. Ruttenberg, T. Pisitkun, M. A. Knepper, and J. D. Hoffert, "PhosphoScore: an open-source phosphorylation site assignment Tool for MS$^n$ data," *Journal of Proteome Research*, vol. 7, no. 7, pp. 3054–3059, 2008.
[16] J. K. Eng, A. L. McCormack, and J. R. Yates III, "An approach to correlate tandem mass spectral data of peptides with amino acid sequences in a protein database," *Journal of the American Society for Mass Spectrometry*, vol. 5, no. 11, pp. 976–989, 1994.
[17] S. A. Beausoleil, J. Villén, S. A. Gerber, J. Rush, and S. P. Gygi, "A probability-based approach for high-throughput protein phosphorylation analysis and site localization," *Nature Biotechnology*, vol. 24, no. 10, pp. 1285–1292, 2006.
[18] T. Taus, T. Kocher, P. Pichler et al., "Universal and confident phosphorylation site localization using phosphoRS," *Journal of Proteome Research*, vol. 10, no. 12, pp. 5354–5362, 2011.
[19] J. Cox, I. Matic, M. Hilger et al., "A practical guide to the MaxQuant computational platform for SILAC-based quantitative proteomics," *Nature protocols*, vol. 4, no. 5, pp. 698–705, 2009.
[20] B. Lu, C. Ruse, T. Xu, S. K. Park, and J. Yates III, "Automatic validation of phosphopeptide identifications from tandem mass spectra," *Analytical Chemistry*, vol. 79, no. 4, pp. 1301–1310, 2007.
[21] B. Lu, C. I. Ruse, and J. R. Yates III, "Colander: a probability-based support vector machine algorithm for automatic screening for CID spectra of phosphopeptides prior to database

search," *Journal of Proteome Research*, vol. 7, no. 8, pp. 3628–3634, 2008.

[22] D. Tsur, S. Tanner, E. Zandi, V. Bafna, and P. A. Pevzner, "Identification of post-translational modifications by blind search of mass spectra," *Nature Biotechnology*, vol. 23, no. 12, pp. 1562–1567, 2005.

[23] N. Echols, P. Harrison, S. Balasubramanian et al., "Comprehensive analysis of amino acid and nucleotide composition in eukaryotic genomes, comparing genes and pseudogenes," *Nucleic Acids Research*, vol. 30, no. 11, pp. 2515–2523, 2002.

[24] Q. Hu, R. J. Noll, H. Li, A. Makarov, M. Hardman, and R. G. Cooks, "The Orbitrap: a new mass spectrometer," *Journal of Mass Spectrometry*, vol. 40, no. 4, pp. 430–443, 2005.

[25] B. Lu, D. B. McClatchy, Y. K. Jin, and J. R. Yates III, "Strategies for shotgun identification of integral membrane proteins by tandem mass spectrometry," *Proteomics*, vol. 8, no. 19, pp. 3947–3955, 2008.

[26] R. G. Sadygov, J. Shofstahl, and A. Humer, "Improvements to the database search algorithm SEQUEST for accurate mass support and improved phosphorylation searching," in *Proceedings of the Annual Conference on Mass Spectrometry and Allied Topics*, 2005.

[27] B. Paizs and S. Suhai, "Fragmentation pathways of protonated peptides," *Mass Spectrometry Reviews*, vol. 24, no. 4, pp. 508–548, 2005.

[28] R. G. Sadygov, F. M. Maroto, and A. F. R. Hühmer, "ChromAlign: a two-step algorithmic procedure for time alignment of three-dimensional LC-MS chromatographic surfaces," *Analytical Chemistry*, vol. 78, no. 24, pp. 8207–8217, 2006.

# 2

# Identification of a Novel Biomarker for Biliary Tract Cancer Using Matrix-Assisted Laser Desorption/Ionization Time-of-Flight Mass Spectrometry

**Shintaro Kikkawa,[1] Kazuyuki Sogawa,[2] Mamoru Satoh,[2] Hiroshi Umemura,[3] Yoshio Kodera,[2, 4] Kazuyuki Matsushita,[3] Takeshi Tomonaga,[2, 5] Masaru Miyazaki,[6] Osamu Yokosuka,[1] and Fumio Nomura[2, 3]**

[1] Department of Medicine and Clinical Oncology, Graduate School of Medicine, Chiba University, 1-8-1 Inohana, Chuo-ku, Chiba, Chiba City 260-8670, Japan
[2] Clinical Proteomics Center, Chiba University Hospital, 1-8-1 Inohana, Chuo-ku, Chiba, Chiba City 260-8670, Japan
[3] Department of Molecular Diagnosis, Graduate School of Medicine, Chiba University, 1-8-1 Inohana, Chuo-ku, Chiba, Chiba City 260-8670, Japan
[4] Department of Physics, School of Science, Kitasato University, 1-15-1 Kitasato, Minami-ku, Kanagawa, Sagamihara City 228-8555, Japan
[5] Laboratory of Proteome Research, National Institute of Biomedical Innovation, 7-6-8 Saito Asagi, Osaka, Ibaraki City 567-0085, Japan
[6] Department of General Surgery, Graduate School of Medicine, Chiba University, 1-8-1 Inohana, Chuo-ku, Chiba, Chiba City 260-8670, Japan

Correspondence should be addressed to Fumio Nomura, fnomura@faculty.chiba-u.jp

Academic Editor: Terence C. W. Poon

Early diagnosis of biliary tract cancer (BTC) is important for curative surgical resection. Current tumor markers of BTC are unsatisfactory in terms of sensitivity and specificity. In a search for novel biomarkers for BTC, serum samples obtained from 62 patients with BTC were compared with those from patients with benign biliary diseases and from healthy controls, using the MALDI-TOF/TOF ClinProt system. Initial screening and further validation identified a peak at 4204 Da with significantly greater intensity in the BTC samples. The 4204 Da peak was partially purified and identified as a fragment of prothrombin by amino acid sequencing. The sensitivity of the 4204 Da peptide for detection of stage I BTC cancer was greater than those for CEA and CA19-9. Also, serum levels of the 4204 Da peptide were above the cut-off level in 15 (79%) of 19 cases in which the CEA and CA19-9 levels were both within their cut-off values. Receiver operating characteristic analysis showed that the combination of the 4204 Da peptide and CA19-9 was significantly more sensitive for detection of stage I BTC cancer compared to CEA and CA19-9. These results suggest that this protein fragment may be a promising biomarker for biliary tract cancer.

## 1. Introduction

Biliary tract cancer (BTC) is a neoplasm that accounts for 3% of all gastrointestinal cancers and 15% of all primary liver cancers. Over the last two decades, the incidence of BTC has risen, mainly due to an increase in the intrahepatic form [1, 2], which has a particularly high incidence in Northern Thailand [3]. Surgical resection is the only curative treatment and this requires an early diagnosis. Even in cases in which surgical resection with negative histological margins is achieved, the 5-year survival rates range from 20% to 40% [4, 5]. The mean one-year survival rate for unresectable cases is only 6 months [4]. Therefore, there is a need to establish a tool for early diagnosis of BTC. Currently, diagnosis of

Identification of a Novel Biomarker for Biliary Tract Cancer Using Matrix-Assisted Laser Desorption/Ionization Time-of-Flight Mass Spectrometry

7

TABLE 1: Clinical characteristics of patients with biliary tract cancer or benign biliary disease and healthy volunteers.

| Healthy volunteers | Benign biliary disease | Biliary tract cancer | |
| --- | --- | --- | --- |
| No. of patients | 30 | 30 | 62 |
| Male/female | 18/12 | 18/12 | 36/26 |
| Mean age | 65.5 ± 4.5 | 64.4 ± 39.0 | 64.7 ± 37.4 |
| CEA (ng/mL) | 3.1 ± 3.5 | 2.5 ± 3.6 | 37.6 ± 1577.3 |
| CA19-9 (U/mL) | 15.3 ± 18.6 | 70.4 ± 603.9 | 5033.6 ± 190367.2 |

BTC depends on imaging of the biliary tree using computed tomography (CT), ultrasonography, and endoscopic retrograde cholangiography (ERC) in symptomatic subjects. Brush cytology during ERC can lead to morphological diagnosis, but the sensitivity is limited because of the highly desmoplastic reaction of BTC [5, 6]. For these reasons, tumor markers that can detect BTC with high diagnostic efficiency are urgently needed. Carcinoembryonic antigen (CEA) and carbohydrate antigen 19.9 (CA19-9) are tumor markers that are used for diagnosis of BTC, but their sensitivity and specificity are unsatisfactory [2, 7].

Proteome analysis is increasingly being applied to cancer biomarker discovery. Surface enhanced laser desorption/ionization time-of-flight mass spectrometry (SELDI-TOF MS) is a proteomics technique used for high-throughput fingerprinting of serum proteins [8]. We have used this technology to identify diagnostic markers for alcohol abuse [9] and a prognostic marker for pancreatic cancer [10]. SELDI-TOF MS can be used to analyze many samples rapidly and simultaneously, but has drawbacks of high cost and difficulty with protein identification. More recently, high-throughput workflow with matrix-assisted laser desorption/ionization-time of flight/time of flight-mass spectrometry (MALDI-TOF/TOF MS) has been established for discovery and identification of serum peptides [11]. This method uses magnetic beads with different chemical chromatographic surfaces, instead of ProteinChip arrays. Proteins selectively bound to the magnetic beads are eluted and analyzed by MALDI-TOF/TOF MS. Compared with the SELDI-TOF MS ProteinChip system, the cost is low, and subsequent protein identification is relatively easy. We recently used the ClinProt system for MALDI-TOF/TOF MS to detect novel biomarkers for alcohol abuse that could not be detected using SELDI [12]. In the present study, we carried out a serum peptidome study to identify novel biomarkers for biliary tract cancer using the MALDI-TOF/TOF MS ClinProt system.

## 2. Methods

*2.1. Patients and Samples.* Serum samples were obtained from 62 patients with BTC (36 males, 26 females; median age 64.7 years old, range 27–81 years old), 30 age-matched healthy controls (18 males, 12 females; median age 65.5 years old, range 61–69 years old), and 30 age-matched patients with benign biliary disease (18 males, 12 females; median age 64.4 years old, range 27–90 years old). Clinicopathological data for all the subjects are shown in Tables 1, 2, and 3. The

TABLE 2: Characteristics of patients with biliary tract cancer.

| Item | Number of patients |
| --- | --- |
| Location | |
| Extrahepatic | 16 |
| Intrahepatic | 17 |
| Klatskin | 7 |
| Ampulla of Vater | 6 |
| Gallbladder | 16 |
| UICC stage | |
| Stage I | 6 |
| Stage II | 10 |
| Stage III | 16 |
| Stage IV | 30 |

TABLE 3: Characteristics of patients with benign biliary disease ($n = 30$).

| Item | Number of patients |
| --- | --- |
| Cholelithiasis | 26 |
| Benign fibrous stricture | 2 |
| Primary sclerosing cholangitis | 2 |

BTC group (Table 2) included cases of intrahepatic cholangiocarcinoma ($n = 17$), Klatskin tumor ($n = 7$), extrahepatic cholangiocarcinoma ($n = 16$), tumor of the ampulla of Vater ($n = 6$), and gallbladder tumor ($n = 16$). The pathological stages of the BTC patients were defined according to the Union Internationale Contre le Cancer tumor node metastasis classification [13]. The patients with benign biliary diseases (Table 3) included cases of cholelithiasis ($n = 24$), benign fibrous stricture ($n = 4$), and primary sclerosing cholangitis ($n = 2$). All the cases of BTC were diagnosed by radiological imaging. In 58 cases, cytology was also compatible with the diagnosis. All of the patients with benign biliary disease were diagnosed by endoscopic retrograde cholangiopancreatography and were followed up for more than 12 months to confirm that they had no malignancy. Serum samples were obtained and processed under standardized conditions that we have described elsewhere [14] and were stored at $-80°C$ until analysis. Written informed consent was obtained from all the subjects. The study was approved by the Ethics Committee of Chiba University School of Medicine.

*2.2. Serum Pretreatment with Magnetic Beads Using the ClinProt Robot.* We used weak cation exchange (WCX) magnetic beads (Bruker Daltonics) and performed serum peptidome fractionation according to the manufacturer's protocol. A $5\,\mu L$ serum sample was mixed with $10\,\mu L$ of binding buffer to which $5\,\mu L$ of WCX beads was added, and the solution was carefully mixed. The peptides in the serum were then allowed to bind to the WCX beads for 5 min. The tube was then placed in a magnetic bead separator (Bruker Daltonics) for separating unbound beads, and the supernatant was removed. The beads were washed three times with $100\,\mu L$ of washing buffer, and the proteins as well as peptides were then eluted from the magnetic beads with $10\,\mu L$ each of elution and stabilization buffer. Thereafter, $2\,\mu L$ of peptide elution solution was mixed with $20\,\mu L$ of alpha-cyano-4-hydroxy-cinnamic acid matrix (Bruker Daltonics). Then $0.8\,\mu L$ of this mixture was spotted onto an AnchorChip target plate (Bruker Daltonics) and crystallized. Each sample was duplicated, and quadruplicate spotting was performed using each eluate; eight spots were developed from each sample. The mean spectra from these eight spots were used for data analyses. These procedures from bead fractionation to spotting were performed automatically using the ClinProt robot (Bruker Daltonics) under strictly controlled humidity, as we previously described [14].

### 2.3. Mass Spectrometry.
*2.3. Mass Spectrometry.* The AnchorChip target plate was placed in an AutoFlex II TOF/TOF mass spectrometer (Bruker Daltonics) controlled by Flexcontrol 2.4 software (Bruker Daltonics). The instrument was equipped with a 337 nm nitrogen laser, delayed-extraction electronics, and a 25 Hz digitizer. All acquisitions were generated by an automated method included in the instrument software and based on averaging of 1000 randomized shots. Spectra were acquired in positive linear mode in the mass range of 600–10000 Da. Peak clusters were completed using second pass peak sections (signal to noise ratio > 5). The relative peak intensities of $m/z$ between 600 and 10000 normalized to a total ion current were expressed in arbitrary units. Calibration was performed using Peptide Calibration Standard II (Bruker Daltonics). All MALDI-TOF MS spectra from $m/z$ 1000 to 10000 were analyzed with FlexAnalysis 2.1 and ClinProtools 2.1 software (Bruker Daltonics).

*2.4. Protein Identification.* A CM ceramic Hyper DF spin column (Bio-Rad Laboratories, Irvine, CA, USA) was washed 3 times with $400\,\mu L$ of MB-WCX binding solution (Bruker Daltonics). Serum samples ($320\,\mu L$) were diluted 5-fold with binding buffer and the diluted sample ($1600\,\mu L$) was applied to the spin column. The sample was allowed to bind at $4°C$ for 1 h on a shaker and then the spin column was washed 3 times with $400\,\mu L$ of binding buffer. Finally, $320\,\mu L$ of MB-WCX stabilization solution (Bruker Daltonics) was added to the spin column for elution. Ten volumes of ice cold acetone were added to the eluate. Peptides/proteins were allowed to precipitate at $-20°C$ for 2 h and then obtained by centrifugation at 13000 g for 10 min at $4°C$. After decanting the acetone, the peptides/proteins were allowed to air dry. The dried pellets were resuspended in buffer (0.1% trifluoroacetic

TABLE 4: Discriminatory peaks and *P* values in the training set.

| Higher in biliary tract cancer group | | Lower in biliary tract cancer group | |
|---|---|---|---|
| *m/z* | *P* value | *m/z* | *P* value |
| 1207 | <0.0001 | 1944 | <0.0001 |
| 1466 | <0.0001 | 2669 | <0.001 |
| 3261 | <0.001 | 2931 | <0.0001 |
| 3950 | <0.001 | 3239 | <0.0001 |
| 4202 | <0.001 | 3272 | <0.001 |
| 4635 | <0.001 | 3878 | <0.01 |
| 4654 | <0.001 | 4051 | <0.001 |
| 5791 | <0.0001 | 4086 | <0.0001 |
| 5890 | <0.0001 | 4276 | <0.0001 |
| 5929 | <0.001 | 6414 | <0.001 |
| 9246 | <0.001 | | |
| 9285 | <0.0001 | | |

acid in water, vol/vol) and further separated by reversed-phase HPLC in an automated HPLC system (Shiseido Nanospace SI-2, Shiseido Fine Chemicals, Tokyo, Japan). The concentrated flow-through sample ($75\,\mu L$) was directly loaded onto an Intrada WP-RP column (Imtakt, Kyoto, Japan). The reversed-phase separations for each flow-through fraction were performed using a multisegment eluk tion gradient with eluent A (0.1% trifluoroacetic acid in water, vol/vol) and eluent B (0.08% trifluoroacetic acid in 90% acetonitrile, vol/vol). The gradient elution program consisted of three steps with increasing concentrations of eluent B (5% B for 5 min, 5% to 95% B for 23 min, and 95% B for 11 min) followed by 5% B for 21 min for reequilibration of the column at a flow rate of 0.40 mL/min for a total run time of 60 min. Based on the chromatogram recorded by measuring the absorbance of the eluate at 280 nm, fractions eluted at retention times between 19.1 and 39.1 min were collected in 40 0.2 mL aliquots at a fraction size setting of 0.5 min. Fractions including objective peaks were confirmed by MALDI-TOF MS. N-terminal amino acid sequence analysis was performed using a Procise 494 cLC protein sequencing system (Applied Biosystems, Foster City, CA, USA).

*2.5. Statistical Analysis.* Univariate analysis of individual peaks was performed using a nonparametric Mann-Whitney *U* test, with $P < 0.05$ considered significant. Discriminatory power for putative markers was further evaluated by receiver operating characteristic (ROC) analysis and the area under the curve (AUC) using IBM SPSS Statistics 18 (SPSS Inc., Ill, USA).

## 3. Results

*3.1. MALDI-TOF-MS Analysis of Peptides in BTC Sera.* As a first step, we compared the peptide profiles of serum samples obtained from BTC patients ($n = 30$) with those from healthy controls ($n = 12$) as a training set (Table 4). Totally 134 peaks were detected and compared in the MALDI

Identification of a Novel Biomarker for Biliary Tract Cancer Using Matrix-Assisted Laser Desorption/Ionization Time-of-Flight Mass Spectrometry

9

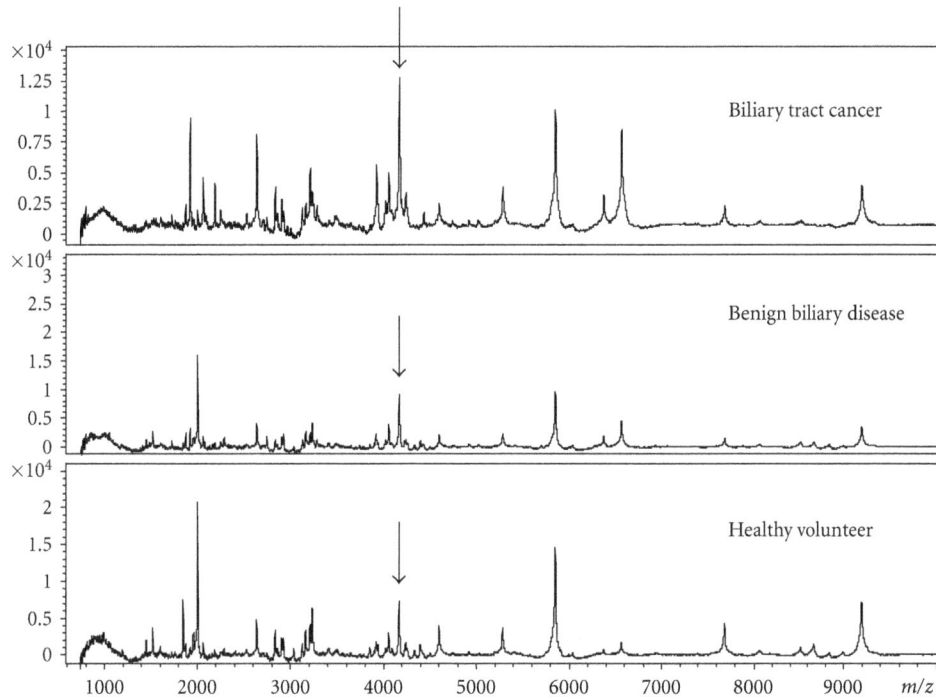

FIGURE 1: The protein mass profile between $m/z$ 0 and 10000 highlighting the differentially expressed peaks in serum from healthy volunteers, patients with benign biliary disease, and BTC patients. The $m/z$ 4204 peak (indicated by arrows) intensity was higher in cancer patients compared with patients with benign disease and healthy volunteers.

proteomic profile. Total of 22 peak intensities differed significantly between the BTC patients and healthy controls, including 12 that were higher and 10 that were lower in the BTC group. In typical spectra for serum samples from each group (Figure 1), the intensity of the 4204 $m/z$ peak was higher in BTC samples compared with those from patients with benign biliary diseases and from healthy controls.

Next, we tested whether the differences observed in the 22 peaks in the training set were reproducible in another set of samples (test set) (Table 5). 32 BTC patients, 30 benign biliary patients, and 18 healthy controls were included in the test set. Out of these 22 peaks, the intensities of 2 peaks (3272 $m/z$ and 4204 $m/z$) were again significantly different between the BTC and control groups. Out of these 2 peaks, the intensity of one peak (4204 $m/z$) was also found to be significantly higher in the BTC group compared to the benign disease group. The relative intensities of the 4204 $m/z$ peak in sera obtained from the three groups of subjects are summarized in Figure 2.

### 3.2. Identification of the 4204 Da Peptide as a Fragment of Prothrombin.
Partial purification of the peptide corresponding to the 4204 $m/z$ peak was conducted as outlined in Section 2. N-terminal amino acid sequencing of trypsin digests of the final preparation containing the 4204 Da peptide revealed that it was a fragment of prothrombin (Figure 3).

### 3.3. Diagnostic Value of the 4204 Da Peptide Compared with Conventional Markers.
Patients with BTC were divided into 4 groups based on clinical stage. The sensitivities of CEA,

TABLE 5: Discriminatory peaks detected in the training set and test set.

| Higher in biliary tract cancer group | | Lower in biliary tract cancer group | |
|---|---|---|---|
| $m/z$ | $P$ value | $m/z$ | $P$ value |
| 4202 | <0.001 | 3272 | <0.001 |

CA19-9, and 4204 Da in the BTC patients were determined (Figure 4). The optimal cut-off point for CEA, CA19-9, and the 4204 Da peptide were selected based on mean + 2SD in healthy subjects. The cut-off levels for CEA, CA19-9, and the 4204 Da peptide were set at 6.4 ng/mL, 33.5 U/mL, and 372.1 AU, respectively. The sensitivities of CEA, CA19-9, and the 4204 Da peptide in stage IV patients were 33.3%, 80.0%, and 66.7%, and the specificities of CEA, CA19-9, and the 4204 Da peptide were 93.3%, 93.3%, 96.7%, respectively. In contrast, these sensitivities in stage I patients were 0.0%, 16.7%, and 50.0%, and these specificities were 93.3%, 93.3%, 96.7%. The sensitivity of the 4204 Da peptide was also greater than those of CEA and CA19-9 in stage II patients.

The ROC curves for the 4204 Da peptide, CEA, and CA19-9 as single markers and combinations are shown in Figure 5. The sensitivities were determined from the results for the 62 patients with BTC and specificities were based on the 60 non-BTC subjects. The AUCs for the 4204 Da peptide, CEA, and CA19-9 as single markers were 0.75, 0.60, and 0.732, respectively. The AUC for the combination of the 4204 Da peptide and CA19-9 was significantly greater than

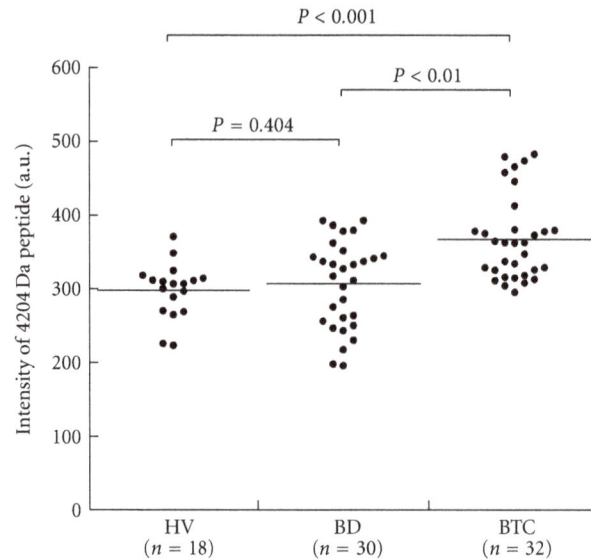

FIGURE 2: Normalized intensities of the peak corresponding to the 4204 Da peptide in serum from healthy volunteers ($n = 18$), patients with benign biliary disease ($n = 30$), and BTC patients ($n = 32$). The peak intensity was significantly higher in sera obtained from patients with biliary tract cancer (BTC) compared with sera from healthy volunteers. There was no significant difference in intensity between BTC sera and benign biliary disease (BB) sera (Mann-Whitney $U$ test).

Prothrombin fragment detected in this study

```
MAHVRGLQLP GCLALAALCS LVHSQHVFLA PQQARSLLQR VRRANTFLEE VRKGNLEREC
VEETCSYEEA FEALESSTAT DVFWAKYTAC ETARTPRDKL AACLEGNCAE GLGTNYRGHV
NITRSGIECQ LWRSRYPHKP EINSTTHPGA DLQENFCRNP DSSTTGPWCY TTDPTVRRQE
CSIPVCGQDQ VTVAMTPRSE GSSVNLSPPL EQCVPDRGQQ YQGRLAVTTH GLPCLAWASA
QAKALSKHQD FNSAVQLVEN FCRNPDGDEE GVWCYVAGKP GDFGYCDLNY CEEAVEEETG
DGLDEDSDRA IEGRTATSEY QTFFNPRTFG SGEADCGLRP LFEKKSLEDK TERELLESYI
DGRIVEGSDA EIGMSPWQVM LFRKSPQELL CGASLISDRW VLTAAHCLLY PPWDKNFTEN
DLLVRIGKHS RTRYERNIEK ISMLEKIYIH PRYNWRENLD RDIALMKLKK PVAFSDYIHP
VCLPDRETAA SLLQAGYKGR VTGWGNLKET WTANVGKGQP SVLQVVNLPI VERPVCKDST
RIRITDNMFC AGYKPDEGKR GDACEGDSGG PFVMKSPFNN RWYQMGIVSW GEGCDRDGKY
GFYTHVFRLK KWIQKVIDQF GE
```

FIGURE 3: N-terminal amino acid sequence of the purified fraction. Red letters identify the N-terminal sequence. The molecular weight of the underlined region is 4204 Da.

that for CEA and CA19-9 ($P < 0.01$). The sensitivity and specificity for combination of the 4204 Da and CA19-9 were 59.8% and 84.0%.

The 62 patients with BTC were also classified into 8 groups based on their tumor marker status, as shown in Table 6. The cut-off values for CEA and CA19-9 were set at 5 ng/mL and 37 U/mL, respectively. The optimal cut-off point for the 4204 Da peptide was selected based on the ROC analysis. The 4204 Da peptide level was greater than the cut-off value in 15 (79%) of 19 cases in which the CEA and CA19-9 levels were within their respective cut-off values.

## 4. Discussion

The sequencing of the human genome has opened the door for comprehensive analysis of all mRNAs (transcriptome) and proteins (proteome). However, the levels of mRNAs are not necessarily predictive of the corresponding protein levels. Indeed, a recent report indicated that the consistency

TABLE 6: Positive or negative status of 4204 Da peptide, CEA, and CA19-9 in patients with biliary tract cancer.

| CEA (≥5 ng/mL) | CA19-9 (≥37 U/mL) | 4204 Da peptide (≥322 A.U.) | Number of patients |
|---|---|---|---|
| − | − | − | 4 |
| − | − | + | 15 |
| − | + | − | 6 |
| + | − | − | 2 |
| + | + | − | 3 |
| + | − | + | 2 |
| − | + | + | 18 |
| + | + | + | 12 |

between cDNA microarray and proteome-based profiles is limited for identification of candidate biomarkers in renal cell carcinoma [15]. Therefore, proteome analysis is a prerequisite for identification of novel biomarkers.

Identification of a Novel Biomarker for Biliary Tract Cancer Using Matrix-Assisted Laser Desorption/Ionization
Time-of-Flight Mass Spectrometry

11

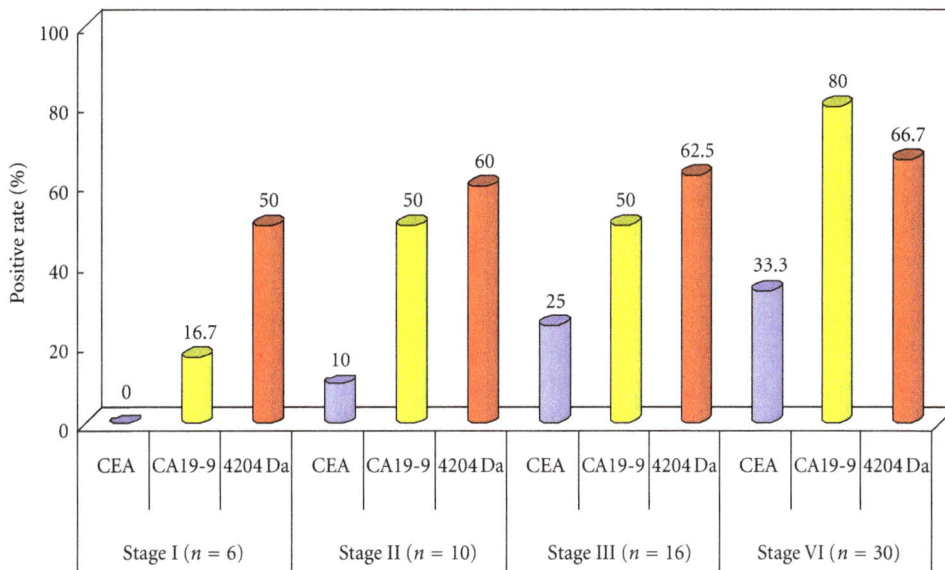

FIGURE 4: Positive rates of detection of the 4204 Da peptide, CEA, and CA19-9 in each UICC stage.

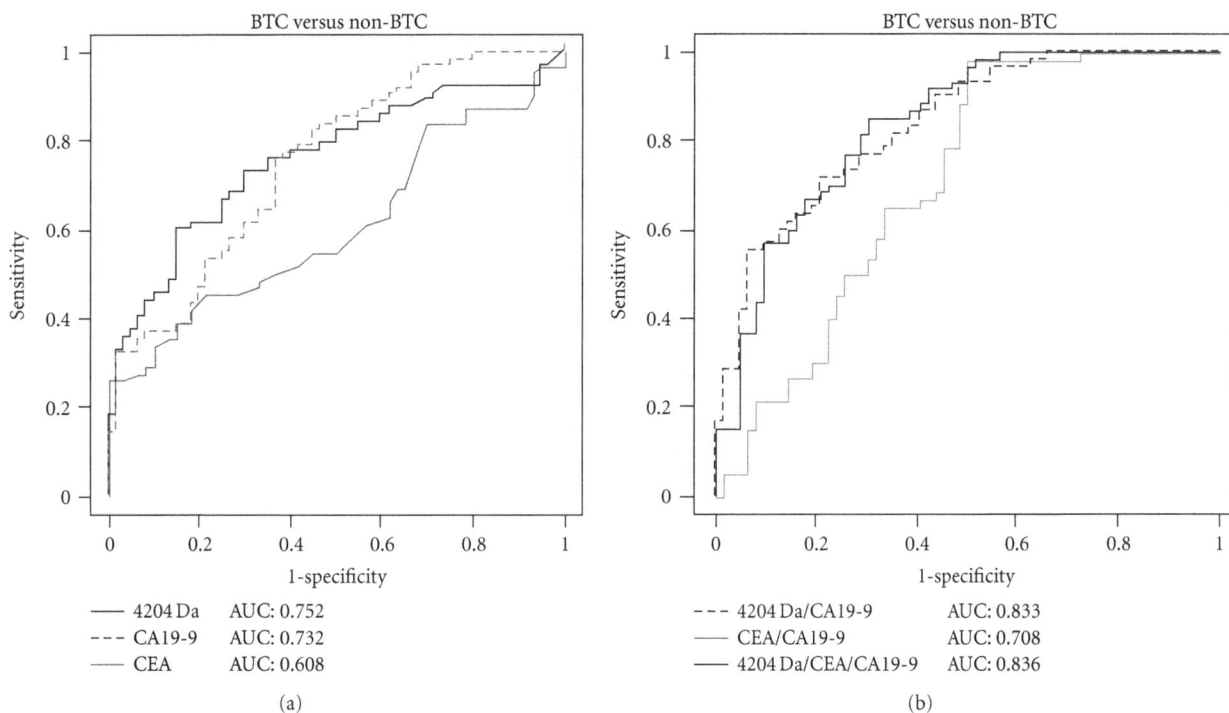

FIGURE 5: ROC analyses of the performance of the 4204 Da peptide, CA19-9, and CEA. (a) AUC was 0.752 for 4204 Da, 0.732 for CA19-9, and 0.608 for CEA. (b) AUC was 0.833 for 4204 Da + CA19-9, 0.708 for CA19-9 + CEA, and 0.836 for 4204 Da + CEA + CA19-9.

Biliary tract cancer is a particularly lethal malignancy with a mean 1-year survival of only 6% for unresectable cases [4]. The lack of a sensitive and specific biomarker for early detection of BTC is one of the reasons for this limited survival. Cholangiocarcinoma often grows along the bile duct without forming a mass, and thus is often missed in CT and ultrasound. Serum biomarkers with satisfactory sensitivity and specificity are likely to be beneficial in the clinical management of this malignancy. There have been previous attempts to discover biomarkers for cholangiocarcinoma. Scarlett et al. conducted proteomic profiling of sera from cases of cholangiocarcinoma using SELDI-TOF MS and found that a serum peptide corresponding to a 4463 $m/z$ peak had superior discriminatory ability to CA19-9 and CEA, but did not identify the peak [16]. More recently, a membrane protein enrichment strategy coupled with $^{18}O$ labeling-based quantitative proteomics was used to identify proteins that are highly expressed in cholangiocarcinoma tissues [17]. Golgi membrane protein, annexin IV, and epidermal growth factor were proposed as candidate markers. However, their diagnostic roles at the serum level were not described.

CEA and CA19-9 are tumor markers for BTC with average sensitivity and specificity for detecting cholangiocarcinoma of 51% and 88%, respectively, for CEA, and 71% and 78%, respectively, for CA19-9 [7]. In the present study, the sensitivities of CEA, CA19-9, and the 4204 Da peptide for detection of all BTC cases were 50%, 61.3%, and 75.8%, respectively. It was of note that the sensitivity of 4204 Da in stages I and II patients was far greater than those of the conventional markers and that serum 4204 Da peptide levels were elevated in 79% of cases in which both CA19-9 and CEA were within their reference intervals. These findings suggest that this novel peptide is complementary to conventional markers in diagnosis of BTC. This is supported by the greater AUC with the combination of CEA, CA19-9, and the 4204 Da peptide, compared to individual AUCs.

The result obtained in identification of the 4204 Da peptide was unexpected. The peptide was identified as a fragment of prothrombin, which makes it unlikely that the fragment originated from cancer tissues. It is possible that production of the fragment occurred in the cancer-tissue microenvironment. Alternatively, the 4204 Da peptide might have been generated ex vivo by undefined degradative proteases during the clotting process [18]. The exact mechanism for production of the 4204 Da peptide remains to be clarified. We also note that intrahepatic cholangiocellular carcinoma, extrahepatic cholangiocellular carcinoma, and gall bladder carcinoma were analyzed together as BTCs in the present study. Separate analyses of these diseases on a larger scale are needed to discover biomarkers that are more specific for each form of BTC. Also, antibody-based verification will be necessary to further confirm the findings obtained in this study. Bile samples may also be an alternative for discovery of disease markers leaking from the biliary tree [19].

## Abbreviations

| | |
|---|---|
| AU: | Arbitrary unit |
| BTC: | Biliary tract cancer |
| MALDI-TOF MS: | Matrix-associated laser desorption ionization time-of-flight mass spectrometry |
| CEA: | Carcinoembryonic antigen |
| CA19-9: | Carbon hydrate antigen 19.9 |
| ROC: | Receiver operation characteristic |
| AUC: | Area under the curve. |

## Authors' Contribution

These authors contributed equally to the work.

## Acknowledgments

The authors thank Fumie Iida and Manami Miura for their technical support.

## References

[1] S. A. Khan, H. C. Thomas, B. R. Davidson, and S. D. Taylor-Robinson, "Cholangiocarcinoma," *Lancet*, vol. 366, no. 9493, pp. 1303–1314, 2005.

[2] T. Patel, "Cholangiocarcinoma," *Nature Clinical Practice Gastroenterology and Hepatology*, vol. 3, no. 1, pp. 33–42, 2006.

[3] Y. Shaib and H. B. El-Serag, "The epidemiology of cholangiocarcinoma," *Seminars in Liver Disease*, vol. 24, no. 2, pp. 115–125, 2004.

[4] W. R. Jarnagin and M. Shoup, "Surgical management of cholangiocarcinoma," *Seminars in Liver Disease*, vol. 24, no. 2, pp. 189–199, 2004.

[5] G. J. Gores, "Cholangiocarcinoma: current concepts and insights," *Hepatology*, vol. 37, no. 5, pp. 961–969, 2003.

[6] E. M. Abu-Hamda and T. H. Baron, "Endoscopic management of cholangiocarcinoma," *Seminars in Liver Disease*, vol. 24, no. 2, pp. 165–175, 2004.

[7] O. Nehls, M. Gregor, and B. Klump, "Serum and bile markers for cholangiocarcinoma," *Seminars in Liver Disease*, vol. 24, no. 2, pp. 139–154, 2004.

[8] H. J. Issaq, T. P. Conrads, D. A. Prieto, R. Tirumalai, and T. D. Veenstra, "SELDI-TOF MS for diagnostic proteomics," *Analytical Chemistry*, vol. 75, no. 7, pp. 148A–155A, 2003.

[9] F. Nomura, T. Tomonaga, K. Sogawa et al., "Identification of novel and downregulated biomarkers for alcoholism by surface enhanced laser desorption/ionization-mass spectrometry," *Proteomics*, vol. 4, no. 4, pp. 1187–1194, 2004.

[10] S. Takano, H. Yoshitomi, A. Togawa et al., "Apolipoprotein C-1 maintains cell survival by preventing from apoptosis in pancreatic cancer cells," *Oncogene*, vol. 27, no. 20, pp. 2810–2822, 2008.

[11] J. Villanueva, J. Philip, D. Entenberg et al., "Serum peptide profiling by magnetic particle-assisted, automated sample processing and MALDI-TOF mass spectrometry," *Analytical Chemistry*, vol. 76, no. 6, pp. 1560–1570, 2004.

[12] K. Sogawa, M. Satoh, Y. Kodera, T. Tomonaga, M. Iyo, and F. Nomura, "A search for novel markers of alcohol abuse using magnetic beads and MALDI-TOF/TOF mass spectrometry," *Proteomics*, vol. 3, no. 7, pp. 821–828, 2009.

[13] L. H. Sobin and I. D. Fleming, "TNM classification of malignant tumors, 5th edition (1997)," *Union Internationale Contre le Cancer and American Joint Committee on Cancer*, vol. 80, no. 9, pp. 1803–1804, 1997.

[14] H. Umemura, M. Nezu, Y. Kodera et al., "Effects of the time intervals between venipuncture and serum preparation for serum peptidome analysis by matrix-assisted laser desorption/ionization time-of-flight mass spectrometry," *Clinica Chimica Acta*, vol. 406, no. 1-2, pp. 179–180, 2009.

[15] B. Seliger, S. P. Dressler, E. Wang et al., "Combined analysis of transcriptome and proteome data as a tool for the identification of candidate biomarkers in renal cell carcinoma," *Proteomics*, vol. 9, no. 6, pp. 1567–1581, 2009.

[16] C. J. Scarlett, A. J. Saxby, A. Nielsen et al., "Proteomic profiling of cholangiocarcinoma: diagnostic potential of SELDI-TOF MS in malignant bile duct stricture," *Hepatology*, vol. 44, no. 3, pp. 658–666, 2006.

[17] T. Z. Kristiansen, H. C. Harsha, M. Grønborg, A. Maitra, and A. Pandey, "Differential membrane proteomics using 18O-labeling to identify biomarkers for cholangiocarcinoma," *Journal of Proteome Research*, vol. 7, no. 11, pp. 4670–4677, 2008.

[18] J. Villanueva, D. R. Shaffer, J. Philip et al., "Differential exoprotease activities confer tumor-specific serum peptidome

patterns," *Journal of Clinical Investigation*, vol. 116, no. 1, pp.
271–284, 2006.

[19] A. Farina, J. M. Dumonceau, and P. Lescuyer, "Proteomic
analysis of human bile and potential applications for cancer
diagnosis," *Expert Review of Proteomics*, vol. 6, no. 3, pp. 285–
301, 2009.

# 3

# A Comprehensive Subcellular Proteomic Survey of *Salmonella* Grown under Phagosome-Mimicking versus Standard Laboratory Conditions

**Roslyn N. Brown,[1] James A. Sanford,[2] Jea H. Park,[1] Brooke L. Deatherage,[1] Boyd L. Champion,[1] Richard D. Smith,[1] Fred Heffron,[3] and Joshua N. Adkins[1]**

[1] *Biological Sciences Division, Pacific Northwest National Laboratory, 902 Battelle Boulevard, Richland, WA 99352, USA*
[2] *Biomedical Sciences Graduate Program, University of California San Diego, 9500 Gilman Dive, La Jolla, CA 92063, USA*
[3] *Department of Molecular Microbiology and Immunology, Oregon Health and Science University, 3181 SW Sam Jackson Park Road, Portland, OR 97239, USA*

Correspondence should be addressed to Joshua N. Adkins, joshua.adkins@pnnl.gov

Academic Editor: Gary B. Smejkal

Towards developing a systems-level pathobiological understanding of *Salmonella enterica*, we performed a subcellular proteomic analysis of this pathogen grown under standard laboratory and phagosome-mimicking conditions *in vitro*. Analysis of proteins from cytoplasmic, inner membrane, periplasmic, and outer membrane fractions yielded coverage of 25% of the theoretical proteome. Confident subcellular location could be assigned to over 1000 proteins, with good agreement between experimentally observed location and predicted/known protein properties. Comparison of protein location under the different environmental conditions provided insight into dynamic protein localization and possible moonlighting (multiple function) activities. Notable examples of dynamic localization were the response regulators of two-component regulatory systems (e.g., ArcB and PhoQ). The DNA-binding protein Dps that is generally regarded as cytoplasmic was significantly enriched in the outer membrane for all growth conditions examined, suggestive of moonlighting activities. These observations imply the existence of unknown transport mechanisms and novel functions for a subset of *Salmonella* proteins. Overall, this work provides a catalog of experimentally verified subcellular protein locations for *Salmonella* and a framework for further investigations using computational modeling.

## 1. Introduction

The pursuit of a systems-level understanding of bacterial physiology requires not only knowledge about the identity, function, and relative abundance of proteins, but also insight into the subcellular localization of these proteins. Subcellular protein localization is linked to protein function, potential protein-protein interactions, and to interactions between a cell and its exterior environment. The observation of proteins in unexpected cellular compartments gives clues about the presence of possible alternate functions. Hence, there is a growing appreciation for the presence of bacterial "moonlighting proteins," that is, those proteins that have a secondary function depending on subcellular location

[1–3]. Experimentally verified localization also provides a foundation for describing proteins that are "hypothetical," uncharacterized, or that contain domains of unknown function. Furthermore, with the increasing use of systems biology approaches, including genome-scale models of metabolism [4] and regulation to study microbial functions, experimentally founded protein localization on a global scale is necessary to produce more accurate model constraints.

Subcellular proteomics has emerged as a powerful tool for large-scale profiling of protein subcellular location [5–9]. Unlike traditional Western blot or high-resolution microscopy methods that rely on the use of antibodies or molecular tags to identify individual proteins, proteomic methods enable high-throughput, unbiased, and large-scale

A Comprehensive Subcellular Proteomic Survey of Salmonella Grown under Phagosome-Mimicking versus
Standard Laboratory Conditions

15

identification of the protein complement of subcellular fractions [5, 6, 10]. Moreover, interrogation of the subcellular proteome under different growth or environmental conditions allows for the investigation of changes in protein abundance and possibly protein location.

Subcellular proteomic analysis of bacterial pathogens holds promise for identifying novel virulence determinants and potential therapeutic targets [11–13]. For Gram-negative pathogens such as *Salmonella enterica*, each of the four main protein-containing compartments—the outer and inner membranes, periplasm, and cytoplasm—is a potential source of virulence determinants. Outer membrane/cell surface proteins mediate adhesion, cell-cell communication, immune evasion, sequestration, transport (including antibiotic efflux), and secretion, whereas inner membrane proteins accomplish transport and assembly of complex structures, such as flagella and secretion apparati. Periplasmic proteins sense and respond to the host environment, and cytoplasmic proteins include secretion substrates, chaperones, and housekeeping proteins important in maintaining the pathogenic lifestyle. Comprehensive characterization of these subcellular fractions can provide insight into the potential for virulence-related interactions with the host as well as fundamental information on the subcellular architecture of this organism.

Our present goals were twofold: (1) to survey the localization of proteins in *Salmonella* cells as a reference of protein localization in this bacterium and (2) to observe changes in protein abundance or location upon growth under phagosome-mimicking conditions relative to standard laboratory conditions to generate new biological insights, as well as improved data for computational modeling. Towards this end, cytoplasmic (CYT), inner membrane (IM), periplasmic (PERI), and outer membrane (OM) fractions were analyzed using liquid chromatography-tandem mass spectrometry (LC-MS/MS). We did not analyze the secretome as we recently completed an extensive analysis of the proteins secreted by *Salmonella* under phagosome-mimicking conditions [14]. In the present study, over 25% of the theoretical *Salmonella* proteome was represented, and confident assignment of subcellular locations was achieved for most proteins. In addition, we assigned subcellular-level localization to the response of the bacteria to growth under conditions that mimic the host macrophage intracellular environment. This study represents the most comprehensive global survey of subcellular localization in *Salmonella* to date and affords a resource to others interested in protein location, improving location predictions and systems computational models.

## 2. Methods

*2.1. Rationale for Media and Strains Used in This Study.* Growth to mid-logarithmic phase in Luria-Bertani broth represents a standard laboratory growth condition in this study and is noninducing for *Salmonella* pathogenicity island 2 (SPI-2) gene expression [15]. Growth of *Salmonella* in defined, acidic media with low concentrations of phosphate and magnesium induces expression of SPI-2 genes that are

required for intracellular survival and replication [15–19]. mLPM has been shown to induce expression and secretion of SPI2-related virulence factors [14] and was used in this study to mimic the environment of a macrophage phagosome.

We previously identified flagellin (especially FliC) as one of the most abundant proteins secreted by *Salmonella* into culture media [14] and also in cell envelope fractions (Supplemental Table 1, supplementary material available online at doi: no# 10.1155/2012/123076). *Salmonella* flagellins are downregulated during the intracellular stage of infection, and SPI-2-expressing bacteria are not motile [20]. Since flagella are not relevant to the stage of infection we intended to mimic, we deleted flagellin genes *fliC* and *fljB* from wildtype *Salmonella enterica* serovar Typhimurium (*S.* Typhimurium) ATCC 14028 in an attempt to achieve better sensitivity by depleting these abundant proteins.

*2.2. Bacterial Strains, Media, and Chemicals.* Bacteria were maintained in LB broth (Difco, Franklin Lakes, NJ, USA) or on LB plates. Unless otherwise noted, components of mLPM [14] and other chemicals were purchased from Sigma (St. Louis, MO, USA). Protein concentrations were determined by bicinchoninic acid (BCA) assay (Pierce, Rockford, IL, USA) using bovine serum albumin as standards. Trypsin used for protein digestions was purchased from Promega (Madison, WI, USA).

*2.3. Deletion of Flagellin Genes.* In an attempt to achieve better sensitivity by depleting a nonessential abundant protein (Supplemental Table 1), a double-flagellin mutant ($\Delta fliC\Delta fljB$) was created using $\lambda$ Red recombination [21]. *fliC* was deleted using oligos FliC P1: AGCCCA-ATAACATCAAGTTGTAATTGATAAGGAAAAGAT-CGTGTAGGCTGGAGCTGCTTC and FliC P2: CCTTGA-TTGTGTACCACGTGTCGGTGAATCAATCGCCGG-ACATATGAATATCCTCCTTAG.

For deleting *fljB*, oligos FljB P1: GATTTTCTC-CTTTACATCAGATAAGGAAGAATTTTAGTC-GGTGTAGGCTGGAGCTGCTTC and FljB P2: CTC-GCCCGTAGGAAATATCATTTACAGCCATACATTCCA-TCATATGAATATCCTCCTTAG were used. Underlined portions of the above oligos represent pKD4 sequences. Insertion of the kanamycin resistance cassette was confirmed using oligos FliC test1: AATGATGAAATTGAAGCCAT and K1: CAGTCATAGCCGAATAGCCT for *fliC* and using FljB test1: AACGCCACCAGGTTTTTCAC and K1 for *fljB*. The kanamycin resistance gene was removed using pCP20 as previously described [21]. The flagellin mutant was tested for lack of motility, compared to the wildtype, using 0.4% agar plates.

*2.4. Subcellular Fractionation.* Overnight starter cultures of WT and the $\Delta fliC\Delta fljB$ mutant were grown in LB broth at 37°C with shaking at 200 rpm. The cultures were diluted 1 : 100 into LB and grown to mid-log phase (OD600 ∼ 0.6) for the "LB-log" condition or diluted 1 : 10 into mLPM and grown for 4 or 20 h for "LPM4" and "LPM20," respectively.

The cell fractionation protocol was adapted from that described by Brown et al. [9]. Unless otherwise noted, centrifugation steps were performed at 4°C. Cells were collected via centrifugation (10,000 ×g, 10 min) and washed with 10 mL of 50 mM Tris-HCl (pH 8.0). PERI fractions were generated by suspending cell pellets in 10 mL spheroplasting buffer (50 mM Tris-HCl, pH 8, 250 mM sucrose, 2.5 mM EDTA) and incubating at room temperature for 5 min, after which they were centrifuged at 11,500 ×g for 10 min. Pellets were then suspended in 1.3 mL cold 5 mM $MgSO_4$ and kept on ice for 10 min with occasional mixing. After centrifugation (11,500 ×g, 10 min), the supernatant was retained as the soluble PERI fraction, while the pelleted spheroplasts were suspended in 1.0 mL 20 mM $NaH_2PO_4$.

Half of the spheroplasts from each condition were then used to perform fractionation into CYT, IM, and OM fractions. The volumes were adjusted to 3.0 mL in 20 mM $NaH_2PO_4$ and lysed by passing three times through a prechilled French Press (8,000 PSI). Cell lysate suspensions were adjusted to 10 mL using 20 mM $NaH_2PO_4$ and centrifuged at 5,000 ×g for 30 min to pellet unbroken cells. Supernatants were then centrifuged at 45,000 ×g for 60 min to separate the soluble CYT fraction from the crude membrane pellet. The CYT fractions were centrifuged again to remove residual membrane contaminants. After tubes containing membrane pellets were inverted to dry, the pellets were suspended in 10 mL 20 mM $NaH_2PO_4$ containing 0.5% Sarkosyl and shaken at 200 rpm for 30 min at room temperature. This mixture was then centrifuged at 45,000 ×g for 60 min to pellet the OM fraction, and the supernatant containing the IM fraction was removed. OM fractions were washed once by suspending in 5 mL $NaH_2PO_4$ and repeating the centrifugation.

*2.5. Tryptic Digests.* Tryptic digests of the soluble CYT and PERI fractions were prepared as follows. To 75 μg of protein from each sample, urea and DTT were added to final concentrations of 7 M and 5 mM, respectively, followed by incubation at 60°C for 30 min. Samples were then diluted 7-fold with 100 mM $NH_4HCO_3$, and $CaCl_2$ was added to a final concentration of 1 mM. Trypsin was then added in a 1 : 50 trypsin : protein ratio, and digestions were performed at 37°C with shaking at 600 rpm for 3 hours. Following digestion, samples were cleaned using 1 mL, 50 mg Discovery DSC-18 solid phase extraction (SPE) columns (Supelco, St. Louis, MO, USA). Briefly, each column was conditioned with methanol and then rinsed with 0.1% TFA in water. Digested samples were run through the columns under vacuum and rinsed with 95 : 5 $H_2O$ : ACN with 0.1% TFA. Excess liquid was removed from the columns, and peptides were eluted using 80 : 20 ACN : $H_2O$ containing 0.1% TFA. Peptides were collected and concentrated using a SpeedVac (Thermo-Savant) to a final volume of 50–100 μL, after which final peptide concentrations were determined by BCA protein assay.

Tryptic digests of the insoluble IM and OM fractions were prepared as follows. To 75 μg of protein from each

sample, urea, DTT, and CHAPS were added to final concentrations of 7 M, 10 mM, and 1%, respectively, followed by incubation at 60°C for 30 min. Samples were then diluted 7-fold with 100 mM $NH_4HCO_3$, and $CaCl_2$ was added to a final concentration of 1 mM. Digestion was performed as described for the soluble fractions. Digested samples were then cleaned using 1 mL, 50 mg Discovery SCX strong cation exchange SPE columns (Supelco, St. Louis, MO, USA). Briefly, columns were conditioned with methanol and then rinsed in varying sequences and amounts of 10 mM ammonium formate in 25% ACN (pH 3.0), 500 mM ammonium formate in 25% ACN (pH 6.8), and nanopure water. Peptide samples were acidified to pH < 4 with formic acid, centrifuged at 10,000 ×g for 5 min, applied to the columns, and washed with 10 mM ammonium formate in 25% ACN (pH 3.0). Peptides were eluted using 80 : 15 : 5 MeOH : $H_2O$ : $NH_4OH$ and concentrated to a final volume of 50–100 μL using a SpeedVac. Final peptide concentrations were calculated by BCA protein assay.

*2.6. SDS-PAGE.* For visualization of the protein fractions, 5 μg of each protein sample was suspended in NuPAGE LDS sample buffer (Invitrogen, Carlsbad, CA, USA), heated at 70°C for 10 min, and resolved on NuPAGE Novex 4–12% Bis-Tris gradient gels (Invitrogen). Gels were run at a constant voltage of 200 V for 35 min and subsequently stained with GelCode Blue stain (Pierce) to observe protein profiles.

*2.7. Capillary LC-MS/MS Analysis.* The high-performance liquid chromatography (HPLC) system and method used for nanocapillary liquid chromatography have been described in detail elsewhere [19, 22]. Analysis was performed using an LTQ-Orbitrap mass spectrometer (Thermo Fisher Scientific, San Jose, CA, USA) with electrospray ionization. The HPLC column was coupled to the mass spectrometer using an in-house manufactured interface. The heated capillary temperature and spray voltage were 200°C and 2.2 kV, respectively. Data acquisition began 20 min after the sample was injected and continued for 100 min over an $m/z$ range of 400–2000. For each cycle, the six most abundant ions from MS analysis were selected for MS/MS analysis, using a collision energy setting of 35 eV. A dynamic exclusion time of 60 s was used to discriminate against previously analyzed ions. All subcellular fractions from the ΔfliCΔfljB mutant were analyzed in addition to the PERI of the WT (Supplemental Table 2) to ensure that the loss of flagellins did not alter periplasmic proteome expression. Each sample was analyzed in triplicate.

*2.8. Data Analysis.* Peptides were identified by using SEQUEST to search the mass spectra from LC-MS/MS analyses. These searches were performed using the annotated *S.* Typhimurium 14028 FASTA file, containing 5590 protein sequences [23]. Porcine trypsin protein sequences were included in the search to detect trypsin autocleavage contaminants. The SEQUEST parameter file contained no modifications to amino acid residues and a mass error window of 3 $m/z$ units for precursor mass and 0 $m/z$ units

A Comprehensive Subcellular Proteomic Survey of Salmonella Grown under Phagosome-Mimicking versus
Standard Laboratory Conditions

17

for fragmentation mass. The searches allowed for all possible peptide termini, that is, not limited by tryptic terminus state. Results were filtered using the MS-Generating Function [24], a software tool that assigns $P$ values (spectral probabilities) to spectral interpretations. The prescribed spectral probability cutoff ($1E^{-10}$) was used. This corresponded to a false-positive rate of 0.88% at the unique peptide level and 0.16% and the spectrum level using a traditional decoy approach, that is, searching against a reversed FASTA database [25].

The number of peptide observations from each protein (spectral count) was used as a measure of relative abundance. Multiple charge states of a single peptide were considered as individual observations, as were the same peptides detected in different mass spectral analyses. Similar approaches for quantitation have been described previously [9, 14, 19, 26]. A protein was considered present in a sample (subcellular fraction) only if observed in at least 2 of 3 technical replicates, and means of triplicate samples were adjusted to zero if this rule was not satisfied.

Statistical analyses were performed using Microsoft Excel and R (http://www.r-project.org/). K-means clustering and construction of heat maps were done using OmniViz 6.0.

## 3. Results

### 3.1. Protein Identification in Salmonella Subcellular Fractions.
To survey the localization of proteins in *Salmonella* cells as a reference of protein localization and to observe changes in protein abundance upon growth under phagosome-mimicking conditions relative to standard laboratory conditions, *S.* Typhimurium 14028 flagellin mutant (see Section 2 for rationale) was grown in Luria-Bertani broth (LB) to log phase or in a low-phosphate, low-magnesium, low-pH minimal medium (LPM) for 4 or 20 h. Subcellular fractionation based on osmotic shock, differential centrifugation, and differential detergent solubilization yielded CYT, IM, PERI, and OM fractions (Figure 1) from which tryptic peptides were identified using LC-MS/MS (see Section 2). The total number of peptide observations from each protein (spectral count) was used as an estimate of relative abundance, and a protein was considered present in a sample only if observed in at least two of three replicates. This step served the dual purpose of globally removing proteins with only one peptide observation and increasing confidence in peptide identifications within each subcellular fraction. The average sequence coverage for each protein was ~30%. Similar numbers of proteins were identified in LB (993), LPM-4h (1102), and LPM-20h (1006) growth conditions.

### 3.2. Subcellular Fraction Enrichment.
Each subcellular fraction contained a unique protein profile (Supplemental Figure 1), although the IM contained a larger proportion of cofractionating CYT proteins, as noted previously [9]. We avoided high-pH treatment of membrane fractions [27] in an attempt to maintain physiologically relevant protein-protein and protein-membrane interactions; thus, peripheral membrane proteins were not removed in our protocol.

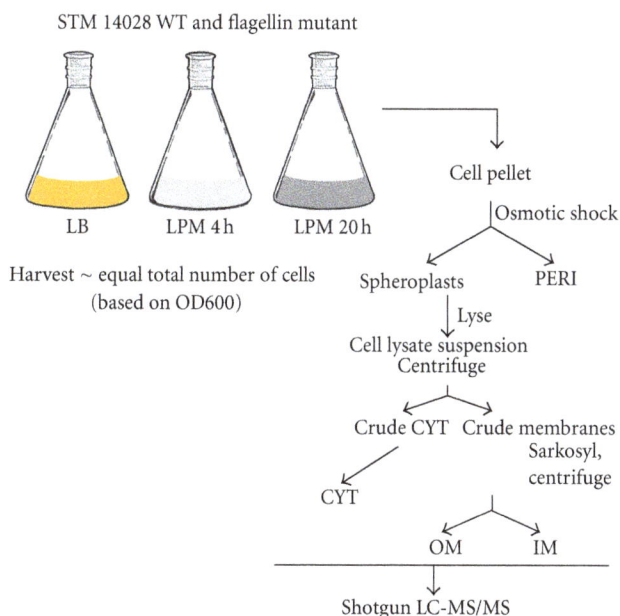

FIGURE 1: Experimental workflow. A fractionation scheme based on differential centrifugation and Sarkosyl solubilization of membranes was combined with spheroplasting to obtain PERI, CYT, OM, and IM samples from *S.* Typhimurium strain 14028. Subcellular fractions were further processed prior to high-resolution LC-MS/MS analysis.

Agreement between observed and computationally predicted protein localization was assessed. Subcellular predictions were computed using PSORTb [28], with the caveat that ~17% of the observed proteins had no PSORTb subcellular assignment (unknown or unknown with multiple possible localizations). Each subcellular fractionation was enriched in the types of proteins expected to reside there (Figure 2(a); Table 1). Both the IM and OM contained a large number of predicted CYT proteins. Since many proteins were likely observed in multiple fractions as minor contaminants due to cofractionation, protein abundance contributions were more informative than the absolute number of proteins observed [9]. From this analysis, predicted CYT proteins contributed to 80–86% of the total protein abundance in CYT fractions, predicted OM proteins, to 65–80% in OM fractions, and predicted PERI proteins, to 68–75% in PERI fractions. In contrast to the expected agreement between predicted and observed enrichment, predicted IM proteins contributed to only ~25% of the total protein abundance observed in IM fractions (Figure 2(b)). This relatively limited enrichment was due largely to cofractionation of abundant CYT proteins and to the general low observability of integral membrane proteins by proteomics [29, 30].

As many proteins involved in bacterial pathogenesis are located outside the cytoplasm where they may more readily target and respond to the host environment, we assessed our success in enriching envelope proteins in the appropriate fractions. Cell envelope (IM, PERI, and OM) proteins can be distinguished by physicochemical properties, such as hydrophobicity (IM proteins), amphipathic beta sheets (OM

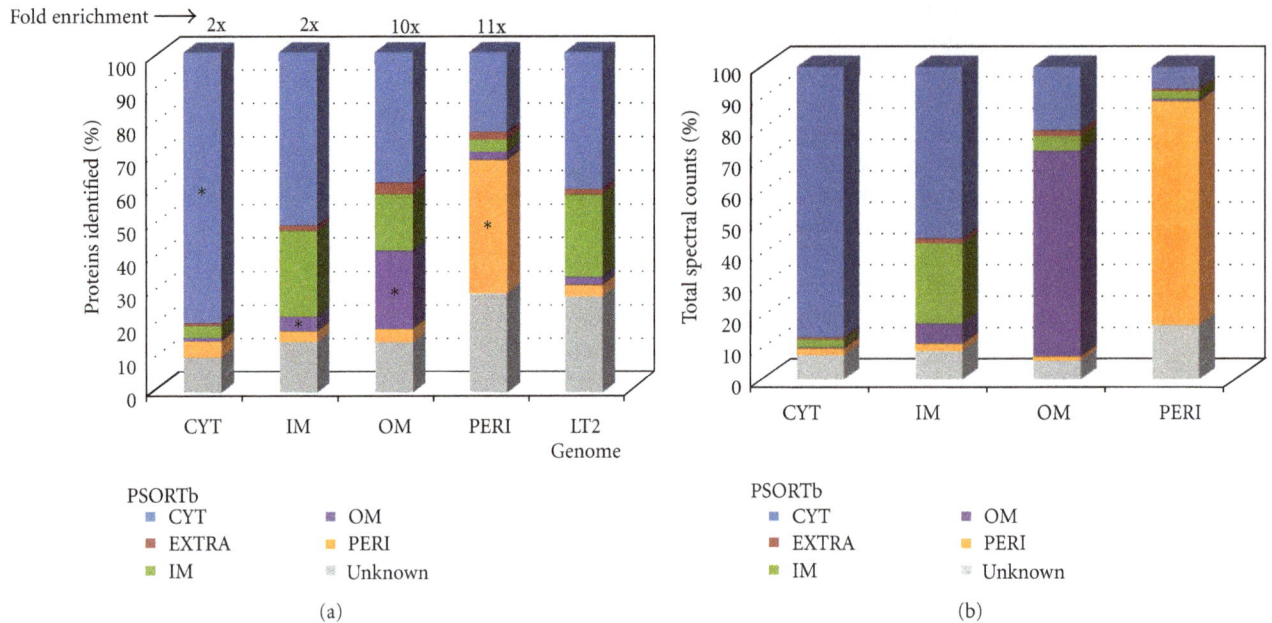

FIGURE 2: Distribution of proteins observed in subcellular fractions via LC-MS/MS (a). Protein composition of each subcellular fraction, based on number of proteins observed in each fraction sorted according to predicted subcellular location [16]. Data are percentage of proteins observed in each fraction. The fold-enrichment in proteins compared to the genomic potential is noted above each bar. *$P \leq 0.002$, $\chi^2$ test, compared to genome (b). Summed spectral counts (total abundance) of proteins observed in subcellular fractions.

TABLE 1: Enrichment of proteins with expected physicochemical properties.

| Protein type | CYT | IM | PERI | OM | All observed | In genome | Percentage observed |
|---|---|---|---|---|---|---|---|
| OM beta barrel | 8 | 27 | 7 | **44** | 51 | 99 | 52% |
| Signal Peps | 81 | 120 | **130** | 100 | 239 | 532 | 45% |
| TMD > 0 | 26 | **196** | 6 | 54 | 204 | 1167 | 18% |
| TMD > 1 | 10 | **130** | 2 | 33 | 130 | 812 | 16% |
| TMD > 2 | 4 | **97** | 2 | 21 | 97 | 683 | 14% |
| TMD > 3 | 3 | **88** | 2 | 19 | 88 | 619 | 14% |
| GRAVY > 0 | 258 | **488** | 54 | 140 | 611 | 2882 | 21% |
| GRAVY > 0.1 | 158 | **335** | 31 | 88 | 413 | 2201 | 19% |
| GRAVY > 0.2 | 69 | **194** | 16 | 47 | 231 | 1637 | 14% |
| GRAVY > 0.3 | 24 | **133** | 8 | 34 | 145 | 1276 | 11% |
| GRAVY ≥ 0.5 | 3 | **66** | 0 | 17 | 66 | 890 | 7% |

proteins), and signal peptides (many envelope proteins). The IM, PERI, and OM were significantly enriched in envelope proteins based on observed physicochemical properties. For example, 239 proteins with predicted signal peptides (using PSORTb) were observed (45% of genomic potential). These proteins were mainly identified in the IM, OM, and PERI fractions, with the highest number (130) observed in the PERI (Table 1). Of 51 predicted outer membrane $\beta$-barrel proteins [31] observed (51% of genomic potential), 44 of these were in outer membrane fractions. Similarly, proteins with predicted transmembrane $\alpha$-helices [32], a feature of integral membrane proteins, were concentrated in the IM, as expected. Of 97 proteins with ≥3 transmembrane domains, all were observed in the IM, while only 24 were observed in the other three fractions combined (Table 1). Hydrophob-

icity, another hallmark of integral membrane proteins [33], correlated well with proteins observed in IM samples. For the 66 proteins that could be considered very hydrophobic (hydrophobicity average ≥0.5) [33], all were observed in the IM with high abundance values (not shown), while 3, 0, and 17 were observed in the CYT, PERI, and OM, respectively.

### 3.3. Determination of Primary Observed Localization.
For proteins observed in multiple subcellular fractions, it was useful to identify the fraction in which each protein was observed at its highest level (i.e., the likely true subcellular location of the protein). Primary localization was determined within each growth condition by calculating the Z-score of protein abundance in each subcellular fraction. Z-scores

A Comprehensive Subcellular Proteomic Survey of Salmonella Grown under Phagosome-Mimicking versus
Standard Laboratory Conditions

19

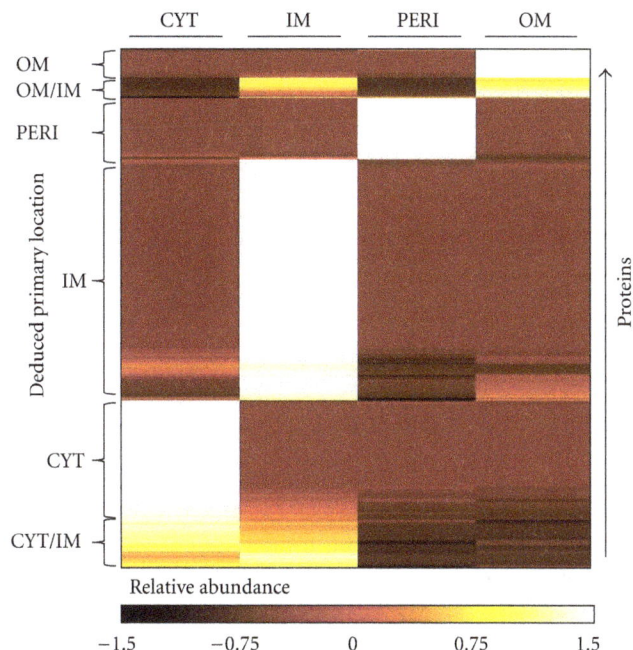

FIGURE 3: Use of Z-scores and K-means clustering to assign primary subcellular locations to proteins. Since many proteins were observed in two or more subcellular fractions, Z-scores of spectral counts across the four fractions were calculated to highlight the primary observed localization of each protein. K-means clustering was used to group proteins with similar profiles.

TABLE 2: Two-component regulators showing localization changes depending on growth conditions.

| Protein Description (PhoP/Q) | Gene | PSORTb v3 | LB 1°Loc | LPM4 1°Loc | LPM20 1°Loc |
|---|---|---|---|---|---|
| Sensor protein PhoQ | PhoQ | IM | IM* | IM* | IM* |
| DNA-binding transcriptional regulator PhoP | PhoP | Cyt | IM/CYT | CYT | CYT |
| Protein Description (ArcA/B) | Gene | PSORTb v3 | LB 1°Loc | LPM4 1°Loc | LPM20 1°Loc |
| Aerobic respiration control sensor protein ArcB | ArcB | IM | IM* | IM* | IM* |
| Two-component response regulator | ArcA | Cyt | IM* | IM | CYT |

*Indicates that a protein is observed exclusively in one location.

were clustered using the K-means algorithm to group similar profiles of subcellular localization (Figure 3; Supplemental Table 3). Note that similar approaches have been described previously [7, 9]. Using the LB culture as an example, 91% of proteins could be assigned a single primary localization using this scheme.

Some proteins (∼9%) were highly observed in two or more subcellular fractions and usually occurred between the CYT and IM or IM and OM. It is noteworthy that six of the 22 IM/OM proteins were lipoproteins, which likely reflects the increased hydrophobicity and tendency to partition with the Sarkosyl-soluble IM. Other members of the IM/OM class included membrane-bound portions of type 3 secretion systems (T3SS): PrgH and PrgK of the invasion-related T3SS and FliF, FliG, and FlgE that represent the ring, basal body, and hook of the flagellar T3SS. In these cases, cofractionation reflects the association of these supramolecular structures with both membranes.

Of the proteins that were multilocalized or had secondary locations, several have been implicated in strong physiologically relevant protein-protein and protein-membrane interactions that can influence localization. For example, seven of the eight subunits of ATP synthase were observed primarily in the IM fraction (Figure 4). While only two subunits are integral to the IM, close protein-protein interactions likely mediated the cofractionation of the entire complex to the IM. Peripheral membrane proteins and multisubunit cytoplasmic proteins made up a majority of the known CYT proteins that had IM or IM/CYT as their primary observed location. Using a combination of available subunit information in Uniprot (http://www.uniprot.org/) and published literature, 45 of the 50 IM/CYT proteins were justified in their observed location due to their multimeric forms or peripheral membrane association that are tied to protein function (Supplemental Table 4).

Another group of proteins in this class were the two-component regulatory systems. These systems consist of a membrane-bound sensor-kinase protein and a cytoplasmic response regulator that interacts with, and is phosphorylated by, the sensor-kinase at the membrane, which promotes DNA binding and regulation of gene expression [34]. In both the PhoP/PhoQ and ArcA/ArcB systems, the sensor-kinases

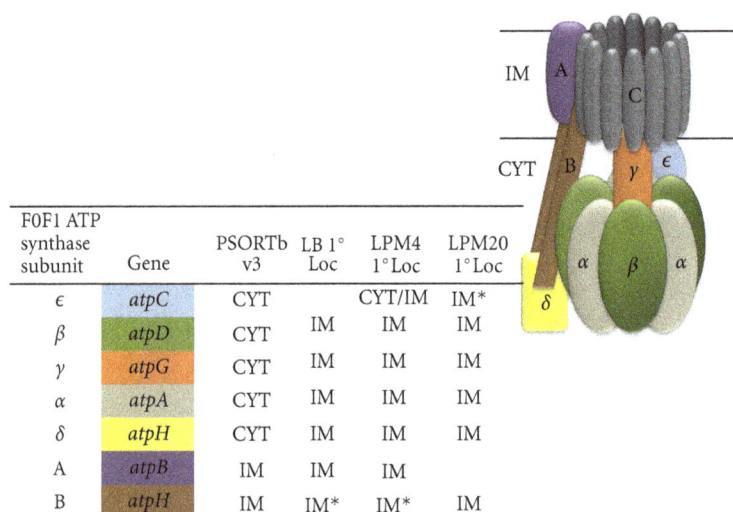

| F0F1 ATP synthase subunit | Gene | PSORTb v3 | LB 1° Loc | LPM4 1°Loc | LPM20 1°Loc |
|---|---|---|---|---|---|
| ε | atpC | CYT | | CYT/IM | IM* |
| β | atpD | CYT | IM | IM | IM |
| γ | atpG | CYT | IM | IM | IM |
| α | atpA | CYT | IM | IM | IM |
| δ | atpH | CYT | IM | IM | IM |
| A | atpB | IM | IM | IM | |
| B | atpH | IM | IM* | IM* | IM |

FIGURE 4: ATP synthase complex exemplifies observed protein-protein and protein-membrane interactions. Schematic representation of membrane-bound ATP synthase, modeled after KEGG Bacterial F-type ATPase Color-coded table shows protein observations in subcellular fractions. n/a: protein not observed. C: AtpE (not observed in this study). *Protein was exclusive to one subcellular fraction in a given growth condition.

were observed exclusively in the IM, while the response regulators were observed either in the IM (i.e., presumably bound to the kinase) or in the CYT (i.e., presumably interacting with DNA), depending on growth condition (Table 2). Our results iterate that PhoP is bound to DNA during growth in LPM (for either 4 or 20 h), which is supported by known activation of the PhoP regulon within acidified macrophage phagosomes [35] and during growth under phagosome-mimicking conditions [26]. Conversely, the response regulator ArcA is IM-localized in cells grown in LB or those grown in LPM for a short duration, but is CYT-localized in cells grown overnight in LPM. These results provide insight into the function of this regulatory system under these specific growth conditions.

We note that some instances of multilocalized proteins may be due to the inability of our methods to perfectly resolve subcellular fractions, or may be artifacts of fractionation. As an example of the latter, DnaK and Ef-Tu can be translocated out of the cytoplasm during osmotic shock [36]. In our study, Ef-Tu was observed at high levels in both the IM and CYT. While DnaK was observed primarily in the CYT, DnaJ, a cochaperone with DnaK, was observed primarily in the IM in all growth conditions in this study.

For those proteins annotated as "putative" ($n = 274$) or "hypothetical" ($n = 92$), we were able to confidently assign localization to a majority based on protein abundances in subcellular fractions (Supplemental Table 5). For many of these proteins, the assignment of subcellular localization as well as data on relative expression levels in different growth conditions represents the most extensive characterization available to date.

*3.4. Putative Moonlighting Proteins.* Some proteins were observed in unexpected subcellular locations regardless of growth condition, while the location of other proteins appeared to be influenced by growth condition. Several proteins with well-characterized housekeeping roles (e.g., enolase and glyceraldehyde-3-phosphate dehydrogenase) have been observed on the cell surfaces of pathogens, where they have secondary functions such as adhesion and immune modulation [3]. The term "moonlighting" refers to proteins that exhibit more than one biological function [1–3]. Here too, proteins that were observed in unexpected locations based on predictions, annotations, and known functions could point to novel interactions or functions yet to be characterized. In these cases, proteins with higher spectral counts (relative abundance) and greater numbers of unique peptides (more confident identifications) were considered more reliable candidates for assignment of localization.

One of the best moonlighting protein candidates observed in this study is Dps (DNA protection during starvation). This protein has been well characterized as a cytoplasmic DNA-binding protein (reviewed in [37]) and has no predicted signal peptide. In each growth condition tested, we observed Dps significantly enriched in the OM fraction (Figure 5), which shows for the first time that this protein is OM-localized in *Salmonella*. Dps is a known virulence determinant of *Salmonella* [38], but how it translocates to the OM and its role(s) at the cell surface remain to be investigated. Interestingly, Dps was recently observed on the cell surface of *Escherichia coli* [38, 39], where it may play a role in attachment to abiotic surfaces [38]. We observed >2-fold increase in the relative abundance of OM-localized Dps between LB and LPM20 growth conditions, which indicates that *Salmonella* Dps is responsive to growth under phagosome-mimicking conditions (Figure 5).

Because the CYT and OM are the two most physically separated subcellular locations studied here and contain proteins with fairly distinct physicochemical properties, we considered known cytoplasmic proteins observed in the OM as the most promising moonlighting candidates.

A Comprehensive Subcellular Proteomic Survey of Salmonella Grown under Phagosome-Mimicking versus
Standard Laboratory Conditions

21

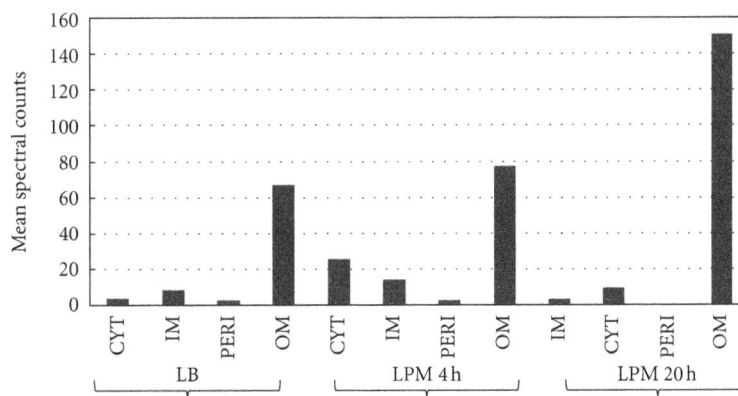

FIGURE 5: Localization and relative abundance of potential moonlighting protein, Dps. Spectral counts of Dps in each subcellular fraction in each growth condition. Values are means of 3 replicates.

These candidates included a (3R)-hydroxymyristoyl-ACP dehydratase (FabZ), a curved DNA-binding protein (CbpA), an imidazole glycerol-phosphate dehydratase/histidinol phosphatase (HisB), and an ATP-dependent RNA helicase (SrmB). All of these cases included proteins generally accepted to be cytoplasmic, with no detectable signal peptides, transmembrane helices, or beta barrel predictions that were confidently observed in OM or in a mix of OM and IM fractions (Supplemental Table 6). These proteins represent the first candidates for an investigation of moonlighting activities in *Salmonella*.

*3.5. Subcellular Responses to Growth Conditions.* Although not a perfect replica of the *in vivo* environment, defined *in vitro* synthetic growth media provide valuable insights into the pathogenic strategies of *Salmonella* [40, 41]. Growth in LB to mid-exponential phase induces genes of the *Salmonella* pathogenicity island 1 (SPI-1) involved in host cell invasion [42–44], while genes of the *Salmonella* pathogenicity island 2 (SPI-2) can be induced by growth in LPM that simulates the environment of the *Salmonella*-containing vacuole (phagosome) [45, 46]. We used these growth conditions to probe the subcellular-level responses of *Salmonella* to phagosome-mimicking conditions.

When qualitatively assessed, similar numbers of proteins were observed in the three growth conditions: 993 in LB, 1102 in LPM-4h, and 1006 in LPM-20h. Approximately 10% of the proteins identified in each growth condition were unique to a given culture: 175 in LB, 100 in LPM-4h, and 92 in LPM-20h (Supplemental Figure 2), and less than half of all identified proteins (688) were observed in all growth conditions, which underscores the utility of using multiple growth conditions for improved coverage of a bacterial proteome.

We have previously investigated the proteome response of *Salmonella* to phagosome-mimicking *in vitro* conditions [19, 26]; however, the use of subcellular fractionation presented an opportunity for obtaining better proteome coverage, especially of proteins that are typically under-represented in global proteomic strategies, in addition to highlighting the subcellular location of proteins of interest.

Based on studies of *Salmonella* grown in acidic minimal media [19, 26], we confirmed the expected increases in abundance of proteins associated with the SPI-2 T3SS (SsaC, SseA, and SsaJ), the SsrB regulon (SsrA, SsrB, and SrfN), and the PhoP regulon (PhoP, PhoQ, PagC, MgtA, and MgtB) during growth in LPM (Supplemental Table 7). Conversely, proteins related to the invasion-associated SPI-1 T3SS (SipA, B, C, D, SopB, SicA, InvG, PrgK, and PrgL) decreased in abundance with growth in LPM. Further analyses focused on envelope proteins because the proteins primarily detected in previous global analyses were cytoplasmic proteins and because envelope proteins have high potential for host-pathogen interactions.

OM proteins whose abundance increased during growth in LPM included iron transporters (FepA, FhuA, IroN, and FoxA), ABC transporters, and virulence-related proteins (PagC; T3SS-related SsaC and SseC), which reflects the nutrient-limited and virulence gene-inducing nature of LPM (Figure 6(a)). A notable OM protein was the outer membrane protease PgtE that was increased 13- and 89-fold in LPM4 and LPM20, respectively ($P < 0.001$). PgtE is involved in cleavage of serum complement during the extracellular phase of *Salmonella* systemic infection [47], but its induction under phagosome-mimicking conditions suggests an intracellular role as well. In addition to the importance of OM proteins that increase in abundance in LPM, those that decrease in abundance may be indicative of immune evasion or virulence-related OM remodeling. For example, putative outer membrane lipoprotein maltoporin and outer membrane protein N were significantly decreased during growth in LPM for 20 h (Supplemental Table 7). Known SPI-1 T3SS-related surface proteins such as PrgK, PrgI, and InvG were also significantly decreased in the OM during growth in LPM, indicating the expected shift away from SPI-1 T3SS expression during growth in LPM.

Notable in the IM was a decrease in chemotaxis-related proteins (CheA, B, M, and Z; Tsr, Trg, and Tcp) and motility-related proteins (FliF, FliI, FliN, and MotA) in LPM compared to LB. A range of IM-integral and peripheral IM proteins of various functions were enriched during growth in LPM, including expected functions such as magnesium

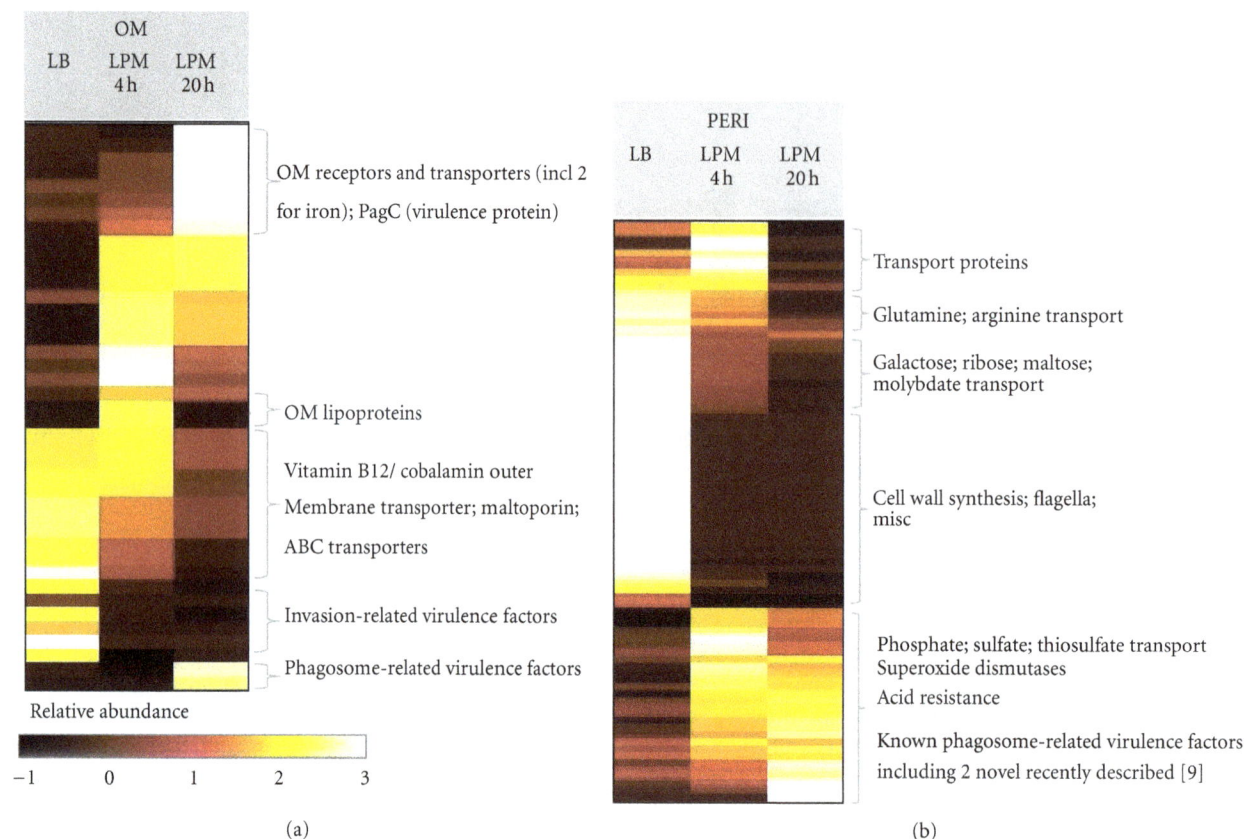

FIGURE 6: Heat map representation of differentially expressed OM and PERI proteins. Z-scores of protein abundance were calculated across the 3 growth conditions for proteins observed at their highest levels in the OM (a) and PERI (b) fractions. Each protein showed $\geq$ 2-fold difference in abundance in any two growth conditions.

transport (MgtA and MgtB), virulence proteins (PhoQ and SsaC), and various transporters, enzymes, and proteins of unknown function (Supplemental Table 7).

The PERI shifted from transport of sugars (galactose, ribose, and maltose), oligopeptides, dipeptides, aminoacids, and related compounds (arginine and putrescine) in LB to transport of phosphate, sulfate, and thiosulfate in LPM (Figure 6(b)). Also showing increased abundance in LPM were PERI proteins involved in superoxide and acid resistance (SodC and PhoN) and known secreted factors CigR [14] and SrfN [48, 49] for which the subcellular location prior to being translocated into infected mammalian cells was previously unknown.

## 4. Discussion

Comparative proteomics is an emerging tool for studying bacterial pathogenesis both *in vitro* and during infection [19, 26, 50]. Subcellular fractionation complements such analyses by providing a means to resolve physiologically relevant protein location in the bacterium. Our analysis of CYT, IM, PERI, and OM fractions of S. Typhimurium grown under laboratory and phagosome-mimicking conditions yielded ~1400 unique proteins, most of which could be confidently localized to a single subcellular fraction in a

given growth condition. Each subcellular fraction contained a unique protein profile (Figures 1 and 3 and Supplemental Figure 1) and protein physicochemical properties generally agreed well with their observed localization (Table 1).

To our knowledge, this study represents the most comprehensive global survey of subcellular localization in *Salmonella* to date. In earlier work, Coldham and Woodward [51] assessed cytosolic, cell envelope, and outer membrane protein preparations of *Salmonella* by extensive chromatographic fractionation followed by mass spectrometry. They observed 816 proteins, with 371 in the CYT, 565 in the envelope, and 262 in the OM samples. Of the latter 262, only 20 were OM proteins. Recently, the OM proteome of *S. enterica* was identified using a lipid-based method [52]. In that study, 54 OM proteins were identified with $\geq$2 peptides, using a multistep digest procedure on outer membrane vesicle preparations. In an early attempt to catalogue the OM proteome of *Escherichia coli*, Molloy and colleagues [27] identified ~30 proteins in the OM fraction, using 2D gel electrophoresis and MS approach. In our present study, at least 74 OM proteins were identified in OM fractions (deduced by PSORTb prediction, annotation, or by the presence of OM $\beta$-sheets). In addition to high coverage of OM proteins, confident assignment of CYT, IM, and PERI proteins was presented (Supplemental Table 3).

A Comprehensive Subcellular Proteomic Survey of Salmonella Grown under Phagosome-Mimicking versus Standard Laboratory Conditions

23

Among the challenges in any subcellular fractionation endeavor are to maximize fraction purity and correctly assign proteins to a subcellular location. Due to the close proximity of fractions, protein-protein interactions between fractions, or to the presence of protein domains that span multiple fractions, proteins sometimes copurify to two or more fractions. These biological phenomena are difficult to distinguish from experimental noise. In our analysis, large multi-subunit cytoplasmic complexes often concentrated in the membrane fractions (particularly the IM); likewise, many protein complexes that are known to be peripherally IM-associated also co-fractionated with the IM (e.g., ATP synthase). In cases where a protein was observed in multiple fractions, we were able to use relative abundance data to deduce the primary observed localization (Figure 3). However, some fractions posed more of a challenge than others; for example, the IM was more ambiguous than the OM, PERI, or CYT. Over 40% of proteins whose primary observed location was the IM were predicted by PSORTb to be cytoplasmic. It is important to note that this localization prediction does not take into account the many potential IM-interacting proteins. While the IM fraction is a good potential source of novel protein-protein and protein-membrane interactions, a clearer picture of the integral IM landscape could emerge upon high-pH buffer treatment of the IM fraction to remove peripherally bound proteins [27].

An aspect of this study that may be helpful to others interested in subcellular proteome characterization was our use of a mutant that was depleted in an abundant cell envelope component, flagellin (ΔfliCΔfljB). Because flagellin was one of the most abundant proteins observed in the PERI (and contaminated all envelope fractions) in a preliminary subcellular proteomic analysis (Supplemental Table 1), we hypothesized that deleting flagellin genes would enable better detection of low abundance of PERI proteins and likely increase the signal of most other proteins in the PERI fraction. Flagella are not essential for survival in macrophage phagosomes [53] and are downregulated under the environmental conditions simulated by our mLPM culture condition [20]. Thus, deleting flagellins should not interfere with the physiological responses we were interested in. In addition, flagella are nonessential for growth in LB (not shown). Proteomic analysis of the wild type versus ΔfliCΔfljB mutant PERI fractions showed no differences in presence of "housekeeping" proteins such as elongation factor Tu, elongation factor G, chaperonin GroEL, and ribosomal proteins that co-fractionated with the PERI (Supplemental Table 2). Also, IM and OM proteins that co-fractionated with the PERI were observed at similar (low) levels in both the wild type and mutant. Most importantly, we observed higher spectral counts of PERI proteins in the mutant relative to wild type, and several PERI proteins were detected only in the flagellin mutant (Supplemental Table 2). Thus, we advocate the use of relevant mutations in abundant nonessential proteins for improved subcellular proteome coverage.

The availability of experimentally observed subcellular localization data for such a large number of Salmonella proteins provides opportunities for further study. Among these opportunities are using high-confidence localization information for training subcellular localization prediction tools and for computationally predicting Salmonella function in host cells through the use of genome-scale models [4]. In addition, localization data for hypothetical or uncharacterized proteins (Supplemental Table 5) is a first step towards functional characterization of these unknown proteins. To extend the utility of these data, our future study will focus on multilocalized proteins and those that changed localization depending on growth condition. Both categories present the possibility of exciting discoveries in terms of protein function. Moonlighting protein candidates are included in this class; determining the transport mechanism and secondary function of our candidates are challenges for future study.

In summary, we presented a comparative subcellular proteomic analysis of Salmonella representative of laboratory growth and infection-like states. We cataloged the confident localization of over 1000 proteins and provided evidence of differential protein movement and the appearance of some proteins in unexpected subcellular compartments. These results imply the existence of unknown transport mechanisms and novel functions for a subset of Salmonella proteins.

## Abbreviations

| | |
|---|---|
| IM: | Inner membrane |
| OM: | Outer membrane |
| CYT: | Cytoplasmic |
| PERI: | Periplasmic |
| WCL: | Whole cell lysate |
| LC-MS/MS: | Liquid chromatography-tandem mass spectrometry |
| TTSS: | Type III secretion system |
| SPI: | Salmonella pathogenicity island. |

## Acknowledgments

The authors thank Robbie Heegel and Dr. Joseph Brown for their contributions to this research. Support for this work was provided by the National Institute of Allergy and Infectious Diseases NIH/DHHS through interagency agreement Y1-A1-8401-01. Proteomic analyses were performed in the Environmental Molecular Sciences Laboratory, a U.S Department of Energy Office of Biological and Environmental Research (DOE/BER) national scientific user facility on the Pacific Northwest National Laboratory (PNNL) campus in Richland, Washington. PNNL is a multiprogram national laboratory operated by Battelle for the DOE under Contract DE-AC05-76RL01830. This work used instrumentation and capabilities developed with funds provided by NIH grants from the National Center for Research Resources (5P41RR018522-10) and the National Institute of General Medical Sciences (8 P41 GM103493-10) and by DOE/BER. Mass spectrometry results are available via http://SysBEP.org/ and http://omics.pnl.gov/.

# References

[1] N. R. Smalheiser, "Proteins in unexpected locations," *Molecular Biology of the Cell*, vol. 7, no. 7, pp. 1003–1014, 1996.

[2] V. Pancholi and G. S. Chhatwal, "Housekeeping enzymes as virulence factors for pathogens," *International Journal of Medical Microbiology*, vol. 293, no. 6, pp. 391–401, 2003.

[3] B. Henderson and A. Martin, "Bacterial virulence in the moonlight: multitasking bacterial moonlighting proteins are virulence determinants in infectious disease," *Infection and Immunity*, vol. 79, no. 9, pp. 3476–3491, 2011.

[4] I. Thiele, D. R. Hyduke, B. Steeb et al., "A community effort towards a knowledge-base and mathematical model of the human pathogen *Salmonella* Typhimurium LT2," *BMC Systems Biology*, vol. 5, p. 8, 2011.

[5] M. Dreger, "Subcellular proteomics," *Mass Spectrometry Reviews*, vol. 22, no. 1, pp. 27–56, 2003.

[6] M. Dreger, "Proteome analysis at the level of subcellular structures," *European Journal of Biochemistry*, vol. 270, no. 4, pp. 589–599, 2003.

[7] S. J. Callister, M. A. Dominguez, C. D. Nicora et al., "Application of the accurate mass and time tag approach to the proteome analysis of sub-cellular fractions obtained from Rhodobacter sphaeroides 2.4.1. aerobic and photosynthetic cell cultures," *Journal of Proteome Research*, vol. 5, no. 8, pp. 1940–1947, 2006.

[8] M. Thein, G. Sauer, N. Paramasivam, I. Grin, and D. Linke, "Efficient subfractionation of gram-negative bacteria for proteomics studies," *Journal of Proteome Research*, vol. 9, no. 12, pp. 6135–6147, 2010.

[9] R. N. Brown, M. F. Romine, A. A. Schepmoes, R. D. Smith, and M. S. Lipton, "Mapping the subcellular proteome of Shewanella oneidensis MR-1 using sarkosyl-based fractionation and LC-MS/MS protein identification," *Journal of Proteome Research*, vol. 9, no. 9, pp. 4454–4463, 2010.

[10] E. Jung, M. Heller, J.-C. Sanchez, and D. F. Hochstrasser, "Proteomics meets cell biology: the establishment of subcellular proteomes," *Electrophoresis*, vol. 21, no. 16, pp. 3369–3377, 2000.

[11] C. Bell, G. T. Smith, M. J. Sweredoski, and S. Hess, "Characterization of the mycobacterium tuberculosis proteome by liquid chromatography mass spectrometry-based proteomics techniques: a comprehensive resource for tuberculosis research," *Journal of Proteome Research*, vol. 11, no. 1, pp. 119–130, 2012.

[12] D. Becher, K. Hempel, S. Sievers et al., "A proteomic view of an important human pathogen-towards the quantification of the entire staphylococcus aureus proteome," *PLoS One*, vol. 4, no. 12, p. e8176, 2009.

[13] E. Carlsohn, J. Nyström, H. Karlsson, A. M. Svennerholm, and C. L. Nilsson, "Characterization of the outer membrane protein profile from disease-related *Helicobacter pylori* isolates by subcellular fractionation and nano-LC FT-ICR MS analysis," *Journal of Proteome Research*, vol. 5, no. 11, pp. 3197–3204, 2006.

[14] G. S. Niemann, R. N. Brown, J. K. Gustin et al., "Discovery of novel secreted virulence factors from *Salmonella enterica* serovar Typhimurium by proteomic analysis of culture supernatants," *Infection and Immunity*, vol. 79, no. 1, pp. 33–43, 2011.

[15] U. Silphaduang, M. Mascarenhas, M. Karmali, and B. K. Coombes, "Repression of intracellular virulence factors in *Salmonella* by the Hha and YdgT nucleoid-associated proteins," *Journal of Bacteriology*, vol. 189, no. 9, pp. 3669–3673, 2007.

[16] D. M. Cirillo, R. H. Valdivia, D. M. Monack, and S. Falkow, "Macrophage-dependent induction of the *Salmonella* pathogenicity island 2 type III secretion system and its role in intracellular survival," *Molecular Microbiology*, vol. 30, no. 1, pp. 175–188, 1998.

[17] J. Delwick, T. Nikolaus, S. Erdogan, and M. Hensel, "Environmental regulation of *Salmonella* pathogenicity island 2 gene expression," *Molecular Microbiology*, vol. 31, no. 6, pp. 1759–1773, 1999.

[18] B. K. Coombes, N. F. Brown, Y. Valdez, J. H. Brumell, and B. B. Finlay, "Expression and secretion of *Salmonella* pathogenicity island-2 virulence genes in response to acidification exhibit differential requirements of a functional type III secretion apparatus and SsaL," *Journal of Biological Chemistry*, vol. 279, no. 48, pp. 49804–49815, 2004.

[19] J. N. Adkins, H. M. Mottaz, A. D. Norbeck et al., "Analysis of the *Salmonella* Typhimurium proteome through environmental response toward infectious conditions," *Molecular and Cellular Proteomics*, vol. 5, no. 8, pp. 1450–1461, 2006.

[20] L. A. Knodler, B. A. Vallance, J. Celli et al., "Dissemination of invasive *Salmonella* via bacterial-induced extrusion of mucosal epithelia," *Proceedings of the National Academy of Sciences of the United States of America*, vol. 107, no. 41, pp. 17733–17738, 2010.

[21] K. A. Datsenko and B. L. Wanner, "One-step inactivation of chromosomal genes in *Escherichia coli* K-12 using PCR products," *Proceedings of the National Academy of Sciences of the United States of America*, vol. 97, no. 12, pp. 6640–6645, 2000.

[22] Y. Shen, N. Tolić, R. Zhao et al., "High-throughput proteomics using high-efficiency multiple-capillary liquid chromatography with on-line high-performance ESI FTICR mass spectrometry," *Analytical Chemistry*, vol. 73, no. 13, pp. 3011–3021, 2001.

[23] T. Jarvik, C. Smillie, E. A. Groisman, and H. Ochman, "Short-term signatures of evolutionary change in the *Salmonella enterica* serovar Typhimurium 14028 genome," *Journal of Bacteriology*, vol. 192, no. 2, pp. 560–567, 2010.

[24] S. Kim, N. Gupta, and P. A. Pevzner, "Spectral probabilities and generating functions of tandem mass spectra: a strike against decoy databases," *Journal of Proteome Research*, vol. 7, no. 8, pp. 3354–3363, 2008.

[25] J. Peng, J. E. Elias, C. C. Thoreen, L. J. Licklider, and S. P. Gygi, "Evaluation of multidimensional chromatography coupled with tandem mass spectrometry (LC/LC-MS/MS) for large-scale protein analysis: the yeast proteome," *Journal of Proteome Research*, vol. 2, no. 1, pp. 43–50, 2003.

[26] L. Shi, C. Ansong, H. Smallwood et al., "Proteome of *Salmonella enterica* serotype Typhimurium grown in a low Mg$^{2+}$/pH medium," *Journal of Proteomics and Bioinformatics*, vol. 2, no. 9, pp. 388–397, 2009.

[27] M. P. Molloy, B. R. Herbert, M. B. Slade et al., "Proteomic analysis of the *Escherichia coli* outer membrane," *European Journal of Biochemistry*, vol. 267, no. 10, pp. 2871–2881, 2000.

[28] N. Y. Yu, J. R. Wagner, M. R. Laird et al., "PSORTb 3.0: improved protein subcellular localization prediction with refined localization subcategories and predictive capabilities for all prokaryotes," *Bioinformatics*, vol. 26, no. 13, pp. 1608–1615, 2010.

[29] V. Santoni, M. Molloy, and T. Rabilloud, "Membrane proteins and proteomics: un amour impossible?" *Electrophoresis*, vol. 21, no. 6, pp. 1054–1070, 2000.

[30] T. Rabilloud, "Membrane proteins and proteomics: love is possible, but so difficult," *Electrophoresis*, vol. 30, supplement

A Comprehensive Subcellular Proteomic Survey of Salmonella Grown under Phagosome-Mimicking versus
Standard Laboratory Conditions

25

1, pp. S174–S180, 2009.

[31] F. S. Berven, K. Flikka, H. B. Jensen, and I. Eidhammer, "BOMP: a program to predict integral $\beta$-barrel outer membrane proteins encoded within genomes of Gram-negative bacteria," *Nucleic Acids Research*, vol. 32, pp. W394–W399, 2004.

[32] L. Käll, A. Krogh, and E. L. L. Sonnhammer, "A combined transmembrane topology and signal peptide prediction method," *Journal of Molecular Biology*, vol. 338, no. 5, pp. 1027–1036, 2004.

[33] J. Kyte and R. F. Doolittle, "A simple method for displaying the hydropathic character of a protein," *Journal of Molecular Biology*, vol. 157, no. 1, pp. 105–132, 1982.

[34] R. B. Bourret, K. A. Borkovich, and M. I. Simon, "Signal transduction pathways involving protein phosphorylation in prokaryotes," *Annual Review of Biochemistry*, vol. 60, pp. 401–441, 1991.

[35] C. M. Alpuche Aranda, J. A. Swanson, W. P. Loomis, and S. I. Miller, "*Salmonella* Typhimurium activates virulence gene transcription within acidified macrophage phagosomes," *Proceedings of the National Academy of Sciences of the United States of America*, vol. 89, no. 21, pp. 10079–10083, 1992.

[36] C. Berrier, A. Garrigues, G. Richarme, and A. Ghazi, "Elongation factor Tu and DnaK are transferred from the cytoplasm to the periplasm of *Escherichia coli* during osmotic downshock presumably via the mechanosensitive channel MscL," *Journal of Bacteriology*, vol. 182, no. 1, pp. 248–251, 2000.

[37] L. N. Calhoun and Y. M. Kwon, "Structure, function and regulation of the DNA-binding protein Dps and its role in acid and oxidative stress resistance in *Escherichia coli*: a review," *Journal of Applied Microbiology*, vol. 110, no. 2, pp. 375–386, 2011.

[38] R. M. Goulter-Thorsen, I. R. Gentle, K. S. Gobius, and G. A. Dykes, "The DNA protection during starvation protein (Dps) influences attachment of *Escherichia colis* to abiotic surfaces," *Foodborne Pathogens and Disease*, vol. 8, no. 8, pp. 939–941, 2011.

[39] A. Lacqua, O. Wanner, T. Colangelo, M. G. Martinotti, and P. Landini, "Emergence of biofilm-forming subpopulations upon exposure of *Escherichia coli* to environmental bacteriophages," *Applied and Environmental Microbiology*, vol. 72, no. 1, pp. 956–959, 2006.

[40] S. Löber, D. Jäckel, N. Kaiser, and M. Hensel, "Regulation of *Salmonella* pathogenicity island 2 genes by independent environmental signals," *International Journal of Medical Microbiology*, vol. 296, no. 7, pp. 435–447, 2006.

[41] C. R. Beuzón, G. Banks, J. Deiwick, M. Hensel, and D. W. Holden, "pH-dependent secretion of SseB, a product of the SPI-2 type III secretion system of *Salmonella* Typhimurium," *Molecular Microbiology*, vol. 33, no. 4, pp. 806–816, 1999.

[42] E. A. Miao and S. I. Miller, "A conserved amino acid sequence directing intracellular type III secretion by *Salmonella* Typhimurium," *Proceedings of the National Academy of Sciences of the United States of America*, vol. 97, no. 13, pp. 7539–7544, 2000.

[43] K. Eichelberg and J. E. Galán, "Differential regulation of *Salmonella* Typhimurium type III secreted proteins by pathogenicity island 1 (SPI-1)-encoded transcriptional activators InvF and HilA," *Infection and Immunity*, vol. 67, no. 8, pp. 4099–4105, 1999.

[44] K. Ehrbar, B. Winnen, and W. D. Hardt, "The chaperone binding domain of SopE inhibits transport via flagellar and SPI-1 TTSS in the absence of InvB," *Molecular Microbiology*, vol. 59, no. 1, pp. 248–264, 2006.

[45] B. K. Coombes, M. J. Lowden, J. L. Bishop et al., "SseL is a *Salmonella*-specific translocated effector integrated into the SsrB-controlled *Salmonella* pathogenicity island 2 type III secretion system," *Infection and Immunity*, vol. 75, no. 2, pp. 574–580, 2007.

[46] X. J. Yu, K. McGourty, M. Liu, K. E. Unsworth, and D. W. Holden, "pH sensing by intracellular *Salmonella* induces effector translocation," *Science*, vol. 328, no. 5981, pp. 1040–1043, 2010.

[47] P. Ramu, R. Tanskanen, M. Holmberg, K. Lähteenmäki, T. K. Korhonen, and S. Meri, "The surface protease PgtE of *Salmonella enterica* affects complement activity by proteolytically cleaving C3b, C4b and C5," *FEBS Letters*, vol. 581, no. 9, pp. 1716–1720, 2007.

[48] S. E. Osborne, D. Walthers, A. M. Tomljenovic et al., "Pathogenic adaptation of intracellular bacteria by rewiring a cis-regulatory input function," *Proceedings of the National Academy of Sciences of the United States of America*, vol. 106, no. 10, pp. 3982–3987, 2009.

[49] H. Yoon, C. Ansong, J. E. McDermott et al., "Systems analysis of multiple regulator perturbations allows discovery of virulence factors in *Salmonella*," *BMC Systems Biology*, vol. 5, p. 100, 2011.

[50] L. Shi, J. N. Adkins, J. R. Coleman et al., "Proteomic analysis of *Salmonella enterica* serovar Typhimurium isolated from RAW 264.7 macrophages: identification of a novel protein that contributes to the replication of serovar Typhimurium inside macrophages," *Journal of Biological Chemistry*, vol. 281, no. 39, pp. 29131–29140, 2006.

[51] N. G. Coldham and M. J. Woodward, "Characterization of the *Salmonella* Typhimurium proteome by semi-automated two dimensional HPLC-mass spectrometry: detection of proteins implicated in multiple antibiotic resistance," *Journal of Proteome Research*, vol. 3, no. 3, pp. 595–603, 2004.

[52] D. Chooneea, R. Karlsson, V. Encheva, C. Arnold, H. Appleton, and H. Shah, "Elucidation of the outer membrane proteome of *Salmonella enterica* serovar *Typhimurium* utilising a lipid-based protein immobilization technique," *BMC Microbiology*, vol. 10, p. 44, 2010.

[53] C. K. Schmitt, J. S. Ikeda, S. C. Darnell et al., "Absence of all components of the flagellar export and synthesis machinery differentially alters virulence of *Salmonella entericaserovar* Typhimurium in models of typhoid fever, survival in macrophages, tissue culture invasiveness, and calf enterocolitis," *Infection and Immunity*, vol. 69, no. 9, pp. 5619–5625, 2001.

# 4

# Serum Biomarkers Identification by Mass Spectrometry in High-Mortality Tumors

**Alessandra Tessitore, Agata Gaggiano, Germana Cicciarelli, Daniela Verzella, Daria Capece, Mariafausta Fischietti, Francesca Zazzeroni, and Edoardo Alesse**

*Department of Biotechnological and Applied Clinical Sciences, University of L'Aquila, Via Vetoio Coppito 2, 67100 L'Aquila, Italy*

Correspondence should be addressed to Edoardo Alesse; edoardo.alesse@univaq.it

Academic Editor: Visith Thongboonkerd

Cancer affects millions of people worldwide. Tumor mortality is substantially due to diagnosis at stages that are too late for therapies to be effective. Advances in screening methods have improved the early diagnosis, prognosis, and survival for some cancers. Several validated biomarkers are currently used to diagnose and monitor the progression of cancer, but none of them shows adequate specificity, sensitivity, and predictive value for population screening. So, there is an urgent need to isolate novel sensitive, specific biomarkers to detect the disease early and improve prognosis, especially in high-mortality tumors. Proteomic techniques are powerful tools to help in diagnosis and monitoring of treatment and progression of the disease. During the last decade, mass spectrometry has assumed a key role in most of the proteomic analyses that are focused on identifying cancer biomarkers in human serum, making it possible to identify and characterize at the molecular level many proteins or peptides differentially expressed. In this paper we summarize the results of mass spectrometry serum profiling and biomarker identification in high mortality tumors, such as ovarian, liver, lung, and pancreatic cancer.

## 1. Introduction

Cancer-related mortality is one of the leading causes of death worldwide. The most effective treatment to fight cancer is still early diagnosis. On the other hand, it is known that the correct classification of the tumor, coupled to a suitable therapy and to a stringent follow-up, helps to prevent and detect relapses. Cancer is a very heterogeneous disease, and, at the diagnostic level, is defined by many indexes such as histological grade, tumor stage, patient age, sex and, more importantly, genetic background and profiles. Histological evaluation of tumor specimens obtained from tissue biopsy is the gold standard of diagnosis, but often tumors with the same histopathological features respond differently to the same therapy. New generation diagnostic platforms, previously unavailable, have enabled to better characterize transcriptomic signatures that predict tumor behaviour, helping to define diagnosis, prognosis, and the most appropriate therapies [1–3]. Tumor biomarker discovery in biological fluids, such as serum, plasma, and urine, is one of the most challenging aspects

of proteomic research [4]. Many researchers have attempted to identify biomarkers in serum that reflect a particular pathophysiological state. Since the expressed proteins, native, fragmented, or posttranslationally modified, quickly change in response to environmental or pathological stimuli, the serum proteome is considered dynamic, oppositely to the stable nature of the genome. Proteins and their functions can determine the phenotypic diversity that arises from a set of common genes. The study of the serum proteome highlights differences in protein expression reflecting a specific pathological state and provides useful information to diagnose a disease, to evaluate prognosis or therapy response [5].

## 2. Biomarker Discovery in Cancer: The Complexity of Human Serum

Single or a small number of serum-based biomarkers indicative of cancer progression, such as prostate-specific antigen (PSA), alpha-fetoprotein (AFP), CA-125, CA-15.3, CA-19-9, or CEA for prostate, liver, ovary, breast, pancreas, or colon cancer, are currently used. Most of these molecules have been isolated from animals immunized with tumor cells extracts or cell lines, with subsequent screening for monoclonal antibodies against cancer-associated antigens [6]. The above-mentioned proteins increase the accuracy of diagnosis, even though there is an urgent need to isolate and use in clinical practice more specific biomarkers, or groups of biomarkers, to precisely characterize the disease at the diagnostic or prognostic level and to monitor its progression [7, 8]. Biomarkers could also help to predict the response of the patient to anticancer therapy and thereby to guide physicians in choosing the best treatment. This research is more appealing due to the simplicity of obtaining blood samples, but, at the same time, shows limits due to the complexity of the serum protein mixtures. A plethora of molecules from almost every tissue of the body can be found in human serum/plasma. Many of the serum proteins are present at very low concentrations (less than pg/mL), while others are present in very large amounts (more than mg/mL). Serum and plasma are very complex mixtures of proteins and exhibit a broad dynamic range of relative abundance (up to 12 orders of magnitude) [9]. They contain thousands of proteins, whose some are very abundant (e.g., albumin, immunoglobulins, apolipoproteins) and constitute approximately the 95% of the total protein content, but only the 0.1% of total protein species [10, 11]. For these reasons, it is thought that many potentially important proteins and markers, if present at low concentrations, can escape the detection. Evidences show that the circulating fragments from unmodified or post-translationally modified proteins generated in the tumor tissue microenvironment can be used as diagnostic or prognostic markers. Proteolysis within the tissue or deregulated post-translational events (e.g. phosphorylation) generate protein fragments that diffuse into the circulation and could give information about the presence or the progression of the disease, then facilitating the management of the tumor. Among these fragments, the fraction with low molecular weight, the peptidome (<20 KDa, LMW peptides), is protected from renal clearance by interaction with abundant serum proteins and, in particular, seems to be an important source of biomarkers [12].

In order to simplify the biomarker discovery process, many prefractionation methods have been developed to remove highly abundant proteins. These useful precursors of proteomic analysis include the use of immobilized dyes (cibacron blue) [13, 14], immunoaffinity-based techniques [15, 16], solid phase fractionation [17], liquid chromatography [18], or low-molecular weight fraction enrichment [19–21]. A promising method under investigation is based on the use of N-isopropylacrylamide (NIPA) nanoparticles which allow a fast one-step capture and concentration of analytes less than 20–25 KDa in molecular weight [22–24]. At the same time, the nanoparticles are able to protect the proteins from the degradation due to the "ex vivo" enzymatic activity of serum proteases, and, when conjugated with suitable chemical baits, show higher capability to sequester and retain many different proteins from whole serum, on the basis of their chemical and physical properties [25]. With this method, the captured analytes can be recovered by a simple electroelution and then analyzed by HPLC-MS/MS, western blotting or immunoassays for complete molecular characterization. To reduce the complexity of serum proteome, several studies are focused on targeting a specific subset of serum proteins [26]. Glycosylation is one of the most common posttranslational modifications in proteins. Approximately 50% of known eukaryotic proteins are glycosylated [27]. It is known that cancer cells express aberrant glycosylation patterns [28]. The analysis of glycosylated proteins (glycoproteome) has received great interest, because of the glycoproteic nature of the currently used cancer biomarkers. Two major methods have been developed to enrich glycoproteins or glycopeptides, based on chemical capture (reaction between aldehyde groups and hydrazide) [29, 30] and lectin-affinity capture (specific recognition of protein glycan moieties by lectins) [31].

## 3. Mass Spectrometry: The Cancer Biomarker Discovery Tool

Many studies have been focused on the identification of new serum biomarkers by mass spectrometry (MS). This powerful method enables to identify a protein without requiring the knowledge of its amino acid sequence. Further improvement of this technology has provided high accuracy to define mass-to-charge ratio ($m/z$) and to generate high-resolution spectra. In addition, the development of tandem mass spectrometry (MS/MS), able to provide de novo protein sequence information, has enhanced the applications of this technology in proteomics [32, 33]. Several MS methods have been used to characterize body fluids. Different combinations of ionization sources (e.g., MALDI, ESI), analysers (e.g., time of flight TOF, quadrupole, Fourier transform and quadrupole ion traps), and fragmentation methods (e.g., CID collision induced dissociation, ETD electron transfer dissociation) can be used. MALDI-TOF (Matrix-Assisted Laser Desorption/Ionization-time of flight) [34] is based on a soft ionization method where a laser beam generates evaporation of a crystallized sample-matrix mixture. MALDI is used in biochemical areas for the analysis of proteins, peptides and oligonucleotides. Surface-Enhanced Laser Desorption/Ionization Time-of-Flight (SELDI-TOF), a modification of MALDI-TOF, allows the identification of proteins differentially expressed in serum by applying a small amount of sample directly on an array surface involving various chromatographic models based on classical chemistries (i.e., normal phase, hydrophobic, cation- and anion-exchange), affinity-coated surfaces (IMAC, immobilized metal affinity capture), or biomolecular affinity probes [5, 35, 36], with minimal requirements for purification and separation [37]. Selectively retained proteins are analyzed by laser desorption and subsequent ionization. The results are shown by a mass spectrum identifying $m/z$ ratios and peak intensities of peptides/proteins. Data preprocessing (i.e., calibration, baseline correction, normalization, peak detection, and alignment), bioinformatic and statistical analyses are performed in order to highlight and characterize any protein differentially expressed. SELDI-TOF allows the detection of many low molecular weight proteins [38].

Liquid chromatography/electrospray ionization tandem mass spectrometry (HPLC/ESI-MS/MS) technology is an alternative approach for serum biomarker identification. The mixtures of analytes are subjected to HPLC then the solution is nebulized under atmospheric pressure and exposed to a high electrical field which generates a charge on the droplets' surface [39]. Due to the evaporation, droplets become much smaller and enter into the analyzer. HPLC-ESI-MS/MS couples protein fractionation with mass spectrometry, where peptide sequence tags can be produced from peptide fragments. Tandem mass spectrometry and data analysis by suitable bioinformatic tools, algorithms and databases, provides a powerful method to characterize peptides at the aminoacidic level, allowing a highly refined analysis [40].

The use of mass spectrometry for serum biomarker discovery is quite simple: spectral peaks (plots representing on the $x$-axis the $m/z$ ratios of ions, and on the $y$-axis the detected ion abundance), are identified in a pathological group and compared with those obtained from normal control groups. Differences in spectral profiles detect putative biomarkers. Essentially four strategies have been used in MS biomarker discovery: analysis of polypeptides separated by electrophoresis or chromatography, with or without prior fragmentation; analysis of enzymatic peptide fragments separated by HPLC and then analyzed by ESI or MALDI; analysis of proteins adsorbed on a solid surface; and analysis of specific serum fractions, such as the peptidome or the glycoproteic fraction [4].

Due to the lack of effective screening test, these methods have been applied to the serum biomarker discovery in many tumors, including those with high mortality, such as ovary, lung, liver, and pancreatic cancer. Here we report the results of studies focused on serum biomarker identification in the above-mentioned types of cancer. Several protein profiles and specific proteins (Tables 1, 2, 3, and 4) have been characterized and classified as putative biomarkers.

### 3.1. Serum Biomarkers in Ovarian Cancer.

Despite advances in cancer therapy, mortality due to ovarian cancer is almost unmodified during the last decades [50]. Ovarian cancer is usually diagnosed at a late clinical stage in more than 80% of patients [51]. In this group, the 5-year survival is approximately 35%. By contrast, the 5-year survival for patients with stage I ovarian cancer is more than 90% and surgery alone can be used as elective therapy [52]. Cancer antigen 125 (CA-125) is the most widely used biomarker for ovarian cancer. Elevated levels of CA-125 are detected in about 80% of patients with advanced-stage disease, but they are increased in only 50–60% of patients with early stage ovarian cancer [53]. The calculated positive predictive value for CA-125, considered as a single marker, is less than 10%, but this value was improved by ultrasound screening methods [54].

### 3.1.1. Proteomic Patterns for Ovarian Cancer Detection.

Analysis of sera by TOF-MS provided specific signature patterns which can be compared to distinguish ovarian cancer from benign disease or normal individuals. A training set of 50 sera from unaffected women and 50 from patients with ovarian cancer were analyzed by SELDI/TOF by using

TABLE 1: ovarian cancer serum biomarkers identified after MS-based studies.

| Putative ovarian cancer biomarker | Expression in ovarian ca sera | References |
| --- | --- | --- |
| Apolipoprotein I | Decreased | [36, 41] |
| Transthyretin | Decreased | [36, 41] |
| Inter-α-trypsin inhibitor heavy chain H4 (cleavage fragment) | Increased | [36] |
| Haptoglobin-α chain | Increased | [42] |
| Haptoglobin I precursor | Increased | [43] |
| Fibrinopeptide A | Increased | [44] |
| Serum amyloid A1 | Increased | [45] |
| Hemoglobin $\alpha$ and $\beta$ chain | Increased | [41, 46] |
| Transferrin | Decreased | [41] |
| Keratin 2a | Increased | [47] |
| Glycosyltransferase-like 1B | Increased | [47] |
| Complement component 3 precursor | Decreased | [47] |
| Complement component 4A preprotein | Decreased | [47] |
| Casein kinase II alpha 1 subunit isoform a | Decreased | [47] |
| D-amino-acid oxidase | Decreased | [47] |
| Transgelin 2 | Increased | [47] |
| Inter-alpha (globulin) inhibitor H4 | Increased | [47] |
| Fibrinogen, alpha chain isoform alpha preprotein | Increased | [47] |
| CC2 motif ligand 18 (CCL18) | Increased | [48] |
| CXC motif ligand 1 (CXCL1) | Increased | [48] |
| Connective tissue-activating peptide III (CTAPIII) | Decreased | [49] |
| Platelet factor 4 (PF4) | Decreased | [49] |

a C16 hydrophobic interaction protein chip. An algorithm identified a proteomic pattern able to distinguish cancer from non-cancer subjects. The pattern was then used to differentiate an independent set of 116 samples (50 ovarian cancer, 66 non-malignant disease or unaffected), yielding 100% sensitivity and 95% specificity [67]. Some of the characteristic ion peaks in the previous signature have been sequenced and described in an independent cohort of ovarian cancer sera [36]. One hundred nine serum samples from ovarian cancer patients, 19 sera from individuals with benign tumors, and 56 from healthy donors were analyzed on strong anion-exchange surface using SELDI-TOF. Three panels of candidate protein biomarkers were obtained (1st: 4.4 kDa, 15.9 kDa, 18.9 kDa, 23 kDa, 30.1 kDa—95.7% sensitivity, 82.6% specificity; 2nd: 3.1 kDa, 13.9 kDa, 21.0 kDa, 79.0 kDa, and 106.7 kDa—81.5% sensitivity, 94.9% specificity; 3rd: 5.1 kDa, 16.9 kDa, 28 kDa, 93 kDa—72.8% sensitivity, 94.9% specificity). The protein panels correctly diagnosed 41/44 blind test samples: 21/22 malignant ovarian cancers, 6/6 low malignant potential tumors, 5/6 benign tumors, 9/10 normal individuals [68]. A four-peak model ($m/z$ 6195, 6311, 6366, 11498), performing better than CA-125, has been identified for diagnosis or monitoring of the therapy in ovarian cancer. This study

TABLE 2: Liver cancer serum biomarkers identified after MS-based studies.

| Putative liver cancer biomarker | Expression in liver ca sera | References |
| --- | --- | --- |
| Complement c3 | Increased | [55] |
| Histidine rich glycoprotein | Increased | [55] |
| CD14 | Increased | [55] |
| Hepatocyte growth factor | Increased | [55] |
| C-terminal part of vitronectin V10 fragment | Increased | [56] |
| Protein complement C3a | Increased | [57] |
| Annexin VI isoform | Increased | [58] |
| Complement component 9 | Increased | [58] |
| Ceruloplasmin | Increased | [55, 58] |
| Serum amyloid A4 | Increased | [58] |
| Serum amyloid A2 | Increased | [58] |
| Serum amyloid A1 isoform 2 | Increased | [58] |
| Cystatin C | Increased | [59] |
| Neutrophil-activating peptide 2 | Increased | [60] |
| Thrombin light chain | Increased | [61] |
| Growth-related oncogene alpha (GRO-alpha) | Increased | [61] |
| Alpha-1 acid glycoprotein | Increased | [62] |
| Haptoglobin | Increased | [63] |
| N terminus complement C3f | Decreased | [64] |
| Fibrinopeptide | Decreased | [64] |
| Complement C4 alpha peptides | Increased | [64] |
| Zyxin peptide | Increased | [64] |
| Coagulation factor XIII peptide | Increased | [64] |
| Biliverdin diglucuronide | Increased | [64] |
| Heat-shock protein 27 | Increased | [65] |
| MYH2 protein | Increased | [66] |
| Mitochondrial ATP synthase | Increased | [66] |
| Sulphated glycoprotein-2 | Increased | [66] |
| Glial fibrillary acidic protein | Increased | [66] |

TABLE 3: Lung cancer serum biomarkers identified after MS-based studies.

| Putative lung cancer biomarker | Expression in lung ca sera | References |
| --- | --- | --- |
| Serum amyloid protein A | Increased | [71–76] |
| Haptoglobin alpha subunit | Increased | [77] |
| Hepatocyte growth factor | Increased | [77] |
| Transthyretin | Decreased | [75, 78] |
| Alpha-1 acid glycoprotein 1 and 2 | Increased | [74] |
| Apolipoprotein A4 peptides | Increased/decreased | [79] |
| Fibrinogen alpha chain | Increased | [79] |
| Limbin | Increased | [79] |

was conducted on a training (31 primary cancer, 16 benign ovarian disease, 25 healthy controls) and a blind test set (23

TABLE 4: Pancreatic cancer serum biomarkers identified after MS-based studies.

| Putative pancreatic cancer biomarker | Expression in pancreatic ca sera | References |
| --- | --- | --- |
| Apolipoprotein CIII | Increased | [82] |
| Serum amyloid protein A | Increased | [83] |
| Apolipoprotein A-II | Decreased | [84] |
| Apolipoprotein A-I | Decreased | [84] |
| Transthyretin | Decreased | [84] |
| Alpha-2 macroglobulin | Increased | [85] |
| Ceruloplasmin | Increased | [85] |
| Complement 3C | Increased | [85] |
| Platelet factor 4 | Decreased | [86] |
| Mannose-binding lectin 2 | Increased | [87] |
| Myosin light chain kinase 2 | Increased | [87] |
| CXC chemokine ligand 7 | Decreased | [88] |
| TIMP1-ICAM1 | Increased | [89] |
| Alpha-1 antitrypsin | Increased | [90] |
| Fibrinogen gamma | Increased | [91] |
| C14orf166 | Increased | [92] |
| Alpha-1 antichymotrypsin | Increased | [93] |

ovarian cancer, 15 benign, 5 normal). The four peak model showed a sensitivity of 90.8% and a specificity of 93.5% in the training set, and a sensitivity of 87% and a specificity of 95% in the blind test set in discriminating cancer from non cancer patients [69]. Hocker et al. [70] analyzed mass spectrometry peak differences in 35 ovarian cancer patients and 16 disease-free subjects. Proteomic profiles distinguished early-stage from advanced cancer with a sensitivity of 80% and a specificity of 93%.

*3.1.2. Glycoproteomics in Ovarian Cancer.* An et al. [80] developed a glycomic approach to identify oligosaccharide markers for ovarian cancer by analyzing ovary cell lines supernatants and then confirming their presence in sera from patients and healthy controls. Changes in glycosilation were monitored by MALDI-Fourier Transform Ion Cyclotron Resonance-MS. Approximately 15 unique serum glycan markers were detected in all patients and were absent in controls. Leiserowitz et al. [81] analyzed glycan markers and CA-125 levels in 48 sera from ovarian cancer women and 24 controls. Oligosaccharides were cleaved from serum glycoproteins and isolated using solid phase extraction. MALDI-Fourier Transformation-MS was used to identify peaks. Sixteen unique oligosaccharide signals were identified in most of the cancer patients (44/48) and just in 1/24 controls (sensitivity 91.6%, specificity 95.8%).

*3.1.3. Ovarian Cancer Biomarker Proteins Involved in Inflammatory Processes.* Several proteins involved in inflammatory processes and acute-phase response have been identified as putative biomarkers for ovarian cancer. In a study by

Zhang et al. [36], sera were fractionated by anion exchange chromatography. Aliquots were bound in triplicate with a randomized chip/spot allocation scheme to IMAC3-Cu, SAX2, H50, and WCX2 protein chip array. For biomarker identification, proteins were purified, separated by SDS-PAGE, and analyzed by MS/MS. Three proteins/protein fragments, described as acute-phase reactants (apolipoprotein A1, $m/z$ 28043, a truncated form of transthyretin, $m/z$ 12828, and a fragment of inter-$\alpha$-trypsin inhibitor heavy chain H4, $m/z$ 3272), were identified as putative biomarkers able to improve the detection of early stage ovarian cancer [36]. Ye et al. [42] identified by SELDI an 11.7 kDa peak showing higher intensity in cancer sera. After protein purification and LC-MS/MS analysis, the alpha chain of haptoglobin, an acute phase reactant, was identified and further validated by western blot and ELISA as a potential biomarker for ovarian cancer, by using a specific polyclonal antibody against the peptide identified by MS/MS analysis. The marker was 2-fold-expressed in cancer sera and had 64% sensitivity and 90% specificity when used alone, or 91% and 95% sensitivity and specificity, if combined with CA-125. After high abundance protein removal and two-dimensional gel electrophoresis (2-DE), Ahmed et al. [43] performed nanoelectrospray quadrupole-quadrupole time of flight mass spectrometry (n-ESIQ-(q)TOF-MS) and MALDI/TOF analysis on six protein spots over-expressed in cancer sera. Protein isoforms of haptoglobin-I precursor (HAPI), a liver glycoprotein present in human serum, were identified as putative novel biomarkers and confirmed by 2-DE and western blotting on the serum from healthy controls and grade 1 and 3 ovarian cancer patients. Bergen III et al. [44] analyzed by nano-LC-ESI-TOF-MS or Fourier Tranform Ion Cyclotron Resonance (FT-ICR) the low molecular weight serum fraction obtained by ultrafiltration and identified several candidate biomarkers. Among these, the fibrinopeptide-A, already described as involved in acute phase reactions and elevated in many cancer including ovary, was detected. Two peaks (11.7 and 11.5 kDa) were identified by SELDI-TOF in thermostable plasma fractions from 27 ovarian cancer and 34 control sera. A method involving cysteine modifications, 2-DE, and HPLC allowed to characterize the peaks corresponding to serum amyloid A1, an acute phase reactant, and its N-terminal arginine truncated form [45]. Kozak et al. [41] identified by micro-LC-MS/MS four biomarkers for early stage ovarian cancer (transthyretin, apolipoprotein A1, transferrin, and beta-hemoglobin), corresponding to already described 13.9-TTR-, 12.9-TTR-, 15.9-Hb-, 28-ApoAI-, 79-Tf-kDa SELDI peaks [68]. Differential expression of these proteins in sera was also confirmed by western blot and ELISA. Lopez et al. [47] set-up a workflow using carrier protein-bound affinity enrichment of serum samples directly coupled with MALDI/TOF. Subsequent tandem MS analysis defined the serum protein-bound peptides' sequence. The procedure was able to identify several specific biomarker panels to differentiate stage I ovarian cancer from unaffected and age-matched women. Among the peptides identified, proteins involved in inflammatory processes (complement component 3 precursor, complement component 4A preprotein, inter-alpha globulin inhibitor H4), as well as the glycosyltransferase-like 1B, a peroxisomal

oxidation enzyme (D-amino-acid oxidase), two proteins involved in cancerogenesis (transgelin 2, casein kinase II alpha 1 subunit isoform a), fibrinogen alpha chain isoform alpha preprotein, and keratin 2a were described as putative biomarkers. In a study on sera from patients with ovarian cancer, compared with sera from patients with benign masses or different type of cancer, it has been shown that the chemokines CC2 motif ligand 18 (CCL18) and CXC motif ligand 1 (CXCL1), identified by MALDI-MS/MS analysis and further validated by ELISA, can be considered as novel circulating tumor markers for differential diagnosis between ovarian cancer and benign masses (sensitivity 92%, specificity 97%) [48]. A nested case-control study performed on 295 sera from women pre-dating their ovarian cancer diagnosis and 585 matched control samples, showed that two peaks identified by MALDI, described as the connective tissue-activating peptide III (CTAPIII) and the platelet factor 4 (PF4), can be associated with CA-125 to improve early diagnosis [49].

### 3.1.4. Ovarian Cancer Biomarker Proteins Involved in Other Functions.
Woong-Shick et al. [46] studied by SELDI-TOF 35 sera from ovarian cancer patients in comparison to 10 from normal women. After protein purification and N-terminal sequencing, hemoglobin-$\alpha$ and $\beta$ chains were described as corresponding to the most distinctive peaks differentially expressed (15.1 and 15.8 kDa). ELISA validation test for intact hemoglobin indicated a sensitivity of 77% in sera from ovarian cancer patients. As above-mentioned, hemoglobin was also described as a putative ovarian cancer biomarker in a study by Kozak et al. [41].

### 3.2. Serum Biomarkers in Liver Cancer.
Liver cancer is often diagnosed at very late stage and is associated to poor prognosis, high recurrence and mortality. Hepatocellular carcinoma (HCC), the most frequent liver neoplasm, is the fifth most common cancer, affecting approximately one million people every year, with an incidence almost corresponding to death rate [94] and a 5 year survival ranging from 17% to 50%. Some predisposing factors, such as viral infections [95], diabetes, metabolic syndromes, exposition to aflatoxin, or alcohol consumption, are frequently related to liver tumor initiation. This allows a management of the patients at risk, making it possible in some cases to diagnose the disease earlier. The most important liver cancer serum biomarker is alpha-fetoprotein (AFP), an oncofetal glycoprotein with elevated levels in patients affected by cirrhosis and HCC. The sensitivity and specificity of AFP range between 60–80% and 70–90%, respectively [96]. For this reason, AFP test utility for screening procedures is questionable. Many studies have been focused on characterizing differentially expressed proteins in sera from patients affected by liver cancer or predisposing diseases and, similarly to ovarian cancer, proteomic profiles and proteins have been identified.

### 3.2.1. Proteomic Patterns for Liver Cancer Detection.
A total number of 117 HCV-positive sera from 39 patients affected by low-grade fibrosis, 44 with cirrhosis without HCC, and 34 with both cirrhosis and HCC were preprocessed by anion-exchange fractionation and analyzed by SELDI-TOF. A four

markers panel (7486, 12843, 44293, 53598 Da) identified HCC with a sensitivity of 100% and a specificity of 85% in a two-way comparison of HCV-cirrhosis versus HCV-HCC training set. Sensitivity and specificity for the correct identification of HCC were 68% and 80% for random test samples. Fibrosis patients were distinguished from cirrhotic using a five marker panel (2873, 6646, 7775, 10525, 67867 Da, sensitivity and specificity 100% and 85%, 80% and 67% in the training and random test samples, resp.). After purification, MS/MS analysis and immunoassay validation, the 6646 Da protein was identified as apolipoprotein C-I and described as a marker to differentiate liver fibrosis from cirrhosis [97]. SELDI-TOF protein-chip technology was applied to analyze sera from patients affected by HCV-associated chronic liver diseases with (64 samples) or without (77 samples) HCC. Samples were randomly split into two analysis groups. Six selected protein peaks ($m/z$ 3444, 3890, 4067, 4435, 4470, 7770) gave information to perform early diagnosis and to distinguish HCC from chronic liver disease in the absence of HCC (sensitivity and specificity 83% and 76%). The model was also applied to the analysis of sera from 5 subjects HCC-free and from 7 HCC patients collected before the diagnosis by ultrasonography. The markers allowed to correctly predict the presence of HCC in 6/7 patients [98]. Wu et al. [99] identified serum proteins and peptide profiles to differentiate HBV-related HCC and HBV-related cirrhosis. Forty-five protein peaks distinguished HCC from LC (liver cirrhosis) samples. The most significant SELDI 3892 $m/z$ peak showed sensitivity and specificity of 69% and 83% and was identified also in six AFP-negative patients. The 3892 peak was considered as a complementary diagnostic marker or a potential marker for positive or negative $\alpha$-fetoprotein HCC. SELDI-TOF analysis of sera from 120 patients affected by HCC and 120 affected by cirrhosis showed five proteomic peaks ($m/z$ 3324, 3994, 4665, 4795, and 5152) able to achieve, especially for early stage HCC, a diagnostic value better than serum AFP (83% sensitivity and 92% specificity in the test set) [100]. Cui et al. [101] formulated classification trees, based on SELDI serum protein profiles, able to distinguish patients affected by chronic hepatitis B, cirrhosis, and HCC from healthy individuals. Samples were divided into training and testing groups, each composed by HBV, liver cirrhosis, HCC patients matched with normal controls. Decision trees distinguished HCC with 90% sensitivity and 89% specificity, cirrhosis with 100% sensitivity and 86% specificity, and HBV patients with 85% sensitivity and 84% specificity.

*3.2.2. Glycoproteomics in Liver Cancer.* Goldman et al. [102] used a glycomic approach to evaluate the abundance of 83 N-glycans in a total of 202 sera from 73 HCC patients, 52 with chronic liver disease and 77 controls. Glycans were enzymatically obtained from serum and permethylated before MALDI-TOF analysis. The abundance of 57 N-glycans resulted significantly altered in HCC samples. Six glycans were used to differentiate HCC cases from controls and showed sensitivity and specificity of 73–90% and 36–91%, respectively. A combination of three N-glycans ($m/z$ 2472.9, 3241.9, 4052.2) was able to classify HCC with 90% and 89% sensitivity and specificity in an independent validation set of

patients with chronic liver disease. Two-hundred and three serum samples collected from 73 HCC cases, 52 chronic liver disease, and 78 healthy subjects were treated for N-glycans releasing and then analyzed by MALDI. Seven glycan peaks achieved good performance in distinguishing HCC from chronic liver disease patients and normal individuals [103]. After depletion of high abundance proteins, Liu et al. [55] analyzed 27 sera from early HCC in comparison to 27 cirrhosis patients in order to identify glycoprotein biomarkers. A lectin array of 16 selected lectins was used to define glycan structures showing changes between the two groups of samples. Samples were then analyzed by exactag labeling, lectin extraction and LC-MS/MS. Complement C3, ceruloplasmin, histidine rich glycoprotein (HRG), CD14 and hepatocyte growth factor (HGF), as validated by western blot, were considered putative biomarkers in differentiating early HCC from cirrhosis with a sensitivity of 72% and a specificity of 79%.

*3.2.3. Liver Cancer Biomarker Proteins Involved in Inflammatory Processes.* Paradis et al. [56] studied by SELDI-TOF eighty-two sera from patients with cirrhosis, either without (38 samples) or with (44 samples) HCC. Thirty protein peaks significantly differentiated cirrhotic patients affected by HCC from those unaffected. An algorithm showing the six highest scoring peaks allowed the correct classification of patients with or without HCC in 92% of individuals in the test set and in 90% in the validation set. After sera fractionation (IMAC-Zn spin column), analysis on NP20 chip array and protein recovery from tricine SDS-PAGE, tandem MS was performed and the highest discriminating peak (8.9 kDa) was described as the C-terminal part of the V10 fragment of vitronectin, a protein involved in cell adhesion, humoral defense mechanism as well as cell invasion. SELDI-TOF was used to identify differentially expressed proteins in hepatocarcinoma (55 samples in total, 31 HBV-related and 24 HCV-related) and chronic hepatitis patients (18 HBV and 30 HCV). After serum fractionation by anionic exchange chromatography, the proteins were characterized by 2-DE separation and LC-MS/MS analysis. The protein complement C3a (about 8.9 kDa), elevated both in chronic HCV and HCV-related HCC patients, was identified as a candidate biomarker and further validated by PS20 chip immunoassay and western blot [57]. Yang et al. [58] used 2-DE combined to nano-HPLC-ESI-MS/MS to identify 14 proteins differentially expressed (12 up and 2 down-regulated) in HCC patients with respect to normal controls. On the other hand, using whole serum trypsin-digested and then analyzed with nano-HPLC-ESI-MS/MS, twenty-nine proteins were identified with high levels of confidence. Six of them (Annexin VI isoform, Complement component 9, Ceruloplasmin-ferroxidase-, Serum amyloid A4, Serum amyloid A2, Serum amyloid A1 isoform 2), playing a role in immune and acute phase response or in membrane dynamics along endocytosis or exocytosis pathways (Annexin VI), were detected only in HCC patients. An 11 peak algorithm, generated by SELDI/TOF analysis, distinguished patients with HCC (41 samples) from those with hepatitis C cirrhosis (51 samples) better than the currently used biomarkers AFP,

AFP L3 (Lens culinaris agglutinin-reactive AFP) and PIVKA-II (prothrombin induced by vitamin K absence-II). Within the 11-protein signature, the 13.4 kDa feature was purified, identified as cystatin C by MS/MS analysis and further validated by ELISA. The cystatin C, a cysteine protease inhibitor marker of inflammation as well as renal function, resulted overexpressed in HCC samples and was described as a marker to distinguish HCC from HCV-related cirrhosis patients [59]. He et al. [60] performed by SELDI-TOF serum profiling on 81 patients with HBV-related HCC and 33 normal controls, randomly split into a training and a testing set. Six proteomic peaks ($m/z$ 3157.33, 4177.02, 4284.79, 4300.80, 7789.87, 7984.14) were considered to construct the best classification tree (sensitivity 95%, specificity 100% in the testing set). Protein fraction corresponding to the 7489 $m/z$ peak was isolated and characterized by MS/MS analysis as the inflammatory cytokine neutrophil activating peptide 2 (NAP-2). NAP-2 was validated by immunohistochemistry in HCC tissues and resulted specifically associated to hepatitis B-related HCC [60]. Sera from eighty-one patients with HBV-related HCC and 80 healthy controls were divided in two sets and analyzed by SELDI-TOF. Candidate biomarkers were purified and identified by MS/MS and database searching. Two proteins, the thrombin light chain ($m/z$ 4096) and the chemokine growth-related oncogene alpha (GRO-alpha) ($m/z$ 7860) were selected as putative biomarkers. A clinical validation set composed by 48 HCC, 54 liver cirrhosis, 151 patients with other cancers and 42 healthy donors was analyzed to confirm data by SELDI-immunoassay. The proteins, when associated to AFP, resulted in a sensitivity of 91.7% and a specificity of 92.7% [61]. Sera from cirrhosis and HCC patients were analyzed by cleavable stable isotope labeling (cICAT) coupled to LC-ESI-MS/MS. Among 31 proteins differentially expressed, the alpha-1 acid glycoprotein (AGP), an acute phase reactant, was chosen for western blot assay and validation in a separate study. AGP was useful for discrimination of HCC from cirrhosis in patients with AFP less than 500 ng/mL [62]. A study based on 2D-gel electrophoresis and MALDI-TOF in patients with hepatocarcinoma or liver cirrhosis revealed five proteins differentially expressed (haptoglobin, Hp2, preprohaptoglobin, SP40 and SAA1). Western blot analysis showed haptoglobin, the most representative protein, as overexpressed in HCC patients. When used in association to AFP, the molecule improved the diagnostic accuracy. Serum haptoglobin also showed diagnostic potential in AFP-negative patients [63]. Several peptides in the serum low molecular weight fraction were identified by MALDI and then characterized by LC-MS/MS. Differentially expressed peptides were described as truncations of N terminus of complement C3f, a fibrinopeptide, complement C4alpha peptides, a zyxin peptide, a coagulation factor XIII peptide, and a biliverdin diglucuronide [64].

*3.2.4. Liver Cancer Biomarker Proteins Involved in Other Functions.* Feng et al. [65] used a strategy based on sonication, albumin and immunoglobulin depletion, 2-DE and MALDI-TOF MS/MS to analyze 20 sera, respectively, from HCC, hepatitis B (HBV) patients and normal subjects. The same number of additional sera from corresponding groups was used for the validation test. Height proteins, involved in inflammatory processes or classified as acute phase reactants (alpha-1 antitrypsin, clusterin, ceruloplasmin, haptoglobin alpha2 chain, transferrin, and transthyretin) as well as alfa-fetoprotein and the heat-shock protein 27, a stress-inducible protein acting in thermotolerance, cell proliferation, and apoptosis, were differentially expressed in the above-mentioned groups. Validation by western blot analysis revealed HSP27 expressed in 90% of HCC, in 10% of HBV and in none of normal sera. Wu et al. [66] compared by 2-DE and mass spectrometry sera from HCC patients and normal controls. Eight protein spots differentially expressed were analyzed and four proteins were identified as putative biomarkers (MYH2 protein, mitochondrial ATP synthase, sulphated glycoprotein-2-clusterin SGP-2-, and glial fibrillary acidic protein (GFAP). SGP-2, known to be involved in inflammation and in the regulation of cellular proliferation, was also confirmed by immunoblotting in an independent set of samples.

*3.3. Serum Biomarkers in Lung Cancer.* Lung cancer is one of the leading causes of cancer-related mortality worldwide and is responsible for 1.3 million deaths worldwide annually [104, 105]. The poor prognosis is evidenced by the 5-year survival rate which is less than 15% [106]. Lung cancers are grouped into small-cell lung cancer (SCLC) and non-small-cell lung cancer (NSCLC), consisting of adenocarcinomas, squamous cell carcinomas, and large-cell carcinomas [107, 108]. NSCLCs comprise approximately 80% of all lung cancers [50], with adenocarcinomas and squamous cell lung cancers each accounting for approximately 30%. Many serologic biomarkers of lung cancer have emerged recently: these include carcinoembryonic antigen (CEA), the cytokeratin 19 fragment CYFRA21-1, cancer antigen CA-125 [109], plasma kallikrein [110], progastrin-releasing peptide (ProGRP), and neuron-specific enolase (NSE) [111].

*3.3.1. Proteomic Patterns for Lung Cancer Detection.* In a study on 208 sera (158 lung cancer, 50 healthy controls), Yang et al. [112] identified a 5 proteins peak pattern (11493, 6429, 8245, 5335, 2538 Da) which, in a blind test, achieved sensitivity of 86.9% (79% for stage I/II lung cancers), specificity of 80% and positive predictive value of 92.4%. In particular, the pattern sensitivity was 91.4% in the detection of NSCLC. In a group of sera (54 SCLC, 24 NSCLC, 32 pneumonia patients, and 40 healthy subjects), SELDI-TOF spectra data analyzed by support vector machine (SVM) gave three patterns able to distinguish SCLC from pneumonia, NSCLC patients and from healthy individuals better than neuron specific enolase (NSE). The sensitivity and specificity ranged from 88% to 83% and from 91% to 75%, respectively [113]. A 17 MS protein signature was identified in a study on 139 lung cancer patients (stage III-IV), 158 healthy individuals and then validated in two sets of 126 (63 lung cancer stage III-IV, 63 controls) and 50 (25 lung cancer stage I-II, 25 controls) individuals. The signature distinguished lung cancer patients from normal subjects and showed sensitivity and specificity of 87.3% and 81.9% in the first validation set, and 90% and 67% in the second one [114]. Du et al. [115]

captured and concentrated serum peptides by using magnetic beads-based weak cation exchange on the ClinProt robotic platform. The peptides were analyzed by MALDI-TOF. A 5 protein fingerprint distinguished SCLC patients (30 samples) from healthy individuals (44 samples) with a specificity of 97% and a sensitivity of 90%. In particular, 89% of stage I/II SCLC were correctly diagnosed.

### 3.3.2. Glycoproteomics in Lung Cancer.
Glycoproteomic approaches have been applied to identify biomarkers for NSCLC early diagnosis. After immunoaffinity depletion of highly abundant serum proteins, glycoproteins were captured and enriched by hydrazide chemistry, recovered and then analyzed by LC-MS/MS. Thirty-eight glycopeptides from 22 proteins distinguished cases from controls. Three of these proteins (alpha-1-antichymotrypsin ACT, insulin-like growth factor-binding protein 3 IGFBP3, lipocalin-type prostaglandin D synthase L-PGDS) were verified by ELISA, showing correlation with MS results [116].

### 3.3.3. Lung Cancer Biomarker Proteins Involved in Inflammatory Processes.
Howard et al. [71] identified by MALDI-TOF a peak ($m/z$ 11702) differentially expressed in lung cancer patients (24 subjects) with respect to individuals with no evidence of cancer (17 subjects). After purification and peptide mapping, the peak was described as the acute phase reactant serum amyloid protein A and further validated by ELISA. Bharti et al. [77] analyzed by MALDI-TOF differentially expressed albumin-depleted serum proteins from SCLC patients and controls, recovered from a silver-stained SDS-PAGE. The peptides were characterized by sequencing. Haptoglobin $\alpha$-subunit, validated by immunoblot, was considered as a biomarker with its level correlating with the disease stage. In addition, they analyzed by ELISA the levels of hepatocyte growth factor (HGF), a multifunctional protein which regulates both cell growth and motility, and described it as a potential SCLC biomarker. A study performed on 218 sera from 175 lung cancer patients and 43 controls by SELDI-TOF showed an 11.6 kDa protein peak significantly elevated in cancer sera and increased in association to the clinical stage. Serum amyloid A protein was identified by tricine SDS-PAGE and MALDI-MS/MS analysis as a biomarker to discriminate lung cancer patients from healthy individuals. The marker was validated by immunoprecipitation and ELISA in the same samples and showed sensitivity of 84% and specificity of 80% [72]. In a study on 227 sera (146 lung cancer, 41 benign lung disease, 40 normal subjects) three peaks differentially expressed ($m/z$ 13780, 13900, 14070) were identified by SELDI. The peaks corresponded to native transthyretin (negative acute-phase reactant) and its two variant, as demonstrated by SDS-PAGE and ESI-MS/MS analysis and further validated by immunoprecipitation and immunoblotting. The transthyretin expression was significantly lower in lung cancer sera compared with sera from normal individuals, but higher compared with those obtained from benign lung disease. Subsequent ELISA assay indicated that the levels of transthyretin were consistent with those obtained by SELDI, showing approximately 65–75% sensitivity and specificity [78]. The diagnostic accuracy of

MALDI in analyzing unfractionated serum was assessed in a study by Yildiz et al. [117] performed on 142 sera form lung cancer patients matched with 146 samples from normal controls. Samples were split into training and test set. A serum proteomic signature of seven features achieved an overall accuracy of 78% and 72.6% in the training and blinded test set, respectively. The peptides around 11500 Da were further analyzed by using SDS-PAGE separation and LC-MS/MS and described as a cluster of truncated forms of serum amyloid A protein. In a study on 154 sera from pre-treated patients (55% early, 45% advanced-stage) an isoform of serum amyloid A, corresponding to an 11.6 kDa SELDI peak and characterized by SDS-PAGE and tandem MS, was found to be elevated in patients with poor prognosis. In this study, sera were prefractionated in six protein fractions on the basis of their isoelectric points [73]. Forty-nine proteins were found to be differentially expressed by LC-ESI-MS/MS in pools of sera from nonsmall cell lung cancer (adeno-carcinoma and squamous cell carcinoma) with respect to healthy controls. Multiple reaction monitoring (MRM) assay was used to confirm the abundance of four selected proteins (serum amyloid A—SSA-, alpha-1 acid glycoproteins 1 and 2—AAG 1 and 2-, clusterin—CLU-). SSA and AAG 1 and 2 showed higher spectral count in lung cancer serum pool [74]. Analysis by SELDI-TOF of 227 sera showed 5 peaks ($m/z$ 11530, 11700, 13780, 13900, 14070) identifying native serum amyloid A protein and transthyretin, and some of their variants as lung cancer biomarkers [75]. Serum amyloid A1 and A2 proteins were identified by LC-ESI-MS/MS in lung cancer pooled sera after SDS-PAGE fractionation. The levels were higher in lung cancer patients with respect both to patients affected by other pulmonary diseases or different cancers and to healthy controls. The results were confirmed by ELISA. Moreover, SSA expression in lung cancer samples was detected by tissue-microarray analysis [76].

### 3.3.4. Lung Cancer Biomarker Proteins Involved in Other Functions.
Ueda et al. [79] described a method based on enrichment of the peptidomic fraction and analysis by nano-LC-MS/MS. After further characterization by MRM-based relative quantification, peptides from apolipoprotein A4 (APOA4), fibrinogen alpha chain (FIBA), and limbin (LBN), a positive regulator of the hedgehog signaling pathway, have been identified as useful biomarkers for early detection and staging of lung tumors.

### 3.4. Serum Biomarkers in Pancreatic Cancer.
Pancreatic cancer is the fourth leading cause of cancer death in both men and women. The high mortality associated with it can be essentially attributed to advanced stage of disease at patient presentation. Few patients with pancreatic cancer are cured without surgical resection. The overall 5-year survival is about 5%, and only 20% of patients are candidates for surgical resection and possible treatment. For this small percentage of patients undergoing resection, even when followed by multimodal therapy, 5-year survival rates are still less than 25% [118–120]. Current methods for diagnosing pancreatic cancer are relatively ineffective to identify small potentially curable lesions. The marker recommended in clinical practice

is serum CA-19-9. However, this marker is of little utility in establishing early diagnosis [121]. To date, there are no efficient modalities to early detect pancreatic cancer and strategies to improve survival have focused just on chemotherapy in the neoadjuvant setting or after resection.

*3.4.1. Proteomic Patterns for Pancreatic Cancer Detection.* In a study on 15 healthy controls, 24 cancer and 11 chronic pancreatitis patients prospectively collected, the low molecular serum fraction was enriched and analyzed by MALDI-TOF. An eight peaks serum signature ($m/z$ 4470, 4792, 8668, 8704, 8838, 9194, 9713, 15958) differentiated cancer patients from normal individuals (sensitivity and specificity of 88% and 93%), cancer from pancreatitis patients (sensitivity and specificity of 88% and 30%), and cancer from healthy plus pancreatitis-affected individuals (sensitivity and specificity of 88% and 66%). The most significant peak, $m/z$ 9713, was described by MS/MS analysis as the apolipoprotein CIII [82]. Liu et al. [122] used SELDI-TOF technology to differentiate cancer from different pancreatic conditions, by studying 118 serum samples, split in training and test set. Two MS patterns, differentiating pancreatic adenocarcinoma from healthy controls and chronic pancreatitis, yielded in the test set sensitivity and specificity of 91.6% (cancer versus controls) and sensitivity of 90.9%, specificity of 80% (cancer versus chronic pancreatitis).

*3.4.2. Pancreatic Cancer Biomarker Proteins Involved in Inflammatory Processes.* An orthotopic nude mouse model of human pancreatic cancer was used to detect serum biomarkers [83]. Mice were injected with a human pancreatic cancer cell line and then were divided in groups treated with anti-cancer drugs for some weeks. Sera were recovered and analyzed by SELDI Proteinchip technology. Plasma from 135 pancreatic cancer patients and 113 healthy individuals were at the same time examined. An 11.7 kDa protein peak, correlating with tumor weight, was detected in mice sera. After purification and separation by SDS-PAGE, the corresponding protein was identified as serum amyloid protein A and confirmed by western blotting. The level of SAA detected in plasma of pancreatic cancer patients correlated with the clinical stage. Ninety-six sera from pancreatic cancer patients undergoing surgery were fractionated by chromatography and analyzed by SELDI in comparison with as many sera from healthy controls. Twenty-four differentially expressed peaks were identified. Twenty-one of them resulted in downregulated pancreatic cancer samples. After purification, several proteins were identified by peptide mapping and postsource decay-matrix-assisted laser desorption ionization-TOF-MS. Down-regulated apolipoprotein A-II, transthyretin, and apolipoprotein A-I were described as potential markers in pancreatic cancer [84]. Hanas et al. [85] studied sera from pancreatic cancer patients by gel electrophoresis in order to highlight protein bands differing quantitatively. The proteins were analyzed and characterized by ESI ion-trap tandem MS. Three high mass proteins ($\alpha$-2 macroglobulin, ceruloplasmin, complement 3C) were elevated in cancer sera with respect to controls. The ESI-MS analysis revealed great heterogeneity especially in the

low mass region. By statistical analysis, twenty low-mass serum peaks correlating to controls and 20 different peaks correlating to cancer sera were found. A study performed by Fiedler et al. [86] on forty sera from patients matched with forty samples from healthy controls was focused on MALDI-TOF peptidome profile analysis after using magnetic beads for protein fractionation. Data were validated by using an additional 20 plus 20 sera set. Two significant peaks ($m/z$ 3884 and 5959) showed 86.3% sensitivity and 97.6% specificity for discriminating patients from controls. The 3884 peak was described and further validated by immunoassay as platelet factor 4 (PF4). PF4, used in combination with CA-19-9, significantly improved sensitivity and specificity for the identification of pancreatic cancer [86]. Rong et al. [87] used immunoaffinity depletion of highly abundant proteins and 2-DE to identify 16 protein spots differentially expressed (8 iper- and 8 ipoexpressed in cancer sera). The proteins were analyzed and sequenced. Mannose-binding lectin 2 and myosin light chain kinase 2, a serine/threonine kinase, were identified as potential biomarkers for the pancreatic cancer diagnosis and further validated by western blot in an independent set of sera from pancreatic cancer patients and normal controls. Low molecular weight (<60 kDa) serum proteome from a training set composed by 24 patients with pancreatic cancer and 21 controls was analyzed by HPLC-ESI-MS/MS. Among many peaks identified, a peptide from CXC chemokine ligand 7 (CXCL7) was significantly reduced in cancer sera. Data were confirmed by high-density protein microarray in a large cohort of 140 patients affected by pancreatic cancer, 10 patients with chronic pancreatitis and 87 healthy controls. Combination of CXCL7 and CA-19-9, improved the discriminatory power for pancreatic cancer [88]. In order to limit the complexity of the plasma proteome, Pan et al. [89] employed multidimensional fractionation followed by HPLC-MS/MS. Many proteins/peptides were identified with this method. A group of differentially expressed proteins was selected and evaluated on a separate cohort of samples from pancreatic cancer, chronic pancreatitis patients, and nonpancreatic disease control. A composite marker of the tissue inhibitor of metalloproteinases TIMP1 and the adhesion molecule ICAM1, as characterized by ELISA, showed significant better performance than CA-19-9 in distinguishing pancreatic cancer, pancreatitis, nonpancreatic diseases and healthy controls. Forty-five samples from patients affected by pancreatic cancer and 20 from healthy controls were analyzed by 2-DE and LC-MS/MS. Seven protein spots were differentially expressed. Serum isoforms of alpha-1-antitrypsin (AAT), also confirmed by western blot, were described as upregulated and potential serum biomarkers for pancreatic cancer [90].

*3.4.3. Pancreatic Cancer Protein Biomarkers Involved in Other Functions.* Bloomston et al. [91] analyzed by high-resolution 2-DE thirty preoperative sera from pancreatic cancer and thirty-two from healthy individuals. Differentially expressed spots were recovered and analyzed by MALDI-TOF and LC-MS/MS. Approximately 150 proteins resulted commonly overexpressed in all cancer patients. Four proteins discriminated 100% of pancreatic cancer and 94% of normal samples. Among them, fibrinogen-$\gamma$ was identified as putative biomarker and further validated by enzymatic analysis in sera and immunohistochemistry in tumor tissues. One hundred and twenty-six sera form pancreatic cancer patients (84 with diabetes) were examined by SELDI-TOF in comparison to 61 sera from chronic pancreatitis (32 with diabetes), 24 from type 2 diabetes mellitus patients, and 12 from healthy controls. Classification algorithms obtained by MS analysis resulted to improve the diagnostic accuracy of CA-19-9 in pancreatic cancer diagnosis and to facilitate the differential diagnosis between pancreatic cancer and type 2 diabetes mellitus. Among the large number of peptides, that described with the $m/z$ 3519 was identified as a member of the EGF-like family [123]. Fifty-eight sera from patients with pancreatic cancer were compared with 18 samples from patients affected by benign disease and 51 healthy controls. Sera were analyzed using a strong anionic exchange chromatography protein-chip and SELDI-TOF. Sixty-one protein peaks were detected to construct multiple classification trees to distinguish the disease groups, reaching 83% sensitivity and almost 100% specificity in discriminating cancer from controls and benign disease. Putative protein biomarkers were identified: one ($m/z$ 4016) showed a downregulated trend in preoperative versus post-operative sera, three ($m/z$ 4155, 4791, 28068) were detected in the differential diagnosis of the 3 test groups. C14orf166, a protein involved in modulation of mRNA transcription by Polymerase II, was identified as corresponding to the 28068 peak by ProteinChip immunoassay. The molecule showed levels significantly higher in pancreatic cancer patients, as confirmed by immunoenzymatic methods. C14orf166 was also iper-expressed in tumor cells [92]. A SELDI-TOF protein panel derived from the study of a training set composed by 38 pancreatic cancer sera, 54 disease controls, and 68 healthy volunteers was further validated on a first validation set (40 pancreatic cancer, 21 disease controls, 19 healthy volunteers) and then, by ELISA, on a second one (33 pancreatic cancer, 28 disease controls, 18 healthy volunteers). Some proteins corresponding to peaks of interest were purified and identified. A simplified diagnostic panel comprising CA-19-9, apolipoprotein C-I, apolipoprotein A-II, additionally validated by ELISA on the second validation set, resulted to improve the diagnostic ability of CA-19-9 [124]. Potential prognostic markers were initially identified by nano-LC-MS/MS in 4 groups of sera, each from 10 patients, selected on the basis of survival (long or short) and therapy (gemcitabine plus bevacizumab, or gemcitabine plus placebo). Alpha1-antichymotrypsin (AACT) was negatively correlated with overall survival and considered as a prognostic marker for pancreatic cancer [93].

## 4. Conclusions

Cancer is one of the leading causes of death worldwide. Advances in screening methods significantly improved early diagnosis with consequent enhancement of prognosis, survival and treatment efficacy. Unfortunately, some tumors are difficult to diagnose before the disease is in advanced or metastasizing state. Therefore, there is an urgent need to discover novel biomarkers which provide sensitive and specific

disease detection. Over the past decade, serum biomarkers have been identified in sera from cancer patients by using powerful high-throughput technologies. Mass spectrometry allowed the identification of hundreds of proteins within complex biological samples such as tissues, serum, plasma, and urine. MS analytical attributes in biomarker discovery are its high mass accuracy, resolution and ability to characterize the peptides at the level of their aminoacidic sequence. Several workflows including methods for serum samples preparation (e.g., high abundance protein removal, serum fractionation), SDS-PAGE and 2D-GE, LC, different MS platforms, and protein chip arrays have been used for biomarker discovery. Many differential MS peak profiles were identified and several proteins were characterized and described as potential biomarkers for high mortality tumors (Tables 1–4), achieving different levels of sensitivity and specificity to diagnose the disease. Many of these proteins are involved in fundamental processes such as inflammation, cellular differentiation and proliferation, and apoptosis. Among them, positive (i.e., serum amyloid A, ceruloplasmin, complement factors, haptoglobin) and negative acute-phase reactants (i.e., transthyretin, transferrin) were differentially expressed in sera from ovary, lung, liver, and pancreatic cancer patients. Some putative ovarian cancer biomarkers described in this paper, such as the keratin 2a, the glycosyltransferase-like 1B, involved in glycosylation processes, and the casein kinase alpha 1, a serine/threonine kinase involved in cellular differentiation, proliferation and apoptosis, have been associated with processes related to cancer [125–128]. Similarly, the mannose binding lectin 2 (MBL2), a mediator of inflammation which results iperexpressed in pancreatic cancer sera, is involved in cancer processes. Genetic alterations of MBL2 can increase colon cancer susceptibility in African Americans and a MBL genetic polymorphism, associated to a reduction of vaginal MBL concentration, may be a risk for development of ovarian cancer [129, 130]. The chemokine CCL18, here described as candidate ovarian cancer serum biomarker, was also considered as a urine biomarker for bladder cancer detection [131]. Likewise, the HCC biomarker cystatin C, an inhibitor of cystein proteinases, showed significantly higher levels also in sera from lung cancer patients [132], and the HCC and lung cancer putative biomarker alpha-1 acid glycoprotein 1, an acute phase protein, was found as well elevated in sera and tumor tissues from patients affected by gastric carcinoma [133]. The here described liver cancer biomarker heat shock protein 27, a protein with cytoprotective and anti-apoptotic activity, measured by immunoenzymatic assay, was confirmed to be elevated in an independent cohort of sera from HCC patients [134].

Despite the great advances in the application of MS in serum biomarker discovery, several challenges remain. The identification of differential serum protein profiles and specific molecules able to discriminate normal from diseased subjects requires a technology able to highlight small differences and to process large series of serum samples. Although MS is the most powerful approach for biomarker identification, there are some boundaries in the analysis of serum. These can be attributable to the complex nature of serum and its tremendous dynamic range, to diurnal variation in protein expression, instability of proteins due to *in vivo* or *ex vivo* protease activity, pre-analytical methods reproducibility as well as to the intrinsic MS sensitivity ($> \mu$g/mL) [135] in detecting analytes which usually range between 50 pg/mL and 10 ng/mL [136]. Accurate selection of cases and controls, standardization of sample collection and storage conditions, utilization of adequate and effective methods focused on reducing the complexity of serum/plasma prior MS analysis, use of different protein array with complementary binding conditions, refined bioinformatic and statistical analysis to process data, and suitable validation workflows by immunoassay on larger sets of independent samples are necessary elements to circumvent criticisms and improve the biomarker discovery process.

## Acknowledgments

This work was supported in part by MIUR FIRB Grant to E. Alesse, by the "Associazione Italiana per la RicercasulCancro" (AIRC) Grant to F. Zazzeroni, and by MIUR PRIN Grant to F. Zazzeroni.

## References

[1] A. M. Glas, A. Floore, L. J. M. J. Delahaye et al., "Converting a breast cancer microarray signature into a high-throughput diagnostic test," *BMC Genomics*, vol. 7, article 278, 2006.

[2] M. E. Straver, A. M. Glas, J. Hannemann et al., "The 70-gene signature as a response predictor for neoadjuvant chemotherapy in breast cancer," *Breast Cancer Research and Treatment*, vol. 119, no. 3, pp. 551–558, 2010.

[3] S. Mook, M. Knauer, J. M. Bueno-De-Mesquita et al., "Metastatic potential of T1 breast cancer can be predicted by the 70-gene MammaPrint signature," *Annals of Surgical Oncology*, vol. 17, no. 5, pp. 1406–1413, 2010.

[4] M. Zhou and T. D. Veenstra, "Mass spectrometry: m/z 1983–2008," *BioTechniques*, vol. 44, no. 5, pp. 667–670, 2008.

[5] F. E. Ahmed, "Utility of mass spectrometry for proteome analysis—part I: conceptual and experimental approaches," *Expert Review of Proteomics*, vol. 5, no. 6, pp. 841–864, 2008.

[6] E. P. Diamandis, "Mass spectrometry as a diagnostic and a cancer biomarker discovery tool: opportunities and potential limitations," *Molecular and Cellular Proteomics*, vol. 3, no. 4, pp. 367–378, 2004.

[7] E. F. Petricoin, C. Belluco, R. P. Araujo, and L. A. Liotta, "The blood peptidome: a higher dimension of information content for cancer biomarker discovery," *Nature Reviews Cancer*, vol. 6, no. 12, pp. 961–967, 2006.

[8] M. De Bock, D. de Seny, M. A. Meuwis et al., "Challenges for biomarker discovery in body fluids using SELDI-TOF-MS," *Journal of Biomedicine & Biotechnology*, vol. 2010, Article ID 906082, 15 pages, 2010.

[9] J. M. Jacobs, J. N. Adkins, W. J. Qian et al., "Utilizing human blood plasma for proteomic biomarker discovery," *Journal of Proteome Research*, vol. 4, no. 4, pp. 1073–1085, 2005.

[10] N. L. Anderson, M. Polanski, R. Pieper et al., "The human plasma proteome: a non redundant list developed by combination of four separate sources," *Molecular and Cellular Proteomics*, vol. 3, no. 4, pp. 311–326, 2004.

[11] Y. Shen, J. Kim, E. F. Strittmatter et al., "Characterization of the human blood plasma proteome," *Proteomics*, vol. 5, no. 15, pp. 4034–4045, 2005.

[12] L. A. Liotta and E. F. Petricoin, "Serum peptidome for cancer detection: spinning biologic trash into diagnostic gold," *Journal of Clinical Investigation*, vol. 116, no. 1, pp. 26–30, 2006.

[13] R. J. Leatherbarrow and P. D. Dean, "Studies on the mechanism of binding of serum albumins to immobilized cibacron blue F3G A," *Biochemical Journal*, vol. 189, no. 1, pp. 27–34, 1980.

[14] F. Di Girolamo, P. G. Righetti, A. D'Amato, and M. C. Chung, "Cibacron Blue and proteomics: the mystery of the platoon missing in action," *Journal of Proteomics*, vol. 74, no. 12, pp. 2856–2865, 2011.

[15] N. Seam, D. A. Gonzales, S. J. Kern, G. L. Hortin, G. T. Hoehn, and A. F. Suffredini, "Quality control of serum albumin depletion for proteomic analysis," *Clinical Chemistry*, vol. 53, no. 11, pp. 1915–1920, 2007.

[16] K. Björhall, T. Miliotis, and P. Davidsson, "Comparison of different depletion strategies for improved resolution in proteomic analysis of human serum samples," *Proteomics*, vol. 5, no. 1, pp. 307–317, 2005.

[17] L. Guerrier, F. Fortis, and E. Boschetti, "Solid-phase fractionation strategies applied to proteomics investigations," *Methods in Molecular Biology*, vol. 818, pp. 11–33, 2011.

[18] L. Guerrier, L. Lomas, and E. Boschetti, "A simplified mono-buffer multidimensional chromatography for high-throughput proteome fractionation," *Journal of Chromatography A*, vol. 1073, no. 1-2, pp. 25–33, 2005.

[19] E. Orvisky, S. K. Drake, B. M. Martin et al., "Enrichment of low molecular weight fraction of serum for MS analysis of peptides associated with hepatocellular carcinoma," *Proteomics*, vol. 6, no. 9, pp. 2895–2902, 2006.

[20] S. Camerini, M. L. Polci, L. A. Liotta, E. F. Petricoin, and W. Zhou, "A method for the selective isolation and enrichment of carrier protein-bound low-molecular weight proteins and peptides in the blood," *Proteomics—Clinical Applications*, vol. 1, no. 2, pp. 176–184, 2007.

[21] A. J. VanMeter, S. Camerini, M. L. Polci et al., "Serum low-molecular-weight protein fractionation for biomarker discovery," *Methods in Molecular Biology*, vol. 823, pp. 237–249, 2012.

[22] A. Luchini, D. H. Geho, B. Bishops et al., "Smart hydrogel particles: biomarker harvesting: one-step affinity purification, size exclusion, and protection against degradation," *Nano Letters*, vol. 8, no. 1, pp. 350–361, 2008.

[23] A. Luchini, C. Longo, V. Espina, E. F. Petricoin III, and L. A. Liotta, "Nanoparticle technology: addressing the fundamental roadblocks to protein biomarker discovery," *Journal of Materials Chemistry*, vol. 19, no. 29, pp. 5071–5077, 2009.

[24] C. Longo, A. Patanarut, T. George et al., "Core-shell hydrogel particles harvest, concentrate and preserve labile low abundance biomarkers," *PLoS One*, vol. 4, no. 3, article e4763, 2009.

[25] D. Tamburro, C. Fredolini, V. Espina et al., "Multifunctional core-shell nanoparticles: discovery of previously invisible biomarkers," *Journal of the American Chemical Society*, vol. 133, no. 47, pp. 19178–19188, 2011.

[26] H. Zhang, X. J. Li, D. B. Martin, and R. Aebersold, "Identification and quantification of N-linked glycoproteins using hydrazide chemistry, stable isotope labeling and mass spectrometry," *Nature Biotechnology*, vol. 21, no. 6, pp. 660–666, 2003.

[27] R. Apweiler, H. Hermjakob, and N. Sharon, "On the frequency of protein glycosylation, as deduced from analysis of the SWISS-PROT database," *Biochimica et Biophysica Acta*, vol. 1473, no. 1, pp. 4–8, 1999.

[28] L. Tong, G. Baskaran, M. B. Jones, J. K. Rhee, and K. J. Yarema, "Glycosylation changes as markers for the diagnosis and treatment of human disease," *Biotechnology and Genetic Engineering Reviews*, vol. 20, pp. 199–244, 2003.

[29] Y. Tian, Y. Zhou, S. Elliott, R. Aebersold, and H. Zhang, "Solid-phase extraction of N-linked glycopeptides," *Nature Protocols*, vol. 2, no. 2, pp. 334–339, 2007.

[30] B. Sun, J. A. Ranish, A. G. Utleg et al., "Shotgun glycopeptide capture approach coupled with mass spectrometry for comprehensive glycoproteomics," *Molecular and Cellular Proteomics*, vol. 6, no. 1, pp. 141–149, 2007.

[31] H. Kaji, H. Saito, Y. Yamauchi et al., "Lectin affinity capture, isotope-coded tagging and mass spectrometry to identify N-linked glycoproteins," *Nature Biotechnology*, vol. 21, no. 6, pp. 667–672, 2003.

[32] R. Aebersold and M. Mann, "Mass spectrometry-based proteomics," *Nature*, vol. 422, no. 6928, pp. 198–207, 2003.

[33] J. R. Yates III, "Mass spectral analysis in proteomics," *Annual Review of Biophysics and Biomolecular Structure*, vol. 33, pp. 297–316, 2004.

[34] F. Hillenkamp, M. Karas, R. C. Beavis, and B. T. Chait, "Matrix-assisted laser desorption/ionization mass spectrometry of biopolymers," *Analytical Chemistry*, vol. 63, no. 24, pp. 1193A–1203A, 1991.

[35] Y. Qu, B. L. Adam, Y. Yasui et al., "Boosted decision tree analysis of surface-enhanced laser desorption/ionization mass spectral serum profiles discriminates prostate cancer from noncancer patients," *Clinical Chemistry*, vol. 48, no. 10, pp. 1835–1843, 2002.

[36] Z. Zhang, R. C. Bast Jr., Y. Yu et al., "Three biomarkers identified from serum proteomic analysis for the detection of early stage ovarian cancer," *Cancer Research*, vol. 64, no. 16, pp. 5882–5890, 2004.

[37] T. W. Hutchens and T. T. Yip, "New desorption strategies for the mass-spectrometric analysis of macromolecules," *Rapid Communications in Mass Spectrometry*, vol. 7, pp. 576–580, 1993.

[38] N. Tang, P. Tornatore, and S. R. Weinberger, "Current developments in SELDI affinity technology," *Mass Spectrometry Reviews*, vol. 23, no. 1, pp. 34–44, 2004.

[39] J. B. Fenn, M. Mann, C. K. Meng, S. F. Wong, and C. M. Whitehouse, "Electrospray ionization for mass spectrometry of large biomolecules," *Science*, vol. 246, no. 4926, pp. 64–71, 1989.

[40] D. F. Hunt, J. R. Yates III, J. Shabanowitz, S. Winston, and C. R. Hauer, "Protein sequencing by tandem mass spectrometry," *Proceedings of the National Academy of Sciences of the United States of America*, vol. 83, no. 17, pp. 6233–6237, 1986.

[41] K. R. Kozak, F. Su, J. P. Whitelegge, K. Faull, S. Reddy, and R. Farias-Eisner, "Characterization of serum biomarkers for detection of early stage ovarian cancer," *Proteomics*, vol. 5, no. 17, pp. 4589–4596, 2005.

[42] B. Ye, D. W. Cramer, S. J. Skates et al., "Haptoglobin-α subunit as potential serum biomarker in ovarian cancer: identification and characterization using proteomic profiling and mass spectrometry," *Clinical Cancer Research*, vol. 9, no. 8, pp. 2904–2911, 2003.

[43] N. Ahmed, G. Barker, K. T. Oliva et al., "Proteomic-based identification of haptoglobin-1 precursor as a novel circulating biomarker of ovarian cancer," *British Journal of Cancer*, vol. 91, no. 1, pp. 129–140, 2004.

[44] H. R. Bergen III, G. Vasmatzis, W. A. Cliby, K. L. Johnson, A. L. Oberg, and D. C. Muddiman, "Discovery of ovarian cancer biomarkers in serum using NanoLC electrospray ionization TOF and FT-ICR mass spectrometry," *Disease Markers*, vol. 19, no. 4-5, pp. 239–249, 2003-2004.

[45] S. A. Moshkovskii, M. V. Serebryakova, K. B. Kuteykin-Teplyakov et al., "Ovarian cancer marker of 11.7 kDa detected by proteomics is a serum amyloid A1," *Proteomics*, vol. 5, no. 14, pp. 3790–3797, 2005.

[46] A. Woong-Shick, P. Sung-Pil, B. Su-Mi et al., "Identification of hemoglobin-α and -β subunits as potential serum biomarkers for the diagnosis and prognosis of ovarian cancer," *Cancer Science*, vol. 96, no. 3, pp. 197–201, 2005.

[47] M. F. Lopez, A. Mikulskis, S. Kuzdzal et al., "A novel, high-throughput workflow for discovery and identification of serum carrier protein-bound peptide biomarker candidates in ovarian cancer samples," *Clinical Chemistry*, vol. 53, no. 6, pp. 1067–1074, 2007.

[48] Q. Wang, D. Li, W. Zhang, B. Tang, Q. Q. Li, and L. Li, "Evaluation of proteomics-identified CCL18 and CXCL1 as circulating tumor markers for differential diagnosis between ovarian carcinomas and benign pelvic masses," *International Journal of Biological Markers*, vol. 26, no. 4, pp. 262–273, 2011.

[49] J. F. Timms, U. Menon, D. Devetyarov et al., "Early detection of ovarian cancer in samples pre-diagnosis using CA125 and MALDI-MS peaks," *Cancer Genomics Proteomics*, vol. 8, no. 6, pp. 289–305, 2011.

[50] A. Jemal, R. Siegel, J. Xu, and E. Ward, "Cancer statistics, 2010," *CA: A Cancer Journal for Clinicians*, vol. 60, no. 5, pp. 277–300, 2010.

[51] U. Menon and I. J. Jacobs, "Recent developments in ovarian cancer screening," *Current Opinion in Obstetrics and Gynecology*, vol. 12, no. 1, pp. 39–42, 2000.

[52] S. A. Cannistra, "Cancer of the ovary," *New England Journal of Medicine*, vol. 351, no. 24, pp. 2519–2565, 2004.

[53] V. Nossov, M. Amneus, F. Su et al., "The early detection of ovarian cancer: from traditional methods to proteomics. Can we really do better than serum CA-125?" *American Journal of Obstetrics and Gynecology*, vol. 199, no. 3, pp. 215–223, 2008.

[54] L. S. Cohen, P. F. Escobar, C. Scharm, B. Glimco, and D. A. Fishman, "Three-dimensional power doppler ultrasound improves the diagnostic accuracy for ovarian cancer prediction," *Gynecologic Oncology*, vol. 82, no. 1, pp. 40–48, 2001.

[55] Y. Liu, J. He, C. Li et al., "Identification and confirmation of biomarkers using an integrated platform for quantitative analysis of glycoproteins and their glycosylations," *Journal of Proteome Research*, vol. 9, no. 2, pp. 798–805, 2010.

[56] V. Paradis, F. Degos, D. Dargère et al., "Identification of a new marker of hepatocellular carcinoma by serum protein profiling of patients with chronic liver diseases," *Hepatology*, vol. 41, no. 1, pp. 40–47, 2005.

[57] I. N. Lee, C. H. Chen, J. C. Sheu et al., "Identification of complement C3a as a candidate biomarker in human chronic hepatitis C and HCV-related hepatocellular carcinoma using a proteomics approach," *Proteomics*, vol. 6, no. 9, pp. 2865–2873, 2006.

[58] M. H. Yang, Y. C. Tyan, S. B. Jong, Y. F. Huang, P. C. Liao, and M. C. Wang, "Identification of human hepatocellular

carcinoma-related proteins by proteomic approaches," *Analytical and Bioanalytical Chemistry*, vol. 388, no. 3, pp. 637–643, 2007.

[59] N. T. Zinkin, F. Grall, K. Bhaskar et al., "Serum proteomics and biomarkers in hepatocellular carcinoma and chronic liver disease," *Clinical Cancer Research*, vol. 14, no. 2, pp. 470–477, 2008.

[60] M. He, J. Qin, R. Zhai et al., "Detection and identification of NAP-2 as a biomarker in hepatitis B-related hepatocellular carcinoma by proteomic approach," *Proteome Science*, vol. 6, article 10, 2008.

[61] F. X. Wu, Q. Wang, Z. M. Zhang et al., "Identifying serological biomarkers of hepatocellular carcinoma using surface-enhanced laser desorption/ionization-time-of-flight mass spectroscopy," *Cancer Letters*, vol. 279, no. 2, pp. 163–170, 2009.

[62] X. Kang, L. Sun, K. Guo et al., "Serum protein biomarkers screening in HCC patients with liver cirrhosis by ICAT-LC-MS/MS," *Journal of Cancer Research and Clinical Oncology*, vol. 136, no. 8, pp. 1151–1159, 2010.

[63] H. Shu, X. Kang, K. Guo et al., "Diagnostic value of serum haptoglobin protein as hepatocellular carcinoma candidate marker complementary to α fetoprotein," *Oncology Reports*, vol. 24, no. 5, pp. 1271–1276, 2010.

[64] Y. An, S. Bekesova, N. Edwards, and R. Goldman, "Peptides in low molecular weight fraction of serum associated with hepatocellular carcinoma," *Disease Markers*, vol. 29, no. 1, pp. 11–20, 2010.

[65] J. T. Feng, Y. K. Liu, H. Y. Song et al., "Heat-shock protein 27: a potential biomarker for hepatocellular carcinoma identified by serum proteome analysis," *Proteomics*, vol. 5, no. 17, pp. 4581–4588, 2005.

[66] W. Wu, J. Li, Y. Liu, C. Zhang, X. Meng, and Z. Zhou, "Comparative proteomic studies of serum from patients with hepatocellular carcinoma," *Journal of Investigative Surgery*, vol. 25, no. 1, pp. 37–42, 2012.

[67] E. F. Petricoin, A. M. Ardekani, B. A. Hitt et al., "Use of proteomic patterns in serum to identify ovarian cancer," *Lancet*, vol. 359, no. 9306, pp. 572–577, 2002.

[68] K. R. Kozak, M. W. Amneus, S. M. Pusey et al., "Identification of biomarkers for ovarian cancer using strong anion-exchange ProteinChips: potential use in diagnosis and prognosis," *Proceedings of the National Academy of Sciences of the United States of America*, vol. 100, no. 21, pp. 12343–12348, 2003.

[69] H. Zhang, B. Kong, X. Qu, L. Jia, B. Deng, and Q. Yang, "Biomarker discovery for ovarian cancer using SELDI-TOF-MS," *Gynecologic Oncology*, vol. 102, no. 1, pp. 61–66, 2006.

[70] J. R. Hocker, E. A. Bishop, S. A. Lightfoot et al., "Serum profiling to distinguish early-and late-stage ovariancancer patients from disease-free individuals," *Cancer Investigation*, vol. 30, no. 2, pp. 189–197, 2012.

[71] B. A. Howard, M. Z. Wang, M. J. Campa, C. Corro, M. C. Fitzgerald, and E. F. Patz Jr., "Identification and validation of a potential lung cancer serum biomarker detected by matrix-assisted laser desorption/ionization-time of flight spectra analysis," *Proteomics*, vol. 3, no. 9, pp. 1720–1724, 2003.

[72] S. Dai, X. Wang, L. Liu et al., "Discovery and identification of Serum Amyloid A protein elevated in lung cancer serum," *Science in China, Series C*, vol. 50, no. 3, pp. 305–311, 2007.

[73] W. C. S. Cho, T. T. Yip, W. W. Cheng, and J. S. K. Au, "Serum amyloid A is elevated in the serum of lung cancer patients with poor prognosis," *British Journal of Cancer*, vol. 102, no. 12, pp. 1731–1735, 2010.

[74] X. Zeng, B. L. Hood, T. Zhao et al., "Lung cancer serum biomarker discovery using label-free liquid chromatography-tandem mass spectrometry," *Journal of Thoracic Oncology*, vol. 6, no. 4, pp. 725–734, 2011.

[75] L. Liu, J. Liu, Y. Wang et al., "A combined biomarker pattern improves the discrimination of lung cancer," *Biomarkers*, vol. 16, no. 1, pp. 20–30, 2011.

[76] H. J. Sung, J. M. Ahn, Y. H. Yoon et al., "Identification and validation of SAA as a potential lung cancer biomarker and its involvement in metastatic pathogenesis of lung cancer," *Journal of Proteome Research*, vol. 10, no. 3, pp. 1383–1395, 2011.

[77] A. Bharti, P. C. Ma, G. Maulik et al., "Haptoglobin α-subunit and hepatocyte growth factor can potentially serve as serum tumor biomarkers in small cell lung cancer," *Anticancer Research*, vol. 24, no. 2C, pp. 1031–1038, 2004.

[78] L. Liu, J. Liu, S. Dai et al., "Reduced transthyretin expression in sera of lung cancer," *Cancer Science*, vol. 98, no. 10, pp. 1617–1624, 2007.

[79] K. Ueda, N. Saichi, S. Takami et al., "A comprehensive peptidome profiling technology for the identification of early detection biomarkers for lung adenocarcinoma," *PLoS One*, vol. 6, no. 4, article e18567, 2011.

[80] H. J. An, S. Miyamoto, K. S. Lancaster et al., "Profiling of glycans in serum for the discovery of potential biomarkers for ovarian cancer," *Journal of Proteome Research*, vol. 5, no. 7, pp. 1626–1635, 2006.

[81] G. S. Leiserowitz, C. Lebrilla, S. Miyamoto et al., "Glycomics analysis of serum: a potential new biomarker for ovarian cancer?" *International Journal of Gynecological Cancer*, vol. 18, no. 3, pp. 470–475, 2008.

[82] K. Kojima, S. Asmellash, C. A. Klug, W. E. Grizzle, J. A. Mobley, and J. D. Christein, "Applying proteomic-based biomarker tools for the accurate diagnosis of pancreatic cancer," *Journal of Gastrointestinal Surgery*, vol. 12, no. 10, pp. 1683–1690, 2008.

[83] K. Yokoi, L. C. Shih, R. Kobayashi et al., "Serum amyloid A as a tumor marker in sera of nude mice with orthotopic human pancreatic cancer and in plasma of patients with pancreatic cancer," *International Journal of Oncology*, vol. 27, no. 5, pp. 1361–1369, 2005.

[84] M. Ehmann, K. Felix, D. Hartmann et al., "Identification of potential markers for the detection of pancreatic cancer through comparative serum protein expression profiling," *Pancreas*, vol. 34, no. 2, pp. 205–214, 2007.

[85] J. S. Hanas, J. R. Hocker, J. Y. Cheung et al., "Biomarker identification in human pancreatic cancer sera," *Pancreas*, vol. 36, no. 1, pp. 61–69, 2008.

[86] G. M. Fiedler, A. B. Leichtle, J. Kase et al., "Serum peptidome profiling revealed platelet factor 4 as a potential discriminating peptide associated with pancreatic cancer," *Clinical Cancer Research*, vol. 15, no. 11, pp. 3812–3819, 2009.

[87] Y. Rong, D. Jin, C. Hou et al., "Proteomics analysis of serum protein profiling in pancreatic cancer patients by DIGE: up-regulation of mannose-binding lectin 2 and myosin light chain kinase 2," *BMC Gastroenterology*, vol. 10, article 68, 2010.

[88] J. Matsubara, K. Honda, M. Ono et al., "Reduced plasma level of CXC chemokine ligand 7 in patients with pancreatic cancer," *Cancer Epidemiology Biomarkers and Prevention*, vol. 20, no. 1, pp. 160–171, 2011.

[89] S. Pan, R. Chen, D. A. Crispin et al., "Protein alterations associated with pancreatic cancer and chronic pancreatitis found in human plasma using global quantitative proteomics

profiling," *Journal of Proteome Research*, vol. 10, no. 5, pp. 2359–2376, 2011.

[90] Y. Wang, Y. Kuramitsu, S. Yoshino et al., "Screening for serological biomarkers of pancreatic cancer by two-dimensional electrophoresis and liquid chromatography-tandem mass spectrometry," *Oncology Reports*, vol. 26, no. 1, pp. 287–292, 2011.

[91] M. Bloomston, J. X. Zhou, A. S. Rosemurgy, W. Frankel, C. A. Muro-Cacho, and T. J. Yeatman, "Fibrinogen γ overexpression in pancreatic cancer identified by large-scale proteomic analysis of serum samples," *Cancer Research*, vol. 66, no. 5, pp. 2592–2599, 2006.

[92] J. Guo, W. Wang, P. Liao et al., "Identification of serum biomarkers for pancreatic adenocarcinoma by proteomic analysis," *Cancer Science*, vol. 100, no. 12, pp. 2292–2301, 2009.

[93] A. S. Roberts, M. J. Campa, E. B. Gottlin et al., "Identification of potential prognostic biomarkers in patients with untreated, advanced pancreatic cancer from a phase 3 trial (Cancer and Leukemia Group B, 80303)," *Cancer*, vol. 118, no. 2, pp. 571–578, 2012.

[94] J. A. Marrero, "Hepatocellular carcinoma," *Current Opinion in Gastroenterology*, vol. 22, no. 3, pp. 248–253, 2006.

[95] H. E. Blum, "Hepatocellular carcinoma: therapy and prevention," *World Journal of Gastroenterology*, vol. 11, no. 47, pp. 7391–7400, 2005.

[96] K. A. Gebo, G. Chander, M. W. Jenckes et al., "Screening tests for hepatocellular carcinoma in patients with chronic hepatitis C: a systematic review," *Hepatology*, vol. 36, no. 5, supplement 1, pp. S84–S92, 2002.

[97] T. Göbel, S. Vorderwülbecke, K. Hauck, H. Fey, D. Häussinger, and A. Erhardt, "New multi protein patterns differentiate liver fibrosis stages and hepatocellular carcinoma in chronic hepatitis C serum samples," *World Journal of Gastroenterology*, vol. 12, no. 47, pp. 7604–7612, 2006.

[98] S. Kanmura, H. Uto, K. Kusumoto et al., "Early diagnostic potential for hepatocellular carcinoma using the SELDI ProteinChip system," *Hepatology*, vol. 45, no. 4, pp. 948–956, 2007.

[99] C. Wu, Z. Wang, L. Liu et al., "Surface enhanced laser desorption/ionization profiling: new diagnostic method of HBV-related hepatocellular carcinoma," *Journal of Gastroenterology and Hepatology*, vol. 24, no. 1, pp. 55–62, 2009.

[100] L. Chen, D. W. Y. Ho, N. P. Y. Lee et al., "Enhanced detection of early hepatocellular carcinoma by serum SELDI-TOF proteomic signature combined with alpha-fetoprotein marker," *Annals of Surgical Oncology*, vol. 17, no. 9, pp. 2518–2525, 2010.

[101] J. Cui, X. Kang, Z. Dai et al., "Prediction of chronic hepatitis B, liver cirrhosis and hepatocellular carcinoma by SELDI-based serum decision tree classification," *Journal of Cancer Research and Clinical Oncology*, vol. 133, no. 11, pp. 825–834, 2007.

[102] R. Goldman, H. W. Ressom, R. S. Varghese et al., "Detection of hepatocellular carcinoma using glycomic analysis," *Clinical Cancer Research*, vol. 15, no. 5, pp. 1808–1813, 2009.

[103] Z. Tang, R. S. Varghese, S. Bekesova et al., "Identification of N-glycan serum markers associated with hepatocellular carcinoma from mass spectrometry data," *Journal of Proteome Research*, vol. 9, no. 1, pp. 104–112, 2010.

[104] R. S. Herbst, J. V. Heymach, and S. M. Lippman, "Molecular origins of cancer: lung cancer," *New England Journal of Medicine*, vol. 359, no. 13, pp. 1367–1380, 2008.

[105] M. V. Infante and J. H. Pedersen, "Screening for lung cancer: are we there yet?" *Current Opinion in Pulmonary Medicine*, vol. 16, no. 4, pp. 301–306, 2010.

[106] C. Reddy, D. Chilla, and J. Boltax, "Lung cancer screening: a review of available data and current guidelines," *Hospital Practice*, vol. 39, pp. 107–112, 2011.

[107] W. D. Travis, E. Brambilla, M. Noguchi et al., "International association for the study of lung cancer/American Thoracic Society/European Respiratory Society: international multidisciplinary classification of lung adenocarcinoma: executive summary," *Proceedings of the American Thoracic Society*, vol. 8, pp. 381–385, 2011.

[108] O. Lababede, M. Meziane, and T. Rice, "Seventh edition of the cancer staging manual and stage grouping of lung cancer: quick reference chart and diagrams," *Chest*, vol. 139, no. 1, pp. 183–189, 2011.

[109] S. Cedrés, I. Nuñez, M. Longo et al., "Serum tumor markers CEA, CYFRA21-1, and CA-125 are associated with worse prognosis in advanced non-small-cell lung cancer (NSCLC)," *Clinical Lung Cancer*, vol. 12, no. 3, pp. 172–179, 2011.

[110] J. Chee, A. Naran, N. L. Misso, P. J. Thompson, and K. D. Bhoola, "Expression of tissue and plasma kallikreins and kinin B1 and B2 receptors in lung cancer," *Biological Chemistry*, vol. 389, pp. 1225–1233, 2008.

[111] E. Wójcik, J. K. Kulpa, B. Sas-Korczyńska, S. Korzeniowski, and J. Jakubowicz, "ProGRP and NSE in therapy monitoring in patients with small cell lung cancer," *Anticancer Research*, vol. 28, no. 5, pp. 3027–3033, 2008.

[112] S. Y. Yang, X. Y. Xiao, W. G. Zhang et al., "Application of serum SELDI proteomic patterns in diagnosis of lung cancer," *BMC Cancer*, vol. 5, article 83, 2005.

[113] M. Han, Q. Liu, J. Yu, and S. Zheng, "Detection and significance of serum protein markers of small-cell lung cancer," *Journal of Clinical Laboratory Analysis*, vol. 22, no. 2, pp. 131–137, 2008.

[114] R. T. Sreseli, H. Binder, M. Kuhn et al., "Identification of a 17-protein signature in the serum of lung cancer patients," *Oncology Reports*, vol. 24, no. 1, pp. 263–270, 2010.

[115] J. Du, S. Yang, X. Lin et al., "Use of anchorchip-time-of-flight spectrometry technology to screen tumor biomarker proteins in serum for small cell lung cancer," *Diagnostic Pathology*, vol. 5, article 60, 2010.

[116] X. Zeng, B. L. Hood, M. Sun et al., "Lung cancer serum biomarker discovery using glycoprotein capture and liquid chromatography mass spectrometry," *Journal of Proteome Research*, vol. 9, no. 12, pp. 6440–6449, 2010.

[117] P. B. Yildiz, Y. Shyr, J. S. M. Rahman et al., "Diagnostic accuracy of MALDI mass spectrometric analysis of unfractionated serum in lung cancer," *Journal of Thoracic Oncology*, vol. 2, no. 10, pp. 893–901, 2007.

[118] C. Sperti, C. Pasquali, A. Piccoli, and S. Pedrazzoli, "Survival after resection for ductal adenocarcinoma of the pancreas," *British Journal of Surgery*, vol. 83, no. 5, pp. 625–631, 1996.

[119] C. J. Yeo, J. L. Cameron, T. A. Sohn et al., "Pancreatico-duodenectomy with or without extended retroperitoneal lymphadenectomy for periampullary adenocarcinoma: comparison of morbidity and mortality and short-term outcome," *Annals of Surgery*, vol. 229, no. 5, pp. 613–624, 1999.

[120] J. D. Christein, M. L. Kendrick, C. W. Iqbal, D. M. Nagorney, and M. B. Farnell, "Distal pancreatectomy for resectable adenocarcinoma of the body and tail of the pancreas," *Journal of Gastrointestinal Surgery*, vol. 9, no. 7, pp. 922–927, 2005.

[121] K. S. Goonetilleke and A. K. Siriwardena, "Systematic review of carbohydrate antigen (CA 19-9) as a biochemical marker in the diagnosis of pancreatic cancer," *European Journal of Surgical Oncology*, vol. 33, no. 3, pp. 266–270, 2007.

[122] D. Liu, L. Cao, J. Yu et al., "Diagnosis of pancreatic adenocarcinoma using protein chip technology," *Pancreatology*, vol. 9, no. 1-2, pp. 127–135, 2009.

[123] F. Navaglia, P. Fogar, D. Basso et al., "Pancreatic cancer biomarkers discovery by surface-enhanced laser desorption and ionization time-of-flight mass spectrometry," *Clinical Chemistry and Laboratory Medicine*, vol. 47, no. 6, pp. 713–723, 2009.

[124] A. Xue, C. J. Scarlett, L. Chung et al., "Discovery of serum biomarkers for pancreatic adenocarcinoma using proteomic analysis," *British Journal of Cancer*, vol. 103, no. 3, pp. 391–400, 2010.

[125] B. K. Bloor, N. Tidman, I. M. Leigh et al., "Expression of keratin K2e in cutaneous and oral lesions: association with keratinocyte activation, proliferation, and keratinization," *American Journal of Pathology*, vol. 162, no. 3, pp. 963–975, 2003.

[126] S. A. Joosse, J. Hannemann, J. Spötter et al., "Changes in keratin expression during metastatic progression of breast cancer: impact on the detection of circulating tumor cells," *Clinical Cancer Research*, vol. 18, no. 4, pp. 993–1003, 2012.

[127] X. L. Jin, S. S. Zheng, B. S. Wang, and H. L. Chen, "Correlation of glycosyltransferases mRNA expression in extrahepatic bile duct carcinoma with clinical pathological characteristics," *Hepatobiliary and Pancreatic Diseases International*, vol. 3, no. 2, pp. 292–295, 2004.

[128] R. Prudent, C. F. Sautel, and C. Cochet, "Structure-based discovery of small molecules targeting different surfaces of protein-kinase CK2," *Biochimica et Biophysica Acta*, vol. 1804, no. 3, pp. 493–498, 2010.

[129] K. A. Zanetti, M. Haznadar, J. A. Welsh et al., "3′-UTR and functional secretor haplotypes in mannose-binding lectin2 are associated with increased colon cancer risk in African Americans," *Cancer Research*, vol. 72, no. 6, pp. 1467–1477, 2012.

[130] N. S. Nevadunsky, I. Korneeva, T. Caputo, and S. S. Witkin, "Mannose-binding lectin codon 54 genetic polymorphism and vaginal protein levels in women with gynecologic malignancies," *European Journal of Obstetrics & Gynecology and Reproductive Biology*, vol. 163, no. 2, pp. 216–218, 2012.

[131] V. Urquidi, J. Kim, M. Chang, Y. Dai, C. J. Rosser, and S. Goodison, "CCL18 in a multiplex urine-based assay for the detection of bladder cancer," *PLoS One*, vol. 7, no. 5, article e37797, 2012.

[132] Q. Chen, J. Fei, L. Wu et al., "Detection of cathepsin B, cathepsin L, cystatin C, urokinase plasminogen activator and urokinase plasminogen activator receptor in the sera of lung cancer patients," *Oncology Letters*, vol. 2, no. 4, pp. 693–699, 2011.

[133] N. Chirwa, D. Govender, B. Ndimba et al., "A 40-50kDa glycoprotein associated with mucus is identified as $\alpha$-1-acid glycoprotein in carcinoma of the stomach," *Journal of Cancer*, vol. 3, pp. 83–92, 2012.

[134] G. Gruden, P. Carucci, V. Lolli et al., "Serum heat shock protein 27 levels in patients with hepatocellular carcinoma," *Cell Stress and Chaperones*. In press.

[135] E. Kuhn, T. Addona, H. Keshishian et al., "Developing multiplexed assays for troponin I and interleukin-33 in plasma by peptide immunoaffinity enrichment and targeted mass spectrometry," *Clinical Chemistry*, vol. 55, no. 6, pp. 1108–1117, 2009.

[136] N. L. Anderson and N. G. Anderson, "The human plasma proteome: history, character, and diagnostic prospects," *Molecular & Cellular Proteomics*, vol. 1, no. 11, pp. 845–867, 2002.

# 5

# Proteomic Analysis of the Ontogenetic Variability in Plasma Composition of Juvenile and Adult *Bothrops jararaca* Snakes

**Karen de Morais-Zani,**[1,2] **Kathleen Fernandes Grego,**[1]
**Aparecida Sadae Tanaka,**[3] **and Anita Mitico Tanaka-Azevedo**[1,2]

[1] *Laboratório de Herpetologia, Instituto Butantan, Avenida Vital Brazil 1500,*
  *05503-900 São Paulo, SP, Brazil*
[2] *Programa de Pós-Graduação Interunidades em Biotecnologia, Universidade de São Paulo,*
  *Avenida Professor Lineu Prestes 2415, 05508-900 São Paulo, SP, Brazil*
[3] *Departamento de Bioquímica, Universidade Federal de São Paulo, Rua Três de Maio 100,*
  *04044-020 São Paulo, SP, Brazil*

Correspondence should be addressed to Anita Mitico Tanaka-Azevedo; amt.azevedo@uol.com.br

Academic Editor: Djuro Josic

The ontogenetic variability in venom composition of some snake genera, including *Bothrops*, as well as the biological implications of such variability and the search of new molecules that can neutralize the toxic components of these venoms have been the subject of many studies. Thus, considering the resistance of *Bothrops jararaca* to the toxic action of its own venom and the ontogenetic variability in venom composition described in this species, a comparative study of the plasma composition of juvenile and adult *B. jararaca* snakes was performed through a proteomic approach based on 2D electrophoresis and mass spectrometry, which allowed the identification of proteins that might be present at different levels during ontogenetic development. Among the proteins identified by mass spectrometry, antihemorrhagic factor Bj46a was found only in adult plasma. Moreover, two spots identified as phospholipase $A_2$ inhibitors were significantly increased in juvenile plasma, which can be related to the higher catalytic $PLA_2$ activity shown by juvenile venom in comparison to that of adult snakes. This work shows the ontogenetic variability of *B. jararaca* plasma, and that these changes can be related to the ontogenetic variability described in its venom.

## 1. Introduction

Poisonous snakes are responsible for around 50,000 deaths among five million cases of ophidian accidents per year in the world, especially in the rural areas of tropical countries in Asia, Africa, and South America [1, 2].

Envenomation by Viperidae snakes causes local tissue damages such as edema, hemorrhage, and myonecrosis, which are not well neutralized by conventional antivenom serotherapy [3]. *Bothrops jararaca* (*B. jararaca*) snake belongs to the Viperidae family and is the main reason for ophidian accidents in the state of São Paulo, Brazil [4]. Its victims usually have, besides systemic reactions of envenomation such as bleeding and blood incoagulability, local effects at the bite site such as edema, ecchymoses, compartmental syndrome, blisters, and necrosis [5]. The envenomation symptomatology has always stimulated researches on snake venom composition and function.

Unfortunately, the same is not observed for snake plasma. Despite extensive biochemical and molecular characterization of blood coagulation in mammals, little information is available about haemostasis in other vertebrates [6], although there is an increasing interest in the "natural resistance" of snakes towards the toxicity of its own venom and towards other snake venoms. The inter- and intraspecies resistibility can contribute to the development of new strategies for\

FIGURE 1: Adult and juvenile *B. jararaca* specimens.

the treatment of snake envenomation and the discovery of proteins that can neutralize the toxic components of these venoms [7], making snake plasma a rich source of bioactive molecules.

It has been proposed that there are two different mechanisms that may account for this "natural immunity" [8]: (i) mutation of the gene encoding the target of the venom toxin, providing target resistance to the toxin [9–12] or (ii) presence of proteins that neutralize hemorrhagins [13–16], neurotoxins [17–22], or myotoxins [23, 24] in the blood of resistant animals. Several studies have shown that these proteins are either metalloproteinase inhibitors (antihemorrhagic factors) or $PLA_2$ inhibitors (PLIs) (antineurotoxic/antimyotoxic factors) [8, 25–28].

Our group has purified and characterized two proteins from the plasma of *B. jararaca* snake, probably involved in its self-defense against accidental envenomation: (i) BjI, a blood coagulation inhibitor that recognizes thrombin-like enzymes present in *B. jararaca* venom by western blotting, suggesting a protective role of this protein [29] and (ii) fibrinogen [30], which showed resistance to hydrolysis caused by snake venoms. Interestingly, while bovine thrombin coagulated both *B. jararaca* and human fibrinogen, *B. jararaca* venom clotted human fibrinogen, but not *B. jararaca* fibrinogen. In addition, *C. durissus terrificus* and *Lachesis* sp. venoms could also clot human fibrinogen, with no action upon *B. jararaca* fibrinogen [31].

Another interesting feature described in some snake species is the ontogenetic variability in venom composition, a well-documented phenomenon that has long been studied [32]. Ontogenetic variation in venom composition has been reported in a number of genera [33–37], including *Bothrops* snakes [32, 38–41], which accounts for the differences in the clinical manifestations and severity of envenomation by juvenile and adult *B. jararaca* [42].

All the peculiarities related to ontogenetic variation in *B. jararaca* venom raised the question of whether the plasma composition of snakes follows the same modifications described in the venom. Therefore, a comparative study of the plasma composition of juvenile and adult *B. jararaca* snake was carried out. The present study focused on the antivenom proteins, considering their importance for the self-protection of these animals and for the search of new proteins for antivenom treatment.

## 2. Material and Methods

*2.1. Blood Collection.* Six specimens of *B. jararaca* (3 juveniles and 3 adults) from the Laboratory of Herpetology, Butantan Institute, São Paulo, Brazil, were used for this analysis. All specimens were females, juveniles (<60 cm snout-vent length) or adults (>82 cm snout-vent length) [43, 44] (Figure 1). Blood was collected by caudal venipuncture. Citrated blood samples were collected in a 9 : 1 ratio of blood to 3.8% sodium citrate solution, and plasma was obtained by blood centrifugation for 15 min at 1,200 g at room temperature and stored at −20°C. The Committee for the Ethical Use of Animals of Butantan Institute approved the experimental protocols (Protocol no. 542/08).

*2.2. Protein Determination.* Protein concentrations were determined using bicinchoninic acid (Sigma, St. Louis, MO, USA) and bovine serum albumin (BSA) (Sigma, St. Louis, MO, USA) as a standard, according to Smith et al. [45].

*2.3. Two-Dimensional Electrophoresis (2D Electrophoresis).* Two-Dimensional electrophoresis was used to separate proteins in the first dimension by isoelectric focusing (IEF) and in the second dimension by molecular weight using SDS-PAGE electrophoresis. IEF was carried out using precast Immobiline DryStrip gels pH 3–10 gradient (24 cm—IPG strip) using an IPGphor unit (GE Healthcare, Uppsala, Sweden). On each gel, 1 mg of protein was loaded. The IPG strip and sample were covered with Dry Strip Cover Fluid (GE Healthcare, Uppsala, Sweden) and run at constant voltage of 100 V for 12 h and 500 V up to the accumulation of 500 Vh, followed by gradient voltage from 500 to 1,000 V up to the accumulation of 800 Vh, another gradient voltage from 1,000 to 10,000 Vh up to the accumulation of 16,500 Vh, and constant voltage of 10,000 V up to the accumulation of 22,200 Vh. IEF was followed by a SDS-PAGE using 10% resolving gels, according to Laemmli protocol [46], and DALTsix system (GE Healthcare, Uppsala, Sweden). The gels were run at constant amperage of 15 mA and constant voltage of 80 V for 1 h and then constant amperage of 60 mA and constant voltage of 500 V. Protein spots were visualized using Coomassie Blue R350 staining procedure according to GE Healthcare (Uppsala, Sweden) protocol. Each sample was analyzed in triplicate. Image acquisition of gels was performed using the ImageScanner III densitometer (GE Healthcare, Uppsala, Sweden) and the gels were analyzed using ImageMaster Platinum 7.0 software (GE Healthcare, Uppsala, Sweden). The spots were quantified using the % of spot volume criterion, which is automatically calculated by the ImageMaster software. The match analysis was performed in an automatic mode, and further manual editing was performed to correct mismatched and unmatched spots. A criterion of $P < 0.05$ was used to define the significant difference when analyzing the paired spots between the two groups ($n = 3$) according to ANOVA.

*2.4. Protein Identification.* For identification of spots with quantitative variation by mass spectrometry, gel spots were

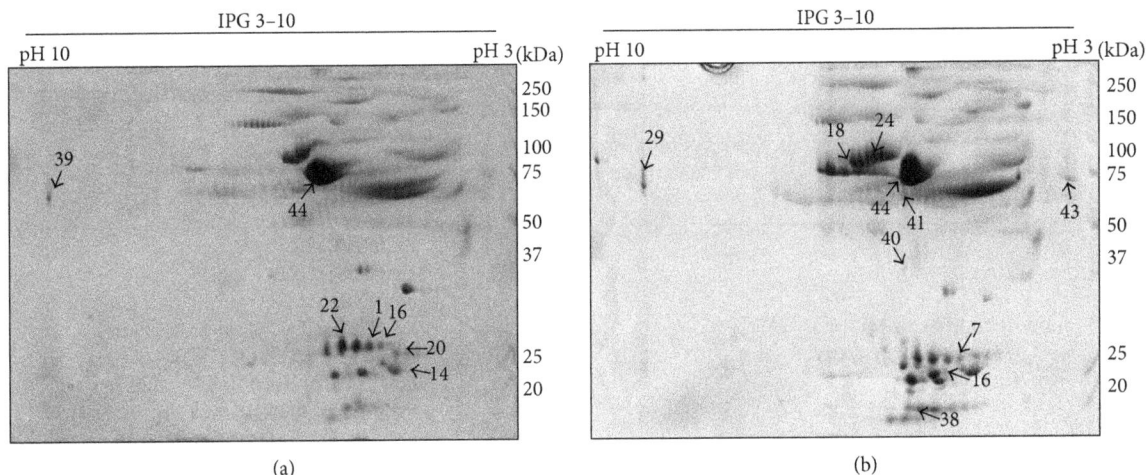

FIGURE 2: Analysis of juvenile and adult *B. jararaca* plasma by 2D electrophoresis. Plasma proteins (1 mg) from (a) juvenile and (b) adult were submitted to isoelectric focusing on 3–10 IPG strips (24 cm) followed by 10% SDS-PAGE. Gels were stained with Coomassie Blue R350. Spots present at different levels were indicated with arrows and identified by nanoESI-Q-TOF.

excised and in-gel trypsin digestion was performed according to Shevchenko et al. [47]. An aliquot (4.5 $\mu$L) of the peptide mixture was separated by $C_{18}$ (100 $\mu$m $\times$ 100 mm) RP-nano UPLC (nanoAcquity UPLC, Waters, Milford, MA, USA) coupled with a Q-TOF Ultima mass spectrometry (Waters, Milford, MA, USA) with a nanoelectrospray source at flow rate of 600 nL/min. The gradient condition was 15–90% acetonitrile in 0.1% formic acid over 10 min. The instrument was operated in the "top three" mode, in which one MS spectrum is acquired followed by MS/MS of the top three most-intense peaks detected. The resulting spectra were acquired using MassLynx v. 4.1 software, and the raw data files were converted into a peak list format (mgf) without summing the scans using Mascot Distiller 2.2.1.0, 2008, Matrix Science (Matrix Science Ltd., London, UK) and searched against nonredundant protein database (NCBI) using Mascot, with carbamidomethylation as fixed modification and oxidation of methionine as variable modifications, one trypsin missed cleavage and tolerance of 20 ppm for both precursor and fragment ions.

## 3. Results and Discussion

It is known that many venomous snakes are resistant to their own venoms and that this natural resistance is due to the neutralizing factors present in their plasma. In the last years, studies on animals that resist the action of snake venoms have led to the discovery and characterization of proteins responsible for this resistance. The result of these studies was the structural and functional characterization of protein inhibitors of hemorrhagic metalloproteinases, as well as myotoxic and neurotoxic $PLA_2$ [7].

Although specific endogenous inhibitors for snake venom have been widely described in the literature and have been the subject of review articles [25, 48, 49], the correlation between venom and plasma ontogenetic development has not been reported yet.

TABLE 1: Number of matches and spots present at different levels in juvenile and adult *B. jararaca* plasma. 2D electrophoresis were analyzed by ImageMaster Platinum 7.0 software (GE Healthcare).

| | Number of matches | Spots showing quantitative variation | |
| --- | --- | --- | --- |
| | | Exclusive spots | Increased spots |
| Juvenile *B. jararaca* | 1,250 | 18 | 5 |
| Adult *B. jararaca* | | 16 | 6 |
| | | Total: 45 | |

Plasma from juvenile and adult *B. jararaca* were analyzed by 2D electrophoresis and were compared using ImageMaster Platinum 7.0 software (see experimental section for details). Figure 2 shows that the proteomic profile of juvenile and adult snakes is similar, suggesting minor ontogenetic differences between the plasma protein content of these two stages of development. The number of matches represents the spots identified in juvenile and adult plasma and exclusive spots were considered those present in only one group, juvenile or adult plasma. The results showed 1,250 matches between juvenile and adult plasma, with only 45 spots showing quantitative variation ($P < 0.05$). Taking into account these 45 spots, 18 are exclusive for juvenile and 16 for adult snakes. In addition, 5 spots were increased in juvenile and 6 in adults (Table 1), suggesting that the ontogenetic development is associated to little changes in the protein content of the plasma.

In order to identify spots present in different levels and correlate these differences to the snake development, the corresponding spots were excised, in-gel digested with trypsin, and submitted to mass spectrometry (MS/MS) (Table 2 and Figure 2). It is important to emphasize that only spots showing quantitative variation were submitted to MS/MS analysis. In addition, out of 45 spots analyzed, only 17 were identified,

TABLE 2: Identification of spots showing quantitative variation in juvenile and adult *B. jararaca* plasma indicated in Figure 2, by ESI-Q-TOF (MS/MS).

(a)

Juvenile *Bothrops jararaca* snake

| Spot number[a] | Protein name (organism) | Score | Protein accession number[b] | Peptide sequences[c] | Volume[d] (%) |
|---|---|---|---|---|---|
| 1 | $\gamma$-phospholipase A$_2$ inhibitor (*Bothrops jararaca*) | 556 | gi\|157885066 | KCIDIVGHR<br>KNCFSSSICKL<br>SCDFCHNIGK<br>VFLEISSASLSVR<br>HEHFPGDIAYNLK<br>LGQIDVNIGHHSYIR<br>DCDGYQQECSSPEDVCGK<br>CIDIVGHRHEHFPGDIAYNLK | 1.6137 |
| 14 | $\gamma$-phospholipase A$_2$ inhibitor subunit B (*Trimeresurus flavoviridis*) | 120 | gi\|155676753 | RACCVGDECK<br>GCATESLCTLLQK | 1.6347 |
| 16 | $\gamma$-phospholipase A$_2$ inhibitor (*Bothrops jararaca*) | 467 | gi\|157885066 | KCIDIVGHR<br>NCFSSSICK<br>VFLEISSASLSVR<br>HEHFPGDIAYNLK<br>TVHKNCFSSSICK<br>LGQIDVNIGHHSYIR<br>DCDGYQQECSSPEDVCGK<br>KCIDIVGHRHEHFPGDIAYNLK | 0.5292 |
| 20 | $\gamma$-phospholipase A$_2$ inhibitor subunit B (*Trimeresurus flavoviridis*) | 284 | gi\|155676753 | ACCVGDECK<br>RACCVGDECK<br>DTENQCLSLTGK<br>GCATESLCTLLQK | 0.5687 |
| 22 | $\gamma$-phospholipase A$_2$ inhibitor (*Bothrops jararaca*) | 496 | gi\|157885068 | INCCEK<br>KCIDIVGHR<br>GRINCCEK<br>NCFSSSICK<br>VFLEISSASLSVR<br>HEHFPGDIAYNLK<br>TVHKNCFSSSICK<br>LGQIDVNIGHHSYIR<br>KCIDIVGHRHEHFPGDIAYNLK | 3.5287 |
| 39 | C3 complement (*Naja naja*) | 221 | gi\|399269 | RVGLVAVDK<br>IWDTIEK<br>IQKPGAAMK<br>GIYTPGSPVR<br>IKLEGDPGAR<br>AVYVLNDKYK<br>EYVLPSFEVR | 0.1644 |
| 44 | Albumin (*Trimeresurus flavoviridis*) | 92 | gi\|56790036 | LVEDIQNDHIIQ<br>IIPQAPTSNLIEITKR | 0.3461 |

(b)

Adult *Bothrops jararaca* snake

| Spot number[a] | Protein name (organism) | Score | Protein accession number[b] | Peptide sequences | Volume[c] (%) |
|---|---|---|---|---|---|
| 7 | $\gamma$-phospholipase A$_2$ inhibitor (*Bothrops jararaca*) | 511 | gi\|157885066 | KCIDIVGHR<br>NCFSSSICK<br>VFLEISSASLSVR<br>HEHFPGDIAYNLK<br>LGQIDVNIGHHSYIR<br>DCDGYQQECSSPEDVCGK<br>KCIDIVGHRHEHFPGDIAYNLK | 0.5813 |

(b) Continued.

| Spot number[a] | Protein name (organism) | Score | Protein accession number[b] | Peptide sequences | Volume[c] (%) |
|---|---|---|---|---|---|
| | Adult *Bothrops jararaca* snake | | | | |
| 16 | $\gamma$-phospholipase A$_2$ inhibitor (*Bothrops jararaca*) | 470 | gi\|157885066 | KCIDIVGHR<br>NCFSSSICK<br>SCDFCHNIGK<br>VFLEISSASLSVR<br>HEHFPGDIAYNLK<br>LGQIDVNIGHHSYIR<br>DCDGYQQECSSPEDVCGK | 0.0990 |
| 18 | Transferrin (*Lamprophis fuliginosus*) | 220 | gi\|108792441 | IVWCAVGK<br>VCTFHTHDW<br>EADAITLDGGHIYTAGK | 0.4829 |
| 24 | Transferrin (*Lamprophis fuliginosus*) | 232 | gi\|108792441 | LVLEQQK<br>IVWCAVGK<br>VCTFHTHDW<br>EADAITLDGGHIYTAGK | 0.2124 |
| 29 | C3 complement (*Naja naja*) | 363 | gi\|399269 | VGLVAVDK<br>LEGDPGAR<br>IWDTIEK<br>IQKPGAAMK<br>GIYTPGSPVR<br>IKLEGDPGAR<br>DTCMGTLVVK<br>AVYVLNDKYK<br>EYVLPSFEVR | 0.2256 |
| 38 | $\alpha$-phospholipase A$_2$ inhibitor precursor (*Bothrops jararaca*) | 383 | gi\|167547111 | LYVTNK<br>REFANLR<br>KNFEALR<br>RSFGSGSER<br>GAFLTVHKA<br>KAFANVLER<br>KVLNSLIDALMHLQRE<br>QICEQAEGHIPSPQLENHNK | 0.1067 |
| 40 | $\beta$-Actin (*Rachycentron canadum*) | 867 | gi\|161376754 | AGFAGDDAPR<br>DLTDYLMK<br>GYSFTTTAER<br>EITALAPSTMK<br>AVFPSIVGRPR<br>DSYVGDEAQSKR<br>IWHHTFYNELR<br>QEYDESGPSIVHR<br>LDLAGRDLTDYLMK<br>SYELPDGQVITIGNER<br>EEEIAALVVDNGSGMCK<br>VAPEEHPVLLTEAPLNPK<br>DLYANTVLSGGTTMYPGIADR<br>TTGIVMDSGDGVTHTVPIYEGYALPHAILR | 0.0060 |
| 41 | Antihemorrhagic factor Bj46a (*Bothrops jararaca*) | 68 | gi\|48428681 | YALNVIKN<br>EGHAHSHLIQQHVEK<br>NCPKCPILLPSNNPQVVDSVEYVLNKHNEK<br>HNEKLSDHVYEVLEISR<br>GDLECDEKDAKEWTDTGVR<br>IMFNVDTFKEDVFAK<br>LSDHVYEVLEISR<br>VPVAFVK<br>ELPKDISDR<br>VHHFEL<br>EWTDTGVR | 0.09264 |

(b) Continued.

| | Adult *Bothrops jararaca* snake | | | | |
|---|---|---|---|---|---|
| Spot number[a] | Protein name (organism) | Score | Protein accession number[b] | Peptide sequences | Volume[c] (%) |
| 43 | Albumin (*Trimeresurus flavoviridis*) | 234 | gi\|56790036 | ECFDTK YGINDCCAK LVEDIQNDHIIQ QLCHCCDSSFISR LEDHVQCLHTGEEQLK | 0.0497 |
| 44 | Albumin (*Trimeresurus flavoviridis*) | 235 | gi\|56790036 | FIETHEK NNCDNYK LVEDIQNDHIIQ QLCHCCDSSFISR LEDHVQCLHTGEEQLK | 0.3964 |

[a] Spot number refers to that shown in Figure 2.
[b] NCBI accession number.
[c] Obtained by MS/MS analysis.
[d] $P < 0.05$.

7 in juvenile and 10 in adult plasma. During this process, we faced the limited available information about reptilian genome and proteome, described by other authors [50, 51]; thus, this study identified proteins by sequence homologies through the National Center for Biotechnology Information database (NCBI).

Among the proteins identified, transferrin was classified as increased in adult plasma (spot no. 18). This could be due to a differential iron transport mechanism across the development stage of snakes, as also reported for humans [52].

The complement system of snakes is of particular interest because the venom of *Naja naja* and related Asian snakes of the genus *Naja* [53] and the venom of *Austrelaps superbus* [54], an Australian elapid, contain a C3 structural and functional analog, cobra venom factor (CVF). Functionally, CVF resembles the C3 activation product C3b as it forms a complex with B factor in the presence of $Mg^{2+}$ [55]. CFV and its analogs have become an important research tool in order to study the role of complement in host defense, immune response, and disease pathogenesis [53].

In *B. jararaca* plasma, C3 complement was identified in juvenile and adult plasma (Figure 2—spots 39 and 29, resp.) with a slight difference concerning the molecular weights. MS/MS analysis identified these spots as C3 complement by sequence homology to C3 from *Naja naja* (gi |399269). This protein, described in *Naja naja*, has molecular weight of 185 kDa and theoretical pI of 5.9. In this work, the two spots identified as C3 complement have molecular mass around 75 kDa and pI around 10, suggesting the presence of fragments in our samples.

Another protein present in adults, according to analysis by 2D electrophoresis, is the antihemorrhagic factor Bj46a (Figure 2—spot 41). Bj46a is a glycoprotein isolated from *B. jararaca* plasma that inhibits the snake venom metalloproteinases atrolysin C and jararhagin and *B. jararaca* venom hemorrhagic activity [56]. Interestingly, Antunes et al. [57] demonstrated that the venom of adult *B. jararaca* specimens was more hemorrhagic than the venom of newborn snakes.

The hemorrhagic activity present in *B. jararaca* venom is generally credited to P-III metalloproteinases, like jararhagin [58, 59]. The reduced hemorrhagic activity present in the newborn *B. jararaca* venom described by Antunes et al. [57] appears to be correlated with the lack of jararhagin in newborn venom. This work showed the sequence of about 35% of Bj46a (data not shown). All of the 122 amino acids identified showed identity to the corresponding sequence present in the databank. However, Bj46a was also identified in juvenile *B. jararaca* plasma submitted to 1D electrophoresis and analyzed by Fourier Transform Ion Cyclotron Resonance mass spectrometry (data not published), suggesting the presence of this inhibitor in juvenile and adult snakes. One hypothesis to explain this finding is that Bj46a might be present in low levels in juvenile *B. jararaca* plasma and could not be identified by 2D electrophoresis. This finding linked to our results suggests a correlation between the ontogenetic development of the venom and the plasma composition of *B. jararaca*.

The high incidence of PLI identified among spots with quantitative variation is noteworthy. This protein corresponds to 71 and 30% of the total proteins identified in juvenile and adult plasma, respectively, and the present work showed that PLIs are increased in juvenile snakes (Figure 2—spots, nos. 16 and 22).

Forty-nine percent of $\gamma$-PLI sequence was obtained in this study (data not shown). Out of 99 amino acids identified, only one has no identity to the corresponding sequence present in the databank. This is the first time that this protein is shown in *B. jararaca*, since the protein sequence described in UNIPROT databank (http://www.uniprot.org) was derived from DNA data. Moreover, $\alpha$-PLI was also possible to be identified, and about 48% of its sequence was obtained (data not shown). As seen for Bj46a, all of the 122 amino acids identified showed identity to the corresponding sequence present in the databank. It is noteworthy the high variability of PLIs found in juvenile or adult plasma, probably not only due to different glycosylation pattern but also to the amino acid sequence, as illustrated by spots 14 and 20

(Figure 2), which are similar to *Trimeresurus flavoviridis* PLIs. This peculiarity is related to the high incidence of PLIs isoforms present in snake plasma, showing the physiological importance of these inhibitors for the physiology of these animals [60, 61].

The role played by PLIs has been the physiological protection of snakes against accidental envenomation or due to the feeding habits of *ophiophagous* specimens [8, 28, 60]. In the last two decades, the number of reports on endogenous PLIs in the plasma of snakes has increased, motivated by the need to develop potentially selective inhibitors for human $PLA_2$.

Snake venom $PLA_2$ exhibits a wide variety of pharmacological effects and is involved in the envenomation pathophysiology, presenting myotoxic and neurotoxic activities [62]. Antunes et al. [57] demonstrated that newborn *B. jararaca* venom shows catalytic $PLA_2$ activity almost twice higher than that of the adult venom, and our results showed that the same occurs regarding $\gamma$-PLI, indicating a connection between venom and plasma components. In addition, besides the antivenom role of PLI, these proteins can be a favorable therapeutic approach in the treatment of inflammatory processes, once $\gamma$-PLI has been studied as a potential model for the development of selective inhibitors of proinflammatory $PLA_2$ in humans [60, 63].

In short, the results showed that there are some differences in plasma protein composition between juvenile and adult *B. jararaca* and that these differences could be related to the ontogenetic variation of the venom composition. This is the first comparative study of protein profiles of juvenile and adult snake plasma. This approach is important for a better understanding of the ontogenetic development of *B. jararaca*. Moreover, associated with the knowledge of ontogenetic changes in venom composition and snakebite clinical reports, the differences identified could be used for the development of more specific antivenoms. It has been suggested that antiophidian serum could be enriched with natural antitoxins in order to increase the serum ability to neutralize snake venom [63]. Thus, the results presented here in this paper can contribute to the knowledge of antivenom mechanisms against ophidian accidents.

## Conflict of Interests

The authors declare that there is no conflict of interests.

## Acknowledgments

The authors acknowledge the Mass Spectrometry Laboratory at Brazilian Biosciences National Laboratory, CNPEM-ABTLuS, Campinas, Brazil for their support with the mass spectrometry analysis. This work was supported by Grants from Fundação de Amparo à Pesquisa do Estado de São Paulo (05/03514-9, 08/08140-8, 09/03484-3, 09/50199-2, 09/54708-9, 11/51857-3) and Conselho Nacional de Desenvolvimento Científico e Tecnológico.

## References

[1] D. A. Warrell, "Clinical features of envenoming from snake bite," in *Envenomings and Their Treatments*, C. Bon and M. Goyffon, Eds., pp. 63–76, Fondation Marcel Mérieux, Lyon, France, 1996.

[2] J. P. Chippaux, "Snake-bites: appraisal of the global situation," *Bulletin of the World Health Organization*, vol. 76, no. 5, pp. 515–524, 1998.

[3] A. Ohsaka, "Hemorragic, necrotizing and edma-forming effects of snake venoms," in *Snake Venoms*, C. Y. Lee, Ed., pp. 480–546, Springer, Berlin, Germany, 1979.

[4] L. A. Ribeiro and M. T. Jorge, "Acidente por serpentes do gênero Bothrops: série de 3.139 casos," *Revista da Sociedade Brasileira de Medicina Tropical*, vol. 30, pp. 475–480, 1997.

[5] M. L. Santoro, I. S. Sano-Martins, H. W. Fan, J. L. Cardoso, R. D. G. Theakston, and D. A. Warrell, "Haematological evaluation of patients bitten by the jararaca, *Bothrops jararaca*, in Brazil," *Toxicon*, vol. 51, no. 8, pp. 1440–1448, 2008.

[6] L. Nahas, A. S. Kamiguti, F. Betti, I. S. Sano Martins, and M. I. Rodrigues, "Blood coagulation mechanism in the snakes *Waglerophis merremii* and *Bothrops jararaca*," *Comparative Biochemistry and Physiology Part A*, vol. 69, no. 4, pp. 739–743, 1981.

[7] S. Lizano, G. Domont, and J. Perales, "Natural phospholipase $A_2$ myotoxin inhibitor proteins from snakes, mammals and plants," *Toxicon*, vol. 42, no. 8, pp. 963–977, 2003.

[8] G. Faure, "Natural inhibitors of toxic phospholipases $A_2$," *Biochimie*, vol. 82, no. 9-10, pp. 833–840, 2000.

[9] S. J. Burden, H. C. Hartzell, and D. Yoshikami, "Acetylcholine receptors at neuromuscular synapses: phylogenetic difference detected by snake $\alpha$ neurotoxins," *Proceedings of the National Academy of Sciences of the United States of America*, vol. 72, no. 8, pp. 3245–3249, 1975.

[10] D. Neumann, D. Barchan, M. Horowitz, E. Kochva, and S. Fuchs, "Snake acetylcholine receptor: cloning of the domain containing the four extracellular cysteines of the $\alpha$ subunit," *Proceedings of the National Academy of Sciences of the United States of America*, vol. 86, no. 18, pp. 7255–7259, 1989.

[11] B. Ohana, Y. Fraenkel, G. Navon, and J. M. Gershoni, "Molecular dissection of cholinergic binding sites: how do snakes escape the effect of their own toxins?" *Biochemical and Biophysical Research Communications*, vol. 179, no. 1, pp. 648–654, 1991.

[12] D. Barchan, S. Kachalsky, D. Neumann et al., "How the mongoose can fight the snake: the binding site of the mongoose acetylcholine receptor," *Proceedings of the National Academy of Sciences of the United States of America*, vol. 89, no. 16, pp. 7717–7721, 1992.

[13] T. Omori-Satoh, S. Sadahiro, A. Ohsaka, and R. Murata, "Purification and characterization of an antihemorrhagic factor in the serum of *Trimeresurus flavoviridis*, a crotalid," *Biochimica et Biophysica Acta*, vol. 285, no. 2, pp. 414–426, 1972.

[14] M. Ovadia, "Purification and characterization of an antihemorrhagic factor from the serum of the snake *Vipera palaestinae*," *Toxicon*, vol. 16, no. 6, pp. 661–672, 1978.

[15] S. Weissenberg, M. Ovadia, G. Fleminger, and E. Kochva, "Antihemorrhagic factors from the blood serum of the western diamondback rattlesnake *Crotalus atrox*," *Toxicon*, vol. 29, no. 7, pp. 807–818, 1991.

[16] S. Pichyangkul and J. C. Perez, "Purification and characterization of a naturally occurring antihemorrhagic factor in the serum of the hispid cotton rat (*Sigmodon hispidus*)," *Toxicon*, vol. 19, no. 2, pp. 205–215, 1981.

[17] M. Ovadia, E. Kochva, and B. Moav, "The neutralization mechanism of *Vipera palaestinae* neurotoxin by a purified factor from homologous serum," *Biochimica et Biophysica Acta*, vol. 491, no. 2, pp. 370–386, 1977.

[18] C. L. Fortes-Dias, B. C. B. Fonseca, E. Kochva, and C. R. Diniz, "Purification and properties of an antivenom factor from the plasma of the South American rattlesnake (*Crotalus durissus terrificus*)," *Toxicon*, vol. 29, no. 8, pp. 997–1008, 1991.

[19] C. L. Fortes-Dias, Y. Lin, J. Ewell, C. R. Diniz, and T. Y. Liu, "A phospholipase A$_2$ inhibitor from the plasma of the South American rattlesnake (*Crotalus durissus terrificus*). Protein structure, genomic structure, and mechanism of action," *Journal of Biological Chemistry*, vol. 269, no. 22, pp. 15646–15651, 1994.

[20] J. Perales, C. Villela, G. B. Domont et al., "Molecular structure and mechanism of action of the crotoxin inhibitor from *Crotalus durissus terrificus* serum," *European Journal of Biochemistry*, vol. 227, no. 1-2, pp. 19–26, 1995.

[21] J. Shao, H. Shen, and B. Havsteen, "Purification, characterization and binding interactions of the Chinese-cobra (*Naja naja atra*) serum antitoxic protein CSAP," *Biochemical Journal*, vol. 293, no. 2, pp. 559–566, 1993.

[22] W. Xiaolu, B. Havsteen, and H. Hansen, "Evidence of the coevolution of a snake toxin and its endogenous antitoxin cloning, sequence and expression of a serum albumin cDNA of the Chinese cobra," *Biological Chemistry Hoppe-Seyler*, vol. 376, no. 9, pp. 545–553, 1995.

[23] S. Lizano, B. Lomonte, J. W. Fox, and J. M. Gutiérrez, "Biochemical characterization and pharmacological properties of a phospholipase A$_2$ myotoxin inhibitor from the plasma of the snake *Bothrops asper*," *Biochemical Journal*, vol. 326, no. 3, pp. 853–859, 1997.

[24] S. Lizano, Y. Angulo, B. Lomonte et al., "Two phospholipase A$_2$ inhibitors from the plasma of *Cerrophidion (Bothrops) godmani* which selectively inhibit two different group-II phospholipase A$_2$ myotoxins from its own venom: isolation, molecular cloning and biological properties," *Biochemical Journal*, vol. 346, no. 3, pp. 631–639, 2000.

[25] J. W. Fox and J. B. Bjarnason, "Metalloproteinase inhibitors," in *Enzymes from Snake Venoms*, G. S. Bailey, Ed., pp. 559–632, Alaken Inc, Fort Collins, Colo, USA, 1998.

[26] J. C. Pérez and E. E. Sánchez, "Natural protease inhibitors to hemorrhagins in snake venoms and their potential use in medicine," *Toxicon*, vol. 37, no. 5, pp. 703–728, 1999.

[27] J. Perales and G. B. Domont, "Are inhibitors of metallopreoteinases, phospholipase A$_2$ and myotoxins members os the innate immune system?" in *Perspectives in Molecular Toxinology*, A. Ménez, Ed., Wiley, Chichester, UK, 2002.

[28] C. L. Fortes-Dias, "Endogenous inhibitors of snake venom phospholipases A$_2$ in the blood plasma of snakes," *Toxicon*, vol. 40, no. 5, pp. 481–484, 2002.

[29] A. M. Tanaka-Azevedo, A. S. Tanaka, and I. S. Sano-Martins, "A new blood coagulation inhibitor from the snake *Bothrops jararaca* plasma: isolation and characterization," *Biochemical and Biophysical Research Communications*, vol. 308, no. 4, pp. 706–712, 2003.

[30] C. O. Vieira, "*Bothrops jararaca* fibrinogen and its resistance to hydrolysis evoked by snake venoms," *Comparative Biochemistry and Physiology*, vol. 151, no. 4, pp. 428–432, 2008.

[31] D. F. Vieira, L. Watanabe, C. D. Sant'ana et al., "Purification and characterization of jararassin-I, a thrombin-like enzyme from *Bothrops jararaca* snake venom," *Acta Biochimica et Biophysica Sinica*, vol. 36, no. 12, pp. 798–802, 2004.

[32] A. Zelanis, A. K. Tashima, M. M. T. Rocha et al., "Analysis of the ontogenetic variation in the venom proteome/peptidome of *Bothrops jararaca* reveals different strategies to deal with prey," *Journal of Proteome Research*, vol. 9, no. 5, pp. 2278–2291, 2010.

[33] S. A. Minton and S. A. Weinstein, "Geographic and ontogenic variation in venom of the western diamondback rattlesnake (*Crotalus atrox*)," *Toxicon*, vol. 24, no. 1, pp. 71–80, 1986.

[34] S. P. Mackessy, "Venom ontogeny in the pacific rattlesnakes *Crotalus viridis* helleri and *C. v. oreganus*," *Copeia*, vol. 1988, pp. 92–101, 1988.

[35] J. M. Gutierrez, C. Avila, Z. Camacho, and B. Lomonte, "Ontogenetic changes in the venom of the snake Lachesis muta stenophrys (bushmaster) from Costa Rica," *Toxicon*, vol. 28, no. 4, pp. 419–426, 1990.

[36] J. M. Gutierrez, M. C. Dos Santos, M. De Fatima Furtado, and G. Rojas, "Biochemical and pharmacological similarities between the venoms of newborn *Crotalus durissus durissus* and adult *Crotalus durissus terrificus* rattlesnakes," *Toxicon*, vol. 29, no. 10, pp. 1273–1277, 1991.

[37] P. Saravia, E. Rojas, V. Arce et al., "Geographic and ontogenic variability in the venom of the neotropical rattlesnake *Crotalus durissus*: pathophysiological and therapeutic implications," *Revista de Biologia Tropical*, vol. 50, no. 1, pp. 337–346, 2002.

[38] M. F. D. Furtado, M. Maruyama, A. S. Kamiguti, and L. C. Antonio, "Comparative study of nine Bothrops snake venoms from adult female snakes and their offspring," *Toxicon*, vol. 29, no. 2, pp. 219–226, 1991.

[39] J. L. López-Lozano, M. V. De Sousa, C. A. O. Ricart et al., "Ontogenetic variation of metalloproteinases and plasma coagulant activity in venoms of wild Bothrops atrox specimens from Amazonian rain forest," *Toxicon*, vol. 40, no. 7, pp. 997–1006, 2002.

[40] R. A. P. Guércio, A. Shevchenko, A. Shevchenko et al., "Ontogenetic variations in the venom proteome of the Amazonian snake *Bothrops atrox*," *Proteome Science*, vol. 4, article 11, 2006.

[41] A. Zelanis, J. de Souza Ventura, A. M. Chudzinski-Tavassi, and M. D. F. D. Furtado, "Variability in expression of Bothrops insularis snake venom proteases: an ontogenetic approach," *Comparative Biochemistry and Physiology C*, vol. 145, no. 4, pp. 601–609, 2007.

[42] J. Monteiro, "Relação da Província do Brasil," in *História da Companhia de Jesus no Brasil*, S. Leite, Ed., Instituto Nacional do Livro, Rio de Janeiro, Brazil, 1949.

[43] I. Sazima, "Natural history of the jararaca pitviper, *Bothrops jararaca*, in southeastern Brazil," in *Biology of the Pitvipers*, J. A. Campbell and E. D. Brodie, Eds., pp. 199–216, Tyler, Selva, Spain, 1992.

[44] T. R. F. Janeiro-Cinquini, "Variação anual do sistema reprodutor de fêmeas de *Bothrops jararaca* (Serpentes, Viperidae)," *Iheringia Série Zoologia*, vol. 94, pp. 325–328, 2004.

[45] P. K. Smith, R. I. Krohn, and G. T. Hermanson, "Measurement of protein using bicinchoninic acid," *Analytical Biochemistry*, vol. 150, no. 1, pp. 76–85, 1985.

[46] U. K. Laemmli, "Cleavage of structural proteins during the assembly of the head of bacteriophage T4," *Nature*, vol. 227, no. 5259, pp. 680–685, 1970.

[47] A. Shevchenko, M. Wilm, O. Vorm, and M. Mann, "Mass spectrometric sequencing of proteins from silver-stained polyacrylamide gels," *Analytical Chemistry*, vol. 68, no. 5, pp. 850–858, 1996.

[48] G. B. Domont, J. Perales, and H. Moussatche, "Natural anti-snake venom proteins," *Toxicon*, vol. 29, no. 10, pp. 1183–1194, 1991.

[49] M. M. Thwin and P. Gopalakrishnakone, "Snake envenomation and protective natural endogenous proteins: a mini review of the recent developments (1991–1997)," *Toxicon*, vol. 36, no. 11, pp. 1471–1482, 1998.

[50] L. N. F. Darville, M. E. Merchant, A. Hasan, and K. K. Murray, "Proteome analysis of the leukocytes from the American alligator (*Alligator mississippiensis*) using mass spectrometry," *Comparative Biochemistry and Physiology Part D*, vol. 5, no. 4, pp. 308–316, 2010.

[51] C. Stegemann, A. Kolobov, Y. F. Leonova et al., "Isolation, purification and de novo sequencing of TBD-1, the first beta-defensin from leukocytes of reptiles," *Proteomics*, vol. 9, no. 5, pp. 1364–1373, 2009.

[52] V. Ignjatovic, C. Lai, R. Summerhayes et al., "Age-related differences in plasma proteins: how plasma proteins change from neonates to adults," *PLoS ONE*, vol. 6, no. 2, Article ID e17213, 2011.

[53] C. W. Vogel, D. C. Fritzinger, B. E. Hew, M. Thorne, and H. Bammert, "Recombinant cobra venom factor," *Molecular Immunology*, vol. 41, no. 2-3, pp. 191–199, 2004.

[54] S. Rehana and R. Manjunatha Kini, "Molecular isoforms of cobra venom factor-like proteins in the venom of *Austrelaps superbus*," *Toxicon*, vol. 50, no. 1, pp. 32–52, 2007.

[55] P. Hensley, M. C. O'Keefe, and C. J. Spangler, "The effects of metal ions and temperature on the interaction of cobra venom factor and human complement Factor B," *Journal of Biological Chemistry*, vol. 261, no. 24, pp. 11038–11044, 1986.

[56] R. H. Valente, B. Dragulev, J. Perales, J. W. Fox, and G. B. Domont, "BJ46a, a snake venom metalloproteinase inhibitor isolation, characterization, cloning and insights into its mechanism of action," *European Journal of Biochemistry*, vol. 268, no. 10, pp. 3042–3052, 2001.

[57] T. C. Antunes, K. M. Yamashita, K. C. Barbaro, M. Saiki, and M. L. Santoro, "Comparative analysis of newborn and adult *Bothrops jararaca* snake venoms," *Toxicon*, vol. 56, no. 8, pp. 1443–1458, 2010.

[58] F. R. Mandelbaum and M. R. Assakura, "Antigenic relationship of hemorrhagic factors and proteases isolated from the venoms of three species of Bothrops snakes," *Toxicon*, vol. 26, no. 4, pp. 379–385, 1988.

[59] M. J. I. Paine, H. P. Desmond, R. D. G. Theakston, and J. M. Crampton, "Purification, cloning, and molecular characterization of a high molecular weight hemorrhagic metalloprotease, jararhagin, from *Bothrops jararaca* venom. Insights into the disintegrin gene family," *Journal of Biological Chemistry*, vol. 267, no. 32, pp. 22869–22876, 1992.

[60] M. I. Estevão-Costa, B. C. Rocha, M. de Alvarenga Mudado, R. Redondo, G. R. Franco, and C. L. Fortes-Dias, "Prospection, structural analysis and phylogenetic relationships of endogenous $\gamma$-phospholipase $A_2$ inhibitors in Brazilian Bothrops snakes (Viperidae, Crotalinae)," *Toxicon*, vol. 52, no. 1, pp. 122–129, 2008.

[61] P. G. Hains and K. W. Broady, "Purification and inhibitory profile of phospholipase $A_2$ inhibitors from Australian elapid sera," *Biochemical Journal*, vol. 346, no. 1, pp. 139–146, 2000.

[62] A. Shimada, N. Ohkura, K. Hayashi et al., "Subunit structure and inhibition specificity of $\alpha$-type phospholipase $A_2$ inhibitor from *Protobothrops flavoviridis*," *Toxicon*, vol. 51, no. 5, pp. 787–796, 2008.

[63] A. M. Soares, S. Marcussi, R. G. Stábeli et al., "Structural and functional analysis of BmjMIP, a phospholipase $A_2$ myotoxin inhibitor protein from Bothrops moojeni snake plasma," *Biochemical and Biophysical Research Communications*, vol. 302, no. 2, pp. 193–200, 2003.

**6**

# Current Status and Advances in Quantitative Proteomic Mass Spectrometry

**Valerie C. Wasinger,[1] Ming Zeng,[1] and Yunki Yau[1,2]**

[1] Bioanalytical Mass Spectrometry Facility, Mark Wainwright Analytical Centre, The University of New South Wales, Sydney, NSW 2052, Australia
[2] Department of Gastroenterology and Liver Services, Concord Repatriation General Hospital, Sydey, NSW 2139, Australia

Correspondence should be addressed to Valerie C. Wasinger; v.wasinger@unsw.edu.au

Academic Editor: Bomie Han

The accurate quantitation of proteins and peptides in complex biological systems is one of the most challenging areas of proteomics. Mass spectrometry-based approaches have forged significant in-roads allowing accurate and sensitive quantitation and the ability to multiplex vastly complex samples through the application of robust bioinformatic tools. These relative and absolute quantitative measures using label-free, tags, or stable isotope labelling have their own strengths and limitations. The continuous development of these methods is vital for increasing reproducibility in the rapidly expanding application of quantitative proteomics in biomarker discovery and validation. This paper provides a critical overview of the primary mass spectrometry-based quantitative approaches and the current status of quantitative proteomics in biomedical research.

## 1. Introduction

Quantification in a proteomics setting relies on the ability to detect small changes in protein and peptide abundance in response to an altered state [1]. Differential analysis is generated from LC-MS experiments and can be carried out using both label and label-free approaches. For trace amounts of proteins within complex proteomes such as plasma, tears, and urine, no singular technique should be used as a stand-alone guarantee of quantitative precision without hypothesis-driven, targeted approaches. Enrichment and fractionation of specific classes of protein is beneficial during the discovery phase of a project, but because these methods can involve numerous steps, they can become a limiting factor for large scale validation. The variability introduced by multiple methods prior to quantitative mass spectrometry should be assessed, and it is paramount that protein measurements reflect the authentic concentration in the original sample. The development of methods for accurate protein quantitation is one of the most challenging areas of proteomics.

Quantitative proteomics comes in two forms: absolute and relative. Relative quantitation compares the levels of a specific protein in different samples with results being expressed as a relative fold change of protein abundance [2]. Absolute quantitation is the determination of the exact amount or mass concentration of a protein, for example, in units of ng/mL of a plasma biomarker.

Traditional proteomic quantitation approaches rely on high-resolution protein separation by 2D gels. The use of dyes, fluorophores, or radioactivity to label proteins allows visualization of spots/bands with differential intensities [3, 4]. These methods facilitate relative abundance comparison but require many replicates and intensive image analysis that can often be quite user subjective. The simplicity of mass spectrometry-based approaches addresses issues of reproducibility [5] and poor representation of low-abundance [6], low-mass, and basic proteins [7, 8], as well as the need for the postdifferential identification by MS [3] as it is inherent in the separation methods. MS-based methods have also come into prominence compared to traditional antibody-based methods due to their higher specificity, good reproducibility and precision, and ability to rapidly analyse hundreds of peptide transitions in one MRM assay [9]. Pragmatically, the course of a biomarker project sees a number of quantitative

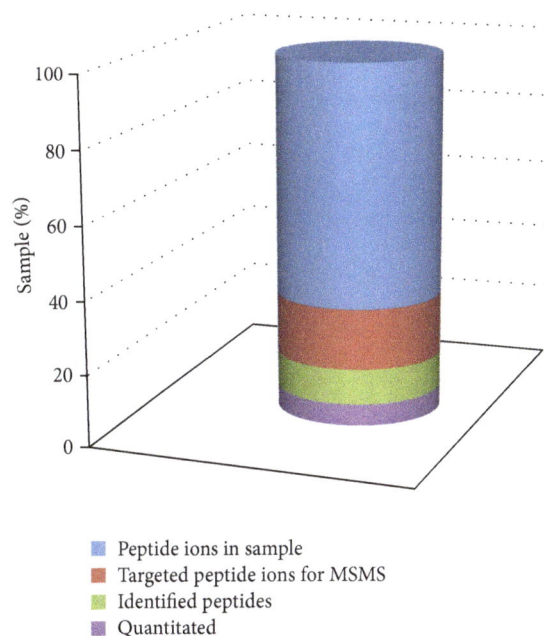

FIGURE 1: Relationship between the peptide ion content and the difficulty in obtaining sufficient MSMS information to both identify and also quantitate those peptides. Adapted from Michalski et al. [10] and Liu et al. [11].

techniques used from discovery-driven low-cost methods such as relative and label-free quantitation to hypothesis-driven quantitation using synthetic standards with complimentary analysis of trends by alternative techniques such as ELISA or Western blot. Here, we provide a critical overview of the main MS-based quantitation approaches and outline the advances and challenges of applying these techniques in protein biomarker discovery and validation.

## 2. Quantitative Proteomics in Biomarker Discovery

The ultimate aim of biomarker discovery is to develop a simple differential test to be used as a clinical evaluation tool. This requires a lengthy and difficult process which involves candidate discovery, verification, validation, and translation to clinical laboratory use [12, 13]. Current discovery studies aim to detect disease-specific markers by analysing and comparing healthy controls and disease-affected subjects [14], and despite the discovery of increasing numbers of potential markers, few have progressed to clinical practice [15, 16]. Much of this dilemma is a reflection of the challenges associated with linking bench to clinic outcomes and providing basic researchers with the opportunity to finance and progress their science past the validation phase [12, 17]. The development of targeted, quantitative approaches that provide accurate and statistically reliable quantitative outcomes for multisite studies may provide a critical bridge to establishing validity of individual or panels of biomarkers.

A challenge facing biomarker development is the sheer complexity and range of concentrations within the human proteome [12, 16]. Human plasma is estimated to contain more than 10,000 core proteins [35], of which only small fractions are effectively characterized with current technology [36]. Proteins in plasma have a $10^{12}$-fold concentration range, from millimolar for albumin, down to attomolar ranges, and further for cytokines [35] and other proteins, hormones, and peptides. This greatly exceeds the ability of current proteomic approaches, which have linearity over ~3 orders of magnitude [16].

Disease-specific proteins, including low-mass peptides, can be low in abundance and difficult to detect amongst a diverse "sea" of proteins [37]. Combined with the immense extent of human and disease variation and the challenges facing the development of sensitive and specific differentiators, developing these technologies to the clinic is a formidable task. Discovery phase quantitative approaches entail the differentiation of as many peptides as possible (rather than the identification of all proteins) from LC-MS experiments and is highly dependent on scan speed, sensitivity, and ability to isolate precursor ions for selection to MS/MS [10]. Figure 1 shows the relationship between peptide ions and quantitation and is adapted from Michalski et al. [10] and Liu et al. [11]. This figure demonstrates the gap between peptide content and ability to quantitate those peptides and proteins comprehensively to provide quantitative coverage. As instruments improve in these areas, there will be an associated increase in depth of coverage and accuracy which is required to discern the very small changes in abundance, peptide modifications, and mass differences that delineate a disease type or process. For targeted approaches, the use of high-resolution instruments has the advantage of relying on the mass accuracy to provide fewer transitions and therefore being able to simultaneously monitor more peptides within the one scheduled experiment. This should assist the reliability and precision of targeted assays to unambiguously identify the target peptide and avoid interfering transitions particularly in complex biological matrices [38]. Indeed, there is a growing consensus that panels of multiple biomarkers are more likely to achieve adequate clinical sensitivity and specificity [12, 37, 39].

There are a number of novel techniques that allow for the fractionation, depletion, enrichment, and equalisation of complex samples to assist in improving the proteome coverage and number of peptide ions targeted for MS/MS within an instrument's detection range. Fractionation techniques can be applied to cut samples into subgroups of fewer proteins [15] and are most commonly in the form of (gel) electrophoresis and liquid chromatography (LC), techniques which exploit a variety of physicochemical properties of proteins to fractionate proteomes [7]. To reduce protein concentration variability, high-abundance proteins such as albumin can be removed from plasma samples through immunodepletion. There is, however, a risk of codepletion of potentially significant biomarkers due to nonspecific binding or loss of biomarkers bound to higher-abundance carrier proteins [40–42]. These techniques in combination effectively allow the detection of trace proteins [7, 15, 16]. However, any additional manipulation during the sample processing

TABLE 1: Overview of the main approaches for quantitative proteomics. Modified from Schulze and Usadel [4] and Ly and Wasinger [4, 7].

| Method | Dynamic range[a] | Coverage | Quant accuracy, (throughput) R = relative, A = absolute | Associated software | Link |
|---|---|---|---|---|---|
| **Label-free** | | | | | |
| 2D gels | 1 to 4, stain dependent | Medium | Medium (low) R Requires MS identification. | PDQuest | http://www.bio-rad.com/ |
| | | | | Progenesis SameSpots | http://www.nonlinear.com/ |
| | | | | Melanie [18] | http://www.genebio.com/ |
| | | | | Phoretix | http://www.perkinelmer.com/ |
| | | | | Progenesis LCMS | http://www.nonlinear.com/ |
| | | | | msInspect [19] | http://proteomics.fhcrc.org |
| | | | | MSight [20] | http://web.expasy.org |
| | | | | TOPP [21] | http://open-ms.sourceforge.net/ |
| Ion intensities MS[1] | 3 | Good | Medium, (medium to high) R, LC dependent. | PEPPeR [22] | http://www.broadinstitute.org/ |
| | | | | SuperHirn [23] | http://www.waters.com/ |
| | | | | DeCyder MS | http://www.gelifesciences.com/ |
| | | | | SIEVE | http://thermo.com/ |
| | | | | ProteinLynx | http://www.waters.com/ |
| Spectrum count MS[2] | 3, Inaccurate for low abundance. | Good | Poor, (medium to high) R LC dependent | Scaffold [24] | http://www.proteomesoftware.com/ |
| | | | | Elucidator | http://www.rosettabio.com/ |
| | | | | ProteoIQ | http://www.bioinquire.com/ |
| APEX, emPAI | 3 or 4 | Good | Poor, (high) R, within sample only. | APEX [25] | http://pfgrc.jcvi.org/index.php/bioinformatics/apex.html |
| | | | | Mascot | http://www.matrixscience.com/ |
| **Metabolic labeling** | | | | | |
| [15]N | 1 to 2 | Medium | Precise, (low). R, between 2 conditions. | Scaffold | http://www.proteomesoftware.com/ |
| | | | | MSQuant [26] | http://msquant.sourceforge.net/ |
| SILAC | 1 to 2 | Medium | Precise, (low). R Between 2 and 3 samples. | Scaffold | http://www.proteomesoftware.com/ |
| | | | | MSQuant | http://msquant.sourceforge.net/ |
| | | | | Elucidator | http://www.rosettabio.com/ |
| | | | | ASAPRatio | http://tools.proteomecenter.org/ |
| **Isotopic labeling** | | | | | |
| ICAT, [18]O, ICPL | 1 to 2 | Poor | Precise, (low). R Between 2 conditions. | Elucidator | http://www.rosettabio.com/ |
| | | | | XPRESS [27] | http://tools.proteomecenter.org/ |
| | | | | MSQuant | http://msquant.sourceforge.net/ |
| | | | | ASAPRatio [28] | http://tools.proteomecenter.org/ |
| | | | | ZoomQuant [29] | http://proteomics.mcw.edu/zoomquant.html |
| **Isobaric labeling** | | | | | |
| iTRAQ, TMT, DIGE | 2 3 | Medium | Medium, (low). R or A Between 2 and 8 conditions. | ProteinPilot | http://www.absciex.com/ |
| | | | | Multi-Q [30] | http://ms.iis.sinica.edu.tw/Multi-Q-Web/ |
| | | | | iTracker [31] | http://www.cranfield.ac.uk/ |
| | | | | MSQuant | http://msquant.sourceforge.net/ |

Table 1: Continued.

| Method | Dynamic range[a] | Coverage | Quant accuracy, (throughput) R = relative, A = absolute | Associated software | Link |
|---|---|---|---|---|---|
| Targeted | | | | | |
| MRM Isotope dilution +/− heavy label | 5 Attomolar detection. | Poor[1] | Precise, (high). R or A Requires intensive method development. | Skyline [32] MaxQuant ATAQS [33] MRMer [34] | https://brendanx-uwl.gs.washington.edu http://maxquant.org/ http://tools.proteomecenter.org/ATAQS/ATAQS.html http://proteomics.fhcrc.org/CPL/MRMer.html |

[a]Orders of magnitude.
APEX: absolute protein expression profiling. emPAI: exponentially modified protein abundance index. SILAC: stable isotope labelling by amino acids. DIGE: Difference Gel Electrophoresis. ICAT: isotope-coded affinity tags. ITRAQ: isobaric tags for absolute and relative quantitation. TMT: tandem mass tags. MRM: multiple reaction monitoring.
[1]Few target proteins can be selected efficiently in a single LC-MS/MS experiment.
MS²: MSMS

can introduce preanalytical variables that cause changes in quantitative peptide amounts [9]. While the previous techniques can improve discovery of trace levels of candidate protein biomarkers, extensive validation and standardization of these steps will be required before they can be used for direct clinical applications [9, 43].

Data analysis is yet another significant challenge associated with MS-based proteomics. With the enormous volumes of proteomic data generated, expert manual analysis would be inconsistent and unfeasible [44]. Thus, bioinformatics tools are crucial in the determination of which proteins and peptides emerge as candidate biomarkers from discovery studies and the interpretation of quantitative data [9, 45]. There is a need for sophisticated yet transparent computational methods and algorithms to allow for consistent analysis and interpretation of proteomic data using statistical principles [45]. The development and validation of such tools is a critical part in the process of developing quality standards for MS experiments and, hence, generating reproducible and accurate data.

## 3. Strengths and Limitations of Mass Spectrometry-Based Quantitative Approaches

Protein mass spectrometry is not inherently quantitative. There are many reasons as to why the amount of analyte compared to the MS signal intensity does not always show a linear relationship [3, 44]. Because of this, accurate comparisons between two samples must be based on the same individual peptide in LC-MS/MS experiments conducted under the same conditions [4], particularly for absolute quantitation. Table 1 presents an overview of the technical parameters of the main quantitative approaches, their strengths and limitations.

### 3.1. Label-Free Approaches.
Two widely used label-free quantitative methods are spectral counting and peptide peak intensity measurement. Spectral counting requires proteins to have sufficient peptides (both in number and abundance) to trigger MS/MS data for quantification and identification. The approach is based on the observation that more abundant proteins will produce more MS/MS spectra than less abundant proteins, and abundant peptides are sampled more often in fragment ion scans than are low abundance peptides. Relative quantitation by spectral count thus involves comparing the number of identified spectra from the same protein between different samples [54]. Spectral counting is a protein-centric approach that is less reliable for trace and/or low mass proteins; and less responsive toward small changes in response (<2 orders of magnitude) [11, 55], favoring higher abundance "average" proteins [2], while lower identification rates for proteins with low sequence coverage and nontryptic or fewer peptides are a consequence of the methods used for identification as much as the dynamic range of the sample and the limited duty cycle of the MS instrument [56]. This approach has been modified into forms such as

the exponentially modified protein abundance index [57] and absolute protein expression profiling [58].

Relative quantitation using peptide peak intensity measurements involves comparing the MS peptide ion intensities belonging to a given protein [59]. The ion chromatograms for every peptide are extracted from an LC-MS run, and their peak areas are integrated over the chromatographic time scale. These values can be compared to respective values in other experiments for relative quantitation, and only the same ion species can be compared between different samples. Hence, this approach requires multiple replicates and correlation of retention time with m/z ion features and charge state to avoid discrepancy in matching common ions detected in each run. The coverage of common ions between different samples is strongly dependent on sample preparation and can be severely affected by column conditions, instrument sensitivity, and calibration. These variables are pronounced when running long-term projects where analysis is carried out over weeks to months and can introduce approximately 40% discrepancy at the peptide level [4]. Label-free techniques have been performed in many studies and are promising alternatives to stable isotope labeling. They are fast, easy to perform, and inexpensive, and they allow higher dynamic range [3]. Furthermore, any soluble biological material can be used, and unlimited numbers of samples can be compared [4].

### 3.2. Stable Isotope Labelling.
Stable isotope labelling techniques are based on the introduction of a differential mass tag which affects only the mass of a protein or peptide without changing the chemical properties during chromatography or MS [2]. Relative or absolute quantitation can be achieved by using MS to compare the abundance of a labeled "heavy" (known concentration) against the endogenous "light" isoforms [60]. Stable isotope labels are introduced metabolically or chemically at either the protein or peptide level during sample preparation.

Metabolic labelling involves the introduction of stable isotopes to whole cells through the growth medium, which enables the labels to be incorporated during normal cell growth and division [61]. Differently labelled samples can be pooled together for subsequent preparation which avoids variability of sample preparation. However, this method is not applicable to samples that are not metabolically active such as plasma [2]. While the original $^{15}N$ labelling can only compare two samples in one experiment, high-throughput quantitation was developed in the form of stable isotope labelling by amino acids (SILAC) [62]. SILAC incorporates heavy and light forms of arginine or lysine *in vivo* and also combines light and heavy samples prior to sample preparation to significantly reduce sample handling and thus quantitative errors, allowing very small changes in protein levels as well as protein modifications to be detected.

In chemical labelling, the isotope label is introduced to proteins or peptides by a chemical reaction, such as with isotope-coded affinity tags (ICAT) [63] and isotope-coded protein labels (ICPL) [64]. ICAT labels specifically bind to cysteine, a relatively rare amino acid, which effectively

TABLE 2: Recent quantitative MS-based studies involving human samples.

| Authors/year | Specimen | Quantitative approach | Sample preparation | Outcomes |
|---|---|---|---|---|
| Yang et al. 2011 [46] | Urine 54 bladder cancer patients, 46 controls | Label-free—spectral count | NIL | Quantified 265 glycoproteins. alpha-1-antitrypsin, 74% sensitivity and 80% specificity for bladder cancer patients. |
| Quintana et al. 2009 [47] | Urine 39 patients kidney chronic allograft dysfunction, 32 controls | Label-free—peak peptide intensity | SCX using magnetic beads | Peptides from uromodulin and kininogen significantly elevated in control compared to CAD patients. |
| Hanas et al. 2008 [48] | Serum 13 pancreatic adenocarcinoma patients, 12 healthy controls | Label-free—peak peptide intensity | NIL | Quantified 20 low-mass serum peaks. Bootstrap analysis showed peaks could differentiate cancer from control sera with 95% accuracy. |
| Xue et al. 2010 [49] | Cell lysates Primary and lymph node metastatic cell lines. 1 patient. | Label-free—peak peptide intensity | NIL | 145 differential proteins. Western blot and ROC curve analysis confirmed that 2 specific proteins could predict colorectal cancer metastasis. |
| Besson et al. 2011 [50] | Colorectal cancer tissue 28 colorectal frozen tissue samples | Stable isotope labeling—iTRAQ | Peptide OFFGEL fractionation | 555 proteins with significant fold change between different cancer stages. Identified a candidate with increased abundance in adenomas and early stage colorectal cancer. |
| Bondar et al. 2007 [51] | Serum 6 healthy male, 20 nonmalignant prostate biopsy patients, 26 malignant prostate cancer patients | Stable isotope labeling | NIL | Higher abundance of Zn-α2 glycoprotein (ZAG) in prostate cancer patients than nonmalignant prostate disease patients and healthy controls. |
| Chaerkady et al. 2008 [52] | Liver tissue 55 samples of hepatocellular carcinoma, 20 samples of adjacent noncancer tissues | Stable isotope labeling—iTRAQ | SCX | 59 proteins increased in abundant, 92 proteins were less abundant in HCC compared to normal tissue. 12 proteins further validated using immunohistochemical labeling. |
| Dayon et al. 2008 [53] | Cerebrospinal fluid 4 postmortem CSF patients, 4 antemortem CSF from living healthy controls | Stable isotope labeling—tandem mass tag isobaric labeling | Immunoaffinity depletion of 6 most abundant proteins and SCX | 78 proteins more abundant in postmortem samples compared to antemortem. |

reduces sample complexity but also limits its use since it cannot track proteins that lack cysteine residues [2]. Another limitation of ICAT is that only two samples can be compared in a single analysis.

The development of isobaric mass tags such as tandem mass tag (TMT) [77] and isobaric tags for relative and absolute quantification (iTRAQ) [78] allows for the comparison of up to eight samples in parallel [79, 80]. iTRAQ involves the introduction of mass-balanced labels at the level of tryptic peptides which produce labelled peptides of the same total mass that coelute in liquid chromatography. The different mass tags are differentiated by the mass spectrometer only upon peptide fragmentation [81]. Despite having disadvantages such as variability in labelling efficiencies and protein digestion [2], TMT and iTRAQ are favourable for quantitative biomarker discovery due to their ability to multiplex up to

eight samples [82]. A summary of some recent projects is demonstrated in Table 2 and shows the variety of techniques applied for quantitation.

*3.3. Multiple Reaction Monitoring.* Multiple reaction monitoring (MRM) is the main current approach for highly confident protein and peptide quantification. MRM targets specific peptides in complex samples by typically using a triple quadrupole mass spectrometer or hybrid triple quadrupole/linear ion trap mass spectrometer. These instruments have two mass filters that can select a predefined peptide ion and a combination of its specific fragment ions to analyse and monitor over time for accurate quantitation [2, 83]. Combinations of peptide mass and product ion masses create a unique signature for a particular peptide with

TABLE 3: Summary of MRM quantitative analysis in blood for a variety disease types.

| Authors/year | Specimen | Target | Sample preparation | MS platform | Outcomes |
|---|---|---|---|---|---|
| Stahl-Zeng et al. 2007 [65] | Plasma | N-glycoproteins | Selective isolation of N-glycosites. Stable isotope $^{13}$C- and/or $^{15}$N-labelled reference peptides. | LC ESI MS/MS Hybrid triple quadrupole linear ion trap | Detection $\leq$ ng/mL concentration range and accurate quantification over a linear range of $\sim 10^5$. LOQ of 50 amol. LOD $\geq$ 10 amol. Protein concentration in plasma of 0.1 ng/mL. |
| Anderson and Hunter 2006 [66] | Plasma 1 healthy donor | 53 plasma proteins | Top six abundant proteins depleted. Stable isotope labeled internal standards. | ESI LC-MS/MS 4000 Q Trap Hybrid triple quadrupole/linear ion trap | Quantitative data for 47 proteins in the $\mu$g/mL level over linear range of $10^4$ |
| Keshishian et al. 2007 [67] | Plasma 1 healthy donor | 6 low abundance plasma proteins | Abundant protein depletion and SCX chromatography. Stable isotope-labeled amino acids. | ESI LC-MS/MS 4000 Q Trap Hybrid triple quadrupole/linear ion trap instrument | LOQ of 1–10 ng/mL range and linearity $\geq 10^2$. LOD in high pg/mL. |
| McKay et al. 2007 [68] | Plasma 4 colorectal cancer patients undergoing chemotherapy | 18 liver-derived proteins in plasma | Immunodepletion (Albumin and IgG removed) | ESI LC-MS/MS 4000 Q Trap Hybrid triple quadrupole/linear ion trap instrument | Increase in target plasma proteins during treatment. Similar trends found in MRM assays and 2-D DIGE |
| Kirsch et al. 2007 [69] | Blood bank pooled serum | 2 human growth hormones (IGFBP-3, IGF-1) | NIL | ESI LC-MS/MS Triple quadrupole mass spectrometer | Detection ranges of 4–10 ng/$\mu$l for IGFBP-3 and 2–8 ng/$\mu$l for IGF-1. |
| Kuhn et al. 2004 [70] | Serum. Pools of healthy, non-erosive RA and erosive RA ( 5 individuals per pool). | C-reactive protein | Immunodepletion of haptoglobin, IgG and HSA, then size exclusion chromatography | ESI LC-MS/MS Triple quadrupole mass spectrometer | Correlation between erosive RA, RA and increased CRP over healthy patients. Results verified using immunoassay. |
| Fortin et al. 2009 [71] | Serum Benign prostate hyperplasia and prostate cancer. | PSA | Immunodepletion of albumin and mixed cation exchange peptide fractionation | ESI LC-MS/MS API 2000 triple quadrupole or 4000 Q Trap hybrid triple quadrupole/linear ion trap | Absolute quant. of PSA to low ng/mL, with good correlation to clinical ELISA tests. |
| Huillet et al. 2012 [72] | Serum Clinical samples from 5 myocardial infarction patients | Clinically validated cardiovascular biomarkers (LDH-B, CKMB, myoglobin, troponin I) | Immunodepletion of six highest abundant proteins and SDS-PAGE Immunocapture prefractionation and SDS-PAGE | ESI LC-MS/MS 5500 Q Trap hybrid triple quadrupole/linear ion trap mass spectrometer | Absolute quant. using Protein Standard Absolute Quantification (PSAQ) and MRM. Demonstrated good correlation with ELISA assay results. |

TABLE 3: Continued.

| Authors/year | Specimen | Target | Sample preparation | MS platform | Outcomes |
|---|---|---|---|---|---|
| Zhao et al. 2010 [73] | Serum 10 hepatocellular carcinoma patients and 10 healthy donors | Candidate biomarkers of hepatocellular carcinoma (vitronectin and clusterin) | NIL | ESI LC-MS/MS 4000 Q Trap Hybrid triple quadrupole/linear ion trap instrument | Stable isotope dilution-MRM using 18O-labelling method demonstrated significant downregulated in HCC compared to healthy group. Results comparable to ELISA. |
| Kuhn et al. 2009 [74] | Plasma 5 patients undergoing PMI and alcohol ablation treatment for HOCM | Troponin I, and Interleukin 33 | Immunoaffinity enrichment SISCAPA | ESI LC-MS/MS Triple quadrupole instrument | Linearity from1.5 to 5000 μg/L and correlated with commercial immunoassay. Demonstrated how SISCAPA-MRM can quantify changes to low μg/L levels. |
| Lopez et al. 2011 [75] | Serum from 24 trisomy 21 and 21 normal first trimester pregnancies | 12 putative markers of Trisomy 21 | NIL | ESI LC-MS/MS LTQ Orbitrap XL mass spectrometer | Developed a workflow for Trisomy 21. Protein biomarkers targeted are high abundance proteins. SRM LOQ of 1–5 femtomoles.. |
| Domanski et al. 2012 [76] | Plasma 90 patients with cardiovascular disease | 67 putative markers of cardiovascular disease | NIL | ESI LC-MS/MS Agilent 6490 triple Quadrupole LC/MS | 117 from 135 peptides with attomolar LOQ for 81 peptides. |

increase in confidence, the more parent and product masses that are detected.

Absolute quantitation can be achieved when MRM is incorporated with isotopically labelled synthetic peptide internal standards, which are designed to be identical to target peptides [84]. For MRM using synthetic internal standards, known concentrations of heavy synthetic peptides are spiked into the sample, and the concentration of the target native peptide can be calculated by measuring the observed MRM response against a standard curve normalised by the internal heavy spike [3, 83].

MRM has a greater sensitivity towards low abundance peptides and relatively good quantitative precision compared to other methods discussed [85]. It is capable of detecting attomole concentrations of peptides across a dynamic range of up to $10^5$ [66, 86]. The main challenge of MRM absolute quantitation is the need for suitable internal standards to be synthesized for each target peptide. Furthermore, absolute MRM quantitation only measures the abundance of individual peptides and makes assumptions on the concentration of the whole protein. Therefore, biomarkers detected and quantified using MRM must be validated using multiple peptides from the same protein (challenging for biofluids) and additional technology to confirm the existence of the actual protein [2]. MRM remains peptide-centric for many biomarker studies.

MRM has been used to quantify major plasma proteins and target biomarkers for a range of diseases. Table 3 lists recent studies conducted using plasma and serum for MRM-based approaches with some quantitation achieving attomolar levels of detection of peptides in one of the most complex human samples available. The MRM approach can also be used for relative quantitation without the use of stable isotopes [87]. A recent multi-site study has confirmed the reproducibility and sensitivity of MRM-based quantitation of plasma proteins [88]. MRM therefore holds great potential to be applied as a specific platform for validation of candidate biomarkers in systematic quantitative studies of clinically relevant peptides.

Further instrument developments have taken advantage of the high resolution and mass accuracy of the TOF and orbitrap analysers and combined them with the selectivity of the quad analysers by replacing the third quad with either an orbitrap or a TOF analyser. These high resolution/accurate mass (HR/AM) instruments are addressing the challenge of eliminating cofiltering interfering ions, while taking advantage of the accuracy afforded by these instruments. In experiments similar to MRM called parallel reaction monitoring (PRM), it is possible to detect all product ions of a peptide in parallel rather than just few transitions per peptide. This allows an increased number of peptides to be quantitated in the one experiment. This combination of analysers firstly uses the quadrupole to select a restricted m/z range (with broad mass filtering window typically 2–100 Th, rather than broad scan of around 700 Th), and the MS/MS mode provides further selectivity and accuracy utilizing the orbitrap or TOF analyser to achieve higher resolution and mass accuracy in both MS and MS/MS scanning modes [89]. A reduced mass filter window as low as 0.2 Th allows reliable discrimination

of targeted ions and increased sensitivity <1 ppm and mass accuracy [90]. These instruments are advancing the reliability and accuracy of quantitative proteomics and are just the beginning to a new era in quantitation that will provide inherent quantitative sampling of all peptides and their product ions in highly complex samples.

## 4. Postdiscovery Validation Phase Platforms

The use of multiparametric assays is becoming an increasing necessity in quantitative studies to overcome a variety of challenges associated with properties of the marker and/or the techniques including immobilisation efficiencies, detection, signal-to-noise [91]. Proteomic-based quantitation of potential biomarkers requires further validation using orthogonal techniques. This is required for both verification as much as for the routine measurement in clinical investigations [12, 13]. The gold standard for validation experiments is by enzyme-linked immunosorbent assays (ELISA). However, alternative techniques such as Western blot, fluorescent bead, chip immunoassay arrays, or Surface Plasmon Resonance (SPR) are also commonly used [91, 92]. Validation by any of these techniques is to complement the onerous requirements for clinical assays: high-throughput, high measurement precision (coefficients of variation of less than 10%) and sufficient sensitivity [92]. The recent developments in multiplexed protein immunoassays such as lateral flow immunoassays and miniaturized microassays [93] hold great promise in advancing panels of biomarkers developed from MS-based proteomics research towards clinical applications. In addition to these orthogonal approaches, parallel validation techniques involving Stable Isotope Standards and Capture by Antipeptide Antibodies (SISCAPA) [94] may also be beneficial.

## 5. Conclusion

Quantitative proteomic analysis has been a point of discussion for the last four decades, with comparative and once limited MS-based techniques heralding the advances that would forge the necessary connection between the dynamic biology of a system and its quantitative proteomic content. The major advances in quantitative MS proteomics have been exceptionally demonstrated over the last decade with the introduction of compatible and reliable label and label-free techniques. These advances now require further developments in bioinformatics and downstream validation, technologies that are required to make sense of complex data and enable researchers to infer more meaningful data that will transform into clinical benefit for years to come.

## References

[1] S. E. Ong and M. Mann, "Mass spectrometry-based proteomics turns quantitative," *Nature Chemical Biology*, vol. 1, no. 5, pp. 252–262, 2005.

[2] M. H. Elliott, D. S. Smith, C. E. Parker, and C. Borchers, "Current trends in quantitative proteomics," *Journal of Mass Spectrometry*, vol. 44, no. 12, pp. 1637–1660, 2009.

[3] M. Bantscheff, M. Schirle, G. Sweetman, J. Rick, and B. Kuster, "Quantitative mass spectrometry in proteomics: a critical review," *Analytical and Bioanalytical Chemistry*, vol. 389, no. 4, pp. 1017–1031, 2007.

[4] W. X. Schulze and B. Usadel, "Quantitation in mass-spectrometry-based proteomics," *Annual Review of Plant Biology*, vol. 61, pp. 491–516, 2010.

[5] K. S. Lilley, A. Razzaq, and P. Dupree, "Two-dimensional gel electrophoresis: recent advances in sample preparation, detection and quantitation," *Current Opinion in Chemical Biology*, vol. 6, no. 1, pp. 46–50, 2002.

[6] S. P. Gygi, G. L. Corthals, Y. Zhang, Y. Rochon, and R. Aebersold, "Evaluation of two-dimensional gel electrophoresis-based proteome analysis technology," *Proceedings of the National Academy of Sciences of the United States of America*, vol. 97, no. 17, pp. 9390–9395, 2000.

[7] L. Ly and V. C. Wasinger, "Protein and peptide fractionation, enrichment and depletion: tools for the complex proteome," *Proteomics*, vol. 11, no. 4, pp. 513–534, 2011.

[8] L. Ly and V. C. Wasinger, "Peptide enrichment and protein fractionation using selective electrophoresis," *Proteomics*, vol. 8, no. 20, pp. 4197–4208, 2008.

[9] P. Findeisen and M. Neumaier, "Mass spectrometry based proteomics profiling as diagnostic tool in oncology: current status and future perspective," *Clinical Chemistry and Laboratory Medicine*, vol. 47, no. 6, pp. 666–684, 2009.

[10] A. Michalski, J. Cox, and M. Mann, "More than 100,000 detectable peptide species elute in single shotgun proteomics runs but the majority is inaccessible to data-dependent LC-MS/MS," *Journal of Proteome Research*, vol. 10, no. 4, pp. 1785–1793, 2011.

[11] H. Liu, R. G. Sadygov, and J. R. Yates III, "A model for random sampling and estimation of relative protein abundance in shotgun proteomics," *Analytical Chemistry*, vol. 76, no. 14, pp. 4193–4201, 2004.

[12] N. Rifai, M. A. Gillette, and S. A. Carr, "Protein biomarker discovery and validation: the long and uncertain path to clinical utility," *Nature Biotechnology*, vol. 24, no. 8, pp. 971–983, 2006.

[13] S. Surinova, R. Schiess, R. Hüttenhain, F. Cerciello, B. Wollscheid, and R. Aebersold, "On the development of plasma protein biomarkers," *Journal of Proteome Research*, vol. 10, no. 1, pp. 5–16, 2011.

[14] R. Schiess, B. Wollscheid, and R. Aebersold, "Targeted proteomic strategy for clinical biomarker discovery," *Molecular Oncology*, vol. 3, no. 1, pp. 33–44, 2009.

[15] E. Boschetti, M. Chung, and P. G. Righetti, "'The quest for biomarkers': are we on the right technical track?" *PROTEOMICS—Clinical Applications*, vol. 6, no. 1-2, pp. 22–41, 2012.

[16] G. L. Hortin, S. A. Jortani, J. C. Ritchie, R. Valdes, and D. W. Chan, "Proteomics: a new diagnostic frontier," *Clinical Chemistry*, vol. 52, no. 7, pp. 1218–1222, 2006.

[17] A. Albalat, H. Mischak, and W. Mullen, "Clinical application of urinary proteomics/peptidomics," *Expert Review of Proteomics*, vol. 8, no. 5, pp. 615–629, 2011.

[18] R. D. Appel, J. R. Vargas, P. M. Palagi, D. Walther, and D. F. Hochstrasser, "Melanie II—a third-generation software package for analysis of two-dimensional electrophoresis images: II. Algorithms," *Electrophoresis*, vol. 18, no. 15, pp. 2735–2748, 1997.

[19] M. Bellew, M. Coram, M. Fitzgibbon et al., "A suite of algorithms for the comprehensive analysis of complex protein mixtures using high-resolution LC-MS," *Bioinformatics*, vol. 22, no. 15, pp. 1902–1909, 2006.

[20] P. M. Palagi, D. Walther, M. Quadroni et al., "MSight: an image analysis software for liquid chromatography-mass spectrometry," *Proteomics*, vol. 5, no. 9, pp. 2381–2384, 2005.

[21] O. Kohlbacher, K. Reinert, C. Gröpl et al., "TOPP—the OpenMS proteomics pipeline," *Bioinformatics*, vol. 23, no. 2, pp. e191–e197, 2007.

[22] J. D. Jaffe, D. R. Mani, K. C. Leptos, G. M. Church, M. A. Gillette, and S. A. Carr, "PEPPeR, a platform for experimental proteomic pattern recognition," *Molecular and Cellular Proteomics*, vol. 5, no. 10, pp. 1927–1941, 2006.

[23] L. N. Mueller, O. Rinner, A. Schmidt et al., "SuperHirn—a novel tool for high resolution LC-MS-based peptide/protein profiling," *Proteomics*, vol. 7, no. 19, pp. 3470–3480, 2007.

[24] B. C. Searle, "Scaffold: a bioinformatic tool for validating MS/MS-based proteomic studies," *Proteomics*, vol. 10, no. 6, pp. 1265–1269, 2010.

[25] J. C. Braisted, S. Kuntumalla, C. Vogel et al., "The APEX quantitative proteomics tool: generating protein quantitation estimates from LC-MS/MS proteomics results," *BMC Bioinformatics*, vol. 9, article 529, 2008.

[26] W. X. Schulze and M. Mann, "A novel proteomic screen for peptide-protein interactions," *The Journal of Biological Chemistry*, vol. 279, no. 11, pp. 10756–10764, 2004.

[27] D. K. Han, J. Eng, H. Zhou, and R. Aebersold, "Quantitative profiling of differentiation-induced microsomal proteins using isotope-coded affinity tags and mass spectrometry," *Nature Biotechnology*, vol. 19, no. 10, pp. 946–951, 2001.

[28] X. J. Li, H. Zhang, J. A. Ranish, and R. Aebersold, "Automated statistical analysis of protein abundance ratios from data generated by stable-isotope dilution and tandem mass spectrometry," *Analytical Chemistry*, vol. 75, no. 23, pp. 6648–6657, 2003.

[29] B. D. Halligan, R. Y. Slyper, S. N. Twigger, W. Hicks, M. Olivier, and A. S. Greene, "ZoomQuant: an application for the quantitation of stable isotope labeled peptides," *Journal of the American Society for Mass Spectrometry*, vol. 16, no. 3, pp. 302–306, 2005.

[30] W. T. Lin, W. N. Hung, Y. H. Yian et al., "Multi-Q: a fully automated tool for multiplexed protein quantitation," *Journal of Proteome Research*, vol. 5, no. 9, pp. 2328–2338, 2006.

[31] I. P. Shadforth, T. P. J. Dunkley, K. S. Lilley, and C. Bessant, "i-Tracker: for quantitative proteomics using iTRAQŮ," *BMC Genomics*, vol. 6, article 145, 2005.

[32] B. MacLean, D. M. Tomazela, N. Shulman et al., "Skyline: an open source document editor for creating and analyzing targeted proteomics experiments," *Bioinformatics*, vol. 26, no. 7, Article ID btq054, pp. 966–968, 2010.

[33] J. Cox and M. Mann, "MaxQuant enables high peptide identification rates, individualized p.p.b.-range mass accuracies and proteome-wide protein quantification," *Nature Biotechnology*, vol. 26, no. 12, pp. 1367–1372, 2008.

[34] D. B. Martin, T. Holzman, D. May et al., "MRMer, an interactive open source and cross-platform system for data extraction and visualization of multiple reaction monitoring experiments," *Molecular and Cellular Proteomics*, vol. 7, no. 11, pp. 2270–2278, 2008.

[35] N. L. Anderson and N. G. Anderson, "The human plasma proteome: history, character, and diagnostic prospects," *Molecular and Cellular Proteomics*, vol. 1, no. 11, pp. 845–867, 2002.

[36] G. S. Omenn, D. J. States, T. W. Blackwell et al., "Challenges in deriving high-confidence protein identifications from data gathered by a HUPO plasma proteome collaborative study," *Nature Biotechnology*, vol. 24, no. 3, pp. 333–338, 2006.

[37] E. F. Petricoin, C. Belluco, R. P. Araujo, and L. A. Liotta, "The blood peptidome: a higher dimension of information content for cancer biomarker discovery," *Nature Reviews Cancer*, vol. 6, no. 12, pp. 961–967, 2006.

[38] J. A. Hewel, S. Phanse, J. Liu, N. Bousette, A. Gramolini, and A. Emili, "Targeted protein identification, quantification and reporting for high-resolution nanoflow targeted peptide monitoring," *Journal of Proteomics*, 2012.

[39] S. M. Hanash, S. J. Pitteri, and V. M. Faca, "Mining the plasma proteome for cancer biomarkers," *Nature*, vol. 452, no. 7187, pp. 571–579, 2008.

[40] W. C. S. Cho and C. H. K. Cheng, "Oncoproteomics: current trends and future perspectives," *Expert Review of Proteomics*, vol. 4, no. 3, pp. 401–410, 2007.

[41] J. Granger, J. Siddiqui, S. Copeland, and D. Remick, "Albumin depletion of human plasma also removes low abundance proteins including the cytokines," *Proteomics*, vol. 5, no. 18, pp. 4713–4718, 2005.

[42] E. Bellei, S. Bergamini, E. Monari et al., "High-abundance proteins depletion for serum proteomic analysis: concomitant removal of non-targeted proteins," *Amino Acids*, vol. 40, no. 1, pp. 145–156, 2011.

[43] A. J. Rai and F. Vitzthum, "Effects of preanalytical variables on peptide and protein measurements in human serum and plasma: implications for clinical proteomics," *Expert Review of Proteomics*, vol. 3, no. 4, pp. 409–426, 2006.

[44] R. Aebersold and M. Mann, "Mass spectrometry-based proteomics," *Nature*, vol. 422, no. 6928, pp. 198–207, 2003.

[45] D. Reker and L. Malmström, "Bioinformatic challenges in targeted proteomics," *Journal of Proteome Research*, vol. 11, no. 9, pp. 4393–4402, 2012.

[46] N. Yang, S. Feng, K. Shedden et al., "Urinary glycoprotein biomarker discovery for bladder cancer detection using LC/MS-MS and label-free quantification," *Clinical Cancer Research*, vol. 17, no. 10, pp. 3349–3359, 2011.

[47] L. F. Quintana, J. M. Campistol, M. P. Alcorea, E. Bañon-Maneus, A. Solé-González, and P. R. Cutillas, "Application of label-free quantitative peptidomics for the identification of urinary biomarkers of kidney chronic allograft dysfunction," *Molecular and Cellular Proteomics*, vol. 8, no. 7, pp. 1658–1673, 2009.

[48] J. S. Hanas, J. R. Hocker, J. Y. Cheung et al., "Biomarker identification in human pancreatic cancer sera," *Pancreas*, vol. 36, no. 1, pp. 61–69, 2008.

[49] H. Xue, B. Lü, J. Zhang et al., "Identification of serum biomarkers for colorectal cancer metastasis using a differential secretome approach," *Journal of Proteome Research*, vol. 9, no. 1, pp. 545–555, 2010.

[50] D. Besson, A. H. Pavageau, I. Valo et al., "A quantitative proteomic approach of the different stages of colorectal cancer establishes OLFM4 as a new nonmetastatic tumor marker," *Molecular and Cellular Proteomics*, vol. 10, no. 12, Article ID M111.009712, 2011.

[51] O. P. Bondar, D. R. Barnidge, E. W. Klee, B. J. Davis, and G. G. Klee, "LC-MS/MS quantification of Zn-α2 glycoprotein: a potential serum biomarker for prostate cancer," *Clinical Chemistry*, vol. 53, no. 4, pp. 673–678, 2007.

[52] R. Chaerkady, H. C. Harsha, A. Nalli et al., "A quantitative proteomic approach for identification of potential biomarkers in hepatocellular carcinoma," *Journal of Proteome Research*, vol. 7, no. 10, pp. 4289–4298, 2008.

[53] L. Dayon, A. Hainard, V. Licker et al., "Relative quantification of proteins in human cerebrospinal fluids by MS/MS using 6-plex isobaric tags," *Analytical Chemistry*, vol. 80, no. 8, pp. 2921–2931, 2008.

[54] M. Wang, J. You, K. G. Bemis, T. J. Tegeler, and D. P. G. Brown, "Label-free mass spectrometry-based protein quantification technologies in proteomic analysis," *Briefings in Functional Genomics and Proteomics*, vol. 7, no. 5, pp. 329–339, 2008.

[55] W. M. Old, K. Meyer-Arendt, L. Aveline-Wolf et al., "Comparison of label-free methods for quantifying human proteins by shotgun proteomics," *Molecular and Cellular Proteomics*, vol. 4, no. 10, pp. 1487–1502, 2005.

[56] A. Prakash, B. Piening, J. Whiteaker et al., "Assessing bias in experiment design for large scale mass spectrometry-based quantitative proteomics," *Molecular and Cellular Proteomics*, vol. 6, no. 10, pp. 1741–1748, 2007.

[57] J. Rappsilber, U. Ryder, A. I. Lamond, and M. Mann, "Large-scale proteomic analysis of the human spliceosome," *Genome Research*, vol. 12, no. 8, pp. 1231–1245, 2002.

[58] E. Xixi, P. Dimitraki, K. Vougas, S. Kossida, G. Lubec, and M. Fountoulakis, "Proteomic analysis of the mouse brain following protein enrichment by preparative electrophoresis," *Electrophoresis*, vol. 27, no. 7, pp. 1424–1431, 2006.

[59] W. Zhu, J. W. Smith, and C. M. Huang, "Mass spectrometry-based label-free quantitative proteomics," *Journal of Biomedicine & Biotechnology*, vol. 2010, Article ID 840518, 6 pages, 2010.

[60] S. E. Ong, L. J. Foster, and M. Mann, "Mass spectrometric-based approaches in quantitative proteomics," *Methods*, vol. 29, no. 2, pp. 124–130, 2003.

[61] S. Julka and F. Regnier, "Quantification in proteomics through stable isotope coding: a review," *Journal of Proteome Research*, vol. 3, no. 3, pp. 350–363, 2004.

[62] S. E. Ong, B. Blagoev, I. Kratchmarova et al., "Stable isotope labeling by amino acids in cell culture, SILAC, as a simple and accurate approach to expression proteomics," *Molecular and Cellular Proteomics*, vol. 1, no. 5, pp. 376–386, 2002.

[63] S. P. Gygi, B. Rist, S. A. Gerber, F. Turecek, M. H. Gelb, and R. Aebersold, "Quantitative analysis of complex protein mixtures using isotope-coded affinity tags," *Nature Biotechnology*, vol. 17, no. 10, pp. 994–999, 1999.

[64] A. Schmidt, J. Kellermann, and F. Lottspeich, "A novel strategy for quantitative proteomics using isotope-coded protein labels," *Proteomics*, vol. 5, no. 1, pp. 4–15, 2005.

[65] J. Stahl-Zeng, V. Lange, R. Ossola et al., "High sensitivity detection of plasma proteins by multiple reaction monitoring of N-glycosites," *Molecular and Cellular Proteomics*, vol. 6, no. 10, pp. 1809–1817, 2007.

[66] L. Anderson and C. L. Hunter, "Quantitative mass spectrometric multiple reaction monitoring assays for major plasma proteins," *Molecular and Cellular Proteomics*, vol. 5, no. 4, pp. 573–588, 2006.

[67] H. Keshishian, T. Addona, M. Burgess, E. Kuhn, and S. A. Carr, "Quantitative, multiplexed assays for low abundance proteins in plasma by targeted mass spectrometry and stable isotope dilution," *Molecular and Cellular Proteomics*, vol. 6, no. 12, pp. 2212–2229, 2007.

[68] M. J. McKay, J. Sherman, M. T. Laver, M. S. Baker, S. J. Clarke, and M. P. Molloy, "The development of multiple reaction monitoring assays for liver-derived plasma proteins," *PROTEOMICS—Clinical Applications*, vol. 1, no. 12, pp. 1570–1581, 2007.

[69] S. Kirsch, J. Widart, J. Louette, J. F. Focant, and E. De Pauw, "Development of an absolute quantification method targeting growth hormone biomarkers using liquid chromatography coupled to isotope dilution mass spectrometry," *Journal of Chromatography A*, vol. 1153, no. 1-2, pp. 300–306, 2007.

[70] E. Kuhn, J. Wu, J. Karl, H. Liao, W. Zolg, and B. Guild, "Quantification of C-reactive protein in the serum of patients with rheumatoid arthritis using multiple reaction monitoring mass spectrometry and 13C-labeled peptide standards," *Proteomics*, vol. 4, no. 4, pp. 1175–1186, 2004.

[71] T. Fortin, A. Salvador, J. P. Charrier et al., "Clinical quantitation of prostate-specific antigen biomarker in the low nanogram/milliliter range by conventional bore liquid chromatography-tandem mass spectrometry (multiple reaction monitoring) coupling and correlation with ELISA tests," *Molecular and Cellular Proteomics*, vol. 8, no. 5, pp. 1006–1015, 2009.

[72] C. Huillet, A. Adrait, D. Lebert et al., "Accurate quantification of cardiovascular biomarkers in serum using protein standard absolute quantification (PSAQ) and selected reaction monitoring," *Molecular and Cellular Proteomics*, vol. 11, no. 2, Article ID M111.008235, 2012.

[73] Y. Zhao, W. Jia, W. Sun et al., "Combination of improved $^{18}O$ incorporation and multiple reaction monitoring: a universal strategy for absolute quantitative verification of serum candidate biomarkers of liver cancer," *Journal of Proteome Research*, vol. 9, no. 6, pp. 3319–3327, 2010.

[74] E. Kuhn, T. Addona, H. Keshishian et al., "Developing multiplexed assays for troponin I and interleukin-33 in plasma by peptide immunoaffinity enrichment and targeted mass spectrometry," *Clinical Chemistry*, vol. 55, no. 6, pp. 1108–1117, 2009.

[75] M. Lopez, R. Kuppusamy, D. Sarracino et al., "Mass spectrometric discovery and selective reaction monitoring (SRM) of putative protein biomarker candidates in first trimester trisomy 21 maternal serum," *Journal of Proteome Research*, vol. 10, no. 1, pp. 133–142, 2011.

[76] D. Domanski, A. J. Percy, J. Yang et al., "MRM–based multiplexed quantitation of 67 putative cardiovascular disease biomarkers in human plasma," *Proteomics*, vol. 12, no. 8, pp. 1222–1243, 2012.

[77] A. Thompson, J. Schäfer, K. Kuhn et al., "Tandem mass tags: a novel quantification strategy for comparative analysis of complex protein mixtures by MS/MS," *Analytical Chemistry*, vol. 75, no. 8, pp. 1895–1904, 2003.

[78] P. L. Ross, Y. N. Huang, J. N. Marchese et al., "Multiplexed protein quantitation in *Saccharomyces cerevisiae* using amine-reactive isobaric tagging reagents," *Molecular and Cellular Proteomics*, vol. 3, no. 12, pp. 1154–1169, 2004.

[79] S. Wiese, K. A. Reidegeld, H. E. Meyer, and B. Warscheid, "Protein labeling by iTRAQ: a new tool for quantitative mass spectrometry in proteome research," *Proteomics*, vol. 7, no. 3, pp. 340–350, 2007.

[80] K. Aggarwal, L. H. Choe, and K. H. Lee, "Shotgun proteomics using the iTRAQ isobaric tags," *Briefings in Functional Genomics and Proteomics*, vol. 5, no. 2, pp. 112–120, 2006.

[81] M. Latterich, M. Abramovitz, and B. Leyland-Jones, "Proteomics: new technologies and clinical applications," *European Journal of Cancer*, vol. 44, no. 18, pp. 2737–2741, 2008.

[82] K. L. Simpson, A. D. Whetton, and C. Dive, "Quantitative mass spectrometry-based techniques for clinical use: biomarker identification and quantification," *Journal of Chromatography B*, vol. 877, no. 13, pp. 1240–1249, 2009.

[83] V. Lange, P. Picotti, B. Domon, and R. Aebersold, "Selected reaction monitoring for quantitative proteomics: a tutorial," *Molecular Systems Biology*, vol. 4, no. 1, article 222, 2008.

[84] S. A. Gerber, J. Rush, O. Stemman, M. W. Kirschner, and S. P. Gygi, "Absolute quantification of proteins and phosphoproteins from cell lysates by tandem MS," *Proceedings of the National Academy of Sciences of the United States of America*, vol. 100, no. 12, pp. 6940–6945, 2003.

[85] W. J. Qian, J. M. Jacobs, T. Liu, D. G. Camp, and R. D. Smith, "Advances and challenges in liquid chromatography-mass spectrometry-based proteomics profiling for clinical applications," *Molecular and Cellular Proteomics*, vol. 5, no. 10, pp. 1727–1744, 2006.

[86] A. Wolf-Yadlin, S. Hautaniemi, D. A. Lauffenburger, and F. M. White, "Multiple reaction monitoring for robust quantitative proteomic analysis of cellular signaling networks," *Proceedings of the National Academy of Sciences of the United States of America*, vol. 104, no. 14, pp. 5860–5865, 2007.

[87] C. Ludwig, M. Claassen, A. Schmidt, and R. Aebersold, "Estimation of absolute protein quantities of unlabeled samples by selected reaction monitoring mass spectrometry," *Molecular and Cellular Proteomics*, vol. 11, no. 3, Article ID M111.013987, 2012.

[88] T. A. Addona, S. E. Abbatiello, B. Schilling et al., "Multi-site assessment of the precision and reproducibility of multiple reaction monitoring-based measurements of proteins in plasma," *Nature Biotechnology*, vol. 27, no. 7, pp. 633–641, 2009.

[89] S. Gallien, E. Duriez, C. Crone, M. Kellmann, T. Moehring, and B. Domon, "Targeted proteomic quantification on quadrupole-orbitrap mass spectrometer," *Molecular and Cellular Proteomics*, vol. 11, no. 12, pp. 1709–1723, 2012.

[90] A. C. Peterson, J. D. Russell, D. J. Bailey, M. S. Westphall, and J. J. Coon, "Parallel reaction monitoring for high resolution and high mass accuracy quantitative, targeted proteomics," *Molecular and Cellular Proteomics*, vol. 11, no. 11, pp. 1475–1488, 2012.

[91] K. Köhler and H. Seitz, "Validation processes of protein biomarkers in serum—a cross platform comparison," *Sensors*, vol. 12, no. 9, pp. 12710–12728, 2012.

[92] S. F. Kingsmore, "Multiplexed protein measurement: technologies and applications of protein and antibody arrays," *Nature Reviews Drug Discovery*, vol. 5, no. 4, pp. 310–321, 2006.

[93] A. A. Ellington, I. J. Kullo, K. R. Bailey, and G. G. Klee, "Antibody-based protein multiplex platforms: technical and operational challenges," *Clinical Chemistry*, vol. 56, no. 2, pp. 186–193, 2010.

[94] N. L. Anderson, N. G. Anderson, L. R. Haines, D. B. Hardie, R. W. Olafson, and T. W. Pearson, "Mass spectrometric quantitation of peptides and proteins using stable isotope standards and capture by anti-peptide antibodies (SISCAPA)," *Journal of Proteome Research*, vol. 3, no. 2, pp. 235–244, 2004.

# 7

# Advantageous Uses of Mass Spectrometry for the Quantification of Proteins

**John E. Hale**

*Hale Biochemical Consulting, 6341 Wyatt Lane, Klamath Falls, OR 97601, USA*

Correspondence should be addressed to John E. Hale; thehales94@gmail.com

Academic Editor: Valerie Wasinger

Quantitative protein measurements by mass spectrometry have gained wide acceptance in research settings. However, clinical uptake of mass spectrometric protein assays has not followed suit. In part, this is due to the long-standing acceptance by regulatory agencies of immunological assays such as ELISA assays. In most cases, ELISAs provide highly accurate, sensitive, relatively inexpensive, and simple assays for many analytes. The barrier to acceptance of mass spectrometry in these situations will remain high. However, mass spectrometry provides solutions to certain protein measurements that are difficult, if not impossible, to accomplish by immunological methods. Cases where mass spectrometry can provide solutions to difficult assay development include distinguishing between very closely related protein species and monitoring biological and analytical variability due to sample handling and very high multiplexing capacity. This paper will highlight several examples where mass spectrometry has made certain protein measurements possible where immunological techniques have had a great difficulty.

## 1. Introduction

Quantitative mass spectrometry of proteins has evolved dramatically over the last decade. Early methods involved labeling proteins with reagents enriched in stable isotopes in order to introduce mass tags into proteins of interest for relative quantification of proteins [1, 2]. These reagents have been refined over the years and have found widespread use in the form of the ITRAQ reagent [3]. Metabolic labeling of proteins with stable isotope-enriched amino acids has also been used as a technique for the relative quantification of proteins in cell culture systems [4]. Additionally, "label-free" methods have been developed for relative quantification of proteins in complex mixtures [5, 6]. As mentioned, these methods were developed for relative quantification of proteins, that is, comparing two or more samples and determining whether levels of proteins increased or decreased in response to some perturbation. Isotopically labeled peptides have been used as standards for the absolute quantification of proteins in complex mixtures. Variations of this approach include the AQUA and SISCAPA methods [7, 8]. Among the advantages of these techniques is that a quantitative assay may be developed for a given protein without the need for an antibody

[7]. Alternatively an antibody to a synthetic peptide may be used [8]. This greatly simplifies the development of assays from immunological formats, such as ELISA, where well-characterized antibodies are needed. However, immunological assays are considered the standard type of assay when quantifying proteins in clinical settings. This is due, in part, to the sensitivity, accuracy, high through put, and relative simplicity of the technology. The implementation of mass spectrometric assays has been slow in this arena since mass spectrometric assays have not provided a clear advantage over ELISA assays. There are certain cases where mass spectrometry is able quantify proteins that are very difficult or impossible to measure by immunological methods. Some of these cases include assaying individual protein isoforms in the presence of all isoforms, measurement of specifically modified proteins, quantification of panels of proteins, and quantification of processed forms of proteins in biological samples. In addition, quantitative mass spectrometric assays may have advantageous dynamic range. While outside the scope of this review, using different combinations of protein analysis and mass spectrometers, dynamic ranges (defined as the lowest level of protein measurable relative to the most abundant protein in a sample) can vary from 1 to 2 orders

TABLE 1: Amino acid substitutions in Apolipoprotein E Isoforms.

| Isoform | Amino acid at position 112 | Amino acid at position 158 |
|---------|----------------------------|----------------------------|
| Apo E2  | Cys                        | Cys                        |
| Apo E3  | Cys                        | Arg                        |
| Apo E4  | Arg                        | Cys                        |

of magnitude to 4-5 orders of magnitude [9]. There are many different types of mass spectrometry available and in some cases, different types of mass spectrometers may be used to develop assays for the same analyte. Selection of a mass spectrometer may depend on several factors including the necessary throughput, complexity of the sample, and method of introduction of the sample to the mass spectrometer. MALDI-Tof mass spectrometry may be adapted to very high throughput applications, as data acquisition requires only a few seconds; however, it may require more sample cleanup prior to mass analysis. Electrospray techniques may be interfaced with liquid chromatographic sample introduction to add an additional sample simplification step prior to mass analysis. Coupling to triple quadrupole or ion-trap instruments allows for the development of assays that can detect the analyte ion directly through extracted ion chromatography (XIC) or that can detect a fragment ion or ions after MS/MS analysis (known as selected reaction monitoring, SRM or multiple reaction monitoring, MRM). In cases where sample complexity is low, XIC may be sufficient for assay development. If complexity is higher, SRM or MRM techniques add an additional level of specificity for a given ion and can reduce overlap with contaminating ions. Finally, in cases of very high sample complexity very high resolution detectors such as FTICR or Orbitrap instruments may be used to isolate very narrow mass ranges for subsequent analysis, thereby reducing contaminating ion overlap. This paper will discuss specific examples of these situations and compare the efficiency and ease of use between mass spectrometry and immunological methods.

## 2. Measurement of Protein Isoforms

Quantification of protein isoforms can be a significant challenge for immunological assay development. Isoforms may result from substitution of only a few amino acids in a protein sequence. Thus, for immunological assay development, highly specific antibodies need to be developed. Mass spectrometry can detect substitutions of single amino acids in proteins and may be used to quantify individual protein isoforms in mixtures.

Using class-specific isolation methods, quantitative assays for protein isoforms may be developed without the need for any antibody. An example of this approach is an assay for the three common apolipoprotein E isoforms, Apo E2, E3, and E4. Amino acid substitutions in positions 112 and 158 of the protein sequence define the isoforms (Table 1). Apo E4 is associated with increased risk for Alzheimer's disease

[10] and Apo E polymorphism has associations with cardiovascular disease [11]. Thus, quantification of the different isoforms is important in understanding the role of Apo E in the various disease states. Immunological methods require the use of different combinations of antibodies for each amino acid substitution and also require multiple assays be performed on a single sample. For instance, an Apo E2 carrier may not be distinguished from an ApoE3/E4 carrier using antibodies recognizing the epitopes with C112 and C158 alone. Antibodies recognizing epitopes with R112 and R158 would need to be included meaning that four separate assays would need to be developed. Recently, a mass spectrometric assay was developed that can measure all three ApoE isoforms simultaneously [12]. By using a lipoprotein absorbant, no antibody was needed to isolate the proteins. Following tryptic digestion, ion-trap-based multiple reaction monitoring assays were designed for the peptides LGADMEDVC$_{112}$GR, LGADMEDVR$_{112}$, LAVYQAGAR, and C$_{158}$LAVYQAGAR. By quantitatively measuring each of these peptides, the concentrations of each isoform could be calculated. This was accomplished with the inclusion of Apo E2 and E4 standards that were metabolically labeled with 13C leucine. This strategy allowed for the quantification of both total Apo E and specific isotope concentrations from a single sample measurement.

Protein isoforms may be isolated using polyclonal or pan-protein antibodies. The protein transthyretin (TTR) provides an example of this approach. Transthyretin is a 127 amino acid protein that tetramerizes and functions as a carrier of T4 and retinol (by binding to retinol binding protein) [13]. Mutations in the protein are associated with a condition known as Familial Transthyretin Amyloidosis [14]. More than one hundred mutant forms of TTR are known and many of these are associated with pathological familial amyloid diseases [15]. At least fifteen nephropathic mutants have been discovered, with a single amino acid substitution of V to M at position 30 as the most common mutation [16]. Diagnosis consists of a combination of DNA testing and IEF analysis [17]. DNA testing can detect amino acid mutations but cannot detect posttranslational modifications (PTMs). IEF may detect some PTMs but may also miss some. Transthyretin monomer is small enough to be directly measured by MALDI-Tof mass spectrometry. Mass spectrometric methods have been developed that involve immunocapture of transthyretin from plasma with polyclonal antibodies followed by MALDI MS analysis of the isolated protein [18]. This technique simultaneously detects amino acid variants, as well as, PTMs, thus negating the need for two tests. By including internal standards (such as isotopically labeled transthyretin), this method may be made quantitative. A more recent refinement of this method incorporated a standard curve in a reference sample yielding a linear working range of thirtyfold [19]. This assay compared very well with an ELISA for TTR but had the advantage of being able to assay multiple TTR variants in a single sample in a high throughput manner which the ELISA cannot do.

Therapeutic monoclonal antibodies (mAb) provide an important challenge for quantification of protein isoforms. In order to monitor drug levels and monitor clearance rates during a therapeutic treatment, assays must measure the levels of a single antibody isoform in the background of host antibody, which consists of thousands of proteins with very high levels of sequence identity with the therapeutic mAb. Immunological methods typically utilize antigenic capture or anti-idiotypic antibodies to isolate the specific antibody of interest. Mass spectrometric methods negate the need for these isolation steps due to the presence of unique peptide sequences in the complementarity determining regions (CDRs) of the antibodies. Individual monoclonal antibodies may be quantified in complex mixtures using SRM or MRM of specific peptides from enzymatic digestions of plasma samples [20–22]. In one example of this approach, a specific, unique peptide from a human monoclonal antibody could be detected in tryptic digests of human plasma [20]. The detectability of this peptide could be increased at least three-fold by a simple two-step solid phase extraction procedure. The assay had a linear range of more than three orders of magnitude and had very good accuracy and precision values in a three-day validation analysis. This assay performed as well or better than an ELISA assay for a monoclonal antibody in a rat pharmacodynamic study. Inclusion of isotopically labeled protein standards allowed for the measurement of absolute levels of mAbs in serum samples [21]. Spiking serum samples with mAb labeled with stable isotopes controls for losses incurred during sample processing and cleanup, since the labeled mAb behaves identically to the unlabeled mAb. Studies monitoring a human mAb spiked into a total IgG fraction from human serum demonstrated that peptides from an individual mAb may be detected by extracted ion chromatography at levels five orders of magnitude below the total IgG concentration [23]. Thus, by isolating the total IgG fraction using protein G, individual antibodies present at 0.01% of the total may be detected. This extends the potential of this approach to the measurement of disease-associated antibodies in biological samples and allows for these assays to be developed rapidly without the need for specific antibody reagents.

Perhaps the most extreme example of isomeric protein quantification is the measurement of epigenetic modification of histones [24–26]. Epigenetics is the study of the regulation of gene expression by cellular modification of DNA and proteins. A major component of this regulation is the various posttranslational modifications that occur on the DNA binding proteins, the histones. This "histone code" controls access to DNA thus regulating gene expression [27]. Modifications that occur include mono-, di-, and tri-methylation of lysine, acetylation of lysine, mono- and di-methylation of arginine, phosphorylation of serine, threonine, and tyrosine. There are literally millions of combinations in which these different modifications may occur, making antibody methods impossible for quantitative analysis. LC/MS/MS methods offer the only practical method of quantifying these modifications and assessing their importance in cellular differentiation and maintenance of phenotype. Studies of the N-terminal 23 amino acid tail of histone 4 using nanoflow-LC coupled to high resolution mass spectrometry indicated multiple different modified forms. Inclusion of ETD MS/MS allowed assignment of positional modifications

and revealed the presence of seventy-four different forms of the N-terminal tail. A label-free quantification technique allows for the assignment of the relative abundance of these forms and makes possible the quantification of changes in the histone epigenome in response to cellular perturbations [26].

## 3. Measurement of Biologically Processed Proteins

In addition to isoforms and posttranslational modification, protein heterogeneity may arise from degradation by proteases, esterases, phosphatases, deacetylases, and many other metabolic enzymes. Degradation of proteins may occur as a part of normal metabolism but may also be artifactual, the result of enzymatic activation during sample collection or processing. Mass spectrometry provides a means to distinguish biological from artifactual processing. Labeling of proteins with stable isotopes creates a standard that can be distinguished by mass spectrometry. These protein standards are chemically and biologically indistinguishable from the unlabeled protein. Introduction of the labeled standard into a biological system prior to sample processing then allows one to monitor the effects of sample handling on the heterogeneity of the protein. An example of this is the peptide hormone, ghrelin. Ghrelin is a 28 amino acid peptide that has an octanoic acid moiety attached to a serine residue at position 3 in the peptide sequence [28]. This modification is essential for the biological activity of ghrelin, which includes increased appetite [29] and increased insulin secretion [30]. The octanoic acid group is easily removed by esterases, which may be activated during sample collection [31, 32]. This may lead to a wide variation in measured levels of octonyl and des octonyl ghrelin. Distinguishing biological from artifactual heterogeneity of ghrelin is difficult if not impossible by immunological methods. By adding 13C labeled ghrelin standards (octonylated and unoctonylated) into sample collection tubes, deacylation upon collection of serum samples may be monitored in real time [33]. Using this assay, it was found that immediate acidification of blood samples was a highly effective method for preserving the acylated version of ghrelin and allowed for much more accurate measurement of its true biological levels.

Another important example of biological heterogeneity of a protein entity is the amyloid beta peptide (A-beta). The association of A-beta with Alzheimer's disease (AD) is well known [34], and its role as a causative agent has recently been strengthened [35]. A-beta exists in multiple forms, the most prevalent of which are 40 and 42 amino acid long peptides. These peptides derive from the transmembrane domain of the Alzheimer's precursor protein (APP) through the action of different proteases. The 42 amino acid peptide (containing two additional C-terminal amino acids from APP) is hydrophobic and aggregates to form plaques in the brain that are the hallmark of AD [34]. A-beta peptides are also present in cerebrospinal fluid, with the 40 amino acid version being more prevalent [36]. Characterization of A-beta from isolated CSF and from brain extracts has demonstrated the presence of additional heterogeneity in the peptide. A-beta immunoprecipitated from CSF exists in multiple C-terminally truncated forms [36]. Conversely, A-beta peptides extracted from brain tissue exhibit heterogeneity at the N-terminus [37]. The differences in the localization of these isoforms may have important implications for the biology of AD and may also provide important biomarkers for therapeutic drug development. The major A-beta isoforms (40 and 42 amino acid versions) have been typically measured by ELISA [38]. ELISA measurement of the multiple forms of the peptides by ELISA is made difficult by the need to develop antibodies that will specifically recognize peptides that differ only by truncation of individual amino acids. Cross reactivity is likely to be high and many antibodies must be developed. Mass spectrometric assays have been developed for A-beta peptides that involve immunoprecipitation with an antibody that cross reacts with multiple A-beta isoforms. Stable isotope labeled peptides, spiked into biological samples, provide standards to control for peptide recovery and to monitor peptide stability through the assay procedure. One such assay employing LC/ESI/MS/MS was developed to simultaneously monitor A-beta 1–40 and 1–42. [39] This stable isotope dilution assay utilized 15N labeled A-beta 1–40 and 1–42 standards spiked into cerebrospinal fluid (CSF) from Alzheimer's disease patients and healthy controls followed by immunoprecipitation with an antibody to the midregion of A-beta [40]. Immunoprecipitated A-beta 1–40 and 1–42 were separated by reversed phase HPLC under basic conditions and sprayed into a linear ion trap mass spectrometer. Peptides were fragmented by MS/MS and quantified by selected reaction monitoring. The labeled internal standards allowed for absolute quantitative analysis, as well as, monitoring sample processing effects. This assay had good sensitivity and strong correlation with the ELISA for A-beta 1–42.

Another assay employing MALDI-Tof mass spectrometry was used to quantify seven different forms of the A-beta peptide in CSF from individuals with Alzheimer's disease and healthy volunteers [41]. The performance of this assay resulted in a quantitative range of nearly two orders of magnitude. Comparison of the mass spectrometric assay to the ELISA assays for the 40 and the 42 amino acid versions of the A-beta peptide showed very good correlations (0.95 and 0.88, resp.). No N- or C-terminal processing of the isotopically labeled peptides was observed ruling out artifactual production of the various forms during sample preparation for the assay.

Yet another multiplex quantitative assay was developed using immunoprecipitation followed by ion trap LCMS analysis [42]. By using an antibody to the mid-region of A-beta, ten different versions of the peptide could be isolated. After positive identification by a combination of accurate mass and MS/MS analyses, quantification was achieved for each of the peptides by extracting ion current for the two most abundant charge states and normalizing the peaks to an isotopically labeled A-beta internal standard. The assays were linear over at least a tenfold concentration range with very low relative standard deviation. This assay has been shown to be useful for *in vitro* drug efficacy and mechanism of action studies [42].

## 4. Conclusion

Several different examples of the advantageous use of mass spectrometry for the assay of closely related protein species have been cited in this paper. Of course there are many additional examples, such as the use of mass spectrometry to quantify protein phosphorylation. The ability of mass spectrometry to accurately distinguish different isoforms or modified forms of proteins, even in mixtures, provides added dimensions in the design of quantitative strategies and simplifies multiplex assay development. Isotopically labeled protein or peptide standards allow for absolute quantification and also provide methods for monitoring stability of analytes throughout sample processing steps involved in preparation for measurement. The sensitivity of mass spectrometers is very high and limitations on the sensitivity of quantitative assays are more often related to the dynamic range of proteins in a sample. Better sample simplification procedures are often the key to increased sensitivity of mass spectrometric protein assays. Perhaps the major limitation to mass-spectrometry-based protein quantification is throughput. Immunological assays can be performed in 96- or 384-well format. Plate readers can measure readouts of entire plates in batch mode. On the other hand, mass spectrometers must analyze samples one at a time. Even under very fast analytical conditions, such as MALDI-Tof analysis, each sample will require a few seconds to process. Future developments in multiplexed mass spectrometric detectors may help to address this limitation [43].

## References

[1] S. P. Gygi, B. Rist, S. A. Gerber, F. Turecek, M. H. Gelb, and R. Aebersold, "Quantitative analysis of complex protein mixtures using isotope-coded affinity tags," *Nature Biotechnology*, vol. 17, no. 10, pp. 994–999, 1999.

[2] A. Chakraborty and F. E. Regnier, "Global internal standard technology for comparative proteomics," *Journal of Chromatography A*, vol. 949, no. 1-2, pp. 173–184, 2002.

[3] L. DeSouza, G. Diehl, M. J. Rodrigues et al., "Search for cancer markers from endometrial tissues using differentially labeled tags iTRAQ and cICAT with multidimensional liquid chromatography and tandem mass spectrometry," *Journal of Proteome Research*, vol. 4, no. 2, pp. 377–386, 2005.

[4] S. E. Ong, B. Blagoev, I. Kratchmarova et al., "Stable isotope labeling by amino acids in cell culture, SILAC, as a simple and accurate approach to expression proteomics," *Molecular & Cellular Proteomics*, vol. 1, no. 5, pp. 376–386, 2002.

[5] P. V. Bondarenko, D. Chelius, and T. A. Shaler, "Identification and relative quantitation of protein mixtures by enzymatic digestion followed by capillary reversed-phase liquid chromatography—tandem mass spectrometry," *Analytical Chemistry*, vol. 74, no. 18, pp. 4741–4749, 2002.

[6] W. M. Old, K. Meyer-Arendt, L. Aveline-Wolf et al., "Comparison of label-free methods for quantifying human proteins by shotgun proteomics," *Molecular & Cellular Proteomics*, vol. 4, no. 10, pp. 1487–1502, 2005.

[7] S. A. Gerber, J. Rush, O. Stemman, M. W. Kirschner, and S. P. Gygi, "Absolute quantification of proteins and phosphoproteins from cell lysates by tandem MS," *Proceedings of the National Academy of Sciences of the United States of America*, vol. 100, no. 12, pp. 6940–6945, 2003.

[8] N. L. Anderson, N. G. Anderson, L. R. Haines, D. B. Hardie, R. W. Olafson, and T. W. Pearson, "Mass spectrometric quantitation of peptides and proteins using Stable Isotope Standards and Capture by Anti-Peptide Antibodies (SISCAPA)," *Journal of Proteome Research*, vol. 3, no. 2, pp. 235–244, 2004.

[9] M. Bantscheff, M. Schirle, G. Sweetman, J. Rick, and B. Kuster, "Quantitative mass spectrometry in proteomics: a critical review," *Analytical and Bioanalytical Chemistry*, vol. 389, no. 4, pp. 1017–1031, 2007.

[10] J. Poirier, J. Davignon, D. Bouthillier, S. Kogan, P. Bertrand, and S. Gauthier, "Apolipoprotein E polymorphism and Alzheimer's disease," *The Lancet*, vol. 342, no. 8873, pp. 697–699, 1993.

[11] J. E. Eichner, S. T. Dunn, G. Perveen, D. M. Thompson, K. E. Stewart, and B. C. Stroehla, "Apolipoprotein E polymorphism and cardiovascular disease: a HuGE review," *American Journal of Epidemiology*, vol. 155, no. 6, pp. 487–495, 2002.

[12] K. R. Wildsmith, B. Han, and R. J. Bateman, "Method for the simultaneous quantitation of apolipoprotein E isoforms using tandem mass spectrometry," *Analytical Biochemistry*, vol. 395, no. 1, pp. 116–118, 2009.

[13] J. R. Murrell, R. G. Schoner, J. J. Liepnieks, H. N. Rosen, A. C. Moses, and M. D. Benson, "Production and functional analysis of normal and variant recombinant human transthyretin proteins," *The Journal of Biological Chemistry*, vol. 267, no. 23, pp. 16595–16600, 1992.

[14] A. Lim, T. Prokaeva, M. E. McComb et al., "Characterization of transthyretin variants in familial transthyretin amyloidosis by mass spectrometric peptide mapping and DNA sequence analysis," *Analytical Chemistry*, vol. 74, no. 4, pp. 741–751, 2002.

[15] Y. Sekijima, R. L. Wiseman, J. Matteson et al., "The biological and chemical basis for tissue-selective amyloid disease. A mutation in a secreted protein can lead to decreased folding efficiency within the ER and reduced exporting, such as the disulfide isomerases," *Cell*, vol. 121, pp. 73–85, 2005.

[16] L. Lobato and A. Rocha, "Transthyretin amyloidosis and the kidney," *Clinical Journal of the American Society of Nephrology*, no. 8, pp. 1337–1346, 2012.

[17] A. Lim, T. Prokaeva, M. E. McComb, L. H. Connors, M. Skinner, and C. E. Costello, "Identification of S-sulfonation and S-thiolation of a novel transthyretin Phe33Cys variant from a patient diagnosed with familial transthyretin amyloidosis," *Protein Science*, vol. 12, no. 8, pp. 1775–1785, 2003.

[18] U. A. Kiernan, K. A. Tubbs, K. Gruber et al., "High-throughput protein characterization using mass spectrometric immunoassay," *Analytical Biochemistry*, vol. 301, no. 1, pp. 49–56, 2002.

[19] O. Trenchevska, E. Kamcheva, and D. Nedelkov, "Mass spectrometric immunoassay for quantitative determination of transthyretin and its variants," *Proteomics*, vol. 11, pp. 3633–3641, 2011.

[20] Z. Yang, M. Hayes, X. Fang, M. P. Daley, S. Effenberg, and F. L. S. Tse, "LC-MS/MS approach for quantification of therapeutic proteins in plasma using a protein internal standard and 2D-solid-phase extraction cleanup," *Analytical Chemistry*, vol. 79, no. 24, pp. 9294–9301, 2007.

[21] O. Heudi, S. Barteau, D. Zimmer et al., "Towards absolute quantification of therapeutic monoclonal antibody in serum by LC-MS/MS using isotope-labeled antibody standard and protein cleavage isotope dilution mass spectrometry," *Analytical Chemistry*, vol. 80, no. 11, pp. 4200–4207, 2008.

[22] C. Hagman, D. Ricke, S. Ewert, S. Bek, R. Falchetto, and F. Bitsch, "Absolute quantification of monoclonal antibodies in biofluids by liquid chromatography-tandem mass spectrometry," *Analytical Chemistry*, vol. 80, no. 4, pp. 1290–1296, 2008.

[23] L. J. M. Dekker, L. Zeneyedpour, E. Brouwer, M. M. van Duijn, P. A. E. Sillevis Smitt, and T. M. Luider, "An antibody-based biomarker discovery method by mass spectrometry sequencing of complementarity determining regions," *Analytical and Bioanalytical Chemistry*, vol. 399, no. 3, pp. 1081–1091, 2011.

[24] J. J. Pesavento, C. A. Mizzen, and N. L. Kelleher, "Quantitative analysis of modified proteins and their positional isomers by tandem mass spectrometry: human histone H4," *Analytical Chemistry*, vol. 78, no. 13, pp. 4271–4280, 2006.

[25] B. A. Garcia, J. J. Pesavento, C. A. Mizzen, and N. L. Kelleher, "Pervasive combinatorial modification of histone H3 in human cells," *Nature Methods*, vol. 4, no. 6, pp. 487–489, 2007.

[26] D. Phanstiel, J. Brumbaugh, W. T. Berggren et al., "Mass spectrometry identifies and quantifies 74 unique histone H4 isoforms in differentiating human embryonic stem cells," *Proceedings of the National Academy of Sciences of the United States of America*, vol. 105, no. 11, pp. 4093–4098, 2008.

[27] T. Jenuwein and C. D. Allis, "Translating the histone code," *Science*, vol. 293, no. 5532, pp. 1074–1080, 2001.

[28] M. Kojima, H. Hosoda, Y. Date, M. Nakazato, H. Matsuo, and K. Kangawa, "Ghrelin is a growth-hormone-releasing acylated peptide from stomach," *Nature*, vol. 402, no. 6762, pp. 656–660, 1999.

[29] M. Tschop, D. L. Smiley, and M. L. Heiman, "Ghrelin induces adiposity in rodents," *Nature*, vol. 407, no. 6806, pp. 908–913, 2000.

[30] Y. Date, M. Nakazato, S. Hashiguchi et al., "Ghrelin is present in pancreatic $\alpha$-cells of humans and rats and stimulates insulin secretion," *Diabetes*, vol. 51, no. 1, pp. 124–129, 2002.

[31] J. Liu, C. E. Prudom, R. Nass et al., "Novel ghrelin assays provide evidence for independent regulation of ghrelin acylation and secretion in healthy young men," *The Journal of Clinical Endocrinology & Metabolism*, vol. 93, no. 5, pp. 1980–1987, 2008.

[32] C. Prudom, J. Liu, J. Patrie et al., "Comparison of competitive radioimmunoassays and two-site sandwich assays for the measurement and interpretation of plasma ghrelin levels," *The Journal of Clinical Endocrinology & Metabolism*, vol. 95, no. 5, pp. 2351–2358, 2010.

[33] J. A. Gutierrez, J. A. Willency, M. D. Knierman et al., "From Ghrelin to Ghrelin's O-acyl transferase," *Methods in Enzymology*, vol. 514, pp. 129–146, 2012.

[34] J. Hardy and D. J. Selkoe, "The amyloid hypothesis of Alzheimer's disease: progress and problems on the road to therapeutics," *Science*, vol. 297, no. 5580, pp. 353–356, 2002.

[35] T. Jonsson, J. K. Atwal, S. Steinberg et al., "A mutation in APP protects against Alzheimer's disease and age-related cognitive decline," *Nature*, vol. 488, no. 7409, pp. 96–99, 2012.

[36] E. Portelius, A. Westman-Brinkmalm, H. Zetterberg, and K. Blennow, "Determination of $\beta$-amyloid peptide signatures in cerebrospinal fluid using immunoprecipitation-mass spectrometry," *Journal of Proteome Research*, vol. 5, no. 4, pp. 1010–1016, 2006.

[37] R. B. DeMattos, M. M. Racke, V. Gelfanova et al., "Identification, characterization, and comparison of amino-terminally truncated A$\beta$42 peptides in Alzheimer's disease brain tissue and in plasma from Alzheimer's patients receiving solanezumab immunotherapy treatment," *Alzheimer's & Dementia*, vol. 5, no. 4, supplement, pp. P156–P157, 2009.

[38] J. R. Cirrito, P. C. May, M. A. O'Dell et al., "*In vivo* assessment of brain interstitial fluid with microdialysis reveals plaque-associated changes in amyloid-$\beta$ metabolism and half-life," *Journal of Neuroscience*, vol. 23, no. 26, pp. 8844–8853, 2003.

[39] T. Oe, B. L. Ackermann, K. Inoue et al., "Quantitative analysis of amyloid $\beta$ peptides in cerebrospinal fluid of Alzheimer's disease patients by immunoaffinity purification and stable isotope dilution liquid chromatography/negative electrospray ionization tandem mass spectrometry," *Rapid Communications in Mass Spectrometry*, vol. 20, no. 24, pp. 3723–3735, 2006.

[40] J. Legleiter, D. L. Czilli, B. Gitter, R. B. DeMattos, D. M. Holtzman, and T. Kowalewski, "Effect of differentanti-Abeta antibodies on Abeta fibrillogenesis asassessed by atomic force microscopy," *Journal of Molecular Biology*, vol. 335, no. 4, pp. 997–1006, 2004.

[41] V. Gelfanova, R. E. Higgs, R. A. Dean et al., "Quantitative analysis of amyloid-$\beta$ peptides in cerebrospinal fluid using immunoprecipitation and MALDI-Tof mass spectrometry," *Briefings in Functional Genomics & Proteomics*, vol. 6, no. 2, pp. 149–158, 2007.

[42] M. J. Ford, J. L. Cantone, C. Polson, J. H. Toyn, J. E. Meredith, and D. M. Drexler, "Qualitative and quantitative characterization of the amyloid $\beta$ peptide (A$\beta$) population in biological matrices using an immunoprecipitation-LC/MS assay," *Journal of Neuroscience Methods*, vol. 168, no. 2, pp. 465–474, 2008.

[43] A. M. Tabert, M. P. Goodwin, J. S. Duncan, C. D. Fico, and R. G. Cooks, "Multiplexed rectilinear ion trap mass spectrometer for high-throughput analysis," *Analytical Chemistry*, vol. 78, no. 14, pp. 4830–4838, 2006.

# 8

# Cathepsin D Expression in Colorectal Cancer: From Proteomic Discovery through Validation Using Western Blotting, Immunohistochemistry, and Tissue Microarrays

**Chandra Kirana,[1] Hongjun Shi,[1] Emma Laing,[2] Kylie Hood,[1] Rose Miller,[3] Peter Bethwaite,[3] John Keating,[4] T. William Jordan,[5] Mark Hayes,[1] and Richard Stubbs[1]**

[1] Wakefield Biomedical Research Unit, University of Otago, Wellington 6242, New Zealand
[2] School of Medicine and Health Sciences, University of Otago, Wellington 6242, New Zealand
[3] Department of Pathology and Molecular Medicine, University of Otago, Wellington 6242, New Zealand
[4] Capital & Coast District Health Board, Wellington Hospital, Wellington 6021, New Zealand
[5] Centre for Biodiscovery, School of Biological Sciences, Victoria University of Wellington, Wellington 6012, New Zealand

Correspondence should be addressed to Chandra Kirana, chandra.kirana@otago.ac.nz

Academic Editor: Fumio Nomura

Despite recent advances in surgical techniques and therapeutic treatments, survival from colorectal cancer (CRC) remains disappointing with some 40–50% of newly diagnosed patients ultimately dying of metastatic disease. Current staging by light microscopy alone is not sufficiently predictive of prognosis and would benefit from additional support from biomarkers in order to stratify patients appropriately for adjuvant therapy. We have identified that cathepsin D expression was significantly greater in cells from invasive front (IF) area and liver metastasis (LM) than those from main tumour body (MTB). Cathepsin D expression was subsequently examined by immunohistochemistry in tissue microarrays from 119 patients with CRC. Strong expression in tumour cells at the IF did not correlate significantly with any clinico-pathological parameters examined or patient survival. However, cathepsin D expression in cells from the MTB was highly elevated in late stage CRC and showed significant correlation with subsequent distant metastasis and shorter cancer-specific survival. We also found that macrophages surrounding tumour cells stained strongly for cathepsin D but there was no significant correlation found between cathepsin D in macrophages at IF and MTB of CRC patient with the clinic-pathological parameters examined.

## 1. Introduction

Despite advances in surgical techniques and therapeutic interventions during the past few decades, colorectal cancer (CRC) remains a major health problem worldwide. The American Cancer Society estimated that some 141,210 people would be diagnosed with colorectal cancer in the US in 2011 and that one-third of them would die of the disease [1]. In New Zealand around 2800 individuals are diagnosed with CRC annually and nearly half of them will die as a result of the disease [2]. Most deaths will result from metastatic spread, most commonly to the liver. Death from CRC is preventable by surgery alone in its

early stages [3]. Adjuvant chemotherapy, which aims to eradicate subclinical tumor deposits after surgical removal of the primary tumor, has been shown to reduce tumor recurrence and improve disease-free survival. While the use of adjuvant chemotherapy for stage III CRC patients has become standard practice, its application for stage II patients is more controversial [4].

Current histological staging methods by light microscopy alone are not sufficiently accurate to predict metastatic spread as there is significant variation with respect to clinical outcomes within currently used stages. Thus, some 20–30% of stage II patients will develop metastases and die of their disease, and some 30% of stage III patients will not develop

Cathepsin D Expression in Colorectal Cancer: From Proteomic Discovery through Validation Using Western Blotting, Immunohistochemistry, and Tissue Microarrays

69

recurrent disease even without adjuvant chemotherapy [4]. Discovery of additional prognostic markers might permit the development of guidelines for better management of CRC in order to improve overall survival. Modern proteomics provides us with the tools to discover new, potentially valuable biomarkers.

Cathepsin D is an aspartic lysosomal endopeptidase present in most mammalian cells. Overexpression of this protease has been associated with the progression of several human cancers including gastric carcinoma [5–7], melanoma [8], and ovarian cancer [9]. Cathepsin D has been comprehensively studied in breast cancer where overexpression of mRNA and protein has been observed [10, 11] and been shown to be an independent marker of poor prognosis [12, 13]. Cathepsin D levels in tumors were reported to be higher than in adjacent normal tissue [14, 15]. The role of cathepsin D in cancer has been postulated to promote tumor growth directly by acting to degrade and remodel the basement membrane and interstitial stroma surrounding the primary tumor [16] and indirectly by stimulation of other enzymes or in cooperation with other cathepsins in the proteolysis process [17]. Previous reports on the clinical significance of cathepsin D in CRC have been variable and inconsistent. On the one hand, cathepsin D expression in tumor and stromal cells at the IF region has been reported to significantly correlate with lymph node metastasis [18] and hence survival. However, another group has reported a study in 48 patients with CRC in which expression of cathepsin D did not differ between MTB and the IF [19].

We used laser microdissection to isolate proteins from CRC tumor cells taken from main tumor body (MTB), invasive front area (IF), and liver metastasis (LM) and then profiled and compared proteins using saturation label dye 2D-DIGE. The concentration of cathepsin D was found to be elevated in tumor cells at the IF area and LM compared to cells at the MTB in tissue from the same patients. This paper explores the expression of cathepsin D in CRC tissue using immunohistochemistry to explore its potential value as a biomarker of metastasis.

## 2. Material and Methods

### 2.1. Identification of Overexpression of Cathepsin D

*2.1.1. Tissue Samples.* Primary colorectal tumor and LM from the same patient were collected from eight patients with sporadic CRC undergoing surgery at Wakefield Hospital, Wellington, New Zealand, and used for proteomic analysis. (See Table 1 for patients' clinico-pathological features.) Tumor specimens were collected directly from the operating theatre and immediately snap-frozen in liquid nitrogen and stored at −80°C. Written, informed consent was obtained from all patients participating in this study and ethics approval was given by the Central Regional Ethics Committee, in accordance with the Helsinki Declaration of 1975.

*2.1.2. Discovery Phase.* We have used a combination of laser microdissection, saturation labeling 2D DIGE, and MALDI

TOF mass spectrometry to compare the protein expression profiles of the main tumor body (MTB), invasive front (IF), and liver metastasis (LM) in sets of tissues from 8 CRC patients, in a biomarker discovery program. The methods in this process have previously been described [20, 21] except that saturation labeling rather than minimal labeling was used. Through this process we identified that cathepsin D was upregulated at both the IF and LM compared to the MTB suggesting that it may play a role in metastasis.

### 2.2. Validation of Overexpression of Cathepsin D in the Invasive Front and Metastases

*2.2.1. Western Blotting.* Saturation-labeled proteins were separated by 2DE as described previously [21] and transferred to PVDF transfer membrane (Amersham Hybond LFP, GE Healthcare, Sweden) at 50 V for 4 h (Bio-Rad, USA) in transfer buffer (3% (w/v) Tris, 14.4% (w/v) glycine, 20% methanol). The blot was blocked overnight in 5% ECL blocking solution (GE Healthcare), rinsed in Tris-buffered saline (TBS) (137 mM NaCl, 20 mM Tris, pH 7.2) containing 0.1% (v/v) Tween-20 (T-TBS), and incubated with a mouse monoclonal antibody [CTD-19] to cathepsin D (Abcam, Cambridge, UK) in T-TBS for 2 h at RT. The blot was rinsed 3 times for 10 min in T-TBS and incubated with an ECL-Plex Cy3-conjugated goat anti-mouse antibody (GE-Healthcare) at 1 : 2500 diluted in T-TBS. The blot was rinsed three times for 10 min in T-TBS, once in TBS, and dried in the dark overnight. Blots were scanned with the Cy3 and Cy5 channel of the Fujifilm FLA-5100 and images overlaid using ImageQuant (GE Healthcare). The primary antibody, mouse anti-cathepsin D antibody condjugated to HRP (CTD-19, ab6313, Abcam, Cambridge, UK) at 1 : 200 dilution was used directly on the gel and detected using ECL-Plus (GE Healthcare) on the Cy2 channel of the Fujifilm FLA5100.

*2.2.2. Immunohistochemistry.* Immunohistochemistry (IHC) was undertaken on formalin-fixed paraffin-embedded (FFPE) tumor specimens. Stains were carried out for cathepsin D, CD45 as a leukocyte marker, and CD68 as a macrophage marker. Sections were deparaffinized in xylene and rehydrated in decreasing concentration of ethanol (100%, 90%, 80%, 70%). Antigen was retrieved using 10 mM sodium citrate buffer (pH 6.8) for 10 min in pressure cooker. Slides were incubated in 3% $H_2O_2$, 50% methanol in wash buffer (phosphate-buffered saline (PBS) (Oxoid), 0.1% Tween-20) for 10 min to quench endogenous peroxidases. Primary antibodies were incubated for 1 h at RT. The ImmPRESS Universal kit (Vector Laboratories, CA, USA) was used to detect primary antibodies and developed with DAB or Nova Red substrate (Labvision, CA, USA). Sections were counterstained with haematoxylin, dehydrated, and mounted. Primary antibodies used in IHC, which were purchased from Abcam (Cambridge, UK), were mouse monoclonal anti-cathepsin D conjugated with HRP (CTD-19, ab6313) at 1 : 7000 dilution, mouse monoclonal anti-CD45 (MEM-28, ab 8216) at 1 : 1000 dilution, and mouse monoclonal anti-CD68 (KP1, ab 955) at 1 : 400 dilution.

TABLE 1: Clinico-pathological details of patients.

| | Patient 1 | Patient 2 | Patient 3 | Patient 4 | Patient 5 | Patient 6 | Patient 7 | Patient 8 |
|---|---|---|---|---|---|---|---|---|
| Site of primary tumor | Recto-sigmoid | Caecum | Ascending colon | Sigmoid colon | Recto-sigmoid | Caecum | Ascending colon | Recto-sigmoid |
| Degree of differentiation | Moderately | Poorly | Poorly | Moderately | Moderately | Poorly | Moderately | Well |
| Age at diagnosis* | 58 | 71 | 63 | 81 | 50 | 54 | 58 | 65 |
| Gender | F | M | F | M | M | M | F | M |
| TNM stage of primary tumor | 2 | 3 | 3 | 2 | 3 | 3 | 3 | 2 |
| LM diagnosis** | Met | Synch | Synch | Synch | Synch | Synch | Synch | Synch |
| Liver involvement*** | Solitary <25% | Multiple 25–50% | Multiple >50% | Multiple 25–50% | Three 25–50% | Multiple >50% | Multiple <25% | Four <25% |

* Mean age at primary tumor diagnosis: 62.9 ± 10 (mean ± SD).
** Met: Metachronous diagnosis, Synch: Synchronous diagnosis.
*** Number of liver metastases (multiple: too many to count), plus percentage of liver involvement.

*2.2.3. Patient Sample for Tissue Microarray.* Out of the 282 patients who had undergone partial colectomy or anterior resection of CRC performed by Dr. John Keating from 1997 to 2005 at Wellington Hospital, 169 patients had archival tissue blocks available for investigation at the time of this study. A representative block from each patient was drawn and sectioned for H&E staining. On histological examination, 42 blocks were excluded from the cohort due to the absence or inadequacy of tumor cells in the sections from the blocks initially chosen from the tissue archives. Consequently a total of 127 CRC cases were finally included in this study. Of the patients chosen, 26 had received chemotherapy (17 × 5FU + leucovorin, 1 oxaliplatin, 1 capecitabine, 2 5FU + mitomycin C, 4 5FU infusion), 13 had received radiotherapy and 6 had received chemoradiation. As chemotherapy and radiation therapy are common treatments before surgery all treated patients were included in the test cohort.

Clinicopathological features of the resected CRC were obtained from a prospective database maintained by Dr. John Keating according to the clinical and pathological reports held at Wellington Hospital. Pathological stages were classified according to the TNM staging system. Histological grading and typing of the tumor were determined according to the World Health Organisation tumor classification system. Cancer-specific survival was defined as the interval between the date of the first operation of the primary tumor to the date when the patient died from recurrent CRC. Cases were censored at the end of the followup or at the time of death due to other causes. Thirty-seven patients died of recurrent CRC during the follow-up period. Medium follow-up time was 61 months (ranging from 2 to 164 months). Construction of tissue microarrays (TMAs) and immunohistochemistry on the TMAs using archival human tissues was conducted with the approval of the New Zealand Central Regional Ethics Committee.

*2.2.4. Tissue Microarray Immunohistochemistry.* Five tissue microarray blocks (TMA) containing a total of 127 CRC cases were constructed. Each TMA consisted of up to

26 tissue cores with a single tissue core per patient's tumor. Formalin-fixed, paraffin-embedded tissue blocks of CRC were obtained from the hospital tumor archive. Before constructing TMAs, a 4 μm section was sliced from each tumor block for a routine H&E inspection by a pathologist. After histological confirmation of the tumors, areas of sampling (AOS) were defined and marked on the microscope slide by the pathologist. These microscope slides with spotted AOS were used later to guide the location of tissue cores for punching. AOS was defined as the area of obvious invasive cancer closest to the lumen, not including any potential adenomatous areas. TMAs were constructed using the Beecher automated tissue arrayer (ATA-27, Beecher Instruments, Sun Prairie, Wisconsin, USA) through the Molecular and Clinical Pathology Research Laboratory, Clinical and Statewide Services, Princess Alexandra Hospital, Queensland, Australia. A tissue core with a diameter of 1 mm was punched from the donor tissue under the guidance of AOS, and transferred to a recipient paraffin block (array margin of 10 × 20 mm). Once the TMAs were made, they were heat cycled from 60°C for 1 h and room temperature (RT) for 1 hr for a total of 5 cycles to aid cutting.

*2.2.5. Scoring of IHC in TMAs.* Expression of cathepsin D and CD68 was graded and scored by two blinded independent pathologists (PB and RM). The intensity of cathepsin D at the MTB and IF was scored from 1 to 3, with 1 for weak or none, 2 for moderate, and 3 for strong. The presence of macrophages, stained with CD68, at the MTB or IF was scored from 1 to 3, with 1 for scanty or none, 2 for moderate, and 3 for plentiful. These were scored separately in order to identify whether expression of cathepsin D was in the tumor cells of the invasive front or in macrophages associated with the invasive front. This expression was then correlated with clinical parameters.

*2.2.6. Statistical Analysis.* Statistical analyses were performed using SPSS (version 17). The association between cathepsin D and CD68 immunoreactivity scores and patient

Cathepsin D Expression in Colorectal Cancer: From Proteomic Discovery through Validation Using Western Blotting, Immunohistochemistry, and Tissue Microarrays

71

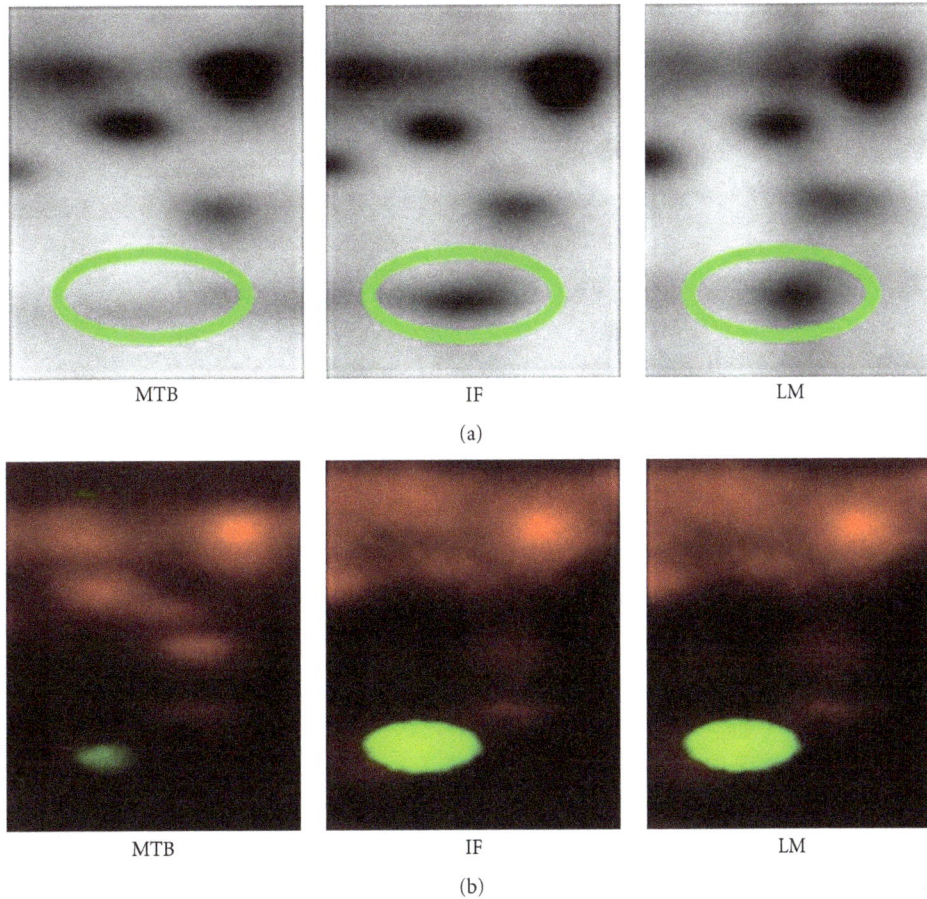

FIGURE 1: (a) Spots of cathepsin D from samples of the main tumor body (MTB), IF area (IF), and liver metastasis (LM) on 2 D gels; (b) validation of cathepsin D profiled by 2DE western blotting of cy5 labeled LMD sample (red) and using Cy3 label antibody (green) (ECL-Plex).

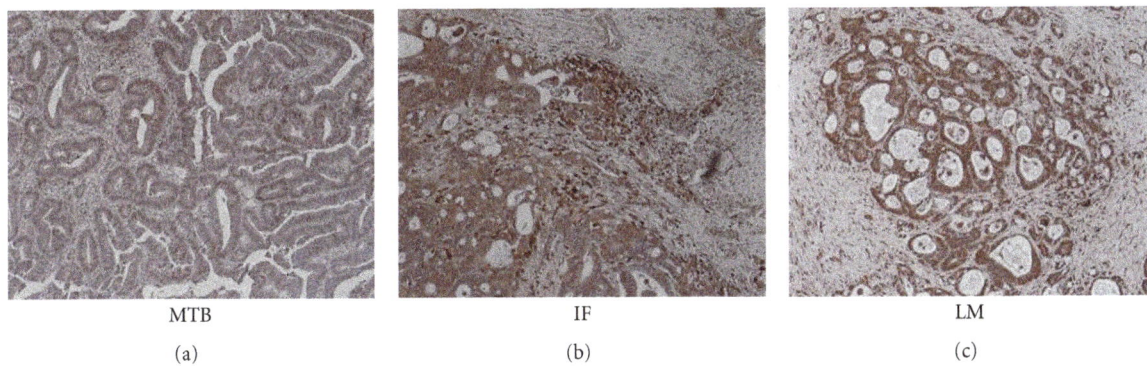

FIGURE 2: Validation of cathepsin D expression at two different regions, main tumor body. (MTB) and invasive front (IF) area of primary tumor and liver metastasis (LM) from the same patient by immunohistochemistry (IHC) (DAB substrate, brown). Sections were counterstained with haematoxylin (blue) (20x objective).

clinico-pathological parameters was assessed by $\chi$-square test. The impact of cathepsin D and CD68 on patient survival was examined by Kaplan-Meier analysis and the statistical significance determined by log-rank test. A multivariate analysis based on Cox proportional hazard regression model was applied to determine independent prognostic factors. Variables included in Cox regression analysis were histological grade, histological types, Dukes stage, vascular invasion, perineural invasion, type of operation, distant metastasis, and cathepsin D and CD68 immunoreactivity scores. A $P$ value less than 0.05 was considered statistically significant.

(a)

(b)

(c)

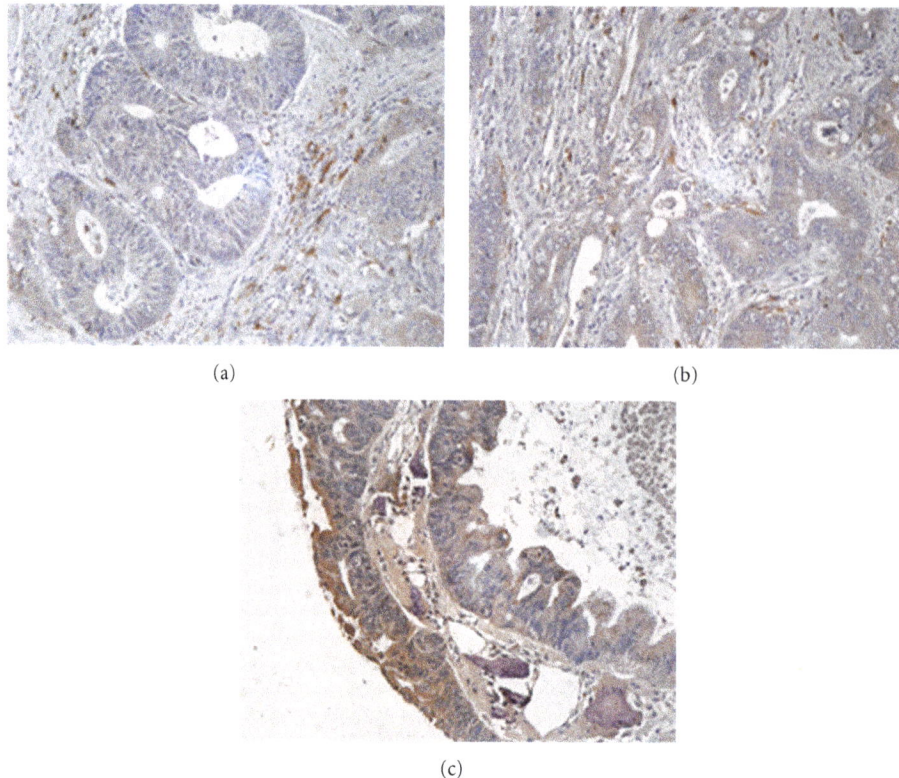

FIGURE 3: Representative CRC tissue for intensity scoring of tissue microarray immunohistochemistry (DAB substrate, brown). None or weak expression of cathepsin D (a) was scored 1. (b) Moderate expression of cathepsin D in tumor cells was scored 2 and strong expression of cathepsin D (c) was scored 3. Tissue sections were counterstained with haematoxylin (blue) (20x objective).

## 3. Results

Cathepsin D was found to have different expressions in three different areas of interest of colorectal tumor tissues in our discovery project. Expression of cathepsin D in tumor cells at the IF and in LM was significantly higher than that in the MTB when profiled using 2 D-DIGE (Figure 1(a)). The gel spot position and concentration of cathepsin D in these three areas of tumor were validated with 2DE western blotting (Figure 1(b)). Validation by IHC also confirmed relatively greater abundance of cathepsin D at the IF and in LM compared to MTB in 9 of 11 (82%) CRC patients. IHC images from a representative set of tissues from the same patient are presented in Figure 2.

Following the creation of TMAs from 127 patients with CRC, tissue damage or loss occurred in 8 patients which left 119 patients available for our validation study. Examples of cathepsin D expression in tissues for TMA IHC for scoring purposes are presented in Figure 3.

High-level expression of cathepsin D at the IF area of colorectal tumor did not correlate significantly with any of the clinico-pathological parameters examined (data not shown). Cathepsin D expression in tumor cells of the MTB did however correlate significantly with distant metastases ($P = 0.038$) and tended to correlate with TNM stage although this did not reach statistical significance ($P = 0.064$). Cathepsin D expression in the MTB did not correlate

with age, gender, tumor location, histological type and grade, vascular invasion, perineural invasion, nodal status, or depth of invasion. This data is shown in Table 2.

Expression of cathepsin D in the MTB was inversely correlated with 5-year cancer-specific survival in univariate analysis using Kaplan-Meier statistics and log-rank test. Those with strong expression of cathepsin D had a 5-year cancer-specific survival of 42%, compared with 63% for those with moderate expression ($P = 0.039$) and 81% for those with weak expression ($P = 0.0001$). 5-year cancer-specific survival was not significantly different between those with weak expression of cathepsin D and those with moderate expression ($P = 0.198$) (Figure 4).

Multivariate Cox regression analysis revealed that cathepsin D is not an independent prognostic factor ($P = 0.958$) unlike TNM stage ($P = 0.0001$) and perineural invasion ($P = 0.039$) (Table 3), which are known to be independent prognostic factors in CRC.

The average immunoreactivity score (score ± SE) of cathepsin D at different stages of CRC is shown in Figure 5. Cathepsin D scores were similar for stage I, II, and III patients but were significantly higher in the patients with stage IV disease. The score of cathepsin D is approximately double in stage IV patients compared to those with earlier stage disease.

Strong cathepsin D staining was noted in a population of cells within stromal tissue at the IF during validation with IHC. Immunohistochemistry of adjacent sections for

Cathepsin D Expression in Colorectal Cancer: From Proteomic Discovery through Validation Using Western Blotting, Immunohistochemistry, and Tissue Microarrays

73

TABLE 2: Correlation of cathepsin D expression at MTB with clinico-pathological features.

| Clinico-pathological parameters | No. of cases (%)[1] | neg/weak (%)[2] | Moderate (%)[2] | Strong (%)[2] | P value[3] |
|---|---|---|---|---|---|
| Age | | | | | 0.139 |
| <65 | 35 (29) | 14 (40) | 14 (40) | 7 (20) | |
| ≥65 | 84 (71) | 50 (60) | 21 (25) | 13 (15) | |
| Gender | | | | | 0.586 |
| F | 59 (50) | 30 (51) | 17 (29) | 12 (20) | |
| M | 60 (50) | 34 (57) | 18 (30) | 8 (13) | |
| Tumor location | | | | | 0.072 |
| Colon | 82 (69) | 49 (60) | 19 (23) | 14 (17) | |
| Rectum | 37 (31) | 15 (41) | 16 (43) | 6 (16) | |
| Histological type | | | | | 0.177 |
| Nonmucinous | 102 (86) | 52 (51) | 33 (32) | 17 (17) | |
| Mucinous | 16 (14) | 12 (75) | 2 (13) | 2 (13) | |
| Histological grade | | | | | 0.211 |
| High grade | 95 (81) | 47 (50) | 31 (33) | 17 (18) | |
| Low grade | 23 (19) | 16 (70) | 4 (17) | 3 (13) | |
| Vascular invasion | | | | | 0.400 |
| Negative | 90 (76) | 51 (57) | 26 (29) | 13 (14) | |
| Positive | 29 (24) | 13 (45) | 9 (31) | 7 (24) | |
| Perineural invasion | | | | | 0.839 |
| Negative | 109 (92) | 59 (54) | 32 (29) | 18 (17) | |
| Positive | 9 (8) | 4 (44) | 3 (33) | 2 (22) | |
| TNM stages | | | | | **0.064** |
| I | 20 (17) | 12 (60) | 5 (25) | 3 (15) | |
| II | 42 (36) | 27 (64) | 10 (24) | 5 (12) | |
| III | 34 (29) | 18 (53) | 13 (38) | 3 (9) | |
| IV | 21 (18) | 7 (33) | 6 (29) | 8 (38) | |
| Distant metastasis | | | | | **0.038*** |
| No | 89 (75) | 52 (58) | 27 (30) | 10 (11) | |
| Yes | 29 (25) | 12 (41) | 8 (28) | 9 (31) | |
| Nodal status | | | | | 0.358 |
| Negative | 66 (58) | 38 (58) | 16 (24) | 12 (18) | |
| Positive | 47 (42) | 24 (51) | 17 (36) | 6 (13) | |
| Depth of invasion | | | | | 0.935 |
| T1/T2 | 23 (20) | 13 (57) | 7 (30) | 3 (13) | |
| T3/T4 | 93 (80) | 51 (55) | 27 (29) | 15 (16) | |
| 5-year recurrence[4] | | | | | *0.090* |
| Recurrence free | 58 (79) | 40 (69) | 14 (24) | 4 (7) | |
| Recurrence | 15 (21) | 6 (40) | 6 (40) | 3 (20) | |

[1] Percentage of the column.
[2] Percentage of the row.
[3] P value based on Pearson's $\chi^2$ test; *$P \leq 0.05$.
[4] Presence or absence of local or distant metachronous recurrence within 5-year followup.
* Significant correlation between expression of cathepsin D at MTB with distant metastasis.

CD45, a leukocyte common antigen, confirmed that cells staining for high levels of cathepsin D also stained for CD45, confirming that they were leukocyte phenotype rather than cancer cells. Using CD68 (a specific marker of monocytes/macrophages), it was demonstrated that the cells at the IF most strongly staining for cathepsin D were of monocytes/macrophage phenotype rather than being cancer cells (Figure 6). The presence of macrophages containing cathepsin D at the IF area of colorectal tumor did not correlate significantly with any of the clinico-pathological parameters examined (data not shown).

## 4. Discussion

The depth of invasion of colorectal cancer through the bowel wall and the presence or not of lymph node involvement have

TABLE 3: Cox regression analysis of tumor characteristics with respect to cancer survival.

| Variable | HR | 95% CI | P value |
|---|---|---|---|
| Histological type (mucinous versus nonmucinous) | 1.230 | 0.328–4.611 | 0.759 |
| Histological grade (high grade versus low grade) | 0.551 | 0.209–1.448 | 0.229 |
| Vascular invasion (positive versus negative) | 0.751 | 0.325–1.735 | 0.503 |
| Operation (emergency versus elective) | 1.079 | 0.467–2.492 | 0.859 |
| Perineural invasion (positive versus negative) | 2.788 | 1.055–7.370 | 0.039 |
| Stage (I, II, III, IV) | 15.333 | 6.030–38.992 | 0.0001 |
| Cathepsin D expression at MTB (strong, moderate, weak) | 0.986 | 0.577–1.684 | 0.958 |

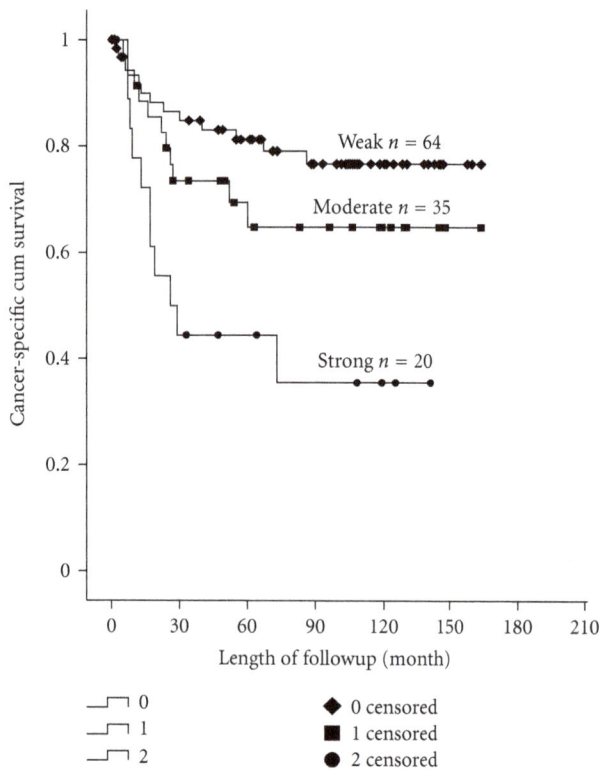

FIGURE 4: Cancer-specific survival (in months) of CRC patients in association with cathepsin D expression in epithelial of main tumor.

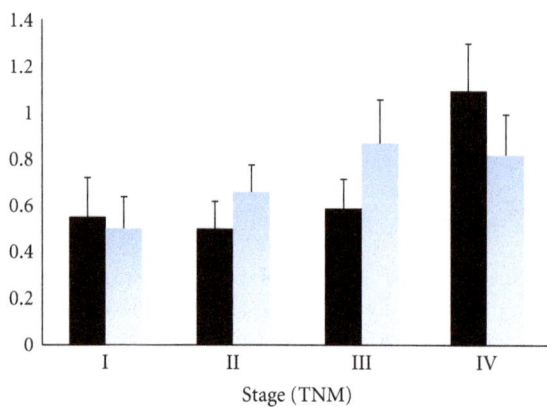

FIGURE 5: The average score of cathepsin D expression in tumor cells at the MTB (solid black) and IF (blue) of CRC patients at different stage (TNM).

provided the basis for pathological staging since it was first described by Dukes some 75 years ago. However useful this is, we know the allocation of patients to stage II (Dukes B) or stage III (Dukes C) disease carries an imprecise estimate of prognosis. This has become problematic following the widespread adoption of adjuvant chemotherapy, as it is now important to identify as accurately as possible those patients who despite a complete surgical clearance of the primary tumor, will develop recurrent or distant disease. There is optimism that molecular markers may be discovered which would help refine assessments of prognosis, allow more precise allocation of adjuvant therapies, and perhaps even point to new drug targets for this disease. Modern proteomics has the capacity to identify such molecular biomarkers. In this paper we briefly describe the initial approach taken by our group, using laser microdissection to precisely identify CRC tissue and specific areas of CRC, for further proteomic analysis using 2D DIGE and MALDI-TOF mass spectrometry.

The IF area of colorectal cancer has been suggested as a critical interface where tumor progression and metastasis begin and may therefore be a critical area in which to search for prognostic markers. Areas of tumor budding have been shown to have overexpressed proteins which are involved in extracellular matrix degradation [22]. Proteins including matrix metalloproteinase-9 (MMP9), cathepsin B [23], matrilysin, and laminin [24] have been identified as being highly expressed in tumor budding. For this reason we were interested in examining and comparing the proteome derived from the MTB and the IF of the primary CRC, and the proteome derived from liver metastases all in the same patient. By so doing we identified that cathepsin D was expressed more strongly at the IF area in both tumor cells and what we identified to be macrophages surrounding tumor glands compared to its expression in the MTB.

Our examination of cathepsin D at the IF by immunohistochemistry suggested that much of the expression in that area was not associated with tumor cells themselves but with what appeared to be leucocytes or macrophages. Using a leukocyte common antigen marker (CD45) we established that much of the positive staining related to leukocytes at the IF, but that not all the leukocytes stained for cathepsin D. Using a specific monocyte/macrophage marker (CD68) we identified that the majority of the cathepsin D staining at the IF was in macrophages which may be considered a marker of the immune response to the tumor. Interestingly,

Cathepsin D Expression in Colorectal Cancer: From Proteomic Discovery through Validation Using Western Blotting, Immunohistochemistry, and Tissue Microarrays

75

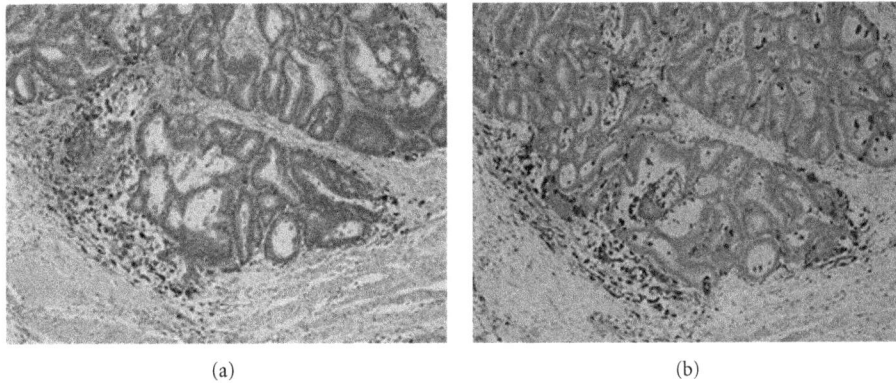

(a)

(b)

FIGURE 6: Adjacent sections of primary colorectal tumor stained for cathepsin D (a) (DAB substrate, brown). CD 68 staining (b) confirmed that very strong expression of cathepsin D in a population of cells within stromal tissue was associated with macrophage (DAB substrate, brown). Sections were counterstained with haematoxylin (blue) (40x objective).

Brujan et al. (2009) [25] also found macrophage-like cells surrounding breast cancer cells which contained strongly positive cathepsin D granules.

Nadji et al. (1996) [26] reported that cathepsin D in stromal cells significantly correlated with disease free and survival but not in tumor cells when examined in node-negative breast cancer patients. Similarly Theodoropoulos et al. (1997) [27] reported that positive cathepsin D staining in stromal cells and negative cathepsin D in tumor cells showed worse prognosis in 60 CRC patients. They suggested that positive CD expression in stromal cells may be used as an important indicator of tumor progression.

Having identified that the cathepsin D staining at the IF could be in either tumor cells or macrophages, we scored the cathepsin D expression at the IF in our TMAs separately in the tumor cells and in the macrophages. Neither cathepsin D expression in tumor tissue nor that in macrophages at the IF significantly correlated with any of the clinico-pathological parameters examined in our 119 patients. Guzińska-Ustymowicz et al.found no correlation between tumor budding and the activity of cathepsin D expression [28]. They concluded that cathepsin D in the tumor cells at the IF area was not involved in tumor progression and metastasis in CRC. On the other hand, the expression of cathepsin D at the MTB was found to be significantly associated with distant metastases and the correlation with TNM stage approached statistical significance ($P = 0.064$). We noted that expression of cathepsin D was highly elevated in late-stage CRC patients (TNM stage IV) compared to the earlier stages (TNM stages II, II, and III). Mayer et al. (1997) [29] noted findings similar to results of our study. They have found that cathepsin D was only elevated in CRC patients with Duke's C and D. They found that the elevation of cathepsin D was not significantly correlated with the clinicopathological parameters examined.

A role for cathepsin D in cancer metastasis was first demonstrated in an *in vitro* study using rat tumor cells in which overexpression of procathepsin D was associated with metastatic potential [30]. The concentration of cathepsin D in chronic ulcerative colitis and familial adenomatous polyposis, which is known to associate with the increase

risk of colorectal carcinoma and colon carcinoma, was higher than that of normal colon [31]. Cathepsin D has been postulated to be secreted from cancer cells and been shown to serve as an autocrine growth factor in several cancer studies conferring proinvasive and prometastatic properties [32]. When our results are taken alongside those of others [18, 22, 28], one is forced to conclude that any role of cathepsin D in CRC progression remains uncertain.

The role of tumor infiltrating lymphocytes (TILs) especially macrophages in solid tumors remains unclear as they have been implicated in both tumor progression and protective host response. There are studies that have reported that pronounced tumor infiltration with TILs is associated with early-stage disease and/or improved survival [33], and yet other compelling evidence has emerged recently to indicate that tumor-associated macrophages (TAMs)— also referred to as alternative M2 macrophages—have an important role in solid tumor progression [34]. In our present IHC microarray study, neither strong expression of cathepsin D in macrophages nor the abundance of macrophages themselves at the IF of colorectal cancer tissue correlated particularly with other important clinico-pathological parameters, survival or metastasis.

In the discovery phase of our study cathepsin D was more highly expressed in LM and at the IF of CRC, relative to the MTB, suggesting it might be associated with tumor progression. When cathepsin D expression was examined in TMAs from 119 patients with CRC, the higher expression at the IF relative to the MTB was confirmed, but this was largely related to its presence in macrophages rather than tumor cells *per se*. However, neither expression of cathepsin D in the CRC cells at the IF nor the presence of cathepsin D staining macrophages at the IF correlated well with other clinico-pathological parameters examined. Paradoxically strong expression of cathepsin D in the cells of the main tumor body was noted in late-stage disease and significantly correlated with distant metastasis and shorter cancer-specific survival. Cathepsin D expression in the main tumor body warrants further consideration as a potential biomarker of prognosis in colorectal cancer.

## Acknowledgments

The authors thank the Cancer Society of New Zealand Wellington Division, NZ Lottery Health Research, Wellington Medical Research Foundation, Wakefield Gastroenterology Research Trust, and Wakefield Hospital for funding in support of this research. The authors are grateful to Mr John Groom at Wakefield Hospital who provided surgical samples for the biodiscovery phase and validation of this study. The proteomic facilities provided by Dr. T. Jordan, School of Biological Sciences, Victoria University of Wellington, are also gratefully acknowledged.

## References

[1] American Cancer Society, "Colorectal cancer facts & figures 2011–2013," Tech. Rep., American Cancer Society, Atlanta, Ga, USA, 2011.

[2] Ministry of Health, "Cancer: new registrations and deaths 2007," Tech. Rep., Ministry of Health, Wellington, New Zealand, 2010.

[3] P. Rougier and E. Mitry, "Epidemiology, treatment and chemoprevention in colorectal cancer," Annals of Oncology, vol. 14, supplement 2, pp. ii3–ii5, 2003.

[4] R. Midgley and D. J. Kerr, "Adjuvant chemotherapy for stage II colorectal cancer: the time is right!," Nature Clinical Practice Oncology, vol. 2, no. 7, pp. 364–369, 2005.

[5] H. Allgayer, R. Babic, K. U. Grützner et al., "An immunohistochemical assessment of cathepsin D in gastric carcinoma: its impact on clinical prognosis," Cancer, vol. 80, no. 2, pp. 179–187, 1997.

[6] J. M. Del Casar, F. J. Vizoso, O. Abdel-Laa et al., "Prognostic value of cytosolyc cathepsin d content in resectable gastric cancer," Journal of Surgical Oncology, vol. 86, no. 1, pp. 16–21, 2004.

[7] T. Saku, H. Sakai, N. Tsuda, H. Okabe, Y. Kato, and K. Yamamoto, "Cathepsins D and E in normal, metaplastic, dysplastic, and carcinomatous gastric tissue: an immunohistochemical study," Gut, vol. 31, no. 11, pp. 1250–1255, 1990.

[8] I. Bartenjev, Z. Rudolf, B. Štabuc, I. Vrhovec, T. Perkovič, and A. Kansky, "Cathepsin D expression in early cutaneous malignant melanoma," International Journal of Dermatology, vol. 39, no. 8, pp. 599–602, 2000.

[9] A. Lösch, M. Schindl, P. Kohlberger et al., "Cathepsin D in ovarian cancer: orognostic value and correlation with p53 expression and microvessel density," Gynecologic Oncology, vol. 92, no. 2, pp. 545–552, 2004.

[10] D. E. Abbott, N. V. Margaryan, J. S. Jeruss et al., "Reevaluating cathepsin D as a biomarker for breast cancer: serum activity levels versus histopathology," Cancer Biology and Therapy, vol. 9, no. 1, pp. 23–30, 2010.

[11] E. Barthell, I. Mylonas, N. Shabani et al., "Immunohistochemical visualisation of cathepsin-D expression in breast cancer," Anticancer Research, vol. 27, no. 4 A, pp. 2035–2039, 2007.

[12] J. A. Foekens, M. P. Look, J. Bolt-De Vries, M. E. Meijer-Van Gelder, W. L. J. Van Putten, and J. G. M. Klijn, "Cathepsin-D in primary breast cancer: prognostic evaluation involving 2810 patients," British Journal of Cancer, vol. 79, no. 2, pp. 300–307, 1999.

[13] G. Jacobson-Raber, I. Lazarev, V. Novack et al., "The prognostic importance of cathepsin D and E-cadherin in early breast cancer: a single-institution experience," Oncology Letters, vol. 2, no. 6, pp. 1183–1190, 2011.

[14] A. Adenis, G. Huet, F. Zerimech, B. Hecquet, M. Balduyck, and J. P. Peyrat, "Cathepsin B, L, and D activities in colorectal carcinomas: relationship with clinico-pathological parameters," Cancer Letters, vol. 96, no. 2, pp. 267–275, 1995.

[15] S. D. Szajda, J. Snarska, A. Jankowska, W. Roszkowska-Jakimiec, Z. Puchalski, and K. Zwierz, "Cathepsin D and carcino-embryonic antigen in serum, urine and tissues of colon adenocarcinoma patients," Hepato-Gastroenterology, vol. 55, no. 82-83, pp. 388–393, 2008.

[16] G. Berchem, M. Glondu, M. Gleizes et al., "Cathepsin-D affects multiple tumor progression steps in vivo: proliferation, angiogenesis and apoptosis," Oncogene, vol. 21, no. 38, pp. 5951–5955, 2002.

[17] E. Křepela, "Cysteine proteinases in tumor cell growth and apoptosis: minireview," Neoplasma, vol. 48, no. 5, pp. 332–349, 2001.

[18] H. Oh-e, S. Tanaka, Y. Kitadai, F. Shimamoto, M. Yoshihara, and K. Haruma, "Cathepsin D expression as a possible predictor of lymph node metastasis in submucosal colorectal cancer," European Journal of Cancer, vol. 37, no. 2, pp. 180–188, 2001.

[19] W. Famulski, K. Guzińska-Ustymowicz, M. Sulkowska et al., "Tumour budding intensity in relation to cathepsin D expression and some clinicopathological features of colorectal cancer," Folia Histochemica et Cytobiologica, vol. 39, supplement 2, pp. 171–172, 2001.

[20] T. Kondo, M. Seike, Y. Mori, K. Fujii, T. Yamada, and S. Hirohashi, "Application of sensitive fluorescent dyes in linkage of laser microdissection and two-dimensional gel electrophoresis as a cancer proteomic study tool," Proteomics, vol. 3, no. 9, pp. 1758–1766, 2003.

[21] H. Shi, K. A. Hood, M. T. Hayes, and R. S. Stubbs, "Proteomic analysis of advanced colorectal cancer by laser capture microdissection and two-dimensional difference gel electrophoresis," Journal of Proteomics, vol. 75, no. 2, pp. 339–351, 2011.

[22] I. Zlobec and A. Lugli, "Invasive front of colorectal cancer: dynamic interface of pro-/anti-tumor factors," World Journal of Gastroenterology, vol. 15, no. 47, pp. 5898–5906, 2009.

[23] K. Guzinska-Ustymowicz, "MMP-9 and cathepsin B expression in tumor budding as an indicator of a more aggressive phenotype of colorectal cancer (CRC)," Anticancer Research, vol. 26, no. 2 B, pp. 1589–1594, 2006.

[24] T. Masaki, M. Sugiyama, H. Matsuoka et al., "Coexpression of matrilysin and laminin-5 $\gamma$2 chain may contribute to tumor cell migration in colorectal carcinomas," Digestive Diseases and Sciences, vol. 48, no. 7, pp. 1262–1267, 2003.

[25] I. Brujan, C. Mărgăritescu, C. Simionescu et al., "Cathepsin-D expression in breast lesion: an immunohistochemical study," Romanian Journal of Morphology and Embryology, vol. 50, no. 1, pp. 31–39, 2009.

[26] M. Nadji, M. Fresno, M. Nassiri, G. Conner, A. Herrero, and A. R. Morales, "Cathepsin D in host stromal cells, but not in tumor cells, is associated with aggressive behavior in node-negative breast cancer," Human Pathology, vol. 27, no. 9, pp. 890–895, 1996.

[27] G. E. Theodoropoulos, D. Panoussopoulos, A. Ch. Lazaris, and B. Ch. Golematis, "Evaluation of Cathepsin D immunostaining in colorectal adenocarcinoma," Journal of Surgical Oncology, vol. 65, no. 4, pp. 242–248, 1997.

[28] K. Guzińska-Ustymowicz, B. Zalewski, I. Kasacka, Z. Piotrowski, and E. Skrzydlewska, "Activity of cathepsin B and D in colorectal cancer: relationships with tumour budding," Anticancer Research, vol. 24, no. 5 A, pp. 2847–2851, 2004.

Cathepsin D Expression in Colorectal Cancer: From Proteomic Discovery through Validation Using Western Blotting, Immunohistochemistry, and Tissue Microarrays

77

[29] A. Mayer, E. Fritz, R. Fortelny, K. Kofler, and H. Ludwig, "Immunohistochemical evaluation of cathepsin D expression in colorectal cancer," *Cancer Investigation*, vol. 15, no. 2, pp. 106–110, 1997.

[30] M. Garcia, D. Derocq, P. Pujol, and H. Rochefort, "Over-expression of transfected cathepsin D in transformed cells increases their malignant phenotype and metastatic potency," *Oncogene*, vol. 5, no. 12, pp. 1809–1814, 1990.

[31] S. Galandiuk, S. Miseljic, A. R. Yang, M. Early, M. D. McCoy, and J. L. Wittliff, "Expression of hormone receptors, cathepsin D, and HER-2/neu oncoprotein in normal colon and colonic disease," *Archives of Surgery*, vol. 128, no. 6, pp. 637–642, 1993.

[32] P. Benes, V. Vetvicka, and M. Fusek, "Cathepsin D—Many functions of one aspartic protease," *Critical Reviews in Oncology/Hematology*, vol. 68, no. 1, pp. 12–28, 2008.

[33] Y. Funada, T. Noguchi, R. Kikuchi, S. Takeno, Y. Uchida, and H. E. Gabbert, "Prognostic significance of CD8+ T cell and macrophage peritumoral infiltration in colorectal cancer," *Oncology Reports*, vol. 10, no. 2, pp. 309–313, 2003.

[34] A. Mantovani, P. Romero, A. K. Palucka, and F. M. Marincola, "Tumour immunity: effector response to tumour and role of the microenvironment," *The Lancet*, vol. 371, no. 9614, pp. 771–783, 2008.

# 9

# A Miniaturized Ligand Binding Assay for EGFR

**Jochen M. Schwenk,[1,2] Oliver Poetz,[1] Robert Zeillinger,[3] and Thomas O. Joos[1]**

[1] NMI Natural and Medical Sciences Institute at the University of Tübingen, Markwiesenstraße 55, 72770 Reutlingen, Germany
[2] Science for Life Laboratory Stockholm, School of Biotechnology, KTH Royal Institute of Technology, P.O. Box 1031, 171 21 Solna, Sweden
[3] Molecular Oncology Group, Department of Obstetrics and Gynecology, Medical University of Vienna, Währinger Gürtel 18-20, 5Q, 1090 Vienna, Austria

Correspondence should be addressed to Jochen M. Schwenk, jochen.schwenk@scilifelab.se

Academic Editor: Ákos Végvári

In order to study receptor abundance and its function in solutions or in homogenates from clinical specimen, methods such as sandwich or radioimmunoassays are most commonly employed. For the determination of epidermal growth factor receptor (EGFR), we describe the development of a miniaturized bead-based ligand binding assay using its ligand EGF as immobilized capture reagent. This assay was used to analyze lysates from cell lines, and the ligand-bound EGFR was detected using an EGFR-specific antibody combined with a fluorescence-based reporter system. In a proof-of concept study with lysates from breast biopsies, the assay allowed to classify breast cancer samples in accordance to clinically the relevant EGFR cut-off level. The study suggests that such a ligand binding receptor assay could become an integral part of protein profiling procedures to provide additional information about receptor functionality in addition to its abundance.

## 1. Introduction

Since their introduction in the 1960s [1], immunoassays based on radiolabeled ligands or antibodies have become well-established technologies in research and clinical chemistry. Today, antibody-based immunoassays have been transferred and applied to various technological platforms such as multiplexed and miniaturized formats used in proteomic experiments [2, 3]. These multiplexed assay systems now allow analyzing hundreds of proteins in a single affinity-based experiment, but, thus far, radioimmunoassay-like measurements of ligand binding active receptors have not yet been transferred in a miniaturized format.

In the context of a detailed analysis of breast cancer, the quantification of active epithelial growth factor receptor (EGFR) in clinical specimen is a prominent example for the application of a radio-ligand binding assay. The investigation of receptor tyrosine kinases such as the EGFR is of interest, as overexpression of EGFR is correlated with poor prognosis [4], and the expression rates of EGFR and hormone receptors were strongly inverse [5]. EGFR itself is a transmembrane-spanning protein, with an extracellular ligand binding and a cytoplasmatic tyrosine kinase domain [6], and its main ligand EGF is a polypeptide consisting of 53 amino acids that binds to domain I and III of the extracellular part of the receptor. Upon binding of EGF to EGFR, the receptor undergoes conformational changes [7]. These lead to a downstream activation of the NF-$\kappa$B transcription factor pathway to induce antiapoptotic and proliferated genes, which eventually results in cell growth. EGFR plays a key role in the regulation of essential normal cellular processes and in the pathophysiology of hyperproliferated diseases such as many epithelial cancer types like breast, ovarian, lung, colorectal or prostate cancer [8]. EGFR is also known to be activated in a variety of tumors via the promotion of its gene c-erbB1, leading to an overexpression [9] by mutation or ligand binding [10]. EGFR expression may even serve as a decision maker for EGFR-targeted therapies [11], and, recently, the use of specific monoclonal antibodies directed against the phosphorylated and mutant form of

EGFR (EGFRvIII) was suggested as a valid predictor of response to therapy with cetuximab, an anti-EGFR antibody [12].

To measure the levels of EGFR in tissue samples, sandwich immunoassays can be used [13]. Complementary to these, radio-ligand binding assays (RLB) perform a more classical approach to measure quantities of functional EGFR by employing $^{125}I$-labeled EGF [14]. This receptor ligand binding assay takes advantage of the strong interaction between the ligand EGF and EGFR, which bind with an affinity of $\leq 12$ nM [15].

In the study presented here, we have developed a miniaturized solid-phase-based ligand binding assay as a possible alternative to assays involving radioactive tracers. Instead of using an EGFR specific antibody, we have employed the ligand EGF as a reagent to capture and profile functional EGFR present in the samples. This assay only requires small amounts of sample to determine relative EGFR level in lysates from cell culture material as well as from patient samples. The study reveals that bead-based ligand binding assay offers a complementary system to sandwich immunoassays and radioimmunoassays.

## 2. Methods and Material

### 2.1. Preparation of Cell Samples.
Frozen pellets of the human breast cancer cells BT-20 (ATCC no. HTB-19) were solubilized on ice with 2% (v/v) NP-40 (Sigma) in a buffer containing 50 mM Tris (Sigma), 120 mM NaCl, 1 mM $CaCl_2$, 1 mM $MgCl_2$, and 2% (v/v) protease inhibitor (Sigma). The cells were sonicated 4 times for 5 s (Branson Sonifier, Korea) and centrifuged at 13,000 rpm for 5 min. The supernatant was collected for further studies.

### 2.2. Tissue Samples.
Membrane fractions of 46 breast cancer samples with EGFR levels between 0 and 600 fmol EGFR per mg protein were included in the study (Supplementary Table 1 available online at doi:10.1155/2012/247059). The EGFR level was analyzed by a radio-ligand binding assay [14]. Seventeen samples had an EGFR level $\leq 10$ fmol/mg, and 29 samples had EGFR levels of 11–246 fmol/mg. One sample had an EGFR concentration of 600 fmol/mg. All samples had been stored at $-80°C$ for about 10 years.

### 2.3. Coupling of Beads.
Biotinylated EGF (EGF-Biot, Molecular Probes) was coupled to avidin-coated beads (LumAvidin beads, Luminex Corporation). In each coupling reaction, $2.5 \times 10^5$ beads were used. Firstly, beads were washed twice in block-and-storage buffer (BSB) containing 1% bovine serum albumin (Carl Roth) in phosphate-buffered saline (pH 7.4). Washing steps were performed as follows: beads were sedimented at 10,000 g for 2 min, supernatant was removed, and beads were resuspended with 250 $\mu$L BSB, vortexed, and sonicated. EGF was coupled onto two differently color-coded bead sets at EGF-Biot concentrations of 1.2 $\mu$M and 0.3 $\mu$M over 30 min under permanent shaking in BSB buffer. Beads were washed twice with BSB and stored in BSB buffer

containing 0.01% sodium azide (Merck) at 4°C in the dark before use.

### 2.4. Miniaturized Ligand Binding Assay.
Assays were performed in a 96-well microtiter plate with a filter-membrane bottom (Millipore) in BSB under permanent shaking at 650 rpm and 23°C in the dark. In each well, 30 $\mu$L containing EGF-Biot beads (1000 per set) and 30 $\mu$L of sample were mixed and incubated for 2 h. The beads were washed with $3 \times 50 \mu$L BSB buffer on a vacuum manifold (Millipore). An anti-EGFR antibody (0.3 $\mu$g/mL, mAb11, LabVision) was selected, and 30 $\mu$L were added to each well. After an incubation of more than 60 min, the beads were washed again as above. A secondary R-phycoerythrin-labeled antibody (2.0 $\mu$g/mL, goat anti-mouse, Dianova) with a volume of 30 $\mu$L/well was incubated for 45 min. Finally, the beads were washed, and a final volume of 75 $\mu$L BSB was added to each well.

For competition experiments, soluble EGF (sEGF, Biomol) was mixed with the sample and coincubated with immobilized EGF-Biot-coupled beads. For detection, anti-EGFR antibody mAb11 was applied at 1.25 $\mu$g/mL and the labeled anti-mouse antibody at 5.0 $\mu$g/mL.

### 2.5. Readout and Data Analysis.
A Luminex LX100 system (Luminex Corporation) was used to determine the bead-bound reporter fluorescence. For each well, at least 100 events per bead ID were counted and the median fluorescence intensity (MFI) of the reporter dye was collected. Data was processed using R, a language and environment for statistical computing and graphics [16]. For the comparison of two groups with differing EGFR expression values, a $P$ value was calculated with Student's $t$-test.

## 3. Results

In a first set of experiments, two concentrations of biotinylated EGF were immobilized onto beads, mixed, and applied to an assay with different amounts of total protein from cell lysates. For this study, BT-20 cells, known to express EGFR at level of 400 fmol/mg or $1.5 \times 10^6$ copies per cell [17], were chosen. For the subsequent detection of bead-bound EGFR, an anti-EGFR antibody was chosen that did not interfere with the EGF-EGFR interaction, according to the data sheet. No cross-reactivity between immobilized EGF-biotin, the anti-EGFR antibody, and the labeled anti-mouse detection antibody could be detected (data not shown). In Figure 1(a), it is shown that both the amount of lysate employed as well as the ligand density on the bead surface influenced the measured signal intensity and no saturation effects were observed for the tested lysate protein concentrations. EGF coupled to the beads at a concentration of 1.2 $\mu$M showed increased signal intensities compared to the bead ID coupled with a 4x lower ligand concentration, in addition to a broader intensity range for the 1.2 $\mu$M loaded EGF beads. As the utilized BT-20 cells are known to express higher levels of EGFR, a sample protein concentration of 500 $\mu$g/mL, reflecting 15 $\mu$g of total protein per well, was chosen for

(a)

(b)

(c)

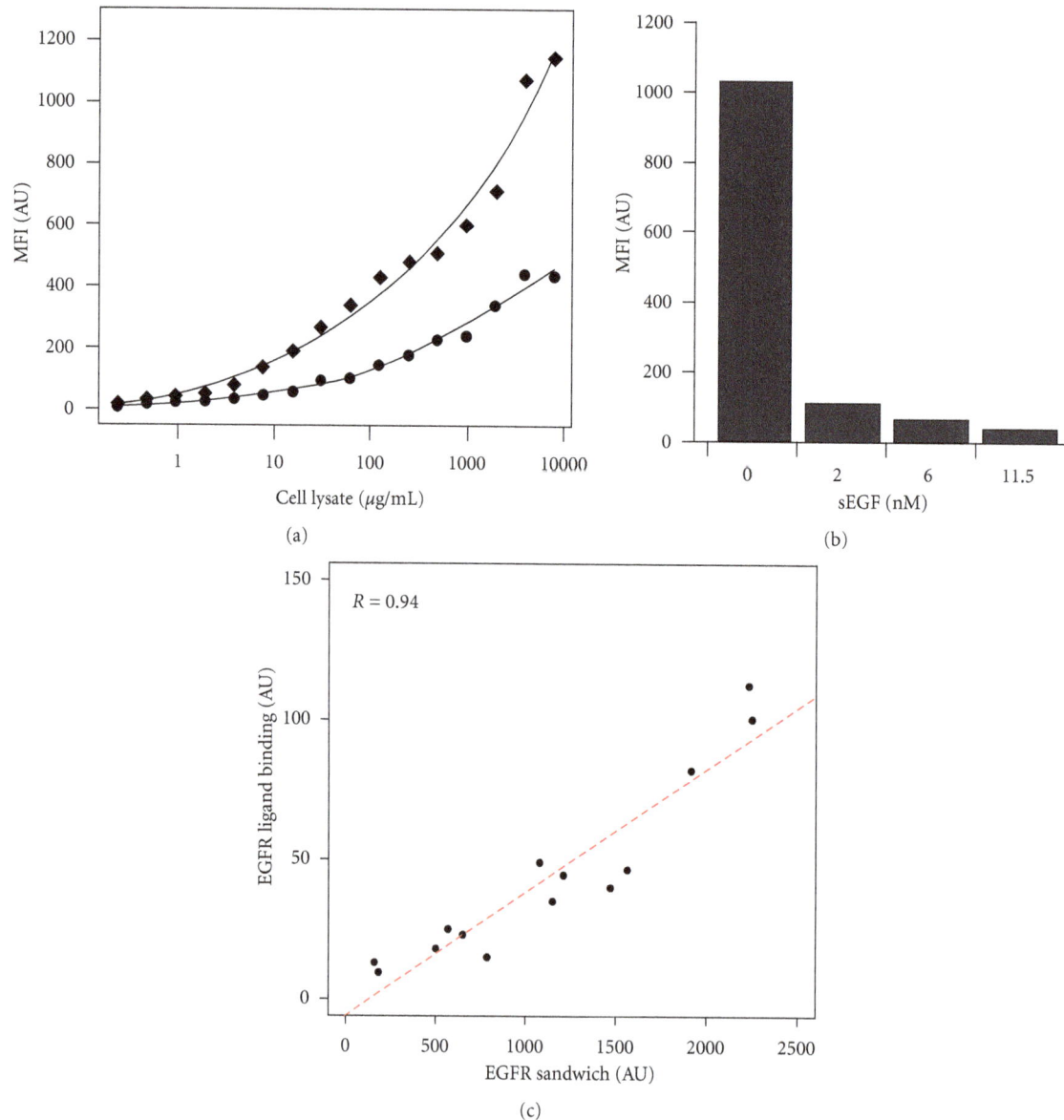

FIGURE 1: (a) Detection of EGFR in cell extracts. Beads coated with different concentrations of EGF were used to capture EGFR from a cell lysate in a concentration-dependent manner. The ligand EGF was immobilized at $1.2\,\mu$M (diamonds) and $0.3\,\mu$M (circles) on different bead types. Both bead sets were mixed and incubated with different amounts of BT-20 cell lysates. Captured EGFR was detected using an anti-EGFR antibody and a labeled secondary antibody. (b) Specificity of capture activity. A BT-20 cell lysate (500 $\mu$g/mL) was coincubated with various concentrations of purified and unbiotinylated sEGF and beads coated with EGF-Biot ($1.2\,\mu$M). The amount of captured EGF receptor was strongly reduced at the presence of sEGF. This indicates that the bead-bound EGF-EGFR complex is captured in a specific fashion by applying an anti-EGFR antibody followed by a labeled secondary antibody. (c) Correlation of ligand binding and sandwich immunoassay. A series of tissue lysates with known EGFR concentrations were analyzed with both, a ligand binding assay and an immunoassay using an antibody as capture reagent. In both assays, the same anti-EGFR detection molecule was used and the profiles obtained reveal a correlation of 0.94, which indicated a good concordance between the two tests.

the following studies to also facilitate the detection of EGFR in specimens with lower expression levels. To validate the specificity of the assay, meaning that EGF is capturing EGFR, BT-20 cell lysates were coincubated with nonbiotinylated soluble EGF (sEGF). In this competition assay, sEGF will bind to free EGFR binding sites and occupy them so that the immobilized EGF cannot bind to such a EGFR molecule. The results from this test (Figure 1(b)) showed a strong

reduction in signal intensity at sEGF levels of 11.5 nM (96% reduction) and 2 nM (89% reduction). This clearly indicated that the observed interactions are due to EGFR-EGF binding. Moreover, we also compared the ligand binding assay to a sandwich immunoassay with both tests being performed on beads and utilizing the same detection system. As shown in Figure 1(c), a concordance of $r = 0.94$ was achieved for the profiles generated with breast cancer tissue samples

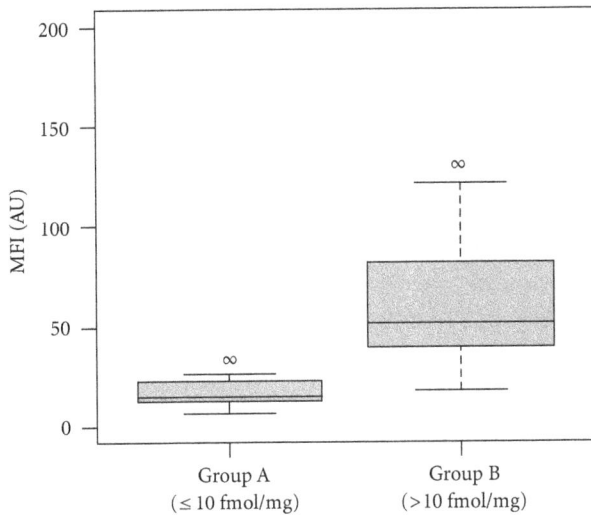

FIGURE 2: Analysis of breast cancer tissue samples. Forty-six breast cancer tissue samples originally analyzed for EGFR expression by a radio-ligand binding assay were reanalyzed in a bead-based ligand binding assay. A cut-off value of 10 fmol EGFR per mg protein, determined by radio-ligand binding assay, was used to divide the samples into group A ($n = 17$) with EGFR values ≤10 fmol/mg and group B ($n = 29$) with EGFR values >10 fmol/mg. Samples were measured in a random order, and a $P$ value of 2.6e-11 was calculated between the two groups.

(see below). This demonstrated that a protocol for a ligand binding assay was developed and that is allowed to provide specific and functional information for profiling the cell surface receptor EGFR in lysates.

Next, the EGFR ligand binding assay was applied in a proof-of-concept study to profile EGFR in 46 tumor samples derived from breast cancer patients. The tissue lysates were measured in random order at a total protein concentration of 0.5 mg/mL with the established protocol. The analyzed samples were grouped based on a clinically relevant cut-off level of 10 fmol/mg [5], which had been previously determined in radio-ligand binding experiments. As shown in Figure 2, 15 of 17 samples (88%) from group A with EGFR ≤10 fmol/mg were measured with MFIs ≤25 AU, while samples with MFI ≥50 AU were not affiliated to this group. For samples in group B with EGFR >10 fmol/mg, 52% had MFI values ≥50 AU, and, for the remaining 48% with MFIs ≤50 AU, the highest EGFR value was 42 fmol/mg. A sample containing 600 fmol/mg EGFR (not shown in Figure 2) was measured with an MFI >1000 AU thus served as a control. Between group A and B, a $P$ value of 2.6e-11 was calculated and demonstrated that the ligand binding assay performed well to provide complementary evidence for separating samples with high and low EGFR values.

## 4. Discussion and Conclusion

Miniaturized ligand binding assays, as described here for EGF and EGFR, offer a radiation-free tool to profile and study target molecules via their in vivo interaction partners compared to conventional radioimmunoassays or other antibody-based methods. In the presented approach, we investigate the potential of the immobilized receptor ligand EGF to capture EGFR in a miniaturized bead-based assay format in which EGF was immobilized to avidin-coated beads via an N-terminal biotin modification. The chosen immobilization strategy enabled to maintain the binding properties of EGF and to achieve high signal intensities. Both, the EGF coupling concentration as well as the amount of the applied sample affected signal intensities. As described elsewhere, the extracellular domain of EGFR undergoes a conformational change upon ligand binding on the cell surface [7] and it seemed likely that such changes should occur even when EGFR is being bound by an immobilized capture ligand. Even though EGF-EGFR interactions take place with a free EGF binding to anchored or soluble EGFR in vivo, attaching EGF to a solid support did not hinder the two proteins from binding to each other.

In competition assays, it was revealed that the binding by detected immobilized EGF was affected by purified and soluble sEGF, thus confirming that the measured interactions were from EGF and EGFR. The strong inhibitory effect of soluble EGF is likely to reflect advantages in conformation adaptation of EGFR when binding in solution. Compared with purely antibody-based sandwich immunoassays, that were described earlier [18], the influence of conformational may have a greater influence on the ligand binding assay. This could be interpreted from the decreased intensity level that was found in ligand binding assays. In addition, this observation may eventually be a consequence of the fact that the total number of receptors accessible to ligand binding is lower due to the applied extraction procedure and sample storage, which may not keep all receptors in nondenatured and functional states.

To investigate the possible potential applicability of this approach in future clinical sample analysis, EGFR-characterized tumor samples from breast cancer patients were studied. The analyzed samples were grouped based on a clinically relevant cut-off level for EGFR, and it was possible to discriminate samples with expression levels below or above cutoff using the described assay setup. Even though this small proof-of-concept study may suggest that such miniaturized ligand binding approaches could have the potential to be a supplementary alternative to radioactivity-based tests, further testing and assay evaluations need to be performed.

The demonstrated miniaturization of receptor ligand binding assays allowed an analysis of potentially functional receptors from minimal sample amounts and would decrease costs by reducing the amount of material and reagents needed. The reduction of sample volume would be of particular importance for applications where only minimal amounts of specimen are available, such as the analysis of multiple tumor markers from biopsies. In addition, the miniaturized ligand binding assays may have the capability to supplement existing single- or multiplexed sandwich immunoassays to further increase the evidence for specific target presence and to add informatory value.

In conclusion, the ligand binding assay system demonstrates that ligand receptor-binding assays can be miniaturized and offer to become part of the variety of protein microarray-based approaches performed today. With new proteins and pathways emerging as indicators of disease, the procedure presented here describes a possibility to profile the interactions of proteins, such as cell surface receptors, in a miniaturized assay. It is possible that such assays are of value for variety of assays in the fields of proteomics and diagnostics.

## Abbreviations

EGFR:      Epidermal growth factor receptor
EGR:       Epidermal growth factor
EGF-Biot:  Biotinylated EGF
sEGF:      Soluble EGF
MFI:       Median fluorescence intensity.

## Acknowledgments

The authors would like to thank Jutta Bachmann for careful reading of this paper. Funding of the project was provided by the Government of Baden-Württemberg through Grant "Adressierbare BioChip arrays für die Diagnostik und Medizin" (no. 720.430-21/15).

## References

[1] R. S. Yalow and S. A. Berson, "Immunoassay of endogenous plasma insulin in man," *The Journal of Clinical Investigation*, vol. 39, pp. 1157–1175, 1960.

[2] B. Ayoglu, A. Häggmark, M. Neiman et al., "Systematic antibody and antigen-based proteomic profiling with microarrays," *Expert Review of Molecular Diagnostics*, vol. 11, no. 2, pp. 219–234, 2011.

[3] M. F. Templin, D. Stoll, J. M. Schwenk, O. Pötz, S. Kramer, and T. O. Joos, "Protein microarrays: promising tools for proteomic research," *Proteomics*, vol. 3, no. 11, pp. 2155–2166, 2003.

[4] J. G. M. Klijn, P. M. J. J. Berns, P. I. M. Schmitz, and J. A. Foekens, "The clinical significance of epidermal growth factor receptor (EGF-R) in human breast cancer: a review on 5232 patients," *Endocrine Reviews*, vol. 13, no. 1, pp. 3–17, 1992.

[5] S. Nicholson, J. R. C. Sainsbury, G. K. Needham, P. Chambers, J. R. Farndon, and A. L. Harris, "Quantitative assays of epidermal growth factor receptor in human breast cancer: cut-off points of clinical relevance," *International Journal of Cancer*, vol. 42, no. 1, pp. 36–41, 1988.

[6] S. R. Hubbard and J. H. Till, "Protein tyrosine kinase structure and function," *Annual Review of Biochemistry*, vol. 69, pp. 373–398, 2000.

[7] A. W. Burgess, H. S. Cho, C. Eigenbrot et al., "An open-and-shut case? recent insights into the activation of EGF/ErbB receptors," *Molecular Cell*, vol. 12, no. 3, pp. 541–552, 2003.

[8] C. L. Arteaga, "Overview of epidermal growth factor receptor biology and its role as a therapeutic target in human neoplasia," *Seminars in Oncology*, vol. 29, no. 5, supplement 14, pp. 3–9, 2002.

[9] A. A. Jungbluth, E. Stockert, H. J. S. Huang et al., "A monoclonal antibody recognizing human cancers with amplification/overexpression of the human epidermal growth factor receptor," *Proceedings of the National Academy of Sciences of the United States of America*, vol. 100, no. 2, pp. 639–644, 2003.

[10] M. V. Grandal and I. H. Madshus, "Epidermal growth factor receptor and cancer: control of oncogenic signalling by endocytosis," *Journal of Cellular and Molecular Medicine*, vol. 12, no. 5A, pp. 1527–1534, 2008.

[11] D. A. Eberhard, G. Giaccone, and B. E. Johnson, "Biomarkers of response to epidermal growth factor receptor inhibitors in non-small-cell lung cancer working group: standardization for use in the clinical trial setting," *Journal of Clinical Oncology*, vol. 26, no. 6, pp. 983–994, 2008.

[12] A. M. Valentini, M. Pirrelli, and M. L. Caruso, "EGFR-targeted therapy in colorectal cancer: does immunohistochemistry deserve a role in predicting the response to cetuximab?" *Current Opinion in Molecular Therapeutics*, vol. 10, no. 2, pp. 124–131, 2008.

[13] C. M. Stoscheck, "Enzyme-linked immunosorbent assay for the epidermal growth factor receptor," *Journal of Cellular Biochemistry*, vol. 43, no. 3, pp. 229–241, 1990.

[14] R. Zeillinger, F. Kury, P. Speiser, G. Sliutz, K. Czerwenka, and E. Kubista, "EGF-R and steroid receptors in breast cancer: a comparison with tumor grading, tumor size, lymph node involvement, and age," *Clinical Biochemistry*, vol. 26, no. 3, pp. 221–227, 1993.

[15] A. Ullrich and J. Schlessinger, "Signal transduction by receptors with tyrosine kinase activity," *Cell*, vol. 61, no. 2, pp. 203–212, 1990.

[16] R. Ihaka and R. Gentleman, "R: a language for data analysis and graphics," *Journal of Computational and Graphical Statistics*, vol. 5, no. 3, pp. 299–314, 1996.

[17] I. Brotherick, T. W. J. Lennard, S. E. Wilkinson, S. Cook, B. Angus, and B. K. Shenton, "Flow cytometric method for the measurement of epidermal growth factor receptor and comparison with the radio-ligand binding assay," *Cytometry*, vol. 16, no. 3, pp. 262–269, 1994.

[18] N. Schneiderhan-Marra, A. Kirn, A. Döttinger et al., "Protein microarrays—a promising tool for cancer diagnosis," *Cancer Genomics and Proteomics*, vol. 2, no. 1, pp. 37–42, 2005.

# 10

# Top-Down Characterization of the Post-Translationally Modified Intact Periplasmic Proteome from the Bacterium *Novosphingobium aromaticivorans*

**Si Wu,**[1] **Roslyn N. Brown,**[2] **Samuel H. Payne,**[3] **Da Meng,**[4] **Rui Zhao,**[1]
**Nikola Tolić,**[1] **Li Cao,**[5] **Anil Shukla,**[3] **Matthew E. Monroe,**[3] **Ronald J. Moore,**[3]
**Mary S. Lipton,**[3] **and Ljiljana Paša-Tolić**[1]

[1] *Environmental Molecular Science Laboratory, Pacific Northwest National Laboratory, P.O. Box 999/MS K8-98, Richland, WA 99352, USA*
[2] *Center for Bioproducts and Bioenergy, Washington State University, Richland, WA, USA*
[3] *Biological Sciences Division, Pacific Northwest National Laboratory, Richland, WA, USA*
[4] *Computational Sciences and Mathematics Division, Pacific Northwest National Laboratory, Richland, WA, USA*
[5] *Department of Neurobiology, 720 Westview Drive SW, Atlanta, GA, USA*

Correspondence should be addressed to Ljiljana Paša-Tolić; ljiljana.pasatolic@pnnl.gov

Academic Editor: Boris Zybailov

The periplasm of Gram-negative bacteria is a dynamic and physiologically important subcellular compartment where the constant exposure to potential environmental insults amplifies the need for proper protein folding and modifications. Top-down proteomics analysis of the periplasmic fraction at the intact protein level provides unrestricted characterization and annotation of the periplasmic proteome, including the post-translational modifications (PTMs) on these proteins. Here, we used single-dimension ultra-high pressure liquid chromatography coupled with the Fourier transform mass spectrometry (FTMS) to investigate the intact periplasmic proteome of *Novosphingobium aromaticivorans*. Our top-down analysis provided the confident identification of 55 proteins in the periplasm and characterized their PTMs including signal peptide removal, N-terminal methionine excision, acetylation, glutathionylation, pyroglutamate, and disulfide bond formation. This study provides the first experimental evidence for the expression and periplasmic localization of many hypothetical and uncharacterized proteins and the first unrestrictive, large-scale data on PTMs in the bacterial periplasm.

## 1. Introduction

The periplasm of Gram-negative bacteria is a hydrated gel located between the cytoplasmic and outer membranes and is comprised of peptidoglycan (cell wall), proteins, carbohydrates, and small solutes [1–3]. The periplasm is a dynamic subcellular compartment important for trafficking of molecules into and out of cells, maintaining cellular osmotic balance, envelope structure, responding to environmental cues and stresses, electron transport, xenobiotic metabolism, and protein folding and modification [4].

The periplasm provides a good model system to study protein biogenesis, composition, sorting, and modification at the molecular level. Indeed, it is analogous in many ways to the endoplasmic reticulum of eukaryotic cells in terms of transport, folding, and quality control [3]. Localization to the periplasm and beyond often involves an N-terminal secretion signal that targets the protein for translocation across the cytoplasmic membrane via the general secretory pathway [5]. These secretion signals (also known as signal peptides) are cleaved by signal peptidases located in the cytoplasmic membrane [6]. Thus, it is expected that signal peptide cleavage is a common modification in the periplasmic proteome.

Compared to the cytoplasm, the periplasm is more vulnerable to changes in pH, temperature, and osmolarity in the external environment [4, 7, 8]. For structural stability in diverse and dynamic environmental conditions, periplasmic proteins often contain disulfide bonds and the periplasm is maintained in an oxidizing state to facilitate this process [9, 10]. Other PTMs, such as the addition of heme groups to cytochromes, may occur in the periplasm [11]. Therefore, the detailed study of bacterial periplasmic proteins not only allows for a better understanding of the physiology of microbial systems, but also provides information toward the complete annotation of mature proteoforms of microbial genomes, and it may give insight into protein sorting and PTMs in more complex systems.

The typical proteomic approach to profile the periplasm is the bottom-up liquid chromatography-tandem mass spectrometry (LC-MS/MS) approach [12, 13]. It has been applied to the periplasmic proteome of the extremophile *Acidithiobacillus ferrooxidans*, where it yielded a total of 131 proteins [14]. A majority of the identified proteins in *A. ferrooxidans* were categorized as periplasmic proteins based on their predicted export signals using software such as SignalP. However, direct evidence for signal peptide removal such as *N*-terminal peptide identifications was not available in that study. The most significant drawback of the bottom-up approach is that it rarely provides complete sequence coverage to ensure the identification of *N*-terminal peptides, which can be essential for understanding localization. Moreover, when multiple PTMs occur in a single protein, the bottom-up approach cannot accurately define proteoforms, as it does not have the ability to determine which combinations of PTMs cooccur in a single proteoform.

To overcome these difficulties, we employ top-down mass spectrometry (MS) to study the periplasmic proteome. Top-down MS measures intact proteins and facilitates the full characterization of proteoforms including PTMs [15]. The top-down approach has been successfully applied for the characterization of various protein PTMs including signal peptide identification [16]. Recent improvements in intact protein LC separations and high performance FTMS instrumentation greatly expand the observable range of proteoforms. Because top-down analysis preserves the mature N-terminus, the proteolytic processing (e.g., N-terminal cleavage) of a protein is evident. Thus top-down MS provides an experimental validation of bioinformatic predictions such as the signal peptide cleavage predicted by SignalP.

As an initial subject for analysis, we focused on the Gram-negative alphaproteobacterium, *Novosphingobium aromaticivorans*. Members of this genus are noted for their remarkable ability to degrade a variety of aromatic hydrocarbons [17]. The genome of only one species, *N. aromaticivorans*, has been completely sequenced. In a genome with 3917 proteins, nearly 30% are annotated as "hypothetical"; moreover, using the subcellular localization predictor PSORTb (http://www.psort.org/psortb/), 33% of proteins have "unknown" localization. Our current goal is to identify protein constituents of the periplasm of this unusual microorganism, to aid in annotation of hypothetical and poorly characterized proteins, and to survey, in an unrestricted manner, the PTMs existing in these proteins. We here report our results on profiling the enriched periplasmic proteome of *N. aromaticivorans* using a high throughput intact protein (top-down) analysis. A total of 55 proteins were confidently identified, and their PTMs were characterized including N-terminal processing (e.g., signal peptide removal), acetylation, glutathionylation, pyroglutamate modification, and disulfide bond formation.

## 2. Experimental Section

*2.1. Periplasmic Protein Extraction.* *N. aromaticivorans* str. DSM 12444 was grown to early stationary phase aerobically in 50% (v/v) Luria Bertani broth. Cells were harvested by centrifugation at 8,500 g for 5 min, washed once with sodium phosphate (pH 7.5), and the periplasm extracted as previously reported [18]. The soluble periplasmic fraction was flash-frozen in liquid nitrogen and concentrated using a SpeedVac (Thermo-Savant) prior to top-down analysis. For peptide-level analysis, 4 volumes of 20% acetonitrile and 3 volumes of water were added, followed by trypsin (trypsin to protein ratio 1 : 50), and incubated at 37°C for 18 h. The sample was concentrated to dryness in a SpeedVac and suspended in 20 μL 0.1% formic acid prior to LC-MS/MS analysis.

*2.2. Intact Protein LC-MS/MS Analysis.* The intact protein RPLC separation was performed on a Waters NanoAcquity system with a column (80 cm × 75 μm i.d.) packed in-house with Phenomenex Jupiter particles (C5 stationary phase, 5 μm particle size, 300 Å pore size). Mobile phase A was composed of 0.5% acetic acid, 0.01% TFA, 5% isopropanol, 10% acetonitrile (ACN), and 84.5% water. Mobile phase B consisted of 0.5% acetic acid, 0.01% TFA, 9.9% water, 45% isopropanol, and 45% ACN. The operating flow rate was 0.3 μL/min. The RPLC system was equilibrated with 100% mobile phase A for 5 minutes and then increased to 20% mobile phase B in 1 minute. A 250 minute linear gradient was set from 20% mobile phase B to 55% mobile phase B. MS analysis was performed using an LTQ-Orbitrap Velos spectrometer (Thermo Scientific, San Jose, CA) outfitted with a custom electrospray ionization (ESI) interface. ESI emitters were custom made using 150 um o.d. × 20 um i.d. chemically etched fused silica [19]. The heated capillary temperature and spray voltage were 275°C and 2.2 kV, respectively. Two LC-MS/MS analyses were performed: one with ETD fragmentation and one with HCD fragmentation. For the LC-MS/MS analysis with ETD fragmentation, a parent spectrum was collected at a 60 K resolution and was followed by high resolution (30 K) ETD MS/MS of the 8 most intense ions from the parent spectrum. The ETD reaction time was fixed at 40 ms. For the LC-MS/MS analysis with HCD fragmentation, a parent spectrum was collected at a 60 K resolution and was followed by high resolution (30 K) HCD MS/MS of the 8 most intense ions from the parent spectrum. FT MS/MS employed 45% normalized collision energy for HCD. Mass calibration

Top-Down Characterization of the Post-Translationally Modified Intact Periplasmic Proteome from the Bacterium Novosphingobium aromaticivorans

85

TABLE 1: Modifications of identified proteins using top-down approach.

| Locus_Tag | Genbank Protein Desc | Export signal | Detected signal peptide[a] | N-terminal | Other modifications |
|---|---|---|---|---|---|
| Saro_2004 | Alkyl hydroperoxide reductase/Thiol specific antioxidant/Mal allergen | SecP | None | Removal of met | |
| Saro_2586 | Cold-shock DNA-binding protein family | SecP | None | Removal of met | |
| Saro_0565 | Glutathione peroxidase | SecP | None | Removal of met | |
| Saro_0483 | Superoxide dismutase | SecP | None | Removal of met | |
| Saro_3290 | Thiamine biosynthesis protein ThiS | SecP | None | Removal of met | |
| Saro_1996 | Thioredoxin | SecP | None | Removal of met | Disulfide bond |
| Saro_1332 | CsbD-like protein | SecP/TatP | None | Removal of met | |
| Saro_1919 | Hypothetical protein | SecP/TatP | Unknown | Proteolytic fragment | |
| Saro_1314 | Conserved hypothetical protein | SecP/TatP | Yes | AXA | Disulfide bond, pyro glu |
| Saro_2385 | Hypothetical protein | SecP/TatP | Yes | AXA | Pyro glu |
| Saro_3257 | Conserved hypothetical protein | SignalP | None | Proteolytic fragment | |
| Saro_3518 | Cupin 2, conserved barrel domain protein | SignalP | None | Removal of met (wrong starting site) | |
| Saro_3173 | OmpA/MotB | SignalP | None | Proteolytic fragment | |
| Saro_1303 | Hypothetical protein | SignalP | Unknown | Proteolytic fragment | |
| Saro_1685 | Amine dehydrogenase | SignalP | Yes | AXA | |
| Saro_2852 | Ankyrin | SignalP | Yes | AXA | Pyro glu |
| Saro_3053 | Beta-Ig-H3/fasciclin | SignalP | Yes | AXA | |
| Saro_0830 | Cell wall surface anchor family protein | SignalP | Yes | VAA, not AXA | |
| Saro_2955 | Conserved hypothetical protein | SignalP | Yes | AXA | |
| Saro_1378 | Conserved hypothetical protein | SignalP | Yes | AXA | |
| Saro_0103 | Conserved hypothetical protein | SignalP | Yes | AXA | Disulfide bond |
| Saro_2067 | Conserved hypothetical protein | SignalP | Yes | SHA, not AXA | Pyro glu |
| Saro_1721 | Conserved hypothetical protein | SignalP | Yes | THA, not AXA | |
| Saro_2522 | Hypothetical protein | SignalP | Yes | ASN, not AXA | Disulfide bond |
| Saro_3326 | Hypothetical protein | SignalP | Yes | AXA | Disulfide bond |
| Saro_2384 | Hypothetical protein | SignalP | Yes | AXA | Pyro glu |
| Saro_1978 | Hypothetical protein | SignalP | Yes | AXA | Pyro glu |
| Saro_1502 | Hypothetical protein | SignalP | Yes | AXA | |
| Saro_1412 | Hypothetical protein | SignalP | Yes | AXA | Disulfide bond |
| Saro_2350 | Peptidase M28 | SignalP | Yes | AXA | |
| Saro_0837 | Peptidyl-prolyl cis-trans isomerase, cyclophilin type | SignalP | Yes | LVA, not AXA | |
| Saro_2251 | Peptidylprolyl isomerase | SignalP | Yes | VAA, not AXA | Pyro glu |
| Saro_0989 | Peptidylprolyl isomerase, FKBP-type | SignalP | Yes | AIS, not AXA | Disulfide bond |
| Saro_0823 | Protein of unknown function DUF192 | SignalP | Yes | AXA | |
| Saro_3075 | TonB-dependent receptor | SignalP | Yes | AXA | Proteolytic fragment |
| Saro_2265 | YceI | SignalP | Yes | MVA, not AXA | Pyro glu |
| Saro_1171 | Hypothetical protein | SignalP/SecP | None | Removal of met | Disulfide bond |
| Saro_1420 | Antibiotic biosynthesis monooxygenase | TatP | None | Removal of met | Disulfide bond |
| Saro_3279 | Arsenate reductase | TatP | None | N/A | S-glutathiolation on cysteine |

TABLE 1: Continued.

| Locus_Tag | Genbank Protein Desc | Export signal | Detected signal peptide[a] | N-terminal | Other modifications |
|---|---|---|---|---|---|
| Saro_1703 | (2Fe-2S)-binding protein | | None | Removal of met | Disulfide bond |
| Saro_1346 | (2Fe-2S)-binding protein | | None | Removal of met | Disulfide bond |
| Saro_1339 | Acyl carrier protein | | None | Removal of met | Modification (382 Da) |
| Saro_2520 | BolA-like protein | | None | Removal of met | |
| Saro_0034 | Chaperonin Cpn10 | | None | Removal of met | |
| Saro_2299 | Conserved hypothetical protein | | None | Removal of met | |
| Saro_2229 | GreA/GreB family elongation factor | | None | Removal of met | Both acetylation and n/a |
| Saro_2403 | H+-transporting two-sector ATPase, delta/epsilon subunit | | None | Removal of met | |
| Saro_1768 | Hypothetical protein | | None | Removal of met | |
| Saro_1177 | Hypothetical protein | | Nonc | Removal of met | |
| Saro_1778 | Molybdopterin binding domain | | None | Removal of met | S-glutathiolation on cysteine |
| Saro_0894 | Nucleoside diphosphate kinase | | None | Removal of met | |
| Saro_1830 | PhnA protein | | None | Removal of met | |
| Saro_1033 | Phosphoribosylformylglycinamidine synthetase PurS | | None | N/A | |
| Saro_0209 | Tetratricopeptide TPR_4 | | None | Removal of met (wrong starting site) | Pyro glu |
| Saro_1194 | Hypothetical protein | | Yes | AXA, unusual[b] | Disulfide bond |

[a] Signal peptide cleavage was annotated as described in the Experimental section.
[b] The unusual cleavage site was further validated. See discussion in text and Figure 3.

was performed prior to analysis according to the method recommended by the instrument manufacturer.

*2.3. Capillary LC-MS/MS Analysis on Trypsin-Digested Peptides.* Bottom-up identification of proteins was achieved through the detection of peptides with LC-MS/MS. The capillary RPLC system used for peptide separations has been previously described [20]. Briefly, the HPLC system consisted of a custom configuration of 100 mL ISCO Model 100DM syringe pumps (Isco, Inc., Lincoln, NE), 2-position Valco valves (Valco Instruments Co., Houston, TX), and a PAL autosampler (Leap Technologies, Carrboro, NC), allowing for fully automated sample analysis across four separate HPLC columns (3 $\mu$m Jupiter C18 stationary phase, Phenomenex, Torrance, CA). Mobile phases consisted of 0.1% formic acid in water (A) and 0.1% formic acid acetonitrile (B). The HPLC system was equilibrated at 10 kpsi with 100% mobile phase A, and a mobile phase selection valve was switched 50 min after injection, which created a near-exponential gradient as mobile phase B displaced A in a 2.5 mL active mixer. A 40 cm length of 360 $\mu$m o.d. × 15 $\mu$m i.d. fused silica tubing was used to split ~17 $\mu$L/min of flow before it reached the injection valve (5 $\mu$L sample loop). The split flow controlled the gradient speed under conditions of constant pressure operation (10 kpsi). Flow through the capillary HPLC column when equilibrated to 100% mobile phase A was ~500 nL/min. ESI using an etched fused-silica tip [19] was employed to interface the RPLC separation to an LTQ mass spectrometer (Thermo Scientific, San Jose, CA). Precursor ion mass spectra

(automatic gain control was set to $1 \times 10^6$) were collected for 400–2000 m/z range at a resolution of 100 K followed by data dependent ion trap CID MS/MS (collision energy 35%, AGC $3 \times 10^4$) of the ten most abundant ions. A dynamic exclusion time of 180 sec was used to discriminate against previously analyzed ions.

*2.4. Data Analysis.* Intact protein MS/MS data were subjected to data analysis and protein identification using MS-Align+[16] (http://bix.ucsd.edu/projects/msalign/) with the following search parameters: minimal precursor mass = 2500 Da; minimal fragment peaks per scan = 10; maximum number of modifications = 2; fragment mass error tolerance = 15 ppm. MS-Align+ reported only the PrSM with the best E-value for each spectrum. LC-MS/MS data were searched against the Genbank protein annotation (accession CP000248). The false discovery rate (FDR) for protein/spectrum matches was estimated by searching all top-down spectra against the human Uniprot database. A final cutoff of E-value $2.7E^{-4}$ was used to achieve FDR 1%. Protein identifications were further manually verified. Peptide-level MS/MS data were searched using SEQUEST and were filtered using MSGF [21] with a spectral probability cutoff of $1E^{-10}$. All the raw datasets and MSAlign+ output results were deposited at http://omics.pnl.gov/view/publication_1074.html.

Signal peptides were determined using the identified peptides and the prokaryotic proteogenomic pipeline [22].

Top-Down Characterization of the Post-Translationally Modified Intact Periplasmic Proteome from the Bacterium
Novosphingobium aromaticivorans

87

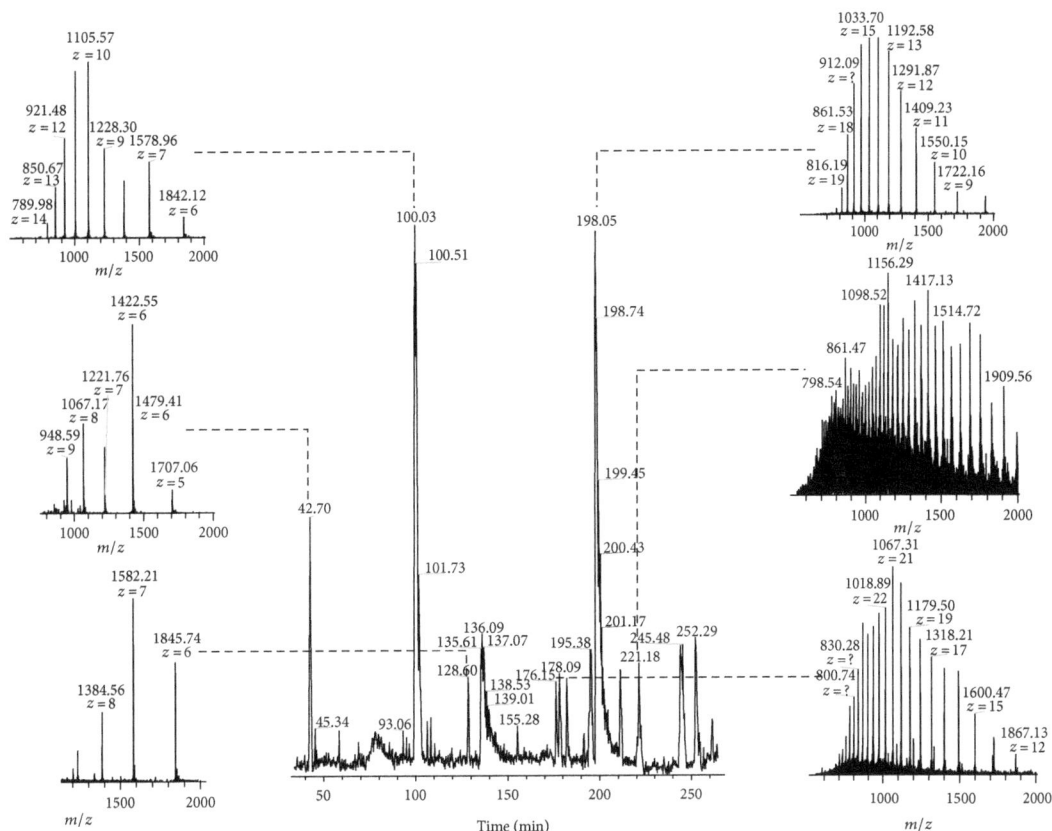

FIGURE 1: Total ion chromatogram (TIC) of an RPLC-MS analysis of intact periplasmic protein from *N. aromaticivorans*. Several representative intact protein spectra are highlighted.

The three criteria were taken from previously recognized signal peptide characteristics [23]. We required a hydrophobic patch of at least eight contiguous amino acids and examined the signal peptide C-terminus for the expected A-X-B cleavage motif (where A = [Ile, Val, Leu, Ala, Gly, Ser, Thr], X = any amino acid, B = [Ala, Gly, Ser]). We also required a basic residue between the start and the hydrophobic patch. Typically the mature protein starts between 15 and 35 residues from the initiator methionine. However, due to the possibility of incorrectly annotated start sites, we allowed for some variance from this requirement.

Subcellular localization and protein functional predictions were made using PSORTb (http://db.psort.org/browse/genome?id=8602), SignalP (http://www.cbs.dtu.dk/services/SignalP/), SecretomeP (http://www.cbs.dtu.dk/services/SecretomeP/), and the Comprehensive Microbial Resource Genome Tools (http://cmr.jcvi.org/tigr-scripts/CMR/shared/Genomes.cgi).

## 3. Results and Discussion

To study the mature proteoforms and PTMs of the *N. aromaticivorans* periplasm, the periplasmic fraction was prepared as previously reported [18]. The enriched intact periplasmic protein fraction was subjected to nano-LC-MS/MS using two different fragmentation methods. Figure 1

shows the base peak chromatogram of a 300-minute LC-MS analysis with several representative intact protein spectra. The detected protein masses varied from 4 kDa to 40 kDa. Intact MS/MS data were analyzed using MS-Align+ [15]. In total, 55 proteins were identified at a 1% FDR (Table 1).

To highlight the specificity and efficiency of the enrichment, we note that abundant cytoplasmic proteins were largely absent in the periplasmic preparation, indicating a low amount of cell lysis during the experiments. For example, none of the ribosomal proteins were detected. Small, highly abundant cytoplasmic proteins such as ribosomal proteins typically dominate in global (whole cell) top-down LC-MS analyses and are often detected in membrane fractions [24, 25]. Several of the proteins identified here were expected to be localized to the periplasm. For example, superoxide dismutase (Saro_0483), a tetratricopeptide repeat protein (Saro_0209), and two peptidyl-prolyl cis-trans isomerases (Saro_0837 and Saro_2251) are known to be localized in the periplasm in other Gram-negative bacteria. Moreover, these proteins were predicted by PSORTb to be periplasmic in *N. aromaticivorans*, and the latter three were enriched in the periplasm of this organism, compared to the cytoplasm, inner membrane, or outer membrane fractions, in a proteomic analysis of multiple subcellular fractions (data not shown). Also, it should be noted that some outer membrane proteins have periplasmic domains. For example, half of

(a)

(b)

(c)

(d)

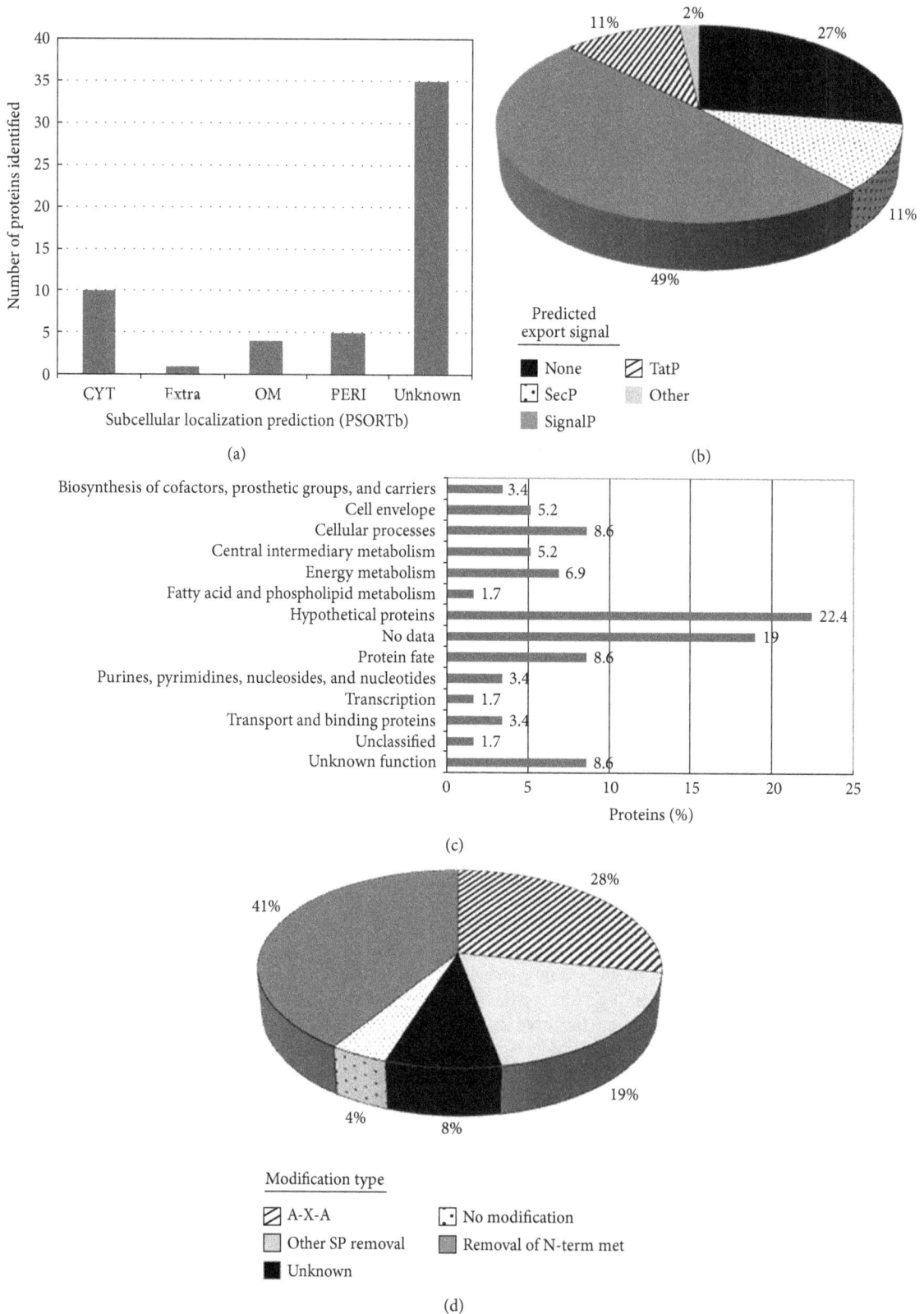

FIGURE 2: (a) Subcellular localization prediction using PSORTb: cytoplasmic (CYT), inner membrane (IM), extracellular (EXTRA), outer membrane (OM), and periplasmic (PERI). (b) Pie chart showing the distribution of predicted export signals among proteins identified by top-down MS. One predicted PERI protein with no detected export signal is designated as "Other." (c) JCVI annotated functional categories of intact proteins identified using top-down analysis. (d) Pie chart representation of protein modifications observed by top-down MS.

Top-Down Characterization of the Post-Translationally Modified Intact Periplasmic Proteome from the Bacterium
Novosphingobium aromaticivorans

89

Hypothetical protein Saro_1194

*E*-value: 1.4 *E*−27; total number of matched peaks: 32

```
MPHAADPHQP STIATSSFGR GWIPAGRDIG LPDDDFPPFN
ANGQADWAGW VSHVAKRSAA EPLSDSGQRP SASAGAPLPF
AGSGIEGSAL PHPNAFTPVR IARFLESLSR CGQVRNAARV
AGVSQQTAYV RRRRDPAFAA GWDAALILAR EAAEQVLAER
ALCGITETIW FRGEAVGERQ RFDGRLLLAH LARLDARVAA
APGAVHHLAE DFDAMLVALA GGEEPAEAVD WPDPARDDHV
EARADAAASA FDHAHPEPED PLDDAAWDAW QSARAAASDA
ARLAAQAEWR AAAQGRDASL ATLLDAPLER KAAGTVVTAA
AAEVGPGSGP QLQRQAECPQ DSVNTVNPAP AAPVPIARGQ
FPPSPHGCRL AAAPESRHAF RPLKGDKIMR RATSTVAIAA
ALAATALASP AVAKPVTLTA SAGAAETGGG DADGVGGFKV
EADDDSGDFC FTLWAEKIAA PTMAHVHEGA AGADGKPVAA
TIEVTGKDSD ACVAMEPELI KKILAAPGDY YVNVHTGDFP
KGAIRGQLQK P
```

FIGURE 3: Fragmentation ion map of the uncharacterized protein Saro_1194 using intact protein MS/MS, indicating that the mature protein contains only the C-terminal portion of the predicted protein sequences starting at residue 414 (the portion of the sequence labeled with grey font was not detected in this experiment).

OmpA (residues 172–325) is periplasmic [26], resulting in its identification in the periplasm here. Subcellular localization predictions of the identified proteins are shown in Figure 2(a). The majority fall into the "unknown" category, making this the first experimental data on subcellular localization for these proteins.

The first and most prevalent type of PTM identified via the top-down approach was proteolytic cleavage. The cleavage events described later were found to be uniformly present; there were no uncleaved forms of the protein detected. We categorized identified proteoforms according to known types of proteolytic maturation. Based on the observed signal peptide cleavage, 25 proteins were localized to the periplasm via Sec-dependent secretion (Figure 2(b)) with detected signal peptide removal. Upstream of the mature protein, the three hallmarks of signal peptides were clearly present: early basic residue(s), a hydrophobic patch of at least 8 residues, and the signal peptidase I cleavage motif. Sixteen proteins were detected with the predominant Ala-Xxx-Ala motif, while 7 of them exhibited tolerated variability at the −3 position [23]. Many of these proteins had poor functional characterization, with 21 lacking any functional annotation (Figure 2(c)) or significant match to protein domain descriptors (e.g., CDD or Pfam; Table 1). Thus, by identifying both their cellular location (periplasm) and their maturation processing, we have significantly added to the annotation of these proteins. We compared the observed signal peptide cleavage to the computational predictions from SignalP4.0 (Figure 2(b)). SignalP correctly predicted 23 of the 25 proteins as containing a signal peptide but did not determine the correct site of cleavage in 6 of the 23 (Table 1) based on top-down analysis. Moreover, SignalP had two false predictions where the identified proteoform lacked a cleaved signal

peptide. Other computational tools were also applied such as TatP and SecP, yielding six more proteins with predicted export signals. Among these six proteins, only two were confirmed with cleaved signal peptides through top-down analysis. Therefore, top-down analysis provided additional information for confident protein categorization, which can be potentially incorporated with currently available software tools to further improve the prediction performance.

We observed that almost all the proteins not exhibiting signal peptide removal had methionine excision (Figure 2(d)) and Table 1). Of the 26 proteins that did not show signal peptide removal or other large N-terminal cleavages, 24 of them began at the second amino acid. The penultimate residue was always consistent with N-terminal methionine excision (NME): alanine, proline, threonine, serine, or glycine [27]. Given the background amino acid frequency and the expected efficiency of methionine amino peptidase [28], the binomial probability of observing such a concentration of NME matured proteins in the periplasm is 4.7 E-6. For comparison, a global top-down analysis of *E. coli* done recently in our lab produced a 1:1 ratio, 69 proteins without methionine excision, and 70 proteins with methionine excision (unpublished results). Additionally, a proteomic and bioinformatic analysis of NME revealed that only a minority of the proteins in a given proteome are subject to NME [29]. The functional significance for pervasive NME in the periplasm is not clear but may be related to protein stability in the potentially hazardous periplasmic environment.

Some of the identified proteins displayed large N-terminal cleavages. For example, the uncharacterized protein Saro_1194 was observed in the data as a mature protein containing only the extreme C-terminal portion of the protein sequence, starting at residue 414, immediately after A-V-A (Figure 3). BLAST analysis showed that the annotated sequence always matched to two separate proteins, well demarcated at the N-terminal and C-terminal extremes of the protein (Figure S1). The C-terminal portion, which was identified from the top-down MS data, also exhibited partial homology to the CHRD domain (pfam07452). Additionally, in two closely related *Erythrobacter* species, the two BLAST hits form a syntenic block in the genome. It is not uncommon for bacteria to combine proteins into multidomain or multifunctional proteins. However, the finding of the mature protein with a perfectly matched and cleaved signal peptide (upstream of the A-V-A is an easily detectable hydrophobic patch and basic residues) suggests that Saro_1194 is actually two separate proteins.

The bacterial periplasm is an oxidizing environment that facilitates disulfide bond formation for correct protein folding and stability [9, 10]. Fifty-three of the 55 proteins identified using our top-down approach contained an even number of cysteines within the detected sequences (i.e., after removal of the signal peptides), including 33 proteins containing no cysteine, 17 proteins containing two cysteines, one protein containing four cysteines, and two proteins containing eight cysteines. Two proteins contain a single cysteine, phosphoribosylformylglycinamidine synthetase PurS, and the uncharacterized protein PhnA. Among the proteins containing two cysteines, only two proteins (arsenate reductase and

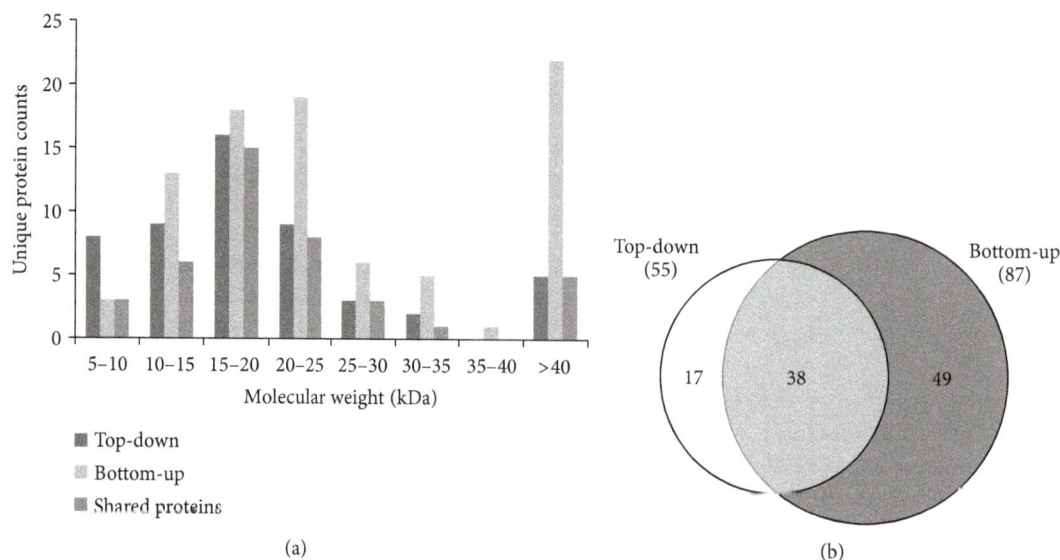

FIGURE 4: (a) Molecular mass distributions of proteins identified using top-down and bottom-up analysis. Theoretical molecular masses were calculated using amino acid sequence. (b) Overlap of proteins identified using top-down and bottom-up analysis (considering proteins identified by at least two unique peptides).

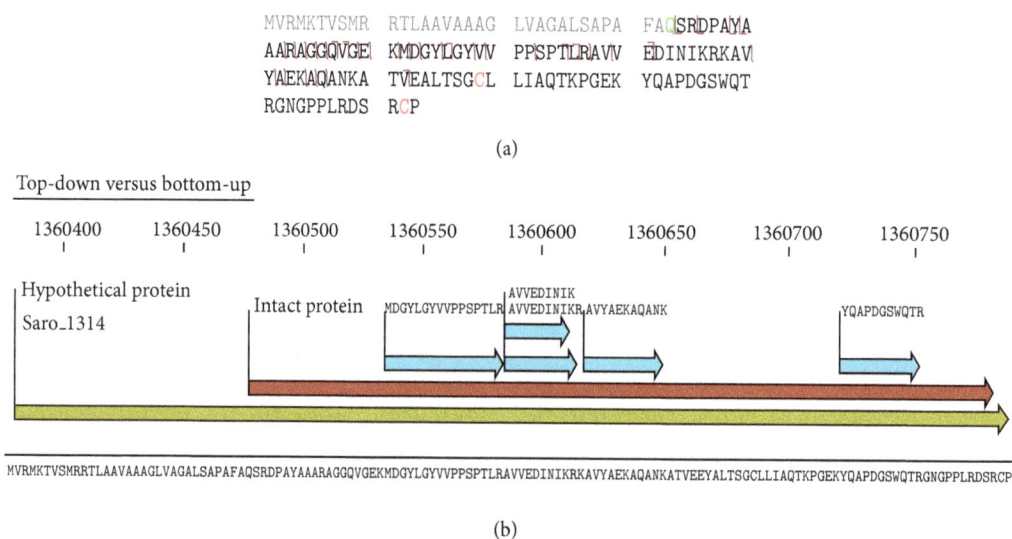

FIGURE 5: Top-down and bottom-up analysis of the hypothetical protein Saro_1314. (a) Fragmentation ion map illustrating high confidence identifications ("Q" highlighted in green font was modified as pyroglutamic acid, and two "C" highlighted in red font formed a disulfide bond). (b) Sequence coverage between top-down approach and bottom-up approach (blue arrows indicate the sequences identified using bottom-up approach).

molybdopterin binding domain, Table 1) did not form a disulfide bond. Instead, both of these proteins contained a glutathionylated adduct (RSSG) on one of the cysteine residues. Although neither of these proteins was detected with signal peptide removal, it has been reported that under certain conditions, 90% of the arsenate reductase activity was found in the periplasmic faction in some bacteria (e.g., *Shewanella* [30]). In other bacteria, several molybdopterin binding proteins (e.g., periplasmic nitrate reductase from *Desulfovibrio desulfuricans* ATCC 27774) were also found in periplasmic fractions [31, 32]. Therefore, these two proteins are likely to be periplasmic proteins, and the observation may indicate the occurrence of cysteine glutathionylation as a form of oxidation in the periplasm other than disulfide bond based oxidation.

Proteoform identifications from top-down also found other PTMs (Table 1). The most common was pyroglutamate, which was very often found on signal peptide cleaved proteins. In nine proteins where the first residue of the mature protein was glutamine, a conversion to pyroglutamate

Top-Down Characterization of the Post-Translationally Modified Intact Periplasmic Proteome from the Bacterium
Novosphingobium aromaticivorans

91

was observed. As mentioned earlier, two proteins were observed with S-glutathiolation, Saro_1778 molybdopterin binding domain protein and Saro_3279 arsenate reductase.

To access the sensitivity and depth of the top-down approach, the same periplasmic enriched protein fraction was analyzed by bottom-up proteomics (Supplementary Table 1 available online at http://dx.doi.org/10.1155/2013/279590). In the bottom-up analysis, 87 proteins were confidently identified with at least two unique peptides. Of these proteins, 37 were also identified in the top-down approach. Fifty proteins were detected only in the bottom-up approach, but most of them have molecular masses larger than 40 kDa (Figure 4(a)), which makes them less amenable to top-down analysis at present. Seventeen proteins were uniquely identified in the top-down approach (Figure 4(b)); eight of these proteins have molecular masses less than 10 kDa. Characterizing small proteins represents a challenge for the bottom-up workflow due to the inability to generate sufficient tryptic peptides for analysis.

We compared the top-down and bottom-up data for their ability to detect mature protein isoforms. The bottom-up data identified 14 proteins with signal peptide cleavage, of which seven were also identified by top-down analysis. Of the remaining seven that were unique to bottom-up, four were large proteins (>45 kDa) and thus largely inaccessible using our current top-down MS platform. We note that for 12 additional proteins, the signal peptide was identified only in the top-down approach, while peptides found in the bottom-up data did not identify a signal peptide cleavage (i.e., none of the peptides captured the mature N-terminus). For example, a hypothetical protein Saro_1314 was confidently identified with a signal peptide removal, a disulfide bond between Cys 99 and Cys 132, and an N-terminal pyroglutamate modification (Figure 5(a)). Only five tryptic peptides were detected for the same protein using the bottom-up approach, and none of them provided evidence for the PTMs (Figure 5(b)). Thus, while the bottom-up approach led to the identification of a larger number of proteins (i.e., a larger survey of periplasmic contents), the top-down analysis provided information on the mature N-terminus and other PTMs.

## 4. Conclusions

Top-down MS analysis of the intact periplasmic fraction of *N. aromaticivorans* indicated the extensive use of sec-dependent signal peptides and disulfide bond formation, as expected for a Gram-negative periplasm. Less expected was the high frequency of NME, which, to our knowledge, has not previously been reported in the bacterial periplasm. Considering these two forms of cleavage and protein maturation, almost all the proteins detected in this study were modified. Moreover, these are cleavage maturation events where no evidence was found of the unmodified protein. Although various modification types were detected, the predominant PTM observed here was proteolysis. Beyond simply showing expression of several "hypothetical" proteins, we have improved the annotation of many genes by providing localization and PTM status, which provides a basis for further functional

annotation of this poorly characterized genus. We propose that top-down MS should be an integral part of efforts towards the characterization of bacterial proteomes in the future.

## Acknowledgments

This research was performed in the W. R. Wiley Environmental Molecular Sciences Laboratory (EMSL, a national scientific user facility sponsored by the US Department of Energy's Office of Biological and Environmental Research and located at Pacific Northwest National Laboratory). Pacific Northwest National Laboratory is operated by Battelle Memorial Institute for the US Department of Energy under Contract DE-AC05-76RLO-1830. Portions of this work were supported by funds from EMSL intramural research projects, and EMSL capability development projects the US Department of Energy's (DOE's) Office of Biological and Environmental Research, and the NIH National Center for Research Resources (Grant RR018522). S. H. Payne was funded by an NSF Grant (EF-0949047). The authors also wish to thank Drs Xiaowen Liu and Pavel Pevzner from UCSD for providing data analysis software MS-Align+.

## References

[1] J. W. Costerton, J. M. Ingram, and K. J. Cheng, "Structure and function of the cell envelope of gram negative bacteria," *Bacteriological Reviews*, vol. 38, no. 1, pp. 87–110, 1974.

[2] J. A. Hobot, E. Carlemalm, W. Villiger, and E. Kellenberger, "Periplasmic gel: new concept resulting from the reinvestigation of bacterial cell envelope ultrastructure by new methods," *Journal of Bacteriology*, vol. 160, no. 1, pp. 143–152, 1984.

[3] M. Ehrmann, *The Periplasm*, ASM press, Washington, DC, USA, 2006.

[4] M. Merdanovic, T. Clausen, M. Kaiser, R. Huber, and M. Ehrmann, "Protein quality control in the bacterial periplasm," *Annual Review of Microbiology*, vol. 65, pp. 149–168, 2011.

[5] J. W. Izard and D. A. Kendall, "Signal peptides: exquisitely designed transport promoters," *Molecular Microbiology*, vol. 13, no. 5, pp. 765–773, 1994.

[6] R. E. Dalbey, M. O. Lively, S. Bron, and J. M. van Dijl, "The chemistry and enzymology of the type I signal peptidases," *Protein Science*, vol. 6, no. 6, pp. 1129–1138, 1997.

[7] D. Missiakas and S. Raina, "Protein folding in the bacterial periplasm," *Journal of Bacteriology*, vol. 179, no. 8, pp. 2465–2471, 1997.

[8] L. M. Stancik, D. M. Stancik, B. Schmidt, D. M. Barnhart, Y. N. Yoncheva, and J. L. Slonczewski, "pH-dependent expression of periplasmic proteins and amino acid catabolism in *Escherichia coli*," *Journal of Bacteriology*, vol. 184, no. 15, pp. 4246–4258, 2002.

[9] H. Kadokura, F. Katzen, and J. Beckwith, "Protein disulfide bond formation in prokaryotes," *Annual Review of Biochemistry*, vol. 72, pp. 111–135, 2003.

[10] H. Kadokura and J. Beckwith, "Mechanisms of oxidative protein folding in the bacterial cell envelope," *Antioxidants and Redox Signaling*, vol. 13, no. 8, pp. 1231–1246, 2010.

[11] Y. Y. Londer, S. E. Giuliani, T. Peppler, and F. R. Collart, "Addressing *Shewanella oneidensis* "cytochromome": the first

step towards high-throughput expression of *cytochromes c*," *Protein Expression and Purification*, vol. 62, no. 1, pp. 128–137, 2008.

[12] M. M. Savitski, M. L. Nielsen, and R. A. Zubarev, "ModifiComb, a new proteomic tool for mapping substoichiometric post-translational modifications, finding novel types of modifications, and fingerprinting complex protein mixtures," *Molecular and Cellular Proteomics*, vol. 5, no. 5, pp. 935–948, 2006.

[13] S. Tanner, S. H. Payne, S. Dasari et al., "Accurate annotation of peptide modifications through unrestrictive database search," *Journal of Proteome Research*, vol. 7, no. 1, pp. 170–181, 2008.

[14] A. Chi, L. Valenzuela, S. Beard et al., "Periplasmic proteins of the extremophile *acidithiobacillus ferrooxidans*: a high throughput proteomics analysis," *Molecular and Cellular Proteomics*, vol. 6, no. 12, pp. 2239–2251, 2007.

[15] X. Liu, Y. Sirotkin, Y. Shen et al., "Protein identification using top-down," *Molecular and Cellular Proteomics*, vol. 11, no. 6, Article ID M111.008524, 2012.

[16] C. M. Ryan, P. Souda, F. Halgand et al., "Confident assignment of intact mass tags to human salivary cystatins using top-down fourier-transform ion cyclotron resonance mass spectrometry," *Journal of the American Society for Mass Spectrometry*, vol. 21, no. 6, pp. 908–917, 2010.

[17] O. Pinyakong, H. Habe, and T. Omori, "The unique aromatic catabolic genes in sphingomonads degrading polycyclic aromatic hydrocarbons (PAHs)," *Journal of General and Applied Microbiology*, vol. 49, no. 1, pp. 1–19, 2003.

[18] R. N. Brown, M. F. Romine, A. A. Schepmoes, R. D. Smith, and M. S. Lipton, "Mapping the subcellular proteome of *Shewanella oneidensis* MR-1 using Sarkosyl-based fractionation and LC-MS/MS protein identification," *Journal of Proteome Research*, vol. 9, no. 9, pp. 4454–4463, 2010.

[19] R. T. Kelly, J. S. Page, Q. Luo et al., "Chemically etched open tubular and monolithic emitters for nanoelectrospray ionization mass spectrometry," *Analytical Chemistry*, vol. 78, no. 22, pp. 7796–7801, 2006.

[20] E. A. Livesay, K. Tang, B. K. Taylor et al., "Fully automated four-column capillary LC-MS system for maximizing throughput in proteomic analyses," *Analytical Chemistry*, vol. 80, no. 1, pp. 294–302, 2008.

[21] S. Kim, N. Gupta, and P. A. Pevzner, "Spectral probabilities and generating functions of tandem mass spectra: a strike against decoy databases," *Journal of Proteome Research*, vol. 7, no. 8, pp. 3354–3363, 2008.

[22] E. Venter, R. D. Smith, and S. H. Payne, "Proteogenomic analysis of bacteria and archaea: a 46 organism case study," *PLoS One*, vol. 6, no. 11, article e27587, 2011.

[23] D. Perlman and H. O. Halvorson, "A putative signal peptidase recognition site and sequence in eukaryotic and prokaryotic signal peptides," *Journal of Molecular Biology*, vol. 167, no. 2, pp. 391–409, 1983.

[24] Y. S. Tsai, A. Scherl, J. L. Shaw et al., "Precursor ion independent algorithm for top-down shotgun proteomics," *Journal of the American Society for Mass Spectrometry*, vol. 20, no. 11, pp. 2154–2166, 2009.

[25] S. M. Patrie, J. T. Ferguson, D. E. Robinson et al., "Top down mass spectrometry of <60-kDa proteins from *Methanosarcina acetivorans* using quadrupole FTMS with automated octopole collisionally activated dissociation," *Molecular and Cellular Proteomics*, vol. 5, no. 1, pp. 14–25, 2006.

[26] E. J. Danoff and K. G. Fleming, "The soluble, periplasmic domain of OmpA folds as an independent unit and displays chaperone activity by reducing the self-association propensity of the unfolded OmpA transmembrane beta-barrel," *Biophysical Chemistry*, vol. 159, no. 1, pp. 194–204, 2011.

[27] C. Flinta, B. Persson, H. Jornvall, and G. von Heijne, "Sequence determinants of cytosolic N-terminal protein processing," *European Journal of Biochemistry*, vol. 154, no. 1, pp. 193–196, 1986.

[28] N. Gupta, J. Benhamida, V. Bhargava et al., "Comparative proteogenomics: combining mass spectrometry and comparative genomics to analyze multiple genomes," *Genome Research*, vol. 18, no. 7, pp. 1133–1142, 2008.

[29] F. Frottin, A. Martinez, P. Peynot et al., "The proteomics of N-terminal methionine cleavage," *Molecular and Cellular Proteomics*, vol. 5, no. 12, pp. 2336–2349, 2006.

[30] D. Malasarn, J. R. Keeffe, and D. K. Newman, "Characterization of the arsenate respiratory reductase from *Shewanella* sp. strain ANA-3," *Journal of Bacteriology*, vol. 190, no. 1, pp. 135–142, 2008.

[31] F. Biaso, B. Burlat, and B. Guigliarelli, "DFT investigation of the molybdenum cofactor in periplasmic nitrate reductases: structure of the Mo(V) EPR-active species," *Inorganic Chemistry*, vol. 51, no. 6, pp. 3409–3419, 2012.

[32] S. Najmudin, P. J. González, J. Trincão et al., "Periplasmic nitrate reductase revisited: a sulfur atom completes the sixth coordination of the catalytic molybdenum," *Journal of Biological Inorganic Chemistry*, vol. 13, no. 5, pp. 737–753, 2008.

# 11

# Miniaturized Mass-Spectrometry-Based Analysis System for Fully Automated Examination of Conditioned Cell Culture Media

Emanuel Weber,[1, 2] Martijn W. H. Pinkse,[1] Eda Bener-Aksam,[1]
Michael J. Vellekoop,[2] and Peter D. E. M. Verhaert[1, 3]

[1] Department of Biotechnology, Netherlands Proteomics Centre, Delft University of Technology, Julianalaan 67,
  2628BC Delft, The Netherlands
[2] Institute of Sensor and Actuator Systems, Vienna University of Technology, Gusshausstrasse 27-29/E366, 1040 Vienna, Austria
[3] Biomedical Research Institute (BIOMED), Hasselt University, Agoralaan building C, 3590 Diepenbeek, Belgium

Correspondence should be addressed to Peter D. E. M. Verhaert, p.d.e.m.verhaert@tudelft.nl

Academic Editor: Paul P. Pevsner

We present a fully automated setup for performing in-line mass spectrometry (MS) analysis of conditioned media in cell cultures, in particular focusing on the peptides therein. The goal is to assess peptides secreted by cells in different culture conditions. The developed system is compatible with MS as analytical technique, as this is one of the most powerful analysis methods for peptide detection and identification. Proof of concept was achieved using the well-known mating-factor signaling in baker's yeast, *Saccharomyces cerevisiae*. Our concept system holds 1 mL of cell culture medium and allows maintaining a yeast culture for, at least, 40 hours with continuous supernatant extraction (and medium replenishing). The device's small dimensions result in reduced costs for reagents and open perspectives towards full integration on-chip. Experimental data that can be obtained are time-resolved peptide profiles in a yeast culture, including information about the appearance of mating-factor-related peptides. We emphasize that the system operates without any manual intervention or pipetting steps, which allows for an improved overall sensitivity compared to non-automated alternatives. MS data confirmed previously reported aspects of the physiology of the yeast-mating process. Moreover, matingfactor breakdown products (as well as evidence for a potentially responsible protease) were found.

## 1. Introduction

In the field of proteomics/peptidomics mass spectrometry has become a well-established tool for protein/peptide sequencing [1–3]. Its steadily increasing performance (sensitivity as well as resolution) enables the analysis of thousands of different molecules at the same time which is of big advantage for "shotgun" approaches, where complex mixtures of unknown samples are targeted for identification. In combination with sophisticated separation methods, protein/peptide analysis has become much faster and more efficient [4–6].

Nevertheless, the whole analysis cycle, starting with peptide extraction from the medium of interest, sample pretreatments (chromatographic purification, digestion) prior to the ultimate injection in the MS instrument requires many time consuming and tedious steps, often done manually. Furthermore, the need for pipetting induces unavoidable sample losses, resulting in a decrease of the overall method sensitivity.

The goal of this work was the design and realization of a system, capable of performing sample extraction, protein/peptide enrichment, purification, and sample preparation for MALDI MS analysis in a fully automated and controlled manner. With the elimination of all previously necessary sample handling steps requiring pipetting, the sensitivity achievable by the system is boosted. Furthermore, using MALDI MS instead of direct connection to an ESI instrument allows for decoupling of the cell cultivation and the actual sample analysis. In that way those two parts can be performed independently from each other, even at different locations.

In addition, sample volumes are kept at a minimum. Reasons to pursue miniaturization include reagent costs. In many studies, different chemicals or additives are needed at certain concentrations to reveal activities of different components in the cell culture. The investment for additives is evidently reduced if the total sample volume is small. Besides these "economy" factors, evolution towards microscale is an essential step to a possible future design of a fully integrated, on-chip analysis system [7]. Once integrated on a single chip, all the advantages of those can be exploited, including (but not limited to) faster analysis cycles, implementation of extrasensing elements (e.g., viability analysis [8]) and on-chip temperature control [9, 10].

As a possible application of this system the analysis of cell-to-cell communication in *Saccharomyces cerevisiae* (baker's yeast) cultures based on peptide secretion was investigated. It is long known that peptides play an important role in cell-to-cell communication in yeast cultures [11]. As best documented example, we selected the mating process as model to evaluate the performance of our novel system. During mating, two yeast cells of opposite haplotype secrete a 13 amino acid pheromone called alpha-mating factor (secreted by alpha-type cells) and a 12 amino acid residue a-mating factor (released by a-haplotypes), respectively. This initiates alpha- and a-haplotype cell fusion to form a diploid cell [12]. In the course of this study the focus was on the detection, accumulation, and analysis of this peptide at different stages during cell culture growth. Therefore, cells were cultivated at small scale (1 mL) while continuously extracting and analyzing the extracellular conditioned medium. As a result a chronological sequence of MS spectra was obtained that could be nicely correlated to the corresponding growth stages. In a second study the effects of an enzyme inhibitor on potential peptidase activity cleaving alpha-mating factor was investigated. This experiment enabled us to collect evidence supporting the hypothesis for the involvement of a yapsin-like protease in an easy and fast way [13].

## 2. Materials and Methods

### 2.1. Strains and Growth Conditions.
A WT *Saccharomyces cerevisiae* strain (CEN.PK 113-1A) mating-type alpha was used [14]. Cells were grown in mineral medium (MM) with addition of glucose (2%, w/v) as sole carbon source [15]. Prior to the transfer into the actual analysis chamber, yeast cells were precultivated in shaker flasks (10 mL MM, 30°C, 200 rpm). Such 24-hour culture has a typical optical density of 19-20 at a measuring wavelength of 600 nm ($OD_{600}$). Dilution to an $OD_{600}$ of 0.1 yielded the initial cell density chosen for all experiments. For every analysis, 1 mL of initial culture was transferred into a custom modified 2 mL glass vial which was prepared for connection to the analysis system.

In the protease inhibition experiments, 10 μM pepstatin (Sigma Aldrich) was added to the culture flask after precultivation.

### 2.2. Miniaturized Cell Culture Chamber.
A modified 2 mL glass vial with cap including septum (Agilent Technologies, USA) was used as a basic module for the miniaturized cell culture chamber. Fused silica tubings (inner diameter 100 μm; BGB Analytik AG, Switzerland) were inserted through the perforated septum, to provide two liquid in- and three liquid outlets (Figure 1). All tubings inside the vial were fitted with a porous glass ending to allow cell culture medium to pass through, while preventing cells to leave the vial and enter the analysis system. Airtight closure of the vial, essential for the functionality of the system, was achieved by deposition of a silicone rubber-based sealant (Bison, Netherlands) on top of the cap. To obtain efficient mixing of the cell suspension the cell culture chamber was equipped with a small magnet and kept on a magnetic stirrer (500 rpm).

Sampling of supernatant from the culture was done by creating an overpressure inside the vial. One of two inlets ("pressure", Figure 1) was connected to a pressurized air system. The second inlet ("MM", Figure 1) was connected to the medium reservoir (via syringe pump 1, SP1, Figure 2) for a constant supply of fresh mineral medium (MM, 0.5 μL/min). In operation only one of the three outlets ("sampling", Figure 1) was opened at a time, with the overpressure inside the chamber resulting in a steady sampling of supernatant.

### 2.3. Automated Setup for In-Line Sampling of Extracellular Medium System Components.
The complete system consists of the cell culture chamber placed on a magnetic stirrer (IKA Labortechnik, Germany), two syringe pumps (Fusion 200; Chemyx, USA), a six-port valve (Rheodyne, USA) controlled by an external interface, three capillary columns packed with 5 μm silica-based C4 beads (300 Å pore size, ReproSil; Dr. Maisch GmbH, Germany) for peptide enrichment/concentration and pressure stabilization, an electrospray unit for sample deposition, a MALDI target plate, and an x-y-z motion controller (MM4005; Newport, USA; Figure 2). The syringe pumps supply solvent for elution (SP2) and MM (SP1), keeping the volume in the vial constant. In the current experiments, no additional glucose or vitamins were supplied via SP1. The MALDI plate was accurately micropositioned by the motion controller. An in-house software program was developed and loaded into the microcontroller of the MM4005 for synchronization with the six-port valve. Real-time determination of optical density was realized with a fiber spectrophotometer (Avaspec-2048) and suitable light source (DH-2000; Avantes, Netherlands). MS analysis was performed directly from the spotted samples in a MALDI Q-TOF mass spectrometer (QTof Premier; Waters, Manchester, UK), equipped with a solid state NdYag laser.

### 2.4. Real-Time Optical Density Measurement inside the Miniaturized Cell Culture Chamber.
The optical density was measured with a fiber spectrophotometer setup especially conceived for use with 2 mL glass vials. Initial linearity of the device at 600 nm was established for ODs between 0.1 up to 1.5. An extended calibration curve was recorded to get

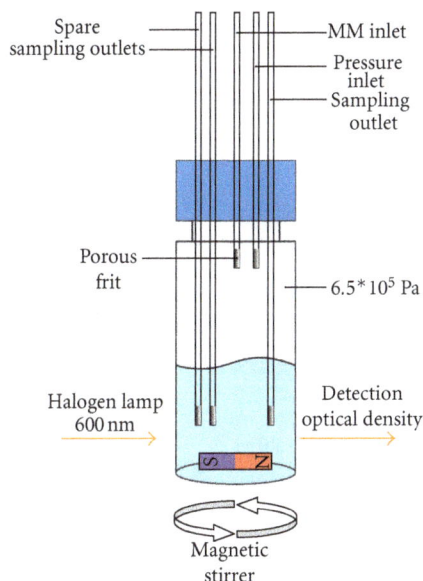

FIGURE 1: Schematic of miniaturized cell culture chamber: a modified 2 mL glass vial. Five fused silica tubings inserted through perforated cap provide liquid in- and outlets. Endings of fused silica tubings are fitted with a porous frit to prevent cells from leaving the vial and contaminating the analysis system. Overpressure of approximately $6.5 * 10^5$ Pa inside chamber operates as pumping system. Efficient mixing of culture is ensured by minimagnet at vial bottom in combination with underlying magnetic stirrer. Light path for OD determination goes straight through vial between magnet and endings of fused silica.

valid data at higher densities as well. After averaging of more than 200 individual measurements for one data point and curve fitting (Matlab, The MathWorks, USA) linearity was obtained for values up to $OD_{600}$ 13, fully covering the range of interest (Figure 3).

*2.5. Sampling Cycle Operation.* Operation of the sampling system was basically divided into two parts: (i) sample accumulation/concentration and (ii) elution. The temporal resolution of the setup in the current configuration is approximately 2 hours. Keeping in mind the life cycle of *S. cerevisiae* (reproduction/cell division each 75–120 minutes [16]) this resolution gives chronological information about the state the whole culture is going through rather than information at the single-cell level. In the accumulation step (90 min) the analysis column C1 was connected directly with the open outlet of the cell culture chamber via valve V1 (Figure 2). The outflow of solution in that time was spotted onto a waste position on the MALDI plate. The flow was adjusted to 0.4–0.6 µL/min resulting in a sampling volume of 36 to 54 µL over the 90 minutes accumulation/concentration period. During this step solvent was pumped through column C2 (Figure 2) connected via the valve to waste. The inclusion of column C2 proved necessary for maintaining constant backpressure inside the system and hence constant (out-) flow. After accumulation, elution followed by switching the valve, which simultaneously

triggered the motion controller to position the first spot on the MALDI plate exactly under the electrospray unit. Solvent (water/acetonitrile/acetic acid; 10 : 90 : 0.6, v/v/v) was pumped through the column C1 at a flow rate of 1 µL/min. Eluates were deposited for 1.5 min per spot (corresponding to 1.5 µL). Ten sample spots were collected in a row to ensure complete elution of the column. Carryover between consecutive analysis runs can be excluded as empty MS spectra (no peptide ion peaks) were obtained for the last sample spots of each series. Extraction of supernatant out of the cell culture chamber continued during the elution step as well. The sample was continuously pushed through a second waste column (C3, Figure 2) for flow stabilization reasons. Both, accumulation and elution step, were repeated up to 11 times, equivalent to more than 19 hours of total analysis time. Throughout the experiment the optical density was measured at a 2 hours interval, and the resulting growth curve was recorded (see e.g., Figure 5).

*2.6. Preparation of Target Plate for MALDI Mass Spectrometry.* Prior to sample spotting, the MALDI plate was ultrasonically cleaned in ammonium bicarbonate solution (10 mM) followed by water/methanol/trifluoroacetic acid (50/50/0.1, v/v/v). Alpha-cyano-4-hydroxycinnamic acid was used as matrix (dissolved at 6 mg/mL in water/acetonitrile/trifluoroacetic acid; 50/50/0.1, v/v/v). After electrospray deposition of the samples, 0.8 µL of matrix solution was added to each spot on the MALDI plate. The plate was analyzed in the MS system, i.c. MALDI Q-TOF after airdrying and complete crystallization of the matrix.

Direct connection of the presented setup to an ESI MS instrument is feasible but requires both parts, cell cultivation as well as sample analysis, to be physically linked which prohibits independent operation and requires all instruments to be placed at the same location.

## 3. Results

*3.1. Detection of S. cerevisiae Mating Factor.* Alpha-pheromone (TrpHisTrpLeuGlnLeuLysProGlyGlnProMetTyr, monoisotopic mass 1682.84 Da) is detected in our MALDI Q-TOF MS as $[M + H]^+$ at *m/z* 1683.85. Associated with this ion often a peak at *m/z* 1699.84 is observed, corresponding to the peptide oxidized at position $Met_{12}$ (a very common posttranslational modification (PTM)). The identity of the peptide could be confirmed by CID of the 1699.84 precursor ion (MS/MS spectrum given in Figure 4).

The alpha-mating-factor-related peptides were detected nearly throughout the whole analysis indicating that alpha factor is expressed and secreted constitutively (also in the absence of opposing-mating-type cells/pheromone). The appearance of alpha-mating factor was more obvious at the late exponential growth phase (corresponding to the diauxic shift, when conditions get less favorable, after nutrient consumption). Two major fragments of the alpha-mating factor (the aminoterminal hexapeptide and the carboxyterminal heptapeptide) were detected in the medium (verified by MS/MS, data not shown) besides the intact pheromone.

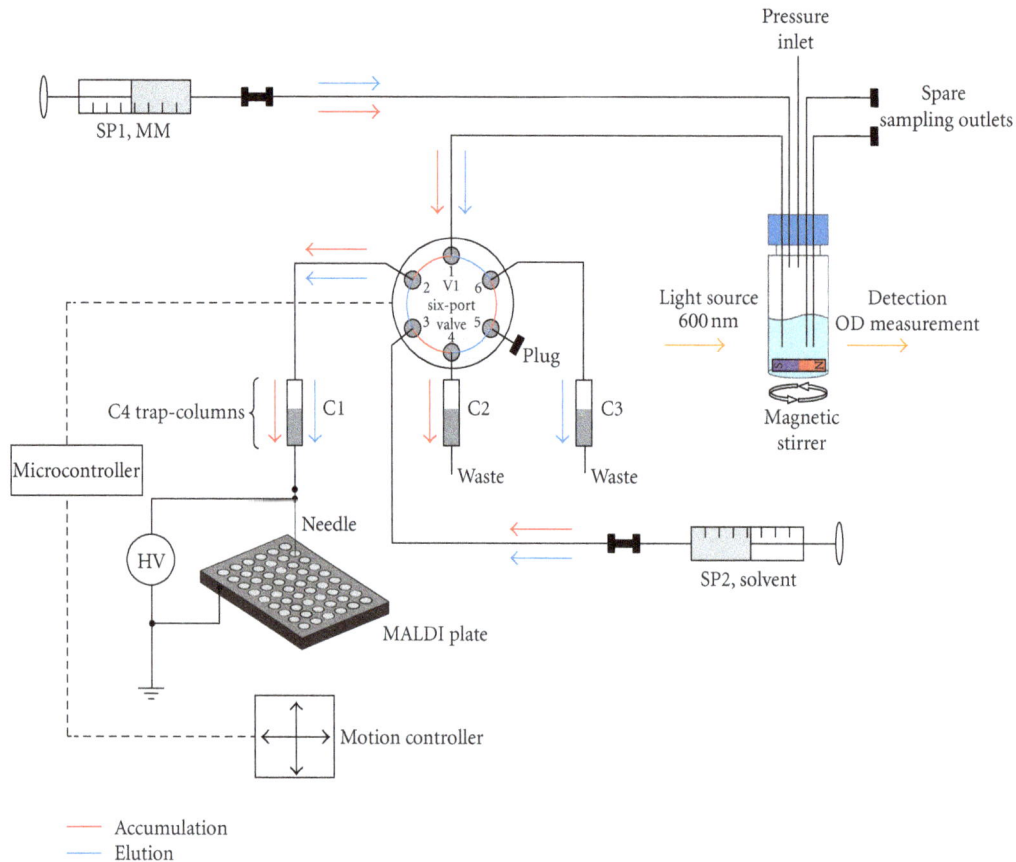

FIGURE 2: Overall analysis setup incorporating two syringes for MM and solvent supply and a switching valve for alternating between concentration/accumulation and elution steps. Cell culture chamber, placed on a magnetic stirrer, is connected via six-port valve with one of three C4 columns, which maintain a stable system backpressure. Via an electrospray needle held at 1.2 kV, sample is deposited onto a MALDI target plate.

These two mating-factor (degradation-) fragments may be products of a protease which cleaves the intact pheromone in two pieces (see Section 3.4 [17]). The fact that their amounts increase in time while intact pheromone decreases at later stages of growth would agree with this.

### 3.2. Real-Time Monitoring of S. cerevisiae Growth in Cell Culture Chamber.

The objective of this study was to develop an automated setup for the analysis of conditioned media of *Saccharomyces cerevisiae* cultures at different growth stages at a miniature scale. To obtain reproducible results and to allow valid comparisons between experiments, it is important to keep the cells at the same physiological/growth state for all experiments. The growth state of the cells was monitored robustly by measuring cell density. For this a fiber spectrophotometer was integrated in our setup, specifically designed for use with 2 mL glass vials. This allowed real-time noninvasive determination of culture ODs. To cover the whole range of ODs a typical yeast culture under the applied conditions goes through (0.1 to 13), a thoroughly elaborated calibration including multiple individual measurements was generated. The real-time recorded growth curves were accurate, as they were in excellent accordance with calibration curves obtained from standard OD determination techniques (diluting the culture to ODs in the linear range of the spectrophotometer and recalculating the actual OD, data not shown). Our analyses confirmed that different batches of yeast cultures show very similar growth behaviors (Figure 5). However, for a meaningful comparison between cultures at various time points/growth stages a perfect match of the two curves is essential. Parameters like small variations in the initial ODs of the inoculated culture or the addition of a test compound may result in slightly delayed or shifted initiation of cell growth. This can be corrected for by software-wise adjustment ("warping") of one of the two curves onto the other (a simple shift along the time axis often being sufficient). The finally obtained diagram shows a perfect match in terms of cell growth of both investigated conditions (Figure 6). All timepoints of both series overlap on a single curve which facilitates a valid comparison.

### 3.3. Time-Resolved Detection of Mating Factor (and Other Peptide-Like Compounds) in Yeast Cell Culture Media under Standard Growth Conditions.

To study the cell culture medium during a standard yeast growth, samples were collected at 2 hours intervals starting at an $OD_{600}$ of approx.

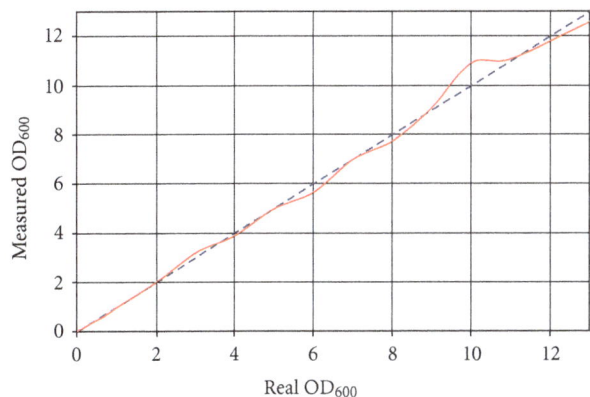

FIGURE 3: Calibration curve for linearization of $OD_{600}$ measurement. Blue dashed line represents optimal behavior. Solid red curve illustrates actually determined $OD_{600}$ after measuring, averaging, and calibration.

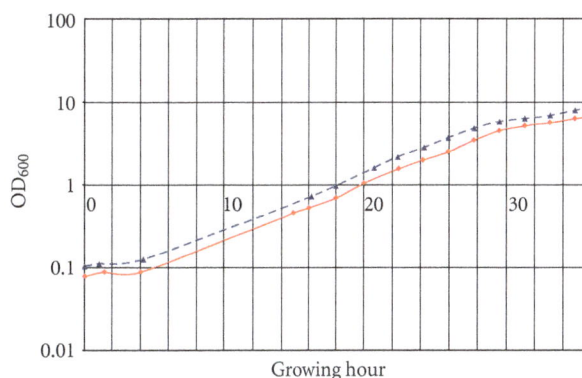

FIGURE 5: Growth curves recorded from two cultures grown under different conditions (blue, dashed curve represents standard growing conditions; red, solid curve with addition of protease inhibitor pepstatin). A small time lag between both curves is evident and makes a comparison based on absolute time points inaccurate.

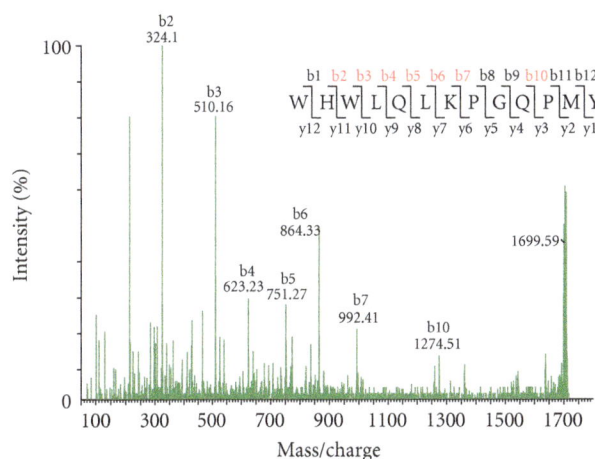

FIGURE 4: MS/MS spectrum of alpha-mating factor (oxidized at $Met_{12}$). Insert: amino-acid sequence with detected sequence ions indicated in red. Note virtually complete b-ions series.

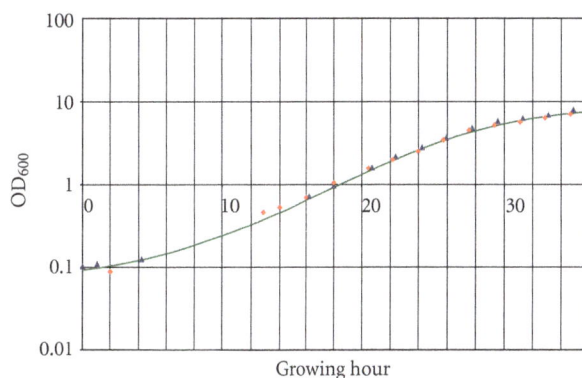

FIGURE 6: After PC-supported adjustment of measurement series of pepstatin-containing culture a perfect matching of both growth curves was obtained. All points lie on fitted sigmoid function (solid, green curve), allowing a valid comparison.

1, which typically is reached 18 to 20 hours after inoculation of the initial culture ($OD_{600}$ of 0.1).

Up to 11 consecutive samples were analyzed by MALDI Q-TOF MS(/MS) yielding chronologically classified MS spectra ("peptide profiles"). Figure 7 shows representative spectra at every second time point sampled (resulting in a difference of 4 hours between each consecutive spectrum displayed). Several different peptide-like signals are evident in the mass spectra acquired. To assist in the interpretation of these profiles, 5 mass over charge ($m/z$) values of interest are highlighted throughout all the spectra, (four between 1520 and 1720 $m/z$ and one at about half the $m/z$). MS/MS analysis confirmed that the peaks at $m/z$ 1683.85, 1699.83, and 1536.75 represent the mating factor, its oxidized version, and a C terminally truncated species (loss of Tyr residue), respectively. The peak in the lower $m/z$ region, at 882.45, represents one of the two mating-factor cleavage products

(the aminoterminal hexapeptide). The complementary peptide fragment (the carboxyterminal heptapeptide) was not readily identified.

One of the most distinct peptide peaks shows up at 1628.74 Da (Figure 7). MS/MS analysis and Mascot database searching (using "no enzyme" as parameter) identified it as the soluble fragment of a cell wall protein; exo-1,3-beta-glucanase (EXG1). EXG1 is known to be involved in cell wall organization by enabling beta-glucan assembly [18]. Literature data [19] and our time-course analysis confirm its presence in the culture medium just before alpha-mating-factor secretion. This suggests that this protein fragment is shed from the membrane just prior to, or simultaneously with, mating-factor release, which may imply a potential role of this protein fragment in mating.

For the time-resolved detection of mating-factor, cells were grown in MM without addition of any special component except those needed for cell growth. It was observed that none of the four mating-factor-related peaks appear early during growth. First unequivocal detection is between hour

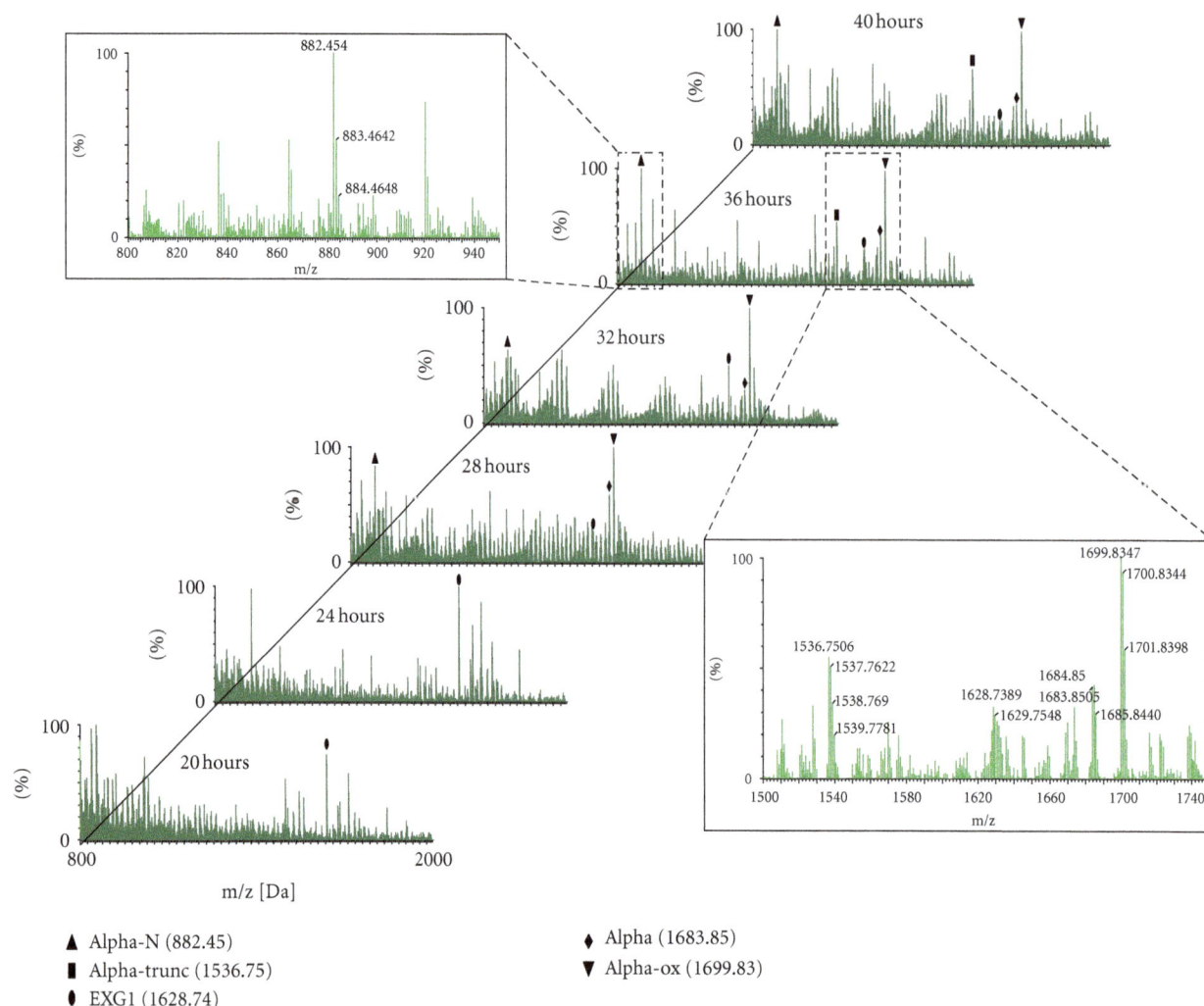

FIGURE 7: MS spectra at 2 hours intervals during standard yeast growth (for figure clarity, every 2nd spectrum is omitted resulting in a 4 hours interval). Marked peaks (legend insert) indicate peptide ions of interest (alpha-N, 882.45 Da: [M + H]$^+$ of aminoterminal mating factor hexapeptide; alpha-trunc, 1536.75 Da: carboxy terminally truncated oxidized mating factor (oxidized alpha factor minus C-terminal tyrosine residue); EXG1, 1628.74 Da: fragment of EXG1 membrane protein; alpha, 1683.85 Da and alpha-ox, 1699.83 Da: alpha-mating factor and its oxidized version). Mating-factor-related peptides do not appear at early time points, whereas they become the most abundant ion peaks at later growth stages, which fits with physiological data. Inserts (top left; bottom right) show magnifications of a spectrum with annotated peaks of interest and their isotopes.

26 and 28, that is, at an OD$_{600}$ of approx. 3.5. Mating-factor concentrations significantly increase in the later growth stages. It is remarkable that both intact mating factor and one of its fragments (the aminoterminal hexapeptide) are detected virtually simultaneously. Multiple repetitions (exceeding 5) confirmed the above-stated behavior. Figure 7 should be understood as an illustrative sequence of MS spectra of a single continuous (40 hours) analysis run.

3.4. Effect of Pepstatin (Inhibition of Aspartic Proteases). To illustrate the usefulness of the miniature culture media analysis system, a simple experiment was designed looking at the effect of a protease inhibitor on the appearance of selected peptide fragments detected above. The proteolytic

cleavage of mating factor at the Leu/Lys peptide bond suggests involvement of an aspartic protease. Hence a general aspartic protease inhibitor was selected to study its effect on the appearance of the peptides/peptide fragments observed [20]. Pepstatin was added to the culture at a concentration of 10 $\mu$M. The anticipated effect reduced appearance of the 881.45 Da fragment ([M + H]$^+$ at m/z 882.45) during the whole experiment. Besides the addition of 10 $\mu$M of pepstatin all conditions were kept strictly the same as for the other experiments. The first appearance of intact mating factor and its oxidized variant was observed at the same time point as for the culture without inhibitor. This indicates that the addition of pepstatin at the chosen concentration had no effect on the actual secretion of alpha factor (MS data not shown). The abundance of the 881.45 Da fragment on the other hand was significantly lower in all spectra (Figure 8, right scale). The red, dashed line in Figure 8 (left scale) depicts the growth curve (in terms of optical density (light absorbance) at 600 nm) at the time of sampling. Data shown represent values from a complete analysis run lasting over 40 hours. Values plotted are representative as biological repetitions of the experiment over shorter time frames showed the very same trend.

## 4. Discussion

*4.1. Alpha-Mating-Factor Profiles at Different Growth Stages.* During the yeast life cycle, mating factor is the trigger for two haploid cells of opposite mating type (alpha and a) to mate and form one single diploid cell. This happens in nature once the growth conditions get unfavorable, for example, lack of nutrition. The strain used in this study is incapable of changing its sex/haplotype [14]. This precludes the formation of diploid cells in the culture flask. However, it is clear that these haploid cells still produce their pheromone in the absence of an opposing mating type or pheromone. The obtained MS spectra in Figure 7 illustrate that the cells "signal for mating" particularly at the later stages of growth, that is, at the end of and after the exponential growth phase. In the early stages of growth no peaks representing the mating factor were identified.

*4.2. Alpha-Mating-Factor Detection in Pepstatin-Containing Cultures.* Comparing the growth curves of the pepstatin containing with those of "standard" cultures confirmed that the growth behavior of the cells was similar in both conditions, justifying a valid time-based peptide profile comparison (Figure 6).

We hypothesized that if mating factor is secreted already at an earlier cell growth stage but readily cleaved by a protease, inhibition or inactivation of this protease could promote the detection of intact mating factor at an earlier time point in the growth curve. This was not observed. The intact mating factor appeared at the same time in both experiments. However, it should be noted that the time resolution of the current setup was 2 hours. Small shifts within this interval may still have been missed.

*4.3. Effect of Pepstatin on Mating-Factor Fragment Appearance.* The comparison of cell cultures with and without pepstatin showed significant differences in terms of the extracellular peptide profiles. The presumed aspartic protease responsible for the formation of the 881.45 Da fragment clearly seems to be inhibited by pepstatin. At all examined time points the ratio between the overall count of the 881.45 Da fragment and that of the intact mating factor (both native (1682.84 Da) and oxidized (1698.83 Da)) was significantly decreased in the culture containing the protease inhibitor (Figure 8). Only at the latest points of inspection, the stationary stage of cell growth, a noticeable count of the 881.45 Da fragment was detected but still far below the intensity level of that in the pepstatin-free culture. Given that the conditions for both cultures were kept identical, the

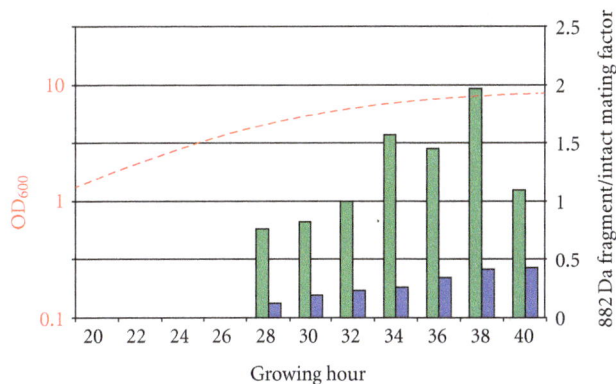

FIGURE 8: Ratio of 881.45 Da fragment to intact mating factor (1682.84 Da plus 1698.83 Da; right scale). No mating factor or fragments are detected in first 4 investigated time points. At later time points ratios between 0.8 and 2 for culture grown under standard conditions were obtained (right scale, green bars). Note significant reduction of relative amount of 881.45 Da fragments in pepstatin-containing culture (blue bars). Dashed, red curve (left, logarithmic scale) plots growth curve of cells.

disappearance or drastic reduction of the fragment in the extracellular medium is clearly to be attributed primarily to the addition of pepstatin. This suggests that pepstatin inhibits the potential protease responsible for "normal" alpha-mating-factor peptide cutting.

*4.4. Additional Mating-Factor-Related Peptide Ions.* In the obtained MS spectra additional mating-factor-related ions were identified. The detection of oxidized mating factor missing the tyrosine at the C-terminus at the later growth stages ($m/z$ 1536.75; Figure 7) may reflect the action of an (carboxyterminal) exopeptidase in the extracellular medium. Concurrently, the amount of intact mating factor inside the culture decreases.

*4.5. Other Nonmating Factor-Related Peptide Ions.* Besides ions related to mating factor, another peptide possibly involved in the secretion process was identified ($m/z$ 1628.74; Figure 7). Database searching identified it as the soluble part of the exo-b-1,3-glucanase EXG1 (a cell wall protein). The role of this protein in peptide secretion remains elusive, but comparing its appearance in both experiments showed marked differences. In standard cultures, the peptide was found most abundant prior to secretion of mating factor and slightly reduced at later growth stage. In-pepstatin containing cultures this peptide was not found prior to mating peptide secretion, and it appeared considerably less prominently present at later growth stages as well (data not shown).

## 5. Conclusion

The presented automated system allows in-line sampling of microliter amounts of extracellular conditioned cell culture media, preparing them for MALDI MS analysis. The minimal amount of cell culture required for this has advantages in terms of handling and cost reduction. For example, compared to standard flask cultivation, an enzyme inhibition study during cell growth could be completed with 10 times less amount of the commercial compound (i.c. pepstatin). In particular when effects on cultured cells of more expensive compounds have to be tested, the experiment cost savings related to the reduced culture chamber volume will become more substantial.

Implementation of the real-time optical density measurement in-line (without disturbing the cell culture) made many tedious extra sample collection and dilution steps redundant and resulted in an overall increase of the practicability of the system.

In summary, we have realized a fully automated setup which eliminates all manual pipetting interventions. This reduces the risk for losses of peptides sticking to microtip or tubing/column wall materials, which often drastically reduces the overall sensitivity of the analysis.

## Acknowledgments

This project was financed by European Marie Curie Research Training Network "CellCheck", Grant no. MCRTN-CT-2006-035854. Also The Netherlands Proteomics Center (Project T3.1b) and The Netherlands Genomics Initiative are gratefully acknowledged for their support.

## References

[1] M. A. Baldwin, "Protein identification by mass spectrometry: issues to be considered," *Molecular and Cellular Proteomics*, vol. 3, no. 1, pp. 1–9, 2004.

[2] T. Nilsson, M. Mann, R. Aebersold, J. R. Yates Jr., A. Bairoch, and J. J. M. Bergeron, "Mass spectrometry in high-throughput proteomics: ready for the big time," *Nature Methods*, vol. 7, no. 9, pp. 681–685, 2010.

[3] B. Ma and R. Johnson, "De novo sequencing and homology searching," *Molecular & Cellular Proteomics*, vol. 11, no. 2, Article ID O111.014902, 2012.

[4] J. R. Wiśniewski, A. Zougman, N. Nagaraj, and M. Mann, "Universal sample preparation method for proteome analysis," *Nature Methods*, vol. 6, no. 5, pp. 359–362, 2009.

[5] H. D. Meiring, E. van der Heeft, G. J. ten Hove, and A. de Jong, "Nanoscale LC-MS(n): technical design and applications to peptide and protein analysis," *Journal of Separation Science*, vol. 25, no. 9, pp. 557–568, 2002.

[6] P. M. Van Midwoud, L. Rieux, R. Bischoff, E. Verpoorte, and H. A. G. Niederländer, "Improvement of recovery and repeatability in liquid chromatography-mass spectrometry analysis of peptides," *Journal of Proteome Research*, vol. 6, no. 2, pp. 781–791, 2007.

[7] M. Stangegaard, S. Petronis, A. M. Jørgensen, C. B. V. Christensen, and M. Dufva, "A biocompatible micro cell culture chamber (μCCC) for the culturing and on-line monitoring of eukaryote cells," *Lab on a Chip*, vol. 6, no. 8, pp. 1045–1051, 2006.

[8] E. Weber, M. Rosenauer, W. Buchegger, P. D. E. M. Verhaert, and M. J. Vellekoop, "Fluorescence based on-chip cell analysis applying standard viability kits," in *Proceedings of the 15th International Conference on Miniaturized Systems for Chemistry and Life Science (microTAS '11)*, pp. 1716–1718, Seattle, Wash, USA, 2011.

[9] A. Jain and K. E. Goodson, "Thermal microdevices for biological and biomedical applications," *Journal of Thermal Biology*, vol. 36, no. 4, pp. 209–218, 2011.

[10] S. Petronis, M. Stangegaard, C. B. V. Christensen, and M. Dufva, "Transparent polymeric cell culture chip with integrated temperature control and uniform media perfusion," *BioTechniques*, vol. 40, no. 3, pp. 368–376, 2006.

[11] T. Tanaka, H. Kita, T. Murakami, and K. Narita, "Purification and amino acid sequence of mating factor from *Saccharomyces cerevisiae*," *Journal of Biochemistry*, vol. 82, no. 6, pp. 1681–1687, 1977.

[12] Y. Wang and H. G. Dohlman, "Pheromone signaling mechanisms in yeast: a prototypical sex machine," *Science*, vol. 306, no. 5701, pp. 1508–1509, 2004.

[13] D. J. Krysan, E. L. Ting, C. Abeijon, L. Kroos, and R. S. Fuller, "Yapsins are a family of aspartyl proteases required for cell wall integrity in *Saccharomyces cerevisiae*," *Eukaryotic Cell*, vol. 4, no. 8, pp. 1364–1374, 2005.

[14] J. P. Van Dijken, J. Bauer, L. Brambilla et al., "An interlaboratory comparison of physiological and genetic properties of four *Saccharomyces cerevisiae* strains," *Enzyme and Microbial Technology*, vol. 26, no. 9-10, pp. 706–714, 2000.

[15] C. Verduyn, E. Postma, W. A. Scheffers, and J. P. Van Dijken, "Effect of benzoic acid on metabolic fluxes in yeasts: a continuous-culture study on the regulation of respiration and alcoholic fermentation," *Yeast*, vol. 8, no. 7, pp. 501–517, 1992.

[16] T. Boekhout and V. Robert, *Yeasts in Food: Beneficial Detrimental Aspects*, Behr's, Hamburg, Germany, 2003.

[17] V. L. MacKay, S. K. Welch, M. Y. Insley et al., "The *Saccharomyces cerevisiae* BAR1 gene encodes an exported protein with homology to pepsin," *Proceedings of the National Academy of Sciences of the United States of America*, vol. 85, no. 1, pp. 55–59, 1988.

[18] A. R. Nebreda, T. G. Villa, J. R. Villanueva, and F. Del Rey, "Cloning of genes related to exo-β-glucanase production in *Saccharomyces cerevisiae*: characterization of an exo-β-glucanase structural gene," *Gene*, vol. 47, no. 2-3, pp. 245–259, 1986.

[19] C. Cappellaro, V. Mrsa, and W. Tanner, "New potential cell wall glucanases of *Saccharomyces cerevisiae* and their involvement in mating," *Journal of Bacteriology*, vol. 180, no. 19, pp. 5030–5037, 1998.

[20] A. V. Azaryan, M. Wong, T. C. Friedman et al., "Purification and characterization of a paired basic residue-specific yeast aspartic protease encoded by the YAP3 gene. Similarity to the mammalian pro- opiomelanocortin-converting enzyme," *Journal of Biological Chemistry*, vol. 268, no. 16, pp. 11968–11975, 1993.

# An Internal Standard-Assisted Synthesis and Degradation Proteomic Approach Reveals the Potential Linkage between VPS4B Depletion and Activation of Fatty Acid β-Oxidation in Breast Cancer Cells

Zhongping Liao,[1] Stefani N. Thomas,[2] Yunhu Wan,[3] H. Helen Lin,[4]
David K. Ann,[4] and Austin J. Yang[1,5]

[1] Greenebaum Cancer Center, University of Maryland School of Medicine, Baltimore, MD 21201, USA
[2] Department of Pharmacology and Molecular Sciences, Johns Hopkins University School of Medicine, Baltimore, MD 21205, USA
[3] Department of Epidemiology and Public Health, University of Maryland School of Medicine, Baltimore, MD 21201, USA
[4] Department of Molecular Pharmacology, Beckman Research Institute, City of Hope Medical Center, Duarte, CA 91010, USA
[5] Department of Anatomy and Neurobiology, University of Maryland School of Medicine, Baltimore, MD 21201, USA

Correspondence should be addressed to Austin J. Yang; ayang@som.umaryland.edu

Academic Editor: Bomie Han

The endosomal/lysosomal system, in particular the endosomal sorting complexes required for transport (ESCRTs), plays an essential role in regulating the trafficking and destination of endocytosed receptors and their associated signaling molecules. Recently, we have shown that dysfunction and down-regulation of vacuolar protein sorting 4B (VPS4B), an ESCRT-III associated protein, under hypoxic conditions can lead to the abnormal accumulation of epidermal growth factor receptor (EGFR) and aberrant EGFR signaling in breast cancer. However, the pathophysiological consequences of VPS4B dysfunction remain largely elusive. In this study, we used an internal standard-assisted synthesis and degradation mass spectrometry (iSDMS) method, which permits the direct measurement of protein synthesis, degradation and protein dynamic expression, to address the effects of VPS4B dysfunction in altering EGF-mediated protein expression. Our initial results indicate that VPS4B down-regulation decreases the expression of many proteins involved in glycolytic pathways, while increased the expression of proteins with roles in mitochondrial fatty acid β-oxidation were up-regulated in VPS4B-depleted cells. This observation is also consistent with our previous finding that hypoxia can induce VPS4B down-regulated, suggesting that the adoption of fatty acid β-oxidation could potentially serve as an alternative energy source and survival mechanism for breast cancer cells in response to hypoxia-mediated VPS4B dysfunction.

## 1. Introduction

VPS4B, a member of the AAA (ATPases associated with diverse cellular activities) protein family, plays an important role in the lysosomal degradation pathway, which functions in ligand-induced membrane receptor downregulation. In lysosomal degradation, endocytosed receptors are sorted into multivesicular bodies (MVBs), which requires the sequential assembly of endosomal sorting complex required for transport I, II, and III (ESCRT-I, -II, and -III) on the endosomal membrane [1]. VPS4B functions to dissociate

ESCRT-III from endosomes for further rounds of endosomal sorting [2]. Expression of functionally inactive VPS4B results in the accumulation of receptors on abnormally enlarged endosomes (class E compartments) and the abolishment of MVB biogenesis [3–6]. As an essential ESCRT-III interacting protein, VPS4B is also involved in protein trafficking among membrane compartments and in signal transduction [4, 5, 7, 8].

Epidermal growth factor- (EGF-) mediated signaling is one of the most important signaling pathways for cell growth, proliferation, invasion, and metastasis in breast cancer [9].

Upon ligand binding, the EGF receptor (EGFR) is phosphorylated, promoting downstream signal transduction, while it is internalized and degraded within the MVB-lysosome [10]. EGFR signaling is initiated at the cell membrane; however, late signaling propagation occurs on the endosomes [11]. We and others have reported that dysfunction of VPS4B not only leads to delayed EGFR degradation, but also to prolonged and altered intracellular EGFR signaling in abnormally enlarged endosomal compartments [4–6, 12]. To further understand the role of VPS4B dysfunction in breast cancer, we set out to determine the consequences of altered EGFR signaling: changes in protein synthesis and degradation, as well as protein dynamics in VPS4B downregulated breast cancer cells. Altered protein synthesis and degradation of cancer-related proteins is involved in cellular transformation and cancer progression [13–18]. Understanding protein synthesis and degradation is essential to fully appreciate cellular dynamics and develop more effective strategies to treat cancer.

Traditional protein synthesis and degradation studies rely on radiolabel tracer labeling (pulse) followed by exposure to nonradioactive medium (chase). However, these approaches are often designed to address the turnover of a specific protein. Recently, high throughput synthesis and degradation mass spectrometry (SDMS) has become a novel approach to study global protein turnover, in which protein degradation and synthesis can be measured simultaneously and unambiguously with minimal cell perturbation [19]. In this approach, proteins are first metabolically labeled by stable isotope labeling with amino acids in cell culture (SILAC) followed by chasing in normal medium. As a result, the preexisting proteins are labeled with stable isotope amino acids, while the newly synthesized proteins only incorporate regular amino acids. These differential stable isotope-labeled proteins can be discriminated from each other and quantified by mass spectrometry based on the differences in masses of their peptides following enzymatic digestion. Since Pratt and coworkers introduced the use of mass spectrometry in protein synthesis and degradation studies [20]; this approach has been applied to numerous systems, such as *Escherichia coli* [21], *Mycobacterium tuberculosis* [22], *Mycoplasma pneumoniae* [23], *Streptomyces coelicolor* [24], *Saccharomyces cerevisiae* [20], HeLa cells [25, 26], human adenocarcinoma cells [27], chicken skeletal muscle [28, 29], and mice [30, 31].

One of the major assumptions of using a pulse-chase labeling experiment to study the dynamics of protein expression is that the level of protein expression is always under the steady state. In SILAC-based protein synthesis and degradation experiments, the degradation rate constant of targeted proteins can be readily derived from the decrease of relative isotope abundance (RIA) or the percentage of preexisting labeled protein [20]. On the contrary, it is more difficult to calculate the rate of protein synthesis in general since proteins are synthesized continuously in cultured proliferating cells, thus the rate of protein accumulation is not likely to be under steady state. In order to bypass such steady state assumption, the incorporation of isobaric tags for relative and absolute quantitation (iTRAQ) [24] or the introduction of an internal control, such as another set of labeled cells [26], permits the simultaneous measurement of protein synthesis,

degradation, and expression under dynamic conditions, such as cell proliferation.

Dysfunction of VPS4B leads to EGFR accumulation and prolonged activation [4–6, 12], which could contribute to cell proliferation and growth. To estimate protein synthesis and degradation rates in the context of altered EGFR signaling caused by downregulation of VPS4B, we independently developed an approach called internal standard-assisted synthesis and degradation mass spectrometry (iSDMS) that normalizes protein abundance across different samples and time points and permits the comparison of protein synthesis, degradation, and expression measurements. A similar approach has also been developed recently by Boisvert et al. to study global protein turnover in HeLa cells [26]. Through this iSDMS analysis, we compared the synthesis and degradation rates of more than 700 proteins between VPS4B downregulated SKBR3 cells and the parental SKBR3 cells. We found that VPS4B downregulation resulted in differential protein expression in energy metabolism pathways by altering the synthesis and degradation of related proteins, in which glycolysis proteins were downregulated, while mitochondrial fatty acid $\beta$-oxidation proteins were upregulated. The adoption of fatty acid $\beta$-oxidation as an alternative energy source could be an unrevealed survival mechanism for breast cancer cells with VPS4B dysfunction.

## 2. Materials and Methods

*2.1. Cell Culture.* SKBR3 cells were obtained from the American Type Culture Collection (Rockville, MD). VPS4B knockdown SKBR3 (SKBR3_shVPS4B) cells were generated by transducing SKBR3 cells with a lentivirus harboring shRNA against VPS4B (SA Biosciences/Qiagen, Frederick, MD) as previously described [32]. To study the role of VPS4B downregulation on the dynamics of protein expression by iSDMS, SKBR3_shVPS4B, and the parental SKBR3 cells were cultured in DMEM SILAC medium (Pierce/Thermo Scientific) supplemented with 28 mg/L $^{13}C_6$-arginine (Arg6, purity 97–99%, Cambridge Isotope Laboratories), 72 mg/L $D_4$-lysine (Lys4, purity 96–98%, Cambridge Isotope Laboratories), 10% dialyzed fetal bovine serum (FBS, Invitrogen), and 1% antibiotic antimycotic solution (Invitrogen). For SKBR3_shVPS4B cells, 2 $\mu$g/mL puromycin were added to the culture medium to maintain the selection pressure. After five passages, cells were starved in serum free SILAC medium for 18 hr. Following three washes with Dulbecco's Phosphate-buffered Saline (DPBS, Invitrogen), cells were stimulated with 100 ng/mL EGF and subsequently "chased" for 0, 2, 6, 12, and 24 hr in the presence of DMEM "light" SILAC medium (Pierce/Thermo Scientific) supplemented with light arginine (Arg0), lysine (Lys0) (Sigma), 10% dialyzed FBS, and 1% antibiotic antimycotic solution. At the end of each chase time period, cell pellets were collected and frozen in liquid nitrogen and stored in $-80°C$ until analysis. To prepare the "heavy-" labeled internal standard, SKBR3_shVPS4B cells were metabolically labeled with $^{13}C_6^{15}N_4$-arginine (Arg10) and $^{13}C_6^{15}N_2$-lysine (Lys8) by culturing in DMEM "heavy" SILAC medium supplemented with 28 mg/L $^{13}C_6^{15}N_4$-arginine (Arg10, purity 97–99%, Cambridge Isotope Laboratories), 72 mg/L $^{13}C_6^{15}N_4$-lysine (Lys8, purity 97–99%, Cambridge Isotope Laboratories), 10% dialyzed FBS, and 1% antibiotic antimycotic solution (Invitrogen). The incorporation of Arg10 and Lys8 in the internal standard SKBR3_shVPS4B cells was ~98% as determined by mass spectrometric analysis of tryptic peptides isolated from the Arg10 and Lys8 labeled SKBR3_shVPS4B cells. Cells were harvested and stored as described above.

### 2.2. Preparation of Cell Lysates and Enzymatic Digestion.
Cell pellets were lysed and sonicated in 4% SDS, 100 mM Tris-HCl (pH 7.6) lysis buffer containing 5 U/mL benzonase nuclease (Novagen) and complete EDTA-free protease inhibitor cocktail (Roche). After removal of cell debris by centrifugation at 10,000 ×g at room temperature for 10 min, protein concentration was measured in triplicate using the BCA protein assay kit (Thermo Scientific/Pierce). 180 μg of protein from each time point were mixed with 60 μg of protein internal standard labeled with $^{13}C_6^{15}N_4$-arginine10 and $^{13}C_6^{15}N_2$-lysine8. Dithiothreitol (DTT, Sigma) and Tris (2-carboxyethyl) phosphine (TCEP, Thermo Scientific/Pierce) were added to the mixture to final concentrations of 100 mM and 10 mM, respectively. Mixtures were kept at 90°C for 10 min. After cooling to room temperature, the cell lysates were processed by the Filter Aided Sample Preparation (FASP) procedure [33] using 30 k VIVACON 500 filtration units (Sartorius Biolab) with modifications. Briefly, cell lysates were first reduced by mixing with 200 μL of UA buffer (8 M urea in 100 mM Tris-HCl pH 8.5) containing 5 mM TCEP, loaded into the filtration units, and centrifuged at 14,000 ×g for 15 min. At the end of the reduction reaction, the samples were washed three times in 200 μL of UA buffer followed by centrifugation at 14,000 ×g for 15 min. Alkylation was performed by adding 100 μL of 50 mM iodoacetamide in UA buffer and incubating at room temperature for 30 min. Excess iodoacetamide was eliminated by centrifugation, followed by three washes with 200 μL of UB buffer (8 M urea in 100 mM Tris-HCl, pH 8.0).

At the end of reductive alkylation reaction, samples were digested by sequential addition of proteases Lys-C and trypsin in the filtration units. Briefly, 2 μg of endoproteinase Lys-C (Roche) in 40 μL water were first added to the filtration units and kept at room temperature in a humidified chamber overnight for 16 hr. After Lys-C digestion, samples were diluted with 200 μL of 50 mM NH$_4$HCO$_3$, and 4 μg of trypsin (Promega) were added to the reaction mixture and incubated at 37°C for another 6–8 hr. After digestion, the peptides were collected by centrifugation of the filtration units, followed by two washes with 50 μL of 50 mM NH$_4$HCO$_3$. All peptides containing filtrates were pooled and acidified by the addition of trifluoroacetic acid (TFA, Sigma) to a final concentration of 1%. Acidified peptides were desalted by a Sep-Pak C18 cartridge (Waters) and dried by SpeedVac (Thermo Scientific). Dried peptides were stored in a −80°C freezer until analysis.

### 2.3. Peptide Fractionation.
Peptides were fractionated using stop and go extraction (STAGE) tips made in house with strong anion exchange (SAX) disks as described in [34–36]. The STAGE tips were assembled by stacking six layers of Empore anion exchange membrane disks (3 M) into a 200 μL micropipet tip. Two stacked tips were used to separate 100 μg of peptides. Britton and Robinson buffer (BR buffer), composed of 20 mM acetic acid, 20 mM phosphoric acid, and 20 mM boric acid and titrated with NaOH to the desired pH, was used in peptide separation. The tip was wet with methanol and washed with 1 M NaOH, followed by equilibration with 100 μL of BR buffer pH 11. Peptides were loaded at pH 11, and fractions were subsequently eluted with BR buffer of pH 11, 8, 6, 5, 4, and 3, respectively. The flow through and all the fractions were acidified by adding TFA to a final concentration of 1% and desalted by a STAGE tip containing three layers of C18 membrane disks (3 M). Peptides were then dried and stored as mentioned above.

### 2.4. LC-MS/MS Analysis.
Peptides were dissolved in 0.5% acetic acid (solvent A) and separated on a 10 cm reverse-phase PicoFrit spray tip (New Objective, Woburn, MA) packed in house with sub-2 μm C18 resin (Prospereon Life Science, IL), using a nanoflow Xtreme simple liquid chromatography system (Microtech/CVC) coupled to a hybrid linear ion trap-Orbitrap mass spectrometer (LTQ Orbitrap, Thermo Scientific). Peptides were loaded onto the column with solvent A at a flow rate of 0.6 μL/min and eluted with a 180 min linear gradient at a flow rate of 0.2 μL/min. A gradient of 2–60% solvent B (40% acetonitrile, 0.5% acetic acid) was applied to the SAX flow through, fractions eluted with pH 11 and pH 8 buffer, a 2–65% solvent B gradient was used for the pH 6 and pH 5 fractions, and a 5–70% solvent B gradient was used for the pH 4 and pH 3 fractions. After the linear gradient, the column was washed with 95% solvent B and reequilibrated with 95% solvent A. Seven fractions from one sample were run sequentially followed by a 40 min wash with 80% acetonitrile with 0.5% acetic acid.

Mass spectra were acquired in the positive ion mode applying a data-dependent automatic switch between the survey scan and MS/MS acquisition. The survey MS1 scans were acquired in the Orbitrap using a mass range of $m/z$ 400–1,600 at a resolution of 60,000 at 400 $m/z$. The MS1 target value was 100,000. MS/MS scans were acquired in the linear ion trap on the 5 most intense ions in each survey scan with dynamic exclusion of previously selected ions; repeat count 1 and exclusion duration 15 seconds. The fragmentation was performed by collision-induced dissociation (normalized collision energy 35%) with a target value of 30,000. Ion selection threshold was 5,000 counts. Charge state screening was enabled, and +1 ions were excluded from fragmentation. Other mass spectrometric parameters that were spray voltage 1.35 kV; no sheath and auxiliary gas flow; ion transfer tube temperature 180°C; activation $q$ = 0.25; activation time of 30 ms were applied in MS2 acquisitions.

### 2.5. Peptide Identification.
Data were searched against a concatenated forward/reverse database, which was built based on a UniProt human database (downloaded on Oct. 18, 2010), using SEQUEST (Bioworks 3.3.1 SP1, Thermo Scientific). Enzyme specificity was set to trypsin and allowed

two missed cleavages. Database search parameters were precursor peptide mass tolerance 50 ppm and fragment ion tolerance 1 amu. Carbamidomethyl cysteine was set as a fixed modification, and oxidized methionine was set as a variable modification. Additional variable modifications included arginine +6.02013, arginine +10.00827, lysine +4.02510, and lysine +8.01420. The peptides were initially filtered by the following criteria: $X_{corr} \geq 2.5$, 3.0, and 3.5 for 2+, 3+, and 4+ peptides, respectively, and $\Delta Cn > 0.1$. The false discovery rate (FDR) was calculated by dividing the number of false positive peptides identified in the reverse database by the number of total identified peptides [37]. In this study, the FDR was 0.94%.

*2.6. Calculation of Peptide Relative Abundance.* By adding the internal standard, three populations of a given peptide were present in the mixture: the peptide containing regular arginine (Arg0) or lysine (Lys0) (light-labeled peptide), the peptide containing $^{13}C_6$-arginine (Arg6) or $D_4$-lysine (Lys4) (medium-labeled peptide), and the peptide containing $^{13}C_6^{15}N_4$-arginine (Arg10) or $^{13}C_6^{15}N_2$-lysine (Lys8) (heavy-labeled internal standard peptide).

The relative abundance of light peptide ($A_l$) was defined as the ratio of the peak intensities of unlabeled peptide ($I_l$) to the intensities of internal standard peptide ($I_h$), (1). Similarly, the relative abundance of labeled peptide ($A_m$) was defined as the ratio of the peak intensities of labeled peptide ($I_m$) to $I_h$, (2). An in-house SILAC-based mass spectrometry quantitation software, IsoQuant (http://www.proteomeumb.org/MZw.html) [38], was used to automatically integrate the isotopic peak intensities of each peptide and calculate $A_l$ and $A_m$. The relative abundance of total peptide ($A_{tot}$) was calculated by summing the relative abundance of light unlabeled peptide ($A_l$) and labeled peptide ($A_m$), (3)

$$A_l = \frac{I_l}{I_h}, \tag{1}$$

$$A_m = \frac{I_m}{I_h}, \tag{2}$$

$$A_{tot} = A_l + A_m. \tag{3}$$

*2.7. Calculation of Peptide Degradation Rate Constant.* Since the SILAC labeled peptides were only subjected to degradation, the degradation rate constant could be derived from the changes in relative abundance of labeled peptides. Since protein degradation rate follows first-order kinetics [39], the relative abundances of the preexisting labeled peptides over time ($A_{m\_t}$) were fit to exponential decay curves to derive the degradation rate constant ($\lambda$). At least four time points were required to find the best fitting curve. Only $A_{m\_t}$ and $t$ that were significantly correlated ($P < 0.05$) with a coefficient of determination ($R^2$) greater than 0.8 were selected to derive the degradation rate constant from the exponential decay equation as follows:

$$A_{m\_t} = a \times e^{-\lambda t}, \tag{4}$$

where $a$ was the corrected-normalized initial peptide amount.

*2.8. Calculation of Peptide Synthesis Rate.* Peptide synthesis rate was defined as the rate of change of the relative abundance of the newly synthesized peptide over time. The relative abundance of the newly synthesized peptide can be calculated by the relative abundance of the unlabeled peptide directly, if the basal label efficiency reaches 100%. However, due to different cell types, cell culture conditions, and the purity of the stable isotope-labeled amino acids, the basal labeling efficiency was not 100%, which consequently resulted in the detection of unlabeled peptides at the beginning of the chase period. In this study, at the beginning of the chase period, ~10% and 30% of the peptides were unlabeled in SKBR3_shVPS4B and SKBR3 cells, respectively. To calculate the relative abundance of the newly synthesized peptide, the relative abundance of the unlabeled peptide has to be corrected by subtracting the relative abundance of the preexisting unlabeled peptide at 0 hr, as well as the later time points. Since both the preexisting unlabeled and labeled peptides were subjected to degradation, the preexisting unlabeled peptide at each time point ($A_{pre\_l\_t}$) was calculated by

$$A_{pre\_l\_t} = A_{l\_0} \times e^{-\lambda t}, \tag{5}$$

where $A_{l\_0}$ was the relative abundance of the unlabeled peptide at 0 hr.

The relative abundance of the newly synthesized peptide at each time point ($A'_{l\_t}$) was calculated by

$$A'_{l\_t} = A_{l\_t} - A_{pre\_l\_t}, \tag{6}$$

where $A_{l\_t}$ was the abundance of the total unlabeled peptide at each time point.

Protein synthesis follows zero-order kinetics [39]. The relative abundances of newly synthesized peptides over time $A'_{l\_t}$ were fit to linear curves to derive the synthesis rates ($\gamma$). At least four time points were required to find the best fitting curve. Only $A'_{l\_t}$ and $t$ that were significantly correlated ($P < 0.05$) with a $R^2$ greater than 0.8 were selected to derive the synthesis rate from the following linear equation:

$$A'_{l\_t} = \gamma \times t + b, \tag{7}$$

where $b$ was the corrected initial peptide amount.

*2.9. Protein Synthesis Rate, Degradation Rate Constant, and Relative Abundance.* The protein synthesis rate, degradation rate constant, and relative protein abundance were calculated by the mean of their identified peptides' synthesis rates, degradation rate constants, and total relative peptide abundance.

## 3. Results

*3.1. iSDMS Analysis of Dynamic Protein Profiles.* Our earlier report has shown that hypoxia leads to the abnormal

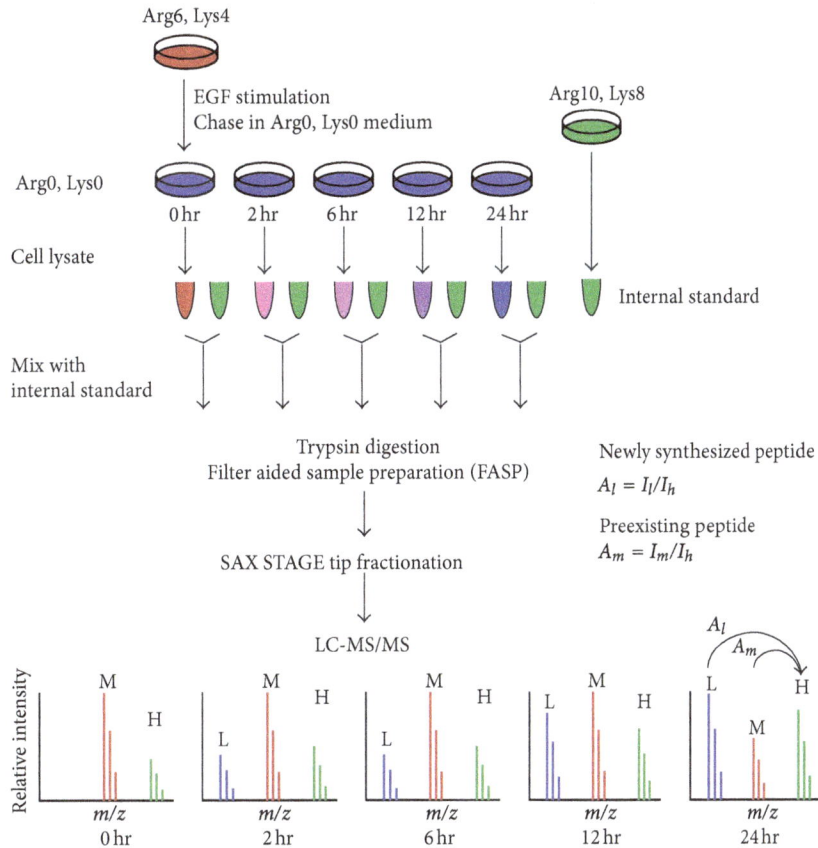

FIGURE 1: Using internal standard-assisted synthesis and degradation mass spectrometry (iSDMS) to study the roles of EGF on global protein synthesis and degradation. SKBR3_shVPS4B and the parental SKBR3 cells were first labeled with $^{13}C_6$-arginine (Arg6) and $D_4$-lysine (Lys4) medium (labeled in red). After overnight serum starvation, Arg6/Lys4-labeled cells were stimulated with 100 ng/mL EGF in medium supplemented with regular arginine (Arg0) and lysine (Lys0) for 0, 2, 6, 12, and 24 hr (labeled in blue). Protein isolated from SKBR3_shVPS4B cells labeled with $^{13}C_6^{15}N_4$-arginine (Arg10) and $^{13}C_6^{15}N_2$-lysine (Lys8)—internal standard (labeled in green)—was spiked into each sample at a ratio of 1 : 3 (wt/wt). The mixtures were digested by the Filter Aided Sample Preparation (FASP) procedure, followed by strong anion exchange (SAX) peptide fractionation. Peptides were analyzed by online LC-MS/MS using an LTQ-Orbitrap mass spectrometer. The relative abundance of the newly synthesized ($A_l$) or preexisting peptides ($A_m$) was defined as the ratio of mass spectrometric peak intensities of the unlabeled peptides ($I_l$) or Arg6/Lys4-labeled peptides ($I_m$) to the intensities of the Arg10/Lys8-labeled peptides ($I_h$), respectively.

accumulation of EGFR and subsequent alteration of cell signaling in breast cancer. To further understand the role of VPS4B dysfunction in global protein dynamics upon EGF treatment, we decided to measure the rates of protein synthesis and degradation in VPS4B downregulated SKBR3 (SKBR3_shVPS4B) and the parental SKBR3 cells using an internal standard assisted synthesis and degradation mass spectrometry (iSDMS) approach. As shown in Figure 1, both cultured SKBR3_shVPS4B and SKBR3 cells were first metabolically labeled with stable isotopic amino acids arginine, and lysine (Arg6/Lys4, labeled in red), and then chased in medium containing regular arginine and lysine amino acids (Arg0/Lys0, labeled in blue) in the presence of EGF. As a result, one can monitor the rates of protein degradation simply by measuring the decrease of Arg6/Lys4-labeled proteins, while newly synthesized proteins can only be monitored by the incorporation of Arg0 and Lys0 amino acids. Since the depletion of VPS4B expression is likely

going to have some effects on global protein expression, it is necessary to normalize the steady state protein profiling and differential gene expression between VPS4B ablation and the parental cells. To overcome the issue of differential gene expression between the two cell lines, an internal standard-cell lysates metabolically labeled with Arg10/Lys8 (labeled in green) was added to each time point in order to normalize the relative abundance of Arg6/Lys4 or Arg10/Lys6-labeled peptides between SKBR3_shVPS4B and SKBR3 cells. The detailed calculation of peptide relative abundance is described in the Section 2.

Figures 2(a) and 2(b) are representative MS spectra of tryptic peptides, SLLVNPEGPTLMR, derived from chain A of human fatty acid synthase (FASN) in SKBR3 (Figure 2(a)) and SKBR3_shVPS4B cells (Figure 2(b)) identified in the chase time periods of 0, 2, 6, 12, and 24 hr. The decreasing abundance of the preexisting Arg6/Lys4-labeled peptides (labeled in red) between the 12 and 24 hr chase periods

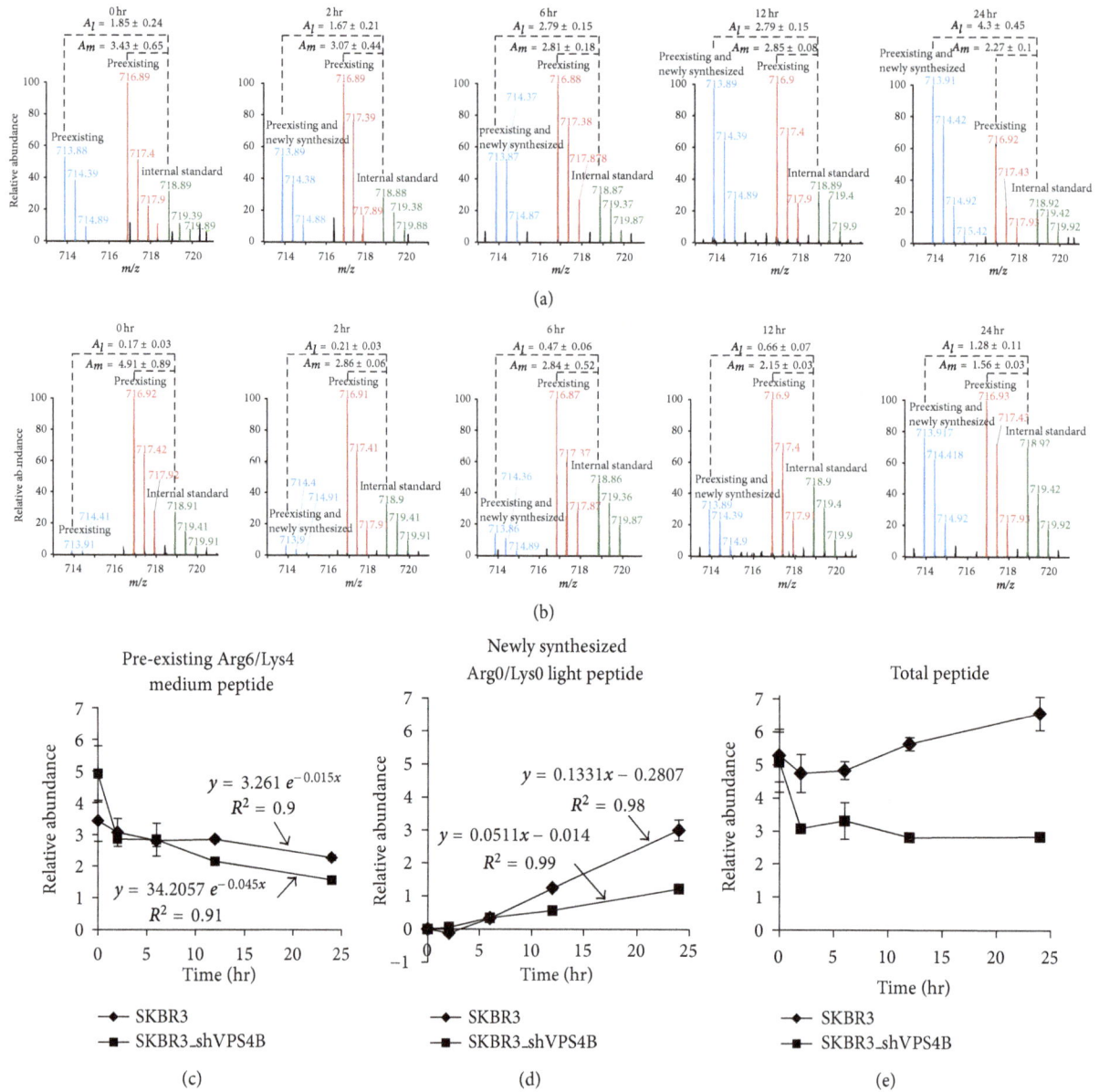

FIGURE 2: VPS4B downregulation decreases the expression of fatty acid synthase (FASN) in SKBR3_shVPS4B cells. Representative MS1 spectra of fatty acid synthase peptide SLLVNPEGPTLMR in SKBR3 (a) and SKBR3_shVPS4B cells (b). The relative abundance of unlabeled peptides ($A_l$, labeled in blue) or labeled peptides ($A_m$, label in red), expressed as mean ± standard deviation, was calculated by our in-house software, IsoQuant [38]. VPS4B downregulation increased the degradation rate of the SLLVNPEGPTLMR FASN peptide (c) and decreased its synthesis rate (d) in SKBR3 cells, which was related to the decrease of its relative abundance after EGF treatment (e).

in both cell lines clearly indicates that the degradation of this peptide can be monitored by this type of approach. To determine the role of VPS4B on protein turnover, a first-order exponential decay curve fitting was constructed (see Section 2.7 for more details). As indicated in Figure 2(c), the peptide derived from fatty acid synthase (SLLVNPEGPTLMR) clearly had a faster degradation rate in SKBR3_shVPS4B cells (labeled in squares, degradation rate constant = 0.015 h$^{-1}$, $R^2$ = 0.91) than SKBR3 cells (labeled in diamonds, degradation rate constant = 0.045 h$^{-1}$, $R^2$ = 0.90).

On the other hand, the calculation of synthesis rate is much more difficult because the pool of unlabeled Arg0/Lys0

(labeled in blue) peptides consisted of both newly synthesized peptides during the chase period, and preexisting unlabeled peptides, which were present at the beginning of the experiment due to incomplete labeling of Arg6/Lys4 amino acids. Since preexisting unlabeled (Arg0/Lys0) peptides were being degraded during the course of the experiment, it is necessary to normalize the relative abundance of the remaining (nondegraded) preexisting unlabeled peptides at each time point based on their degradation rate constant(see Section 2.8 for more details). Figure 2(d) is a representative time-course analysis of newly synthesized human FASN peptides (SLLVNPEGPTLMR). This result suggests that the FASN peptide (SLLVNPEGPTLMR) had a slower synthesis rate in SKBR3_shVPS4B cells (labeled in squares, synthesis rate = $0.055\,\mathrm{h}^{-1}$, $R^2 = 0.99$) as compared to SKBR3 cells (labeled in diamonds, synthesis rate = $0.133\,\mathrm{h}^{-1}$, $R^2 = 0.98$).

Because iSDMS allows one to measure the rates of protein synthesis and degradation simultaneously, it is therefore possible to determine the dynamics of protein expression in VPS4B-depleted cells in response to EGF. Figure 2(e) is a representative time course analysis of FASN dynamic expression. In this study, the relative abundance of a given peptide ($A_{tot}$) of FASN (SLLVNPEGPTLMR) from each time point was calculated and normalized by summing the relative abundance of unlabeled peptide ($A_l$) and labeled peptide ($A_m$) ((3), Section 2). It is clear that the relative abundance of FASN expression in SKBR3_shVPS4B cells (Figure 2(e), labeled in squares) was drastically reduced after two hours of EGF treatment. Together with the result from the degradation study of FASN in Figure 2(c), this observation suggests that the decrease of FASN expression is largely due to the increased turnover rate of FASN in SKBR3_shVPS4B cells (Figure 2(c), labeled in squares). Interestingly, after 6 hr of EGF treatment the rate of FAN degradation was very similar between SKBR3_shVPS4B and SKBR3 cells, indicating that the sudden increase of FASN turnover rate in SKBR3_shVPS4B cells could be potentially regulated by EGF-related cell signaling pathways.

*3.2. Changes of the Dynamic Proteome Profile in Relationship to VPS4B Depletion.* VPS4B is essential for the formation of MVB and has also been documented to play a pivotal role in regulating the degradation of various membrane receptors. Therefore, we decided to examine the consequence of VPS4B ablation on the dynamic proteome profile using our SKBR3_shVPS4B model system. Briefly, both cultured SKBR3_shVPS4B and the parental control SBKR3 cells grown on SILAC "medium," Arg6/Lys4, medium were treated with 100 ng/mL of EGF and chase labeled in SILAC "light," Arg0/Lys0, medium for 2, 6, 12, and 24 hr as described in the Section 2. At the end of each chase period, SKBR3_shVPS4B and SKBR3 cells were collected and analyzed by a FASP-based SAX fractionation and LC-MS/MS analysis using a hybrid Orbitrap mass spectrometer. MS/MS raw files were then subjected to database search and quantification analysis using our in-house software IsoQuant [38]. However, in order to compare the degradation rate constants and synthesis rates of each peptide in SKBR3_shVPS4B and SKBR3 cells at the proteome level, we required that the "same" peptides should be identified and quantified at each time point in both cell lines throughout the entire course of the experiment in all five time points. As indicated in Figures 2(c)-2(d), the relative peptide abundance and the time period used to calculate both degradation rate constants and synthesis rates also need to be highly correlated ($P < 0.05$; $R^2 > 0.8$) with at least four time points in order to derive the rates. These criteria have significantly reduced the depth of proteome; however, we were able to generate a highly statistically relevant dataset on the effects of VPS4B in modulating proteome dynamics. Overall, we have identified

TABLE 1: Increased protein expression is mostly due to increased protein synthesis in SKBR3_shVPS4B cells.

|  | Number of proteins* | Percentage (%) |
| --- | --- | --- |
| Increased synthesis only | 21 | 63.6 |
| Decreased degradation only | 3 | 9.1 |
| Both | 3 | 9.1 |
| Other | 6 | 18.2 |

*Number of proteins was calculated from 33 proteins with increased relative abundance (SKBR3_shVPS4B versus SKBR3 ratio >1.5) at 24 hr.

more than 15,000 total unique peptides assigned to more than 4,500 protein groups in both SKBR3_shVPS4B and SKBR3 cells. Over 70% of the proteins were assigned with a minimum of two peptides. We obtained the degradation rate constants, synthesis rates, and total peptide profiles of 1623 peptides and 723 proteins. At the global level, the overall mean protein synthesis rate in SKBR3_shVPS4B cells was $0.058 \pm 0.010\,\mathrm{h}^{-1}$, which was similar to that of the SKBR3 cells, $0.057 \pm 0.012\,\mathrm{h}^{-1}$ (see Supplemental Figures 1(a) and 1(b) available online at http://dx.doi.org/10.1155/2013/291415). This result suggests that VPS4B does not affect global protein synthesis in general in response to EGF treatment. Similarly, VPS4B depletion also does not have any systematic effect on global protein degradation, because the mean protein degradation rate constant was $0.033 \pm 0.012\,\mathrm{h}^{-1}$ ($t_{1/2} = 20\,\mathrm{hr}$) in SKBR3_shVPS4B cells, which was only slightly higher than that of the SKBR3 cells, $0.028 \pm 0.010\,\mathrm{h}^{-1}$ ($t_{1/2} = 24\,\mathrm{hr}$) (Supplemental Figures 1(c) and 1(d)).

Because the heavy SILAC Arg10/Lys8-labeled internal protein mixture was spiked into every time point, it is possible to analyze the effect of VPS4B downregulation on global protein expression. To address the effect of VPS4B downregulation on dynamic protein profiles, we defined increased and decreased protein synthesis, degradation, and relative protein abundance in SKBR3_shVPS4B cells as those values greater than 1.5-fold that of the SKBR3 cells (SKBR3_shVPS4B versus SKBR3 ratio <0.67 or >1.5). Among 723 proteins, the majority did not show changes in their rates of protein synthesis (80.9% of proteins), degradation (77.2% of proteins), or relative protein abundance (90.2% of proteins) (Supplemental Table 1). As expected, downregulation of VPS4B has a rather limited effect on the alteration of overall protein synthesis, because only ~10% of proteins displayed changes in their rates of protein synthesis. On the other hand, ~20% of proteins had increased rates of degradation in SKBR3_shVPS4B cells, suggesting that VPS4B depletion potentially affects and enhances the turnover of certain proteins. At the global level, our data suggest that VPS4B downregulation does not change the dynamic protein profiles of most proteins in SKBR3 cells. Despite the VPS4B-depleted cell line having a rather different growth phenotype and sensitivity to hypoxia as compared to the parental cells, we found that the overall distributions of relative protein abundance at 24 hr in SKBR3 and SKBR3_shVPS4B cells are also very similar (Supplemental Figures 1(e) and 1(f)).

In this study, we found that among 33 proteins with increased relative abundance (Supplemental Table 2), 63.6%

TABLE 2: Decreased protein synthesis and increased protein degradation contribute to decreased protein expression in SKBR3_shVPS4B cells.

| | Number of proteins* | Percentage (%) |
| --- | --- | --- |
| Decreased synthesis only | 7 | 18.4 |
| Increased degradation only | 6 | 15.8 |
| Both | 25 | 65.8 |

* Number of proteins was calculated from 38 proteins with decreased relative abundance (SKBR3_shVPS4B versus SKBR3 ratio <0.66) at 24 hr.

had increased synthesis only, 9.1% had decreased degradation only, and 9.1% had both (Table 1). Among 38 proteins with decreased relative abundance (Supplemental Table 3), 18.4% had decreased synthesis only, 15.8% had increased degradation only, and 65.8% had both (Table 2). Because the steady state accumulation of proteins is determined by their rates of synthesis and degradation, we therefore decided to further examine the relationship between the steady state protein expression and protein synthesis and degradation in response to the downregulation of VPS4B expression. The relative protein abundance ratios between the two cell lines (SKBR3_shVPS4B versus SKBR3) were plotted against the ratios of protein synthesis rates (Supplemental Figure 2(a)) and the ratios of protein degradation rate constants (Supplemental Figure 2(b)). As indicated in Figure 3, it is clear that the VPS4B-mediated decrease of protein expression is largely due to increased protein degradation, while the increase of VPS4B-mediated protein expression is directly related with increased protein synthesis. For instance, proteins with higher relative abundance in SKBR3_shVPS4B cells (Figure 3, red diamonds) were found to have both lower degradation rate constants and higher synthesis rates (Figure 3, 2nd quadrant). Conversely, those proteins with lower relative abundance have higher degradation rate constants and lower synthesis rates (Figure 3, 4th quadrant, labeled in green diamonds).

*3.3. VPS4B Downregulation Alters Energy Metabolism in SKBR3 Cells.* To further understand the biological consequences of VPS4B ablation in EGF-mediated cell signaling, a gene ontology and Kyoto Encyclopedia of Genes and Genome (KEGG) pathway analysis was performed [40]. As indicated in Figure 4, proteins involved in energy metabolism, in particular glycolysis and mitochondrial fatty acid $\beta$-oxidation pathways, are significantly affected in SKBR3_shVPS4B cells upon EGF treatment. In humans, glycolysis and fatty acid $\beta$-oxidation are the two main sources of acetyl-CoA for the tricarboxylic acid cycle (TCA) to generate adenosine triphosphate (ATP). We found that the expression of many key glycolytic enzymes, such as glyceraldehyde-3-phosphate dehydrogenase (GAPDH) and L-lactate dehydrogenase B chain (LDHB), fructose-bisphosphate aldolase A and C (ALDOA, ALDOC), phosphoglycerate kinase 1 (PGK1), alpha-enolase (ENO1), pyruvate kinase isozymes M1/M2 (PKM2), and L-lactate dehydrogenase A chain (LDHA), was drastically reduced in SKBR3_shVPS4B cells. As indicated in Figure 4, the decreased expression of these proteins is mainly

caused by the simultaneously increased degradation and decreased synthesis rates of these proteins as the consequence of VPS4B downregulation.

Interestingly, the expression of mitochondrial trifunctional protein alpha-subunit (HADHA/ECHA), mitochondrial trifunctional protein beta-subunit (HADHB/ECHB), and mitochondrial hydroxyacyl-coenzyme A dehydrogenase (HADH) was increased more than 1.5-fold in SKBR3_shVPS4B cells. Mitochondrial very long-chain specific acyl-CoA dehydrogenase (ACADV) and mitochondrial enoyl-CoA hydratase (ECHM) were increased more than 1.25-fold. The increased expression of these mitochondrial proteins and proteins involved in fatty acid $\beta$-oxidation are primarily caused by the increased protein synthesis rate, rather than caused by decreased protein degradation. On the other hand, fatty acid synthase (FASN) expression was decreased in SKBR3_shVPS4B cells with increased degradation and decreased synthesis rates. Taken together, our results suggest that downregulation of VPS4B expression can potentially alter the energy metabolism in breast cancer, suggesting that under either hypoxia or VPS4B depletion, fatty acid tends to be oxidized to generate energy rather than being stored, consequently replacing glucose as a main energy source.

## 4. Discussion

Dysfunction of VPS4B results in altered endosomal trafficking of membrane receptors, such as EGFR, as well as EGFR-associated signaling molecules [3–8, 12]. Toward gaining a more thorough understanding of the consequences of altered EGFR signaling caused by VPS4B dysfunction, we developed an iSDMS method to identify changes in the protein synthesis, degradation rates, and dynamic protein expression, in breast cancer SKBR3_shVPS4B cells. In this study, we obtained dynamic profiles of more than 700 proteins in both SKBR3_shVPS4B and SKBR3 cells in response to EGF. Most importantly, we have also identified many previously unidentified energy metabolism pathways that were altered as a result of VPS4B downregulation in breast cancer. In particular, we found that downregulation of VPS4B expression has a profoundly negative effect on the expression of several key proteins involved in glycolysis and fatty acid synthesis, suggesting that TCA cycle and ATP energy metabolism could be compromised. In order to overcome this defect in ATP generation, we found that the expression of many proteins involved in fatty acid $\beta$-oxidation is also elevated in the VPS4B-depleted cells, indicating that the activation of fatty acid $\beta$-oxidation could serve as a potential survival mechanism and ultimately lead to its resistance to chemotherapy and hypoxia.

Protein degradation rate constants vary among different cell types and tissues. In the same biological system, the degradation rate constants also vary widely among different proteins. In *E. coli*, protein degradation rate constants range from $0.017\,h^{-1}$ ($t_{1/2} = 40\,hr$) to $0.058\,h^{-1}$ ($t_{1/2} = 11.9\,hr$), with a mean value of $0.03\,h^{-1}$ ($t_{1/2} = 23\,hr$) [23]. In human A549 adenocarcinoma cells, the degradation rate constants

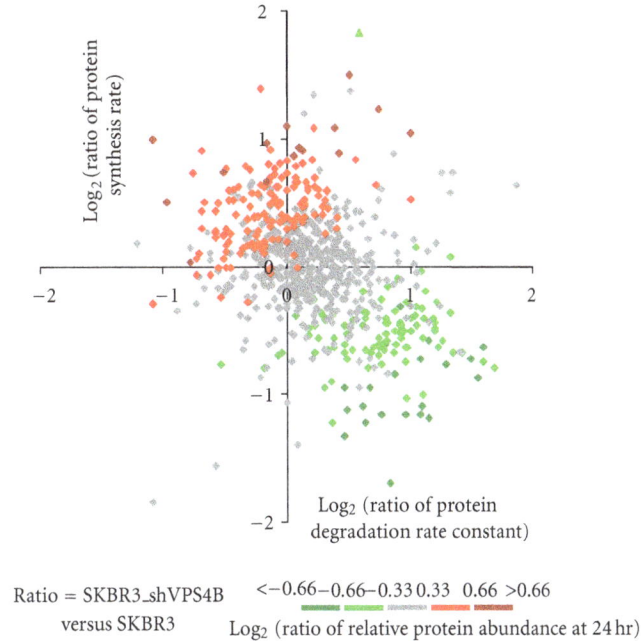

FIGURE 3: The role of VPS4B downregulation on the alteration of global protein synthesis and degradation after EGF treatment. Ratios of synthesis rates of SKBR3_shVPS4B versus SKBR3 cells ($y$ axis) are plotted against ratios of degradation rate constants of SKBR3_shVPS4B versus SKBR3 cells ($x$ axis). Diamonds represent $\log_2$ ratios of protein expression at 24 hr between the two cell lines, and the relative values of which are indicated on the color scale.

range from $2 \times 10^{-5}$ ($t_{1/2} = 69{,}000$ hr) to $5.4\,\text{h}^{-1}$ ($t_{1/2} = 0.13$ hr), with a mean of $0.081\,\text{h}^{-1}$ ($t_{1/2} = 8.5$ hr) [27]. In our study, we found that among more than 700 proteins, protein degradation rate constants ranged from $0.012$ ($t_{1/2} = 57$ hr) to $0.116\,\text{h}^{-1}$ ($t_{1/2} = 5.9$ hr) with a mean of $0.033\,\text{h}^{-1}$ ($t_{1/2} = 20$ hr) in SKBR3_shVPS4B cells and $0.01$ to $0.087\,\text{h}^{-1}$ with a mean of $0.028\,\text{h}^{-1}$ ($t_{1/2} = 24$ hr) in SKBR3 cells. Due to the sensitive of our current iSDMS analysis, we were only able to examine the dynamic profiles of those proteins with $t_{1/2} > 2$ hr.

Protein expression is controlled by the rates of protein synthesis and degradation. Generally, higher protein synthesis rate is related to higher protein expression, and higher protein degradation rate often leads to decreased protein expression. These trends are consistent with our results (Figure 3 and Supplemental Figure 2). Traditionally, increased protein expression has been considered as being mostly attributable to increased gene transcription and translation. However, several recent global dynamic protein profiling studies have indicated that increased protein expression can also be regulated by posttranscriptional and posttranslational control [25–31]. Oksvold et al. have suggested that posttranscriptional control and increased protein half life are two of the main mechanisms to maintain protein homeostasis and buffer cells from various transcriptional and gene expression noise in response to various environmental stimulations [11].

In addition, several lines of evidence have also indicated that the abnormally elevated expression of many important receptors in cancer or drug resistant cancer cells is caused

by decreased or delayed receptor degradation rather than being caused by altered gene expression [13–18]. Our results show that only ~10% of proteins with increased expression in VPS4B have decreased degradation, while the increased expression of the majority of these proteins is attributable to changes in protein synthesis only. These results strongly suggest that increasing protein synthesis or increasing translational efficiency is the main method of increasing protein expression, whereas altered protein degradation is a more protein-specific approach to increasing protein expression in SKBR3 cells. As indicated in Figure 3, we found that VPS4B-dependent protein downregulation (labeled in green diamonds) is largely caused by the combination of decreased protein synthesis and increased protein degradation, suggesting that VPS4B-mediated MVB dysfunction is playing a pivotal role in modulating protein homeostasis in breast cancer. The altered function of MVB-lysosomal degradation caused by VPS4B depletion likely stimulates other cellular degradation pathways.

VPS4B plays an important role in protein degradation, especially in the lysosomal degradation of membrane receptors [2]. It has been found that loss of VPS4B function results in delayed EGFR degradation and prolonged EGFR retention on the limiting membrane of MVBs [4–6]. However, we were not able to obtain the dynamic expression profile for EGFR and its related signaling molecules in this study due to the low abundance of endogenous EGFR and the limited dynamic proteome profile coverage in SKBR3 cells. Surprisingly, we found that overall protein degradation rates at the proteomic

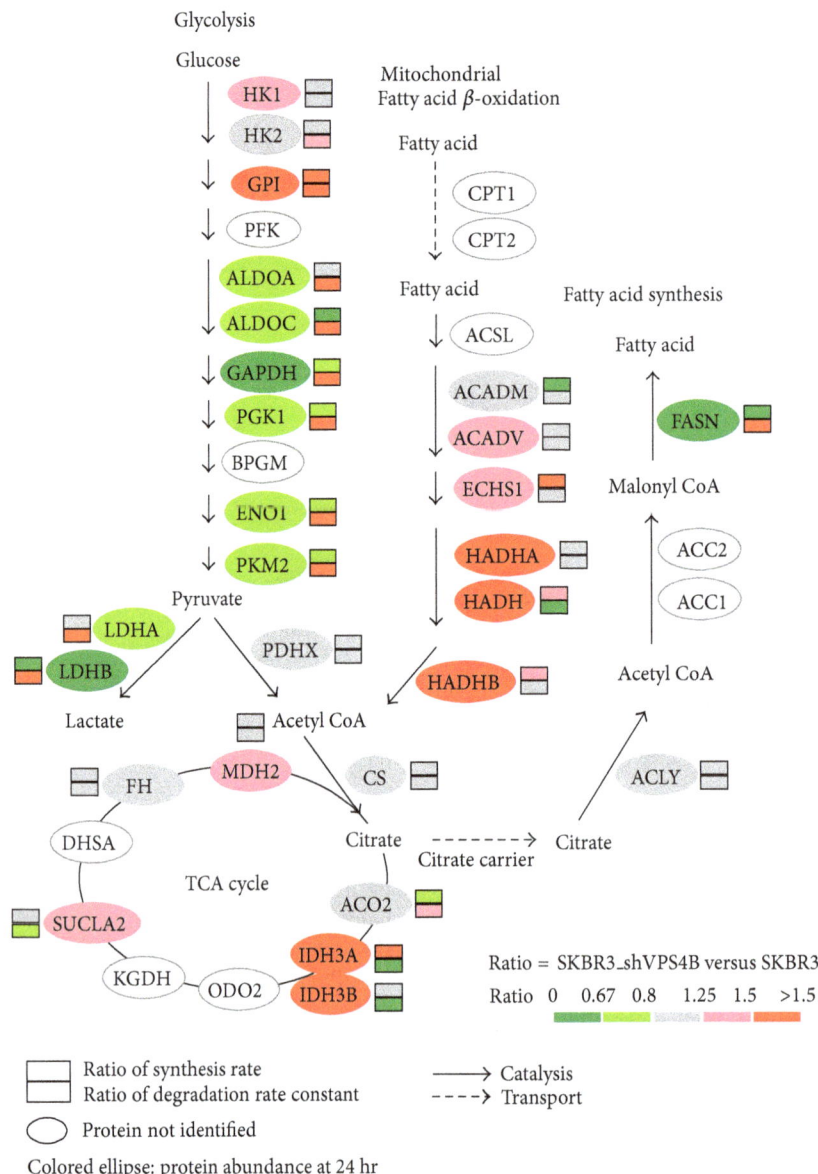

FIGURE 4: VPS4B downregulation increases the expression of mitochondrial fatty acid $\beta$-oxidation related proteins and down regulates the expression of glycolysis related proteins in SKBR3 cells. The energy metabolism pathway was adapted from the Kyoto Encyclopedia of Genes and Genomes (KEGG) [40]. Colors represent the ratios of relative protein abundance at 24 hr (shaded ellipses), protein synthesis rate (shaded upper rectangles), and protein degradation rate constant (shaded lower rectangles) in SKBR3_shVPS4B versus SKBR3 cells. Red color represents protein with increased expression and green color represents protein with decreased expression.

level were similar in SKBR3_shVPS4B and SKBR3 cells. This observation suggests that the VPS4B downregulation-induced delayed lysosomal degradation of endocytosed cargo is likely to be cargo specific and explains why overall protein degradation is not affected. Currently, we are in the process of identifying what protein cargos are specifically targeted and degraded by the Vps4B-dependent MVB-lysosomal degradation system in breast cancer. Identification of specific cargos that are delayed by VPS4B-dependent target degradation is likely going to provide critical information on whether these proteins can be used as potential biomarkers for both the classification and progression of breast cancer.

Glycolysis and fatty acid $\beta$-oxidation are two major metabolic pathways that cells utilize to generate energy. Adaptation to different carbon and energy sources is an important cellular response to environmental or intracellular changes. Our results indicate that dynamic protein expression of many enzymes involved in glycolysis and the TCA cycle are coordinately regulated in response to VPS4B depletion. Recently, Lin et al. have reported that there is a direct correlation between the downregulation of VPS4B and the progression of breast cancer, and that decreased VPS4B expression can be induced by hypoxia [12]. Although the molecular mechanisms underlying the hypoxia-induced

An Internal Standard-Assisted Synthesis and Degradation Proteomic Approach Reveals the Potential Linkage between VPS4B
Depletion and Activation of Fatty Acid β-Oxidation in Breast Cancer Cells

111

VPS4B-associated tumor angiogenesis and metastasis remain unknown, our results clearly indicate that there is a direct link between altered lipid metabolism and VPS4B-mediated MVB dysfunction.

It has been hypothesized that most cancer cells, including breast cancer cells, exhibit increased aerobic glycolysis and lead to the conversion of glucose to lactic acid, also known as the Warburg effect, in part as a result of mitochondrial respiration injury and hypoxia [41]. However, since the initial report of the Warburg hypothesis, numerous reports also indicate that many cancer cells have higher glycolytic activity even under aerobic conditions. In addition, it has been suggested by Gillies and colleagues that up-regulation of glycolysis can significantly provide many growth advantages for various cancers and the proposed conversion of glucose to lactate under aerobic conditions could be an important adaptation step for the cancer cells to survive under the intermittent hypoxic conditions during the early development of cancer (see review by Gatenby and Gillies [42]). Our current study indicated that the glycolytic pathway is downregulated in VPS4B-depleted SKBR3 cells, suggesting a potential cross-talk between the abnormal glycolysis in cancer and MVB dysfunction. Currently, we are exploring whether this attenuation of the glycolytic pathway by VPS4B could also lead to the reduction of lactate-induced acidosis that is commonly associated with many cancers. Interestingly, it has also been recently demonstrated that chronic acid-adapted MDA-MB-231 cells are able to induce the accumulation of LC3 and the possible formation of autophagosomes [43]. However, it is not currently clear whether the accumulation of LC3-positive autophagic vacuoles is due to the dysfunction of VPS4B-MVB-mediated autolysosome formation or enhanced autophagic flux. Although the main molecular and cellular events modulating this acid or environmental-induced autophagy formation in cancers are largely unknown, our systems approach described here provides a potential explanation of the role of VPS4B in regulating the glycolytic pathway and the formation of autophagy under either hypoxia or acidosis.

In addition to the downregulation of the glycolytic pathway and modulating the formation of autophagic vacuole formation, our findings also reveal that fatty acid β-oxidation is significantly upregulated in VPS4B-depleted cells. Specifically, we found that the expression of mitochondrial trifunctional protein alpha subunit (HADHA/ECHA), mitochondrial trifunctional protein beta subunit (HADHB/ECHB), and mitochondrial hydroxyacyl-coenzyme A dehydrogenase (HADH) is significantly elevated. The pathological consequences of abnormal fatty acid β-oxidation activation are unclear. However, it is reasonable to postulate that the activation of fatty acid oxidation is part of the cellular compensatory response to VPS4B-mediated downregulation of the glycolytic pathway. Fatty acid β-oxidation is known to be upregulated in many different cancers and other diseases [44–49]. Since mitochondria are known to be degraded by autophagy, we postulate that the VPS4B-autophagy-mediated mitochondrial degradation could be impaired during the premetastatic stage of cancer or in VPS4B-depleted

cells and ultimately causes the activation of mitochondrial fatty acid β-oxidation.

Finally, another important finding of our study is the decreased expression of fatty acid synthase (FASN) in VPS4B-depleted cells. High levels of FASN are reported in many epithelial cancers, such as breast, colorectal, and prostate, and FASN overexpression is highly associated with a higher risk of both disease recurrence and death [50]. Most significantly, our results show that VPS4B downregulation decreased the expression of FASN by simultaneously decreasing its rates of synthesis and increasing the rate of its degradation, suggesting that both *de novo* fatty acid synthesis and fatty acid β-oxidation are tightly coupled in cancer. Altogether, our study suggests the downregulation of VPS4B causes the alteration of glucose metabolism and the glycolytic pathway, which ultimately leads to the activation of mitochondrial fatty acid β-oxidation.

## 5. Conclusion

Here we have presented an approach, iSDMS, for the direct measurement of protein degradation and synthesis, as well as relative protein expression levels in SKBR3 cells with downregulated VPS4B expression. This approach can be feasibly applied to other cell culture systems to determine global protein dynamics under different genetic manipulations or environmental stimulations. As with other mass spectrometry-based approaches, iSDMS is limited by the depth and reproducibility of protein identification, especially when applied to multiple time points and multiple conditions. However, we are able to establish the dynamic protein profiling of many proteins involved in the central glucose and lipid metabolism that are preferentially regulated by the downregulation of VPS4B expression. As downregulation of VPS4B has been reported to be associated with certain high grade and recurring tumors, the adoption of fatty acid β-oxidation as an alternative energy source could be a distinct feature of breast cancer cells with VPS4B dysfunction. Therapeutic interventions targeting fatty acid ß-oxidation energy metabolism could serve as alternative and complementary strategies to treat breast cancer.

## Acknowledgment

This work was funded by National Institutes of Health Grants AG25323 to A. J. Yang and DE14183 to D. K. Ann.

## References

[1] D. Teis, S. Saksena, and S. D. Emr, "Ordered assembly of the ESCRT-III complex on endosomes is required to sequester cargo during MVB formation," *Developmental Cell*, vol. 15, no. 4, pp. 578–589, 2008.

[2] B. A. Davies, I. F. Azmi, and D. J. Katzmann, "Regulation of Vps4 ATPase activity by ESCRT-III," *Biochemical Society Transactions*, vol. 37, no. 1, pp. 143–145, 2009.

[3] N. Bishop and P. Woodman, "ATPase-defective mammalian VPS4 localizes to aberrant endosomes and impairs cholesterol

trafficking," *Molecular Biology of the Cell*, vol. 11, no. 1, pp. 227–239, 2000.

[4] H. Fujita, M. Yamanaka, K. Imamura et al., "A dominant negative form of the AAA ATPase SKD1/VPS4 impairs membrane trafficking out of endosomal/lysosomal compartments: class E vps phenotype in mammalian cells," *Journal of Cell Science*, vol. 116, no. 2, pp. 401–414, 2003.

[5] T. Yoshimori, F. Yamagata, A. Yamamoto et al., "The mouse SKD1, a homologue of yeast Vps4p, is required for normal endosomal trafficking and morphology in mammalian cells," *Molecular Biology of the Cell*, vol. 11, no. 2, pp. 747–763, 2000.

[6] E. Mizuno, T. Iura, A. Mukai, T. Yoshimori, N. Kitamura, and M. Komada, "Regulation of epidermal growth factor receptor down-regulation by UBPY-mediated deubiquitination at endosomes," *Molecular Biology of the Cell*, vol. 16, no. 11, pp. 5163–5174, 2005.

[7] C. Tu, C. F. Ortega-Cava, P. Winograd et al., "Endosomal-sorting complexes required for transport (ESCRT) pathway-dependent endosomal traffic regulates the localization of active Src at focal adhesions," *Proceedings of the National Academy of Sciences of the United States of America*, vol. 107, no. 37, pp. 16107–16112, 2010.

[8] R. J. Flinn, Y. Yan, S. Goswami, P. J. Parker, and J. M. Backer, "The late endosome is essential for mTORC1 signaling," *Molecular Biology of the Cell*, vol. 21, no. 5, pp. 833–841, 2010.

[9] Y. Yarden and M. X. Sliwkowski, "Untangling the ErbB signalling network," *Nature Reviews Molecular Cell Biology*, vol. 2, no. 2, pp. 127–137, 2001.

[10] A. Sorkin and L. K. Goh, "Endocytosis and intracellular trafficking of ErbBs," *Experimental Cell Research*, vol. 315, no. 4, pp. 683–696, 2009.

[11] M. P. Oksvold, E. Skarpen, L. Wierød, R. E. Paulsen, and H. S. Huitfeldt, "Re-localization of activated EGF receptor and its signal transducers to multivesicular compartments downstream of early endosomes in response to EGF," *European Journal of Cell Biology*, vol. 80, no. 4, pp. 285–294, 2001.

[12] H. H. Lin, X. Li, J.-L. Chen et al., "Identification of an AAA ATPase VPS4B-dependent pathway that modulates epidermal growth factor receptor abundance and signaling during hypoxia," *Molecular and Cellular Biology*, vol. 32, no. 6, pp. 1124–1138, 2012.

[13] A. Ciechanover and A. L. Schwartz, "The ubiquitin system: pathogenesis of human diseases and drug targeting," *Biochimica et Biophysica Acta*, vol. 1695, no. 1–3, pp. 3–17, 2004.

[14] K. I. Nakayama and K. Nakayama, "Ubiquitin ligases: cell-cycle control and cancer," *Nature Reviews Cancer*, vol. 6, no. 5, pp. 369–381, 2006.

[15] D. S. Hirsch, Y. Shen, and W. J. Wu, "Growth and motility inhibition of breast cancer cells by epidermal growth factor receptor degradation is correlated with inactivation of Cdc42," *Cancer Research*, vol. 66, no. 7, pp. 3523–3530, 2006.

[16] N. E. Willmarth, A. Baillo, M. L. Dziubinski, K. Wilson, D. J. Riese, and S. P. Ethier, "Altered EGFR localization and degradation in human breast cancer cells with an amphiregulin/EGFR autocrine loop," *Cellular Signalling*, vol. 21, no. 2, pp. 212–219, 2009.

[17] S. Chandarlapaty, M. Scaltriti, P. Angelini et al., "Inhibitors of HSP90 block p95-HER2 signaling in Trastuzumab-resistant tumors and suppress their growth," *Oncogene*, vol. 29, no. 3, pp. 325–334, 2010.

[18] Y. Wang, O. Roche, M. S. Yan et al., "Regulation of endocytosis via the oxygen-sensing pathway," *Nature Medicine*, vol. 15, no. 3, pp. 319–325, 2009.

[19] I. V. Hinkson and J. E. Elias, "The dynamic state of protein turnover: it's about time," *Trends in Cell Biology*, vol. 21, no. 5, pp. 293–303, 2011.

[20] J. M. Pratt, J. Petty, I. Riba-Garcia et al., "Dynamics of protein turnover, a missing dimension in proteomics," *Molecular & Cellular Proteomics*, vol. 1, no. 8, pp. 579–591, 2002.

[21] B. J. Cargile, J. L. Bundy, A. M. Grunden, and J. L. Stephenson, "Synthesis/degradation ratio mass spectrometry for measuring relative dynamic protein turnover," *Analytical Chemistry*, vol. 76, no. 1, pp. 86–97, 2004.

[22] P. K. Rao, G. Marcela Rodriguez, I. Smith, and Q. Li, "Protein dynamics in iron-starved Mycobacterium tuberculosis revealed by turnover and abundance measurement using hybrid-linear ion trap-fourier transform mass spectrometry," *Analytical Chemistry*, vol. 80, no. 18, pp. 6860–6869, 2008.

[23] T. Maier, A. Schmidt, M. Güell et al., "Quantification of mRNA and protein and integration with protein turnover in a bacterium," *Molecular Systems Biology*, vol. 7, article 511, 2011.

[24] K. P. Jayapal, S. Sui, R. J. Philp et al., "Multitagging proteomic strategy to estimate protein turnover rates in dynamic systems," *Journal of Proteome Research*, vol. 9, no. 5, pp. 2087–2097, 2010.

[25] J. S. Andersen, Y. W. Lam, A. K. L. Leung et al., "Nucleolar proteome dynamics," *Nature*, vol. 433, no. 7021, pp. 77–83, 2005.

[26] F. M. Boisvert, Y. Ahmad, M. Gierliński et al., "A quantitative spatial proteomics analysis of proteome turnover in human cells.," *Molecular & Cellular Proteomics*, vol. 11, no. 3, p. M111.011429, 2012.

[27] M. K. Doherty, D. E. Hammond, M. J. Clague, S. J. Gaskell, and R. J. Beynon, "Turnover of the human proteome: determination of protein intracellular stability by dynamic SILAC," *Journal of Proteome Research*, vol. 8, no. 1, pp. 104–112, 2009.

[28] M. K. Doherty, L. McClean, I. Edwards et al., "Protein turnover in chicken skeletal muscle: understanding protein dynamics on a proteome-wide scale," *British Poultry Science*, vol. 45, supplement 1, pp. S27–S28, 2004.

[29] M. K. Doherty, C. Whitehead, H. McCormack, S. J. Gaskell, and R. J. Beynon, "Proteome dynamics in complex organisms: using stable isotopes to monitor individual protein turnover rates," *Proteomics*, vol. 5, no. 2, pp. 522–533, 2005.

[30] Y. Zhang, S. Reckow, C. Webhofer et al., "Proteome scale turnover analysis in live animals using stable isotope metabolic labeling," *Analytical Chemistry*, vol. 83, no. 5, pp. 1665–1672, 2011.

[31] J. N. Savas, B. H. Toyama, T. Xu, J. R. Yates III, and M. W. Hetzer, "Extremely long-lived nuclear pore proteins in the rat brain," *Science*, vol. 335, no. 6071, p. 942, 2012.

[32] H. V. Nguyen, J. L. Chen, J. Zhong et al., "SUMOylation attenuates sensitivity toward hypoxia- or desferroxamine- induced injury by modulating adaptive responses in salivary epithelial cells," *American Journal of Pathology*, vol. 168, no. 5, pp. 1452–1463, 2006.

[33] J. R. Wiśniewski, A. Zougman, N. Nagaraj, and M. Mann, "Universal sample preparation method for proteome analysis," *Nature Methods*, vol. 6, no. 5, pp. 359–362, 2009.

[34] J. Rappsilber, Y. Ishihama, and M. Mann, "Stop And Go Extraction tips for matrix-assisted laser desorption/ionization, nanoelectrospray, and LC/MS sample pretreatment in proteomics," *Analytical Chemistry*, vol. 75, no. 3, pp. 663–670, 2003.

[35] Y. Ishihama, J. Rappsilber, and M. Mann, "Modular stop and go extraction tips with stacked disks for parallel and multidimensional peptide fractionation in proteomics," *Journal of Proteome Research*, vol. 5, no. 4, pp. 988–994, 2006.

[36] J. R. Wiśniewski, A. Zougman, and M. Mann, "Combination of FASP and StageTip-based fractionation allows in-depth analysis of the hippocampal membrane proteome," *Journal of Proteome Research*, vol. 8, no. 12, pp. 5674–5678, 2009.

[37] J. E. Elias and S. P. Gygi, "Target-decoy search strategy for increased confidence in large-scale protein identifications by mass spectrometry," *Nature Methods*, vol. 4, no. 3, pp. 207–214, 2007.

[38] Z. Liao, Y. Wan, S. N. Thomas, and A. J. Yang, "IsoQuant: a software tool for stable isotope labeling by amino acids in cell culture-based mass spectrometry quantitation," *Analytical Chemistry*, vol. 84, no. 10, pp. 4535–4543, 2012.

[39] I. M. Arias, D. Doyle, and R. T. Schimke, "Studies on the synthesis and degradation of proteins of the endoplasmic reticulum of rat liver," *Journal of Biological Chemistry*, vol. 244, no. 12, pp. 3303–3315, 1969.

[40] M. Kanehisa, S. Goto, M. Furumichi, M. Tanabe, and M. Hirakawa, "KEGG for representation and analysis of molecular networks involving diseases and drugs," *Nucleic Acids Research*, vol. 38, no. 1, pp. D355–D360, 2009.

[41] O. WARBURG, "On respiratory impairment in cancer cells," *Science*, vol. 124, no. 3215, pp. 269–270, 1956.

[42] R. A. Gatenby and R. J. Gillies, "Why do cancers have high aerobic glycolysis?" *Nature Reviews Cancer*, vol. 4, no. 11, pp. 891–899, 2004.

[43] J. W. Wojtkowiak and R. J. Gillies, "Autophagy on acid," *Autophagy*, vol. 8, no. 11, 2012.

[44] S. Zha, S. Ferdinandusse, J. L. Hicks et al., "Peroxisomal branched chain fatty acid β-oxidation pathway is upregulated in prostate cancer," *Prostate*, vol. 63, no. 4, pp. 316–323, 2005.

[45] Y. Liu, "Fatty acid oxidation is a dominant bioenergetic pathway in prostate cancer," *Prostate Cancer and Prostatic Diseases*, vol. 9, no. 3, pp. 230–234, 2006.

[46] I. Samudio, M. Fiegl, T. McQueen, K. Clise-Dwyer, and M. Andreeff, "The warburg effect in leukemia-stroma cocultures is mediated by mitochondrial uncoupling associated with uncoupling protein 2 activation," *Cancer Research*, vol. 68, no. 13, pp. 5198–5205, 2008.

[47] F. Wang, M. Kumagai-Braesch, M. K. Herrington, J. Larsson, and J. Permert, "Increased lipid metabolism and cell turnover of MiaPaCa2 cells induced by high-fat diet in an orthotopic system," *Metabolism*, vol. 58, no. 8, pp. 1131–1136, 2009.

[48] J. Khasawneh, M. D. Schulz, A. Walch et al., "Inflammation and mitochondrial fatty acid β-oxidation link obesity to early tumor promotion," *Proceedings of the National Academy of Sciences of the United States of America*, vol. 106, no. 9, pp. 3354–3359, 2009.

[49] V. R. Holla, H. Wu, Q. Shi, D. G. Menter, and R. N. DuBois, "Nuclear orphan receptor nr4a2 modulates fatty acid oxidation pathways in colorectal cancer," *Journal of Biological Chemistry*, vol. 286, no. 34, pp. 30003–30009, 2011.

[50] J. A. Menendez and R. Lupu, "Fatty acid synthase and the lipogenic phenotype in cancer pathogenesis," *Nature Reviews Cancer*, vol. 7, no. 10, pp. 763–777, 2007.

# 13

# Proteomic Analysis and Label-Free Quantification of the Large *Clostridium difficile* Toxins

**Hercules Moura,**[1] **Rebecca R. Terilli,**[1,2] **Adrian R. Woolfitt,**[1] **Yulanda M. Williamson,**[1] **Glauber Wagner,**[1,3] **Thomas A. Blake,**[1] **Maria I. Solano,**[1] **and John R. Barr**[1]

[1] *Division of Laboratory Sciences, National Center for Environmental Health, Centers for Disease Control and Prevention (CDC), MS F-50, 4770 Buford Hwy NE, Atlanta, GA 30341, USA*

[2] *Association of Public Health Laboratories, Silver Spring, MD 20910, and Oak Ridge Institute for Scientific Education, Oak Ridge, TN 37380, USA*

[3] *Universidade do Oeste de Santa Catarina, 89600 Joacaba, SC, Brazil*

Correspondence should be addressed to John R. Barr; JBarr@cdc.gov

Academic Editor: Jen-Fu Chiu

*Clostridium difficile* is the leading cause of antibiotic-associated diarrhea in hospitals worldwide, due to hypervirulent epidemic strains with the ability to produce increased quantities of the large toxins TcdA and TcdB. Unfortunately, accurate quantification of TcdA and TcdB from different toxinotypes using small samples has not yet been reported. In the present study, we quantify *C. difficile* toxins in <0.1 mL of culture filtrate by quantitative label-free mass spectrometry (MS) using data-independent analysis (MS$^E$). In addition, analyses of both purified TcdA and TcdB as well as a standard culture filtrate were performed using gel-based and gel-independent proteomic platforms. Gel-based proteomic analysis was then used to generate basic information on toxin integrity and provided sequence confirmation. Gel-independent in-solution digestion of both toxins using five different proteolytic enzymes with MS analysis generated broad amino acid sequence coverage (91% for TcdA and 95% for TcdB). Proteomic analysis of a culture filtrate identified a total of 101 proteins, among them TcdA, TcdB, and S-layer proteins.

## 1. Introduction

*Clostridium difficile* is a gram-positive, anaerobic, spore-forming, rod-shaped bacterium that can produce at least three toxins including two Rho GTPase-glucosylating toxins (TcdA and TcdB) and the binary *C. difficile* transferase (CDT) toxin. The organism can cause *C. difficile* infection (CDI) in humans and animals. *C. difficile* is considered an important cause of healthcare-associated infection in humans and is the leading cause of antibiotic-associated diarrhea in hospitals worldwide [1, 2]. CDIs are toxin-mediated illnesses that range from mild diarrhea to fulminant pseudomembranous colitis and toxic megacolon, which may result in death [3]. Laboratorial confirmation of CDI needs to be rapidly performed. The methods include detection of *C. difficile* through cultivation, detection of Tcd A and Tcd B in stool samples using immunoassay and DNA-based methods [2].

Reflecting its changing epidemiology, the incidence and severity of CDI have increased significantly in the past ten years. Among the possible causes of increasing morbidity is the emergence of strains considered to be more virulent [4]. These strains produce greater amounts of toxin than reference strains and are highly transmissible due to their greater sporulation capacity [4]. One example is the rapid emergence of the highly virulent clone-designated BI/NAP1/027 in multiple countries [5]. Additionally, it is estimated that there are ~500,000 cases of CDI per year in US hospitals and long-term facilities alone, with an estimated ~15,000 to 30,000 deaths [2]. Since the illness may be severe and is difficult to treat and there is currently no available vaccine, preventing individual cases and outbreaks of CDI is a major challenge.

Although it has been demonstrated [4, 5] that strains or toxinotypes associated with outbreaks and high morbidity produce more toxin than historic, nonepidemic isolates, the expression levels of these toxins from different toxinotypes are not completely known. Moreover, the toxin load in clinical samples and culture supernatants has always been roughly estimated [5–8], and the accurate quantification of TcdA and TcdB using small samples has not yet been reported. Accurate quantification of toxins can be accomplished using proteomic strategies which may reveal key information applicable to TcdA and TcdB method development, detection, understanding toxic mechanisms, and production of new therapeutics including polyclonal and monoclonal antibodies [2].

Proteomics has been described as a key technology in the postgenomic era that provides information complementary to that provided by genomics. Proteins can be analyzed rapidly, accurately, and with high sensitivity using mass spectrometry (MS). Proteomic examinations have been performed for C. difficile in which proteins released in vitro during high toxin production [7] were identified for strains 630 and VPI 10461. Additional studies characterizing the subproteomes of C. difficile reference strain 630, including a surface protein and insoluble protein fraction analysis [9, 10], spore protein identification [11], and a protein assessment associated with heat stress responses, have also been reported [12]. In an additional study, culture supernatants of C. difficile reference strain 630 were compared to two hypervirulent strains (CD196 and CDR20291), and five secreted proteins were identified exclusively in the supernatants of the hypervirulent strains [13].

Absolute protein quantification by MS is a well-studied technique typically performed using stable isotope dilution [14–16]. However, applying a data-independent analysis ($MS^E$) that does not require labeled compounds and is amenable to sample-limited experiments [17] is a newly available alternative. In $MS^E$, data are acquired in a data-independent fashion using an alternating low/high-energy scan mass analysis and can be used to perform both protein identification and quantification in one MS experiment [18, 19].

We have previously described MS-based proteomics studies of different bacterial toxins [20], including botulinum neurotoxin [21, 22], anthrax lethal factor [20], and pertussis toxin [23]. In the present study, we describe a gel-based proteomic analysis of TcdA and TcdB. A gel-independent approach was also used in which the toxins were digested by five different proteolytic enzymes to maximize amino acid sequence coverage. Toxin digests were analyzed qualitatively using two different MS instrument platforms and were further analyzed using a label-free quantitative methodology. This study provides the performance characteristics and the basis for future development of improved MS-based detection and quantification methods for TcdA and TcdB and may help to identify protein factors involved in C. difficile toxin production by different isolates. We expect that such efforts will contribute to a better understanding of these toxin-mediated illnesses and will lead to new preventive measures and therapies against C. difficile infection.

## 2. Materials and Methods

*2.1. Clostridium difficile Toxins.* Purified TcdA (one lot—lot 1-15215A1C) and TcdB (two lots—lot 1-1551A1B and lot 2-15518A1B) used in this study were purchased from List Biologicals Laboratories Inc. (Campbell, CA, USA). The toxins were purified from C. difficile VPI 10463—ATCC 43255 and were provided in vials containing 100 μg of TcdA or 20 μg of TcdB lyophilized in 0.05 M Tris. Three vials of each toxin (TcdA lot 1, TcdB lot 2) were assigned numbers in house and designated as distinct biological samples (BioS1, BioS2, and BioS3). The vials containing TcdA and TcdB were reconstituted as per the manufacturer's instructions, aliquoted, and used immediately as the designated BioS. Aliquots (20 μL) from the first biological sample (BioS1) were used to run SDS-PAGE gels and for three separated fast trypsin in-solution digestions to verify method reproducibility [24]. Ten microliters aliquots from samples designated as BioS2 were digested using five different enzymes in parallel to obtain maximum protein sequence coverage. Aliquots of BioS3 were serially diluted (1 : 2 factor, starting at 2 μg and ending at 0.125 μg) and used to determine the sensitivity of the $MS^E$ method used in this study. In addition, a commercially available lyophilized C. difficile culture filtrate (CFil) control reagent (Techlab Clostridium difficile Toxin/Antitoxin Kit—T5000, Blacksburg, VA, USA) was used for both the quantitative method and qualitative proteomic analyses. A schematic flow diagram of the procedures used in this work can be found in Figure 1. All chemicals used in this study were purchased from Sigma-Aldrich (St.Louis, MO, USA) unless otherwise indicated.

*2.2. SDS-PAGE Analysis and In-Gel Digestion.* Duplicates of purified TcdA lot 1 (BioS1) and two lots of TcdB lot 1 and lot 2 (BioS1) were treated with NuPAGE (Invitrogen Carlsbad, CA, USA) sample buffer and the proteins were separated by 1D SDS-PAGE using the NuPAGE Novex system (Bis-tris-gels; 4–12% polyacrylamide gradient) as per the manufacturer's instructions (Invitrogen). In addition, BioS1 in-solution tryptic digests were separated using Tricine gels (Invitrogen). The gels were then either stained with Colloidal Coomassie Blue (GelCode Blue Safe Protein Stain—Thermo Scientific, Pierce, Rockford, IL, USA) or with Pierce Silver Stain Kit for Mass Spectrometry (Thermo Scientific). After being scanned, each gel band was cut into slices of approximately 0.4 cm. In-gel digestion was performed with sequence grade trypsin (Promega, Madison, WI, USA) as previously described [21]. Briefly, the gel slices were dried for 30 min using a Centrivap concentrator (Labconco, Kansas City, MO, USA) and 10 μL of trypsin (0.5 μg/μL) diluted in 25 μL of a 50 mM ammonium bicarbonate solution, pH 8.5, containing 1 mM calcium chloride (digestion buffer), was added to each sample. After 5 min incubation at room temperature (RT), 25 μL of digestion buffer was added and the samples were incubated at 37°C overnight (ON). Following ON incubation, the digests were quenched with 0.1% formic acid (FA), sonicated for 3 min, and centrifuged at 1200 g for 10 min. The supernatants were used for nanoscale ultrapressure liquid chromatography (nUPLC)-MS/MS analysis.

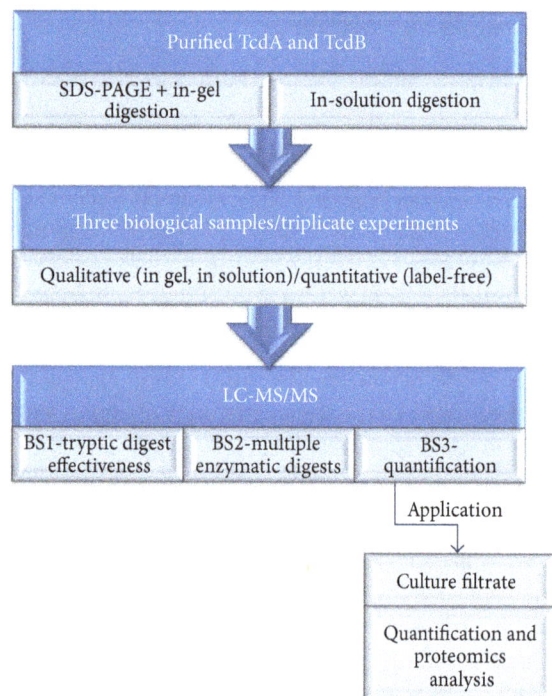

FIGURE 1: Schematic flow diagram of baseline data and biomarker discovery methods to study the large *C. difficile* toxins.

*2.3. In-Solution Enzymatic Digestion.* In-solution detergent-based 3 min tryptic digestions were performed as described [21], using three aliquots of purified TcdA (lot1) and TcdB (lot2) (BioS1, S2, and S3) (1 μg/μL) and CFil (10 μg/μL). Ten microliters of 0.2% RapiGest (RG), an acid-labile surfactant (Waters Corporation, Milford, MA, USA), in-digestion buffer, was added to each 10 μL-aliquot of TcdA, TcdB, and CFil and the tubes were incubated at 99°C for 5 min using a thermocycler (Applied Biosystems, Foster City, CA, USA). The solution was rapidly cooled to RT, and trypsin (~50 pmol) in-digestion buffer, was added. The samples were incubated at 52°C for 3 min. To hydrolyze the RG, 10 μL of 0.45 M HCl was added and the samples were incubated at 37°C for 30 min. The suspension was further diluted to a final volume of 50 μL with 0.1% FA and centrifuged (1200 g) for 10 min. The supernatant containing the peptides was frozen at −70°C if not used immediately. For calibration of the quantification method, yeast alcohol dehydrogenase (ADH) standard tryptic digest solution (Waters Corporation) was added to the supernatant before analysis by MS$^E$, at a concentration required to give 100 fMol ADH on column. Each toxin was digested in triplicate (technical replicates) and submitted to the MS instrument in triplicates (analytical replicates) and the results of all MS runs were compared. To visually evaluate digestion effectiveness, BioS1 tryptic digests were analyzed using SDS-PAGE tricine gels and silver stained. Four additional enzymes, AspN, chymotrypsin, GluC, and LysC (Sigma-Aldrich, San Louis, MO, USA), were separately used to digest aliquots of BioS2 in order to maximize proteome coverage. Enzymatic digestions were performed as per the manufacturer's instructions.

*2.4. MS Analyses and Database Search.* Liquid chromatography-tandem mass spectrometry (LC-MS/MS) was carried out using an nUPLC coupled either to a QTof Premier MS system (Waters Corporation, Milford, MA, USA), or to a linear ion trap (LTQ)-Velos Orbitrap tandem MS instrument (Thermo Scientific, San Jose, CA, USA). All the conditions for nUPLC separation, (including flow rate and solvent concentrations) as well as the MS$^E$ method on the QTof Premier MS, were used as previously described [21].

Each digest was analyzed in triplicate (three analytical replicates per digest sample), and the respective raw data files obtained using data-independent LC-MS$^E$ were further processed using the ProteinLynx Global Server v2.4 software (PLGS, Waters Corporation), for protein identification and quantification. Database searches were performed using the PLGS Identity$^E$ database search algorithm against either a UniProt protein database (November 2009; $6 \times 10^5$ entries) or a modified NCBInr database created with the term "difficile" (December 2010; $3.5 \times 10^3$ entries) to which the ADH 30,030 Da amino acid sequence was added. The PLGS software package provided statistically validated peptide and protein identification along with the determination of the stoichiometry of the protein constituents of the mixture (relative quantification) along with the expected amounts of protein present in the mixture (absolute quantification). The remaining parameters, including mass accuracy for precursor (10 ppm) and product (20 ppm) ions and criteria for protein identifications, were defined as before [21]. Similarly, relative protein quantification for the ADH digest-spiked samples (100 fMol on the column) was obtained using both the PLGS Identity$^E$ and the Expression software [18, 19]. The clustered dataset was exported from PLGS and further analyzed with Microsoft Excel 2010 (Microsoft Corporation, Redmond, WA, USA). Scaffold (v3.01, Proteome Software Inc., Portland, OR, USA) was used to further validate MS/MS based peptide and protein identifications as before [21]. The reported data represent three technical replicates and three analytical replicates.

Additionally, protein digests were analyzed using an LTQ Velos Orbitrap tandem MS instrument. Peptides were separated using an nUPLC system directly coupled online to the MS instrument through an Advance Captive Spray source from Michrom Bioresources (Auburn, CA, USA). The spray voltage was set at 1500 V, and the capillary temperature was 200°C. nUPLC separation was performed as previously described [21]. Briefly, the mobile phase consisted of (solvent A) 0.2% FA, 0.005% trifluoroacetic acid (TFA) in water, and (solvent B) 0.2% FA, 0.005% TFA in ACN. The gradient was set at 5% B for 5 min, followed by a ramp to 40% B over 90 min, and then a ramp up to 95% B in 1 min. The gradient was then held at 95% B for 5 min before returning to 5% B in 2 min, followed by reequilibration at 5% B for 5 min.

The MS was programmed to perform data-dependent acquisition by scanning the mass range from *m/z* 400 to 1400 at a nominal resolution setting of 60,000 FMHM for parent ion acquisition in the Orbitrap. Tandem mass spectra of doubly charged and higher charge state ions were acquired for the top 15 most intense ions in each survey scan. All tandem mass spectra were recorded by use of the linear ion trap. This process cycled continuously throughout the duration of the nUPLC gradient. All tandem mass spectra were extracted from the raw data file using Mascot Distiller (Matrix Science, London, UK; version 2.2.1.0) and searched using Mascot (version 2.2.0). Mascot was setup to search using the entire NCBInr database or a modified NCBInr database created to search "C. difficile" recognized proteins, or to search for C. difficile strain ATCC 43255 in which trypsin is used as the digestion agent. Mascot and Scaffold search parameters were used as described before with stringent parameters so the probability of a wrong assignment was below 0.1% [22, 23]. PSORTb subcellular scores were used to predict and localize identified culture supernatant proteins (http://www.psort.org/psortb/) [24]. NCBI gi accession numbers were employed to assign functions to each of the identified proteins using KEGG identifiers http://www.genome.jp/kegg/kegg3.html as described before [23].

## 3. Results

*3.1. Gel-Based Proteomics Platform.* 1D SDS-PAGE followed by silver stain revealed intense bands corresponding to purified TcdA and TcdB proteins (Figure 2(a)). Interestingly, in the lane containing TcdB (lot 1), an extra, prominent band was present at the ~210 kDa MW region. However, analysis of TcdB (lot 2) revealed only one band at ~270 kDa, as expected. Trypsin digestion and MS analysis of all the protein-extracted gel bands confirmed their amino acid (AA) sequences as TcdA and TcdB. It also indicated that the extra-band in lot 1 represents a truncated TcdB, in which peptides corresponding to AA 1 through 543 are missing. This finding can be observed in Figure 2(b) through the analysis of the matched peptides detected after overlaying the amino acid sequences detected in the two excised gel bands from lot 1.

*3.2. Gel-Independent Proteomics Platform.* First, three separate aliquots of BioS1 (TcdA lot1 and TcdB lot2) were used to verify the effectiveness of the 3 min digestion method for the large C. difficile toxins. As revealed by nUPLC-MS/MS analysis, in-solution digestion of the toxins generated a large peptide pool. Additionally, to confirm completeness of digestion, analysis of the digests using a silver-stained Tricine gel (not shown) revealed no bands, suggesting completeness of digestion.

Secondly, BioS2 aliquots were digested in parallel using multiple enzymes. The amino acid percent coverage, for TcdA and TcdB for each enzyme used, was 37% and 34% for AspN, 28% and 35% for chymotrypsin, 40% and 61% for GluC, 8% and 57% for LysC, and 80% and 66% for trypsin. The two enzymes that delivered the most complementary sequence coverage were trypsin and GluC (combined sequence coverage of 83% for TcdA and 86% for TcdB). Summation of the entire MS peptide analysis revealed broad amino acid sequence coverage for TcdA (91%) and TcdB (95%).

Finally, BioS3 was used for the absolute quantification of TcdA and TcdB in small samples using $MS^E$ (Table 1). The minimum amounts of digested TcdA and TcdB in buffer that

can be loaded on the nUPLC column and still be detected were, respectively, 5 ng (1.6 $\mu$g/mL) and 1.25 ng (0.43 $\mu$g/mL).

A further application of the $MS^E$ method was to successfully determine the relative and absolute amounts of TcdA and TcdB in a commercial lyophilized C. difficile culture filtrate (CFil) control reagent. In addition, because $MS^E$ data analysis can be used to determine the relative abundance of proteins in a complex mixture, data obtained with the CFil were processed and 24 constituents of the filtrate could be quantified (Table 2). Data analysis clearly shows that the S-layer proteins were the most abundant proteins in the mixture, followed by TcdA and TcdB. After proteomic analysis of the CFil using both MS instruments, a total of 101 proteins were identified. With the stringent parameters used for Peptide Prophet and Protein Prophet within the Scaffold software, the false discovery rate was zero. Psort and KEGG localization disclosed that most proteins (72%) were cytoplasmatic (Figure 3).

## 4. Discussion

The large toxins TcdA and TcdB are the major virulence factors of *Clostridium difficile* and are primary markers for diagnosis of CDI. These toxins are glycosyltransferases involved in the inactivation of small GTPases which is a key factor in disease pathogenesis. The present proteomic study is foundational and is aimed at exploring the potential of this methodology for the development of a mass spectrometry-based method for accurate TcdA and TcdB quantification. Gel-based 1D SDS-PAGE analysis of TcdA and TcdB was performed, followed by in-gel trypsin digestion of the proteins and further mass spectrometric analysis. A gel-independent approach was also performed using in-solution multienzymatic digestion of TcdA and TcdB followed by nUPLC-MS/MS, with both data-dependent analysis and LC-$MS^E$. Data-independent analysis is advantageous as it can potentially provide both absolute and relative label-free quantification results.

The gel-based approach provided critical information on TcdA and TcdB protein integrity and amino acid (AA) sequence coverage maps. The expected unique protein band (~300 kDa MW) was visible in the TcdA lane, while two bands were observed in the TcdB lane for the first lot of a commercially available toxin. Further MS analysis of the gel-excised tryptic-digested proteins from this first lot demonstrated that the band at the ~270 kDa region was the complete TcdB, whereas the extra-band (~210 kDa) was a truncated TcdB protein missing the N-terminal peptides (~68 kDa) which are associated with enzymatic activity. Interestingly, the presence of two bands at ~210 and ~68 kDa is normally expected when TcdB is autocleaved by the cysteine-protease present in the toxin molecule during the activation process which can occur *in vivo* or *in vitro* [1, 2]. However, in this case, it is possible that self-cleavage did occur and the 68 kDa portion was lost during the manufacturer's purification process since no band for the N-terminal fragment was observed. Fortunately, analysis of a second lot of TcdB from the same

MW
(kDa)  TcdA    TcdB

Lot 1a Lot 1b  Lot 1  Lot 2

250

150

(a)

TcdA        TcdB

```
TcdA
   1 MSLISKEELI KLAYSIRPRE NEYKTILTNL DEYNKLTTHN NENKYLQLKK
  51 LHESIDVFMN KYKTSSRNRA LSNLKKDILK EVILIKNSHT SPVEKNLHFV
 101 WIGGEVSDIA LEYIKQWADI NAEYNIKLWY DSEAFLVHTL KKAIVESSTT
 151 EALQLLEEEI QNPQFDRMKF YKKRMEFIYD RQKRFINYYK SQIHKPTVPT
 201 IDDIIKSHLV SEYNRDETVL ESYRTNSLRK INSNHGIDIR ANSLFTEQEL
 251 LNIYSQELLN RGHLAAASDI VRLLALKNFG GVYLDVDMLP GIHSDLFKTI
 301 SRPSSIGLDR WEMIKLEADH KYKKYINHYT SENFDKLDQQ LKDNFKLIIE
 351 SKSEKSEIFS KLEMLHVSDL EIKIAFALGS VINQALISKQ GSYLTHLVIE
 401 QVKNRYQFLN QHLNPAIESD NHFTDTTKIF HDSLFNSATA ENSHFLTKIA
 451 PYLQVGHMPE ARSTISLSGP GAYASAYYDF IHLQENTIEK TLKASDLIEF
 501 KFPEHNLSQL TEQEINSLWS FDQASAKYQF EKNVRDYTGG SLSEDNGVDF
 551 HKNTALDKNY LLNNKIPSCH VEEAGSKNYV HYIIQLQGDD ISYEATCNLF
 601 SKHPKNSIII QRHMHESAKS YFLSDDGESI LELHKYRIPE RLKHKEKVKV
 651 TFIGHGKDEF NTSEFARLSV DLSLHEISSF LDTIKLDISP KNVEVNLLGC
 701 HMFSYDFNVE ETYPGKLLLS IMDKITSTLP DVNKNSITIG ANQYEVRINS
 751 EGRKELLAHS GHWIHKEEAI MDLSSKEYI FFDSIDNKLK AKSKNIPGLA
 801 SISEDIKTLL LDASVSPDTK FILNNLKLNI ESSIGDYIYY EKLEPVKNII
 851 HNSIDDLIDE FHLLENVSDE LYELKKLHNL DEKYLISFED ISKNNSTYSV
 901 RFINKSHGES VYVETEKEIP SKYSEHITKE ISTIKHSIIT DVHGNLLDHI
 951 QLDHTSQVNT LNAAFFIQSL IDYSSNKDVL NDLSTSVKVQ LYAQLFSTGL
1001 NTIYDSIQLV HLISHAVHDT INVLPTITEG IPIVSTILDG INLGAAIKEL
1051 LDEHDPLLKK ELEAKVGVLA IHMSLSIAAT VASIVGIGAE VTIFLLPIAG
1101 ISAGIPSLVN HELILHDKAT SVVNYFNHLS ESKKYGPLKT EDDKILVPID
1151 DLVISEIDFN HNSIKLGTCN ILAMEGGSGH TVTGNIDHFF SSPSISSHIP
1201 SLSIYSAIGI ETENLDFSKK IHMLPNAPSR VFNWETGAVP GLRSLENDGT
1251 HLLDSIRDLY PGKFYWRFYA FDKYDAITTLK PVYEDTHIKI KLDKDTRNFI
1301 MPTITTHEIR NKLSYSFDGA GGTYSLLLSS YPISTNIMLS KDDLWIFNID
1351 HEVREISIEN GTIKKGKLIK DVLSKIDINK NKLIIGHQTI DFSGDIDNKD
1401 RYIFLTCELD DKISLIIEIN LVAKSYSLLL SGDKHYLISH LSNTIEKINT
1451 LGLDSKNIAY HYTDESHNKY FGAISKTSQK SIIMYKKDSK HILEFYNDST
1501 LEFNSKDFIA EDIHVFHHDD INTITGKYYV DNNTDKSIDF SISLVSKNQV
1551 KVNGLYLNES VYSSYLDFVK HSDGHHNTSN FMNLFLDNIS FWKLFGFENI
1601 HEVIDKYFTL VGKTHLGYVE FICDHNKHID IYFGEWKTSS SKSTIFSGNG
1651 RHVVVEPIYN PDTGEDISTS LDFSYEPLYG IDRYINKVLI APDLYTSLIN
1701 IHTNYYSNEY EPEILVSNAS TTFHKKVHNL DSSSFEYKWS TEGSDFILVR
1751 YLEESNKKIL QKIRIKGILS HTQSFNKMSI DFKDIKKLSL GYIMSNFKSF
1801 NSENELDRDH LGFKIIDNKT YYYDEDSKLV KGLININHSL FYFDPIEFNL
1851 VTGWQTINGK KYYFDIHTGA ALTSYKIING KHFYFNNDGV HQLGVFKGPD
1901 GFEYFAPANT QHHNIEGQAI VYQSKFLTLH GKKYYFDHHS KAVTGWRIIN
1951 NEKYYFNPHN AIAAVGLQVI DNNKYYFNPD TAIISKGWQT VHGSRYYFDT
2001 DTAIAFHGYK TIDGKHFYFD SDCVVKIGVF STSNGFEYFA PANTYHHNIE
2051 GQAIVYQSKF LTLHGKKYYF DNNSKAVTGL QTIDSKKYYF HTHTAEAATG
2101 WQTIDGKKYY FNTNTAEAAT GWQTIDGKKY YFNTNTAIAS TGYTIINGKH
2151 FYFNTDGIMQ IGVFKGPNGF EYFAPANTDA HNIEGQAILY QNEFLTLHGK
2201 KYYFGSDSKA VTGWRIINHK KYYFNPHNAI AAIHLCTINH DKYYFSYDGI
2251 LQNGYITIER KKYYFNPDNH SKAVTGWFKG PNGFEYFAPA NTHHNNIEGQ
2301 AIVYQHKFLT LNGKKYYFDN DSKAVTGWQT IDGKKYYFNL NTAEAATGWQ
2351 TIDGKKYYFN LNTAEAATGW QTIDGKKYYF NTNTFIASTG YTSINGKHFY
2401 FNTDGIMQIG VFKGPNGFEY FAPANTDAHH IEGQAILYQH KFLTLHGKKY
2451 YFGSDSKAVT GLRTIDGKKY YFNTNTAVAV TGWQTINGKK YYFNTNTSIA
2501 STGYTIISGK HFYFNTDGIM QIGVFKGPDG FEYFAPANTD AHNIEGQAIR
2551 YQNRFLYLHD NIYYFGHNSK AATGWVTIDG NRYYFEPNTA MGANGYKTID
2601 NKHFYFRHGL PQIGVFKGSN GFEYFAPANT DAHNIEGQAI RYQNRFLHLL
2651 GKIYYFGHNS KAVTGWQTIN GKVVYFMPDT AMAAAGGLFE IDGVIYFFGV
2701 DGVKAPGIYG
```

```
TcdB
   1 MSLVNRKQLE KMAHVRFRTQ EDEYVAILDA LEEYHNMSEH TVVEKYLKLK
  51 DINSLTDIYI DTYKKSGRHK ALKKFKEYLV TEVLELKHNH LTPVEKNLHF
 101 VWIGGQINDT AINYINQWKD VHSDYHVNVF YDSHAFLINT LKKIVVESAI
 151 HDTLESFREN LHDPRFDYNK FFRKRMEIIY DKQKNFINYY KAQREENPEL
 201 TIDDIVKTYL SHEYSKEIDE LNTYIEESIN KITQNSGNDV RNFEEFKNGE
 251 SFHLYEQELV ERSWLAAASD ILRISALKEI GGMYLDVDML PGIQPDLFES
 301 IEKPSSVTVD FWEMTKLEAI MKYKEYIPEY TSEHFDMLDE EVQSSFESVL
 351 ASKSDKSEIF SSLGDMEASP LEVKIAFHSK GIIHQGLISV KDSYCSNLIV
 401 KQIEWRYKIL HNSLHPAISE DMDFWTTTNT FIDSIMAEAN ADNGRFMMEL
 451 GKYLRVGFFP DVKTTINLSG PEAYAAAYQD LLMFKEGSMH IHLIEADLRN
 501 FEISKTNISQ STEQHMASLW SFDDARAKAQ FEEYKRNYFE GSIGEDDHLD
 551 FSQNIVVDKE YLLEKISSLA RSSERGYIHY IVQLQGDKIS YEAACHLFAK
 601 TPYDSVLFQK NIEDSEIAYY YHPGDGEIQE IDKYKIPSII SDRPKIKLTF
 651 IGHGKDEFNT DIFAGFDVDS LSTEIEAAID LAKEDISPKS IEIHLLGCHM
 701 FSYSINVEET YPGKLLLKVK DKISELMPSI SQDSIIVSAH QYEVRINSEG
 751 RRELLDHSGE WINKEESIIK DISSKEYISF NKPENKKITVK SKNLPELSTL
 801 LQEIRNHSWS SDIFLEEKVM LTECEINVIS NIDIQIVEER IEEAKHLTSD
 851 SINYIKDEFK LIESISDALC DLKQQHELED SHFISFEDIS ETDEGFSIRF
 901 INKETGESIF VETEKTIFSE YANHITEEIS KIKGTIFDTV NGKLVKKVNL
 951 DTTHEVNTLN AAFFIQSLIE YNSSKESLSN LSVAMKVQVY AQLFSTGLNT
1001 ITDAAKVVEL VSTAIDETID LLPTLSEGIP IIATIIDGVS LGAAIKELSE
1051 TSDPILRQEI EAKIGIMAVH LTTATTAIIT SSLGIASGFS ILLVPLAGIS
1101 AGIPSLVNHE LVLRDKATKV VDYFKHVSLV ETEGVFTLLD DKIMMPQDDL
1151 VISEIDFNHN SIVLGKCEIW RHEGGSGHTV TDDIDHFFSA PSITYREPHL
1201 SIYDVLEVQK EELDLSKDLM VLPHAPHRVF AWETGWTPGL RSLENDGTKL
1251 LDRIRDNYEG EFYWRYFAFI ADALITTLKP RYEDTHIRIH LDSNTRSFIV
1301 PIITTEYIRE KLSYSFYGSG GTYALSLSQY HMGIHIELSE SDVWIIDVDN
1351 VVRDVTIESD KIKKGDLIEG ILSTLSIEEH KIILNSHEIH FSGEVHGSHG
1401 FVSLIFSILE GINAIIEVDL LSKSYKLLIS GELKILMLNS HHIQQKIDYI
1451 GFHSELQKNI PYSFVDSEGK EHGFINGSTK EGLFVSEIPD VVLISKVHMD
1501 DSKPSFGYYS HNLKDVKVIT KDHVHILTGY YLKDDIKISL SLILQDEKTI
1551 KLHSVHLDES GVAEILKFMH RKGHINTSDS LMSFLESNHI KSIFVHFLQS
1601 HIKFILDANF IISGTTSIGQ FEFICDENDH IQPYFIKFNT LETNYTLYVG
1651 HRQHMIVEPN YDLDDSGDIS STVIHFSQKY LYGIDSCVHK VVISPNIYTD
1701 EINITPVYET HNTYPEVIVL DANYIHEKIH VHIHDLSIRY VWSHDGHDFI
1751 LMSTSEEHKV SQVKIRFVHV FKDKTLANKL SFHFSDKQDV PVSEIILSFT
1801 PSYYEDGLIG YDLGVSLYH EKFYINHFGH HVSGLIYIHD SLYYFKPPVH
1851 NLITGFVTVG DDKYYFNPIN GGAASIGETI IDDKNYYFNQ SGVLQTGVFS
1901 TEDGFKYFAP ANTLDEHLEG EAIDFTGKLI IDENIYYFDD HYRGAVEWKE
1951 LDGEMHYFSP ISGKAFKGLN QIGDYKYYFN SDGVMQKGFV SINDHKHYFD
2001 DSGVMKVGYT EIDGKHFYFA ENGEMQIGVF NTEDGFKYFA HHNEDLGNEE
2051 GEEISYSGIL NFNHKIYYFD DSFTAVVGWK DLEDGSKYYF DEDTAEAYIG
2101 LSLINDGQYY FNDDGIMQVG FVTINDKVFY FSDSGIIESG VQWIDDNYFY
2151 IDDNGIVQIG VFDTSDGYKY FAPANTVNDH IYGQAVEYSG LVRWGEDVYY
2201 FGETYTIETG WIYDMENESD KYYFNPETKK ACKGINLIDD IKYYFDEKGI
2251 MRTGLISFEN HNYYFNEHGE MQFGYINIED KMFYFGEDGV MQIGVFNTPD
2301 GFKYFAHQNT LDENFEGESI HYTGWLDLDE KRYYFTDEYI AATGSVIIDG
2351 EEYYFDPDTA QLVISE
```

(b)

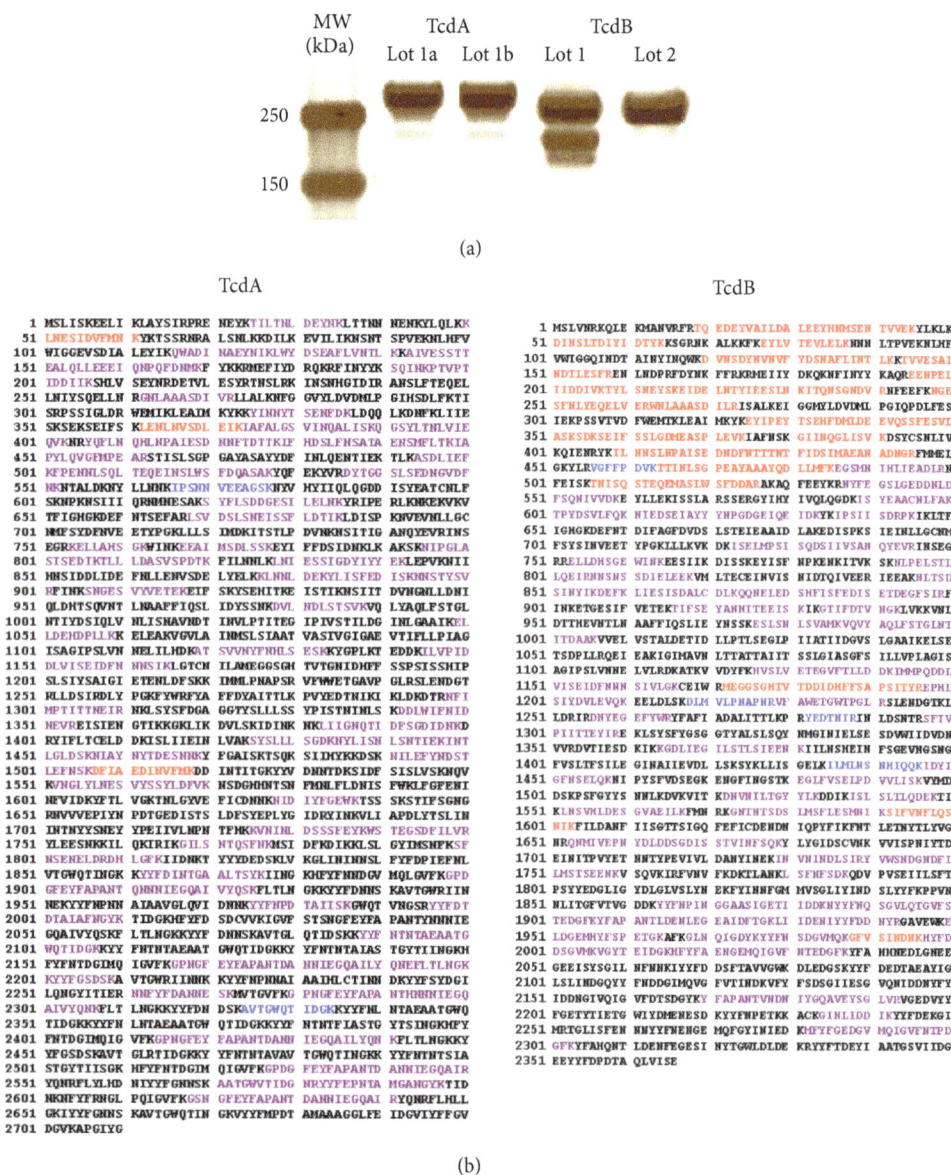

FIGURE 2: Gel-based analysis of purified *C. difficile* toxins. (a) SDS-PAGE of purified TcdA and TcdB. Only one band was observed in the TcdA lanes; two bands were observed in the TcdB lot 1 lane; one band was observed in the TcdB lot 2 lane as expected. (b) Amino acid sequence coverage obtained for the two lots of *C. difficile* toxins. The gel bands were extracted, digested, and MS analyzed. Peptide sequences detected were overlaid. Sequences in Red = peptides from run 1a (TcdA) or band 1 (TcdB); Blue = peptides from run 1b or band 2; Purple = common peptides; Black = not detected.

vendor only revealed the expected ~270 kDa band, and this lot was used in all further experiments.

The gel-independent approach provided unique information related to toxin quantification of small samples (<0.1 mL). The first set of experiments using BioS1 confirmed the robustness and efficiency of our modified rapid trypsin digestion method [21]. While trypsin provided sufficient AA sequence coverage to unambiguously identify TcdA and TcdB, using one enzyme may not provide enough AA sequence information to differentiate toxins produced by different *C. difficile* strains. The use of multiple proteases has been reported before not only to improve AA coverage, but also to differentiate homologous proteins, detect

posttranslational modifications, and identify editing events [25–27]. For initial discovery work, five different enzymes were used to obtain broad *C. difficile* AA coverage. However, since high-quality proteases are expensive, the use of multiple enzymes for each sample can be cost prohibitive. Working towards this, careful analysis of TcdA and TcdB AA sequence coverage maps from all five enzymes proved that using only two enzymes, trypsin and GluC, collectively garnered enough sequence information for most of our applications.

Quantification of TcdA and TcdB in samples irrespective of the volume capacity has typically been performed using enzyme-linked immunosorbent assays (ELISA) [4–8], cytotoxity assays [8, 28, 29], PCR [30], and gel densitometry [6].

TABLE 1: Absolute quantification of TcdA and TcdB in small samples using MS$^E$.

| | Amounts of TcdA and TcdB studied (ng) | | | | | |
|---|---|---|---|---|---|---|
| Digested (ng)* | **63** | **125** | **250** | **500** | **1,000** | **2,000** |
| Expected (ng)** | **1.25** | **2.5** | **5** | **10** | **20** | **40** |
| TcdA | | | | | | |
| Exp1*** | 0 | 0 | 3.08 | 8.7 | 16.46 | 31.7 |
| Exp2 | 0 | 0 | 3.2 | 6.5 | 18.4 | 30.2 |
| Exp3 | 0 | 0 | 3.4 | 6.3 | 16.3 | 32.7 |
| **Average** | 0 | 0 | 3.23 | 7.17 | 17.05 | 31.53 |
| Stdev | 0 | 0 | 0.16 | 1.33 | 1.17 | 1.26 |
| TcdB | | | | | | |
| Exp1*** | 0.84 | 1.1 | 2.05 | 7.9 | 14.1 | 36.05 |
| Exp2 | 0.83 | 1.1 | 2.5 | 4.6 | 13.04 | 33.4 |
| Exp3 | 0.83 | 1.2 | 3.4 | 4.6 | 13.7 | 26.4 |
| **Average** | 0.83 | 2.23 | 2.65 | 5.7 | 13.61 | 31.95 |
| Stdev | 0.01 | 0.06 | 0.69 | 1.91 | 0.54 | 4.99 |

Three samples each of purified toxin were serially diluted by a factor of 2 and digested with trypsin.
Each sample was analyzed three times using MSE. The numbers represent the amounts of toxin digested and the obtained values.
The minimum amounts of digested TcdA and TcdB in buffer that can be loaded on the nUPLC column and still detected were respectively 5 ng (1.6 $\mu$g/mL) and 1.25 ng (0.43 $\mu$g/mL).
*Values were estimated from the theoretical concentration based on values provided by the manufacturer.
**Amounts expected on column based on values provided by the manufacturer.
***Experimental values on column.

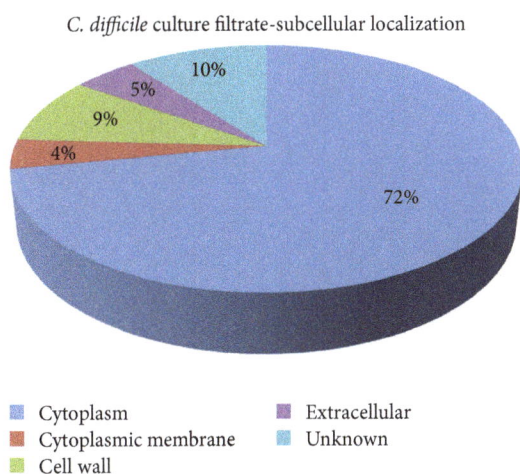

FIGURE 3: Subcellular localization of *C. difficile* proteins identified in a commercial culture filtrate using Psort score. Most proteins were cytoplasmatic (72%).

These methods are limited in that they generally measure total toxin and do not discriminate between TcdA and TcdB.

We report here the use of data-independent nUPLC-MS$^E$ to separately quantify *C. difficile* toxins in small sample volumes. After assessing the MS$^E$ spectrum data, standard curves were constructed for TcdA and TcdB, and the minimum amounts detected were determined for each toxin.

A review of MS$^E$ concepts and applications has been published by others [18, 19]. In addition, a recent systematic evaluation of the MS$^E$ method has been reported. The authors found that there is a linear dynamic range of three orders of magnitude and low limit of quantification when they tested complex mixtures in small volumes [31]. Since the nUPLC-MS$^E$ methodology is expected to be very sensitive [31], the obtained values reported in the present work for both toxins may not be ideal for samples containing small amounts of toxin. Therefore, our current research focuses on improving the sensitivity of our nUPLC-MS$^E$ method, coupling it with other known toxin concentration methods such as antibody capture prior to MS analysis [20].

Previously, we have used nUPLC-MS$^E$ for the quantification of botulinum toxin complexes [21, 22]. These findings taken with the successful quantification of purified *C. difficile* large toxins reported here led us to examine a commercial culture filtrate (CFil) to determine nUPLC-MS$^E$ utility for analysis of a complex *C. difficile* mixture. A proteomic analysis of CFil, normally used as a positive control in toxicity assays, revealed the identification of 101 unique *C. difficile* proteins, of which 24 could be quantified by MS$^E$. The proteins identified are comparable with results from previous *C. difficile* proteomic studies [13]. Most importantly, proteomic quantitative analysis of a culture filtrate demonstrated the potential of these methods for further studies.

Interestingly, label-free quantification of CFil identified four proteins (S-layer protein, TcdA, TcdB, and NAD-specific glutamate dehydrogenase) all detected in significantly large amounts. One possible explanation is that these proteins are traditionally abundant and thus would likely be detected in greater amounts. In addition, they are large proteins, which once digested likely have a larger peptide pool compared to other less abundant proteins, resulting in a greater chance for detection by nUPLC-MS$^E$. Even more, other proteins, such as glycine cleavage system protein H, isoprenoyl-CoA:2-hydroxyisocaproate CoA-transferase, and molecular chaperone DnaK, that have not been previously cited in proteomic studies were also detected in lower but significant

TABLE 2: Summary of the 24 most abundant C. *difficile* ATCC 43255 proteins identified and quantified in a commercial lyophilized culture filtrate.

| | Protein | gi number | MW (Da) | # Unique peptides | # Unique spectra | # Total spectra | % Sequence coverage | Protein.fmolOnColumn | Protein.ngramOnColumn |
|---|---|---|---|---|---|---|---|---|---|
| 1 | 50S ribosomal protein L1 | gi\|255305004 | 24,789 | 4 | 6 | 15 | 28.90% | 6.84 | 0.1698 |
| 2 | 50S ribosomal protein L7/L12 | gi\|255305006 | 12,602 | 10 | 14 | 33 | 75.20% | 36.5 | 0.4605 |
| 3 | Acetyl-CoA acetyltransferase | gi\|255305985 | 40,843 | 2 | 3 | 8 | 65.00% | 100 | 4.1031 |
| 4 | Cell surface protein (S-layer precursor protein) | gi\|255307831 | 79,907 | 7 | 9 | 19 | 78.20% | 376 | 30.0472 |
| 5 | Electron transfer flavoprotein alpha-subunit | gi\|255305398 | 37,131 | 13 | 17 | 39 | 39.70% | 58.7 | 2.1801 |
| 6 | Electron transfer flavoprotein beta-subunit | gi\|255305397 | 28,654 | 7 | 10 | 23 | 38.70% | 54.4 | 1.561 |
| 7 | Enolase | gi\|255308205 | 46,073 | 12 | 14 | 35 | 48.10% | 18.3 | 0.8425 |
| 8 | Gamma-aminobutyrate metabolism dehydratase/isomerase | gi\|255307364 | 55,084 | 6 | 7 | 12 | 17.70% | 10.8 | 0.5948 |
| 9 | Glycine cleavage system protein H | gi\|255305732 | 14,029 | 2 | 3 | 6 | 18.40% | 15.7 | 0.2206 |
| 10 | Isocaprenoyl-CoA: 2-hydroxyisocaproate CoA-transferase | gi\|255305392 | 44,218 | 3 | 3 | 6 | 35.30% | 13.9 | 0.6154 |
| 11 | Molecular chaperone DnaK | gi\|255307491 | 66,452 | 17 | 23 | 50 | 36.60% | 22.3 | 1.4796 |
| 12 | NAD-specific glutamate dehydrogenase | gi\|255305176 | 45,998 | 4 | 4 | 9 | 41.80% | 88.5 | 4.071 |

TABLE 2: Continued.

| Protein | gi number | MW (Da) | # Unique peptides | # Unique spectra | # Total spectra | % Sequence coverage | Protein.fmolOnColumn | Protein.ngramOnColumn |
|---|---|---|---|---|---|---|---|---|
| 13 Oligopeptide ABC transporter, substrate-binding lipoprotein | gi\|255305858 | 58,513 | 2 | 2 | 4 | 5.56% | 9.08 | 0.5315 |
| 14 Oligopeptide ABC transporter, substrate-binding protein | gi\|255307707 | 58,122 | 3 | 3 | 4 | 11.80% | 6.84 | 0.3979 |
| 15 Putative amino acid aminotransferase | gi\|255308702 | 44,781 | 13 | 14 | 29 | 47.00% | 29.4 | 1.3178 |
| 16 Putative aminoacyl-histidine dipeptidase | gi\|255305707 | 52,533 | 2 | 2 | 4 | 18.60% | 15.2 | 0.8 |
| 17 Putative anaerobic nitric oxide reductase flavorubredoxin | gi\|255306060 | 44,820 | 4 | 5 | 11 | 24.20% | 5.66 | 0.2538 |
| 18 Putative propanediol utilization protein | gi\|255307719 | 20,016 | 3 | 3 | 5 | 20.60% | 17.5 | 0.3513 |
| 19 Putative rubrerythrin | gi\|255306441 | 19,990 | 2 | 3 | 9 | 29.30% | 30.7 | 0.6144 |
| 20 Putative translation inhibitor endoribonuclease | gi\|255307543 | 13,582 | 4 | 5 | 14 | 75.40% | 26.2 | 0.356 |
| 21 Subunit of O²-sensit 2-hydroxyisocaproyl-CoA dehydratase | gi\|255305395 | 42,349 | 14 | 16 | 26 | 33.90% | 21.9 | 1.0175 |
| 22 Subunit of O²-sensit 2-hydroxyisocaproyl-CoA dehydratase | gi\|255305394 | 46,345 | 2 | 2 | 5 | 23.50% | 18.8 | 0.796 |
| 23 Toxin A | gi\|255305655 | 308,129 | 55 | 65 | 135 | 29.90% | 41.3 | 12.7356 |
| 24 Toxin B | gi\|255305652 | 269,707 | 41 | 50 | 96 | 28.20% | 41.3 | 7.2551 |

*Concentration based on spectral counting

(−)  (+)

Proteins were identified using PLGS and Mascot search parameters and validated as greater than 95% peptide probability, min 2 peptides, 99% protein probability using Scaffold; FDR was zero. The colors in protein nanogram OnColumn are based on the estimated protein amounts.

amounts. This finding suggests that culture supernatants may harbor a pool of specific proteins and could be a key matrix to search for potential unique *C. difficile* biomarkers. Because of the high TcdA and TcdB toxin concentration detected in the CFil, we initially hypothesized that the filtrate had been enriched by the manufacturer. However, we were assured by the manufacturer that the CFil used in these studies is indeed a filtrate of the culture supernatant of the strain VPI 10463, ATCC43255, which has not undergone any type of enhancement treatment, besides filtration, that would concentrate the protein pool present in this matrix (personal communication). Interestingly, a recent publication describes the proteome examination of culture supernatants of three *C. difficile* strains, two of them hypervirulent isolates [13]. The authors used gel-based analysis followed by MS and identified 5 unique proteins among the hypervirulent strains. They emphasized the usefulness of proteomics to discover and quantify specific biomarkers for hypervirulent strains since at least 234 unique genes have been identified in the strains that cause the majority of hospital outbreaks in North America and Europe [32].

## 5. Conclusions

Proteomic analyses using gel-based and gel-independent platforms were used to further characterize and quantitate the large *C. difficile* toxins. The overall goal was to determine the potential of proteomics using mass spectrometry-based methods to develop a quantification method for these large toxins. Moreover, the use of MS$^E$ for *C. difficile* toxin quantification in small samples, and for multienzymatic digestion to increase protein amino acid sequence coverage, was performed. The gel-based work revealed basic information on toxin integrity and provided sequence confirmation. The gel-independent platform was applied to in-solution digestion of both toxins using five different enzymes followed by analysis using two different mass spectrometer instruments. Broad amino acid sequence coverage for TcdA (91%) and for TcdB (95%) was generated using this approach. These data, if coupled to *in silico* sequencing analysis, suggest that the method has the potential to determine subtle sequence differences of TcdA and TcdB from different *C. difficile* toxinotypes. Moreover, label-free proteomics using MS$^E$ data collection and analysis provided the ability potential to determine the absolute quantity of TcdA and TcdB in small samples and was applied to a culture filtrate. A proteomic study of the culture filtrate demonstrated that the most abundant proteins are S-layer protein, TcdA, TcdB, and NAD-specific glutamate dehydrogenase. Taken together, data presented in this study provide performance characteristics and the basis for future development of improved MS-based detection and quantification methods in determining the factors involved in *C. difficile* toxin production by different isolates.

## Disclaimer

References in this paper to any specific commercial products, processes, services, manufacturers, or companies do not

constitute an endorsement or a recommendation by the US government or the Centers for Disease Control and Prevention (CDC). The findings and conclusions in this report are those of the authors and do not necessarily represent the views of CDC.

## Conflict of Interests

The authors do not have any conflict of interests.

## Acknowledgment

The authors wish to thank the members of the Biological Mass Spectrometry Laboratory at the National Center for Environmental Health, CDC, for helpful discussions.

## References

[1] G. P. Carter, J. I. Rood, and D. Lyras, "The role of toxin A and toxin B in *Clostridium difficile*-associated disease: past and present perspectives," *Gut Microbes*, vol. 1, no. 1, pp. 58–64, 2010.

[2] M. Rupnik, M. H. Wilcox, and D. N. Gerding, "*Clostridium difficile* infection: new developments in epidemiology and pathogenesis," *Nature Reviews Microbiology*, vol. 7, no. 7, pp. 526–536, 2009.

[3] C. V. Gould and L. C. McDonald, "Bench-to-bedside review: *Clostridium difficile* colitis," *Critical Care*, vol. 12, no. 1, p. 203, 2008.

[4] M. Merrigan, A. Venugopal, M. Mallozzi et al., "Human hypervirulent *Clostridium difficile* strains exhibit increased sporulation as well as robust toxin production," *Journal of Bacteriology*, vol. 192, no. 19, pp. 4904–4911, 2010.

[5] M. Warny, J. Pepin, A. Fang et al., "Toxin production by an emerging strain of *Clostridium difficile* associated with outbreaks of severe disease in North America and Europe," *Lancet*, vol. 366, no. 9491, pp. 1079–1084, 2005.

[6] S. Karlsson, L. G. Burman, and T. Åkerlund, "Induction of toxins in *Clostridium difficile* is associated with dramatic changes of its metabolism," *Microbiology*, vol. 154, no. 11, pp. 3430–3436, 2008.

[7] K. Mukherjee, S. Karlsson, L. G. Burman, and T. Åkerlund, "Proteins released during high toxin production in *Clostridium difficile*," *Microbiology*, vol. 148, no. 7, pp. 2245–2253, 2002.

[8] P. Vohra and I. R. Poxton, "Comparison of toxin and spore production in clinically relevant strains of *Clostridium difficile*," *Microbiology*, vol. 157, no. 5, pp. 1343–1353, 2011.

[9] A. Wright, D. Drudy, L. Kyne, K. Brown, and N. F. Fairweather, "Immunoreactive cell wall proteins of *Clostridium difficile* identified by human sera," *Journal of Medical Microbiology*, vol. 57, no. 6, pp. 750–756, 2008.

[10] A. Wright, R. Wait, S. Begum et al., "Proteomic analysis of cell surface proteins from *Clostridium difficile*," *Proteomics*, vol. 5, no. 9, pp. 2443–2452, 2005.

[11] T. D. Lawley, N. J. Croucher, L. Yu et al., "Proteomic and genomic characterization of highly infectious *Clostridium difficile* 630 spores," *Journal of Bacteriology*, vol. 191, no. 17, pp. 5377–5386, 2009.

[12] S. Jain, C. Graham, R. L. J. Graham, G. McMullan, and N. G. Ternan, "Quantitative proteomic analysis of the heat stress response in *Clostridium difficile* strain 630," *Journal of Proteome Research*, vol. 10, no. 9, pp. 3880–3890, 2011.

[13] A. Boetzkes, K. W. Felkel, J. Zeiser, N. Jochim, I. Just, and A. Pich, "Secretome analysis of *Clostridium difficile* strains," *Archives of Microbiology*, 2012.

[14] J. R. Barr, V. L. Maggio, D. G. Patterson Jr. et al., "Isotope dilution—mass spectrometric quantification of specific proteins: model application with apolipoprotein A-I," *Clinical Chemistry*, vol. 42, no. 10, pp. 1676–1682, 1996.

[15] B. Domon and R. Aebersold, "Options and considerations when selecting a quantitative proteomics strategy," *Nature Biotechnology*, vol. 28, no. 7, pp. 710–721, 2010.

[16] S. Pan, R. Aebersold, R. Chen et al., "Mass spectrometry based targeted protein quantification: methods and applications," *Journal of Proteome Research*, vol. 8, no. 2, pp. 787–797, 2009.

[17] W. Zhu, J. W. Smith, and C. M. Huang, "Mass spectrometry-based label-free quantitative proteomics," *Journal of Biomedicine and Biotechnology*, vol. 2010, Article ID 840518, 6 pages, 2010.

[18] S. J. Geromanos, J. P. C. Vissers, J. C. Silva et al., "The detection, correlation, and comparison of peptide precursor and product ions from data independent LC-MS with data dependant LC-MS/MS," *Proteomics*, vol. 9, no. 6, pp. 1683–1695, 2009.

[19] J. C. Silva, R. Denny, C. Dorschel et al., "Simultaneous qualitative and quantitative analysis of the *Escherichia coli* proteome: a sweet tale," *Molecular and Cellular Proteomics*, vol. 5, no. 4, pp. 589–607, 2006.

[20] A. E. Boyer, M. Gallegos-Candela, R. C. Lins et al., "Quantitative mass spectrometry for bacterial protein toxins—a sensitive, specific, high-throughput tool for detection and diagnosis," *Molecules*, vol. 16, no. 3, pp. 2391–2413, 2011.

[21] H. Moura, R. R. Terilli, A. R. Woolfitt et al., "Studies on botulinum neurotoxins type/C1 and mosaic/DC using Endopep-MS and proteomics," *FEMS Immunology and Medical Microbiology*, vol. 61, no. 3, pp. 288–300, 2011.

[22] R. R. Terilli, H. Moura, A. R. Woolfitt, J. Rees, D. M. Schieltz, and J. R. Barr, "A historical and proteomic analysis of botulinum neurotoxin type/G," *BMC Microbiology*, vol. 11, article 232, 2011.

[23] R. West, J. Whitmon, Y. M. Williamson et al., "A rapid method for capture and identification of immunogenic proteins in *Bordetella pertussis* enriched membranes fractions: a fast-track strategy applicable to other microorganisms," *Journal of Proteomics*, vol. 75, no. 6, pp. 1966–1972, 2012.

[24] N. Y. Yu, J. R. Wagner, M. R. Laird et al., "PSORTb 3.0: improved protein subcellular localization prediction with refined localization subcategories and predictive capabilities for all prokaryotes," *Bioinformatics*, vol. 26, no. 13, pp. 1608–1615, 2010.

[25] R. G. Biringer, H. Amato, M. G. Harrington, A. N. Fonteh, J. N. Riggins, and A. F. R. Hühmer, "Enhanced sequence coverage of proteins in human cerebrospinal fluid using multiple enzymatic digestion and linear ion trap LC-MS/MS," *Briefings in Functional Genomics and Proteomics*, vol. 5, no. 2, pp. 144–153, 2006.

[26] F. Fischer and A. Poetsch, "Protein cleavage strategies for an improved analysis of the membrane proteome," *Proteome Science*, vol. 4, article 2, 2006.

[27] D. L. Swaney, C. D. Wenger, and J. J. Coon, "Value of using multiple proteases for large-scale mass spectrometry-based proteomics," *Journal of Proteome Research*, vol. 9, no. 3, pp. 1323–1329, 2010.

[28] J. E. Blake, F. Mitsikosta, and M. A. Metcalfe, "Immunological detection and cytotoxic properties of toxins from toxin A-positive, toxin B-positive *Clostridium difficile* variants," *Journal of Medical Microbiology*, vol. 53, no. 3, pp. 197–205, 2004.

[29] G. Yang, B. Zhou, J. Wang et al., "Expression of recombinant *Clostridium difficile* toxin A and B in *Bacillus megaterium*," *BMC Microbiology*, vol. 8, article 192, 2008.

[30] T. Åkerlund, B. Svenungsson, Å. Lagergren, and L. G. Burman, "Correlation of disease severity with fecal toxin levels in patients with *Clostridium difficile*-associated diarrhea and distribution of PCR ribotypes and toxin yields in vitro of corresponding isolates," *Journal of Clinical Microbiology*, vol. 44, no. 2, pp. 353–358, 2006.

[31] Y. Levin, E. Hradetzky, and S. Bahn, "Quantification of proteins using data-independent analysis ($MS^E$) in simple and complex samples: a systematic evaluation," *Proteomics*, vol. 11, no. 16, pp. 3273–3287, 2011.

[32] R. A. Stabler, M. He, L. Dawson et al., "Comparative genome and phenotypic analysis of *Clostridium difficile* 027 strains provides insight into the evolution of a hypervirulent bacterium," *Genome Biology*, vol. 10, no. 9, article R102, 2009.

# 14

# Human Myocardial Protein Pattern Reveals Cardiac Diseases

**Jonas Bergquist,[1] Gökhan Baykut,[2] Maria Bergquist,[1,3] Matthias Witt,[2] Franz-Josef Mayer,[2] and Doan Baykut[4]**

[1] Analytical Chemistry, Department of Chemistry, Biomedical Center and SciLife Lab, Uppsala University, P.O. Box 599, 751 24 Uppsala, Sweden
[2] Bruker Daltonik GmbH, 28359 Bremen, Germany
[3] Department of Medical Sciences, Hedenstierna Laboratory, Uppsala University, 75185 Uppsala, Sweden
[4] Institute of Biophysics, University of Frankfurt, 60438 Frankfurt/M, Germany

Correspondence should be addressed to Jonas Bergquist, jonas.bergquist@kemi.uu.se

Academic Editor: Ákos Végvári

Proteomic profiles of myocardial tissue in two different etiologies of heart failure were investigated using high performance liquid chromatography (HPLC)/Fourier transform ion cyclotron resonance mass spectrometry (FT-ICR MS). Right atrial appendages from 10 patients with hemodynamically significant isolated aortic valve disease and from 10 patients with isolated symptomatic coronary heart disease were collected during elective cardiac surgery. As presented in an earlier study by our group (Baykut et al., 2006), both disease forms showed clearly different pattern distribution characteristics. Interesting enough, the classification patterns could be used for correctly sorting unknown test samples in their correct categories. However, in order to fully exploit and also validate these findings there is a definite need for unambiguous identification of the differences between different etiologies at molecular level. In this study, samples representative for the aortic valve disease and coronary heart disease were prepared, tryptically digested, and analyzed using an FT-ICR MS that allowed collision-induced dissociation (CID) of selected classifier masses. By using the fragment spectra, proteins were identified by database searches. For comparison and further validation, classifier masses were also fragmented and analyzed using HPLC-/Matrix-assisted laser desorption ionization (MALDI) time-of-flight/time-of-flight (TOF/TOF) mass spectrometry. Desmin and lumican precursor were examples of proteins found in aortic samples at higher abundances than in coronary samples. Similarly, adenylate kinase isoenzyme was found in coronary samples at a higher abundance. The described methodology could also be feasible in search for specific biomarkers in plasma or serum for diagnostic purposes.

## 1. Introduction

Understanding the differences of proteomic profiles has a crucial importance for gaining insight into molecular mechanisms of disease. Although the molecular origin of the cardiac dysfunction is still largely unknown in the majority of heart diseases [1, 2], heart insufficiency is increasingly expected to be a result from alterations in gene and protein expression. Gene expression is a more static process; however, protein patterns may be altered faster in close relationship with the appearance of disease, making up an adequate identification of proteome characteristics important. If different etiologies and pathways of cardiac disease progression can be represented in the form of unequivocal proteomic patterns, these patterns may help simplify and accelerate the diagnostics and be used as an appropriate diagnostic method in daily clinical routine [3–6]. Molecular differences extracted out of these patterns may also reveal potential biomarkers that could be targeted for screening purposes [7, 8]. A possibility to locate some of these molecules in circulating blood, where an easy access for high-throughput assays would be valid, is especially attractive.

The objective of this study is to evaluate selected individual myocardial samples which are representative for aortic valve disease (AVD) and coronary heart disease (CHD), respectively. As presented in our recent study [9], proteomic

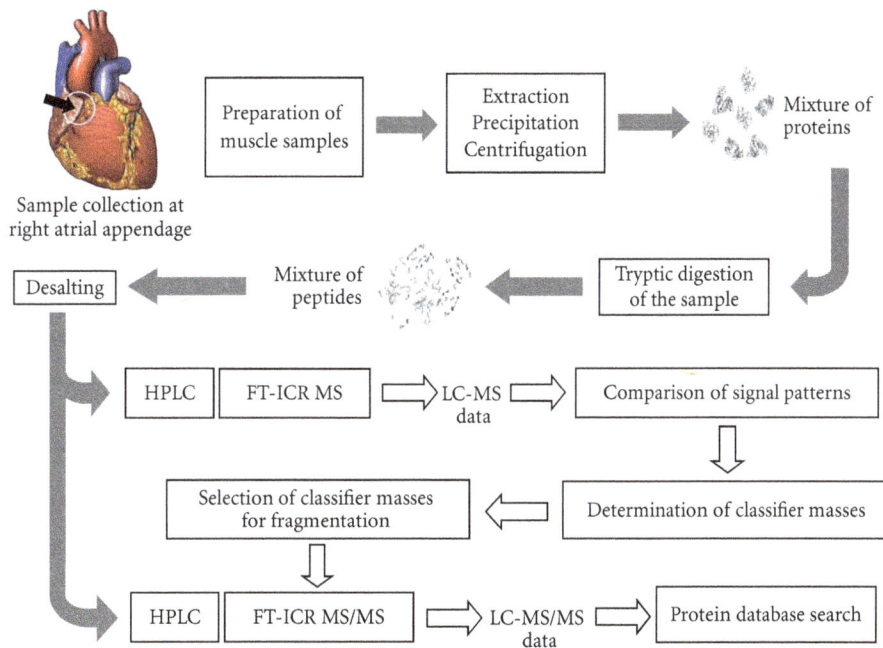

FIGURE 1: Flowchart for the path of the sample preparation and measurements with 9.4 Tesla FT-ICR MS. The measurement of the samples with two different instrument had the reason that the samples were measured with a classical FT-ICR instrument without external MS/MS capability. After the differential mass spectrometric runs followed by the pattern comparison, a quadrupole/hexapole FT-ICR instrument with external MS/MS capability was available. The fragmentation studies for the protein identification were performed with this latter instrument.

pattern distribution characteristics of myocardial tissue in AVD and CHD were found to be clearly different using liquid-chromatography Fourier transform ion cyclotron resonance mass spectrometric (LC-FT-ICR MS) analysis. Classification patterns obtained from the comparison of LC-FT-ICR MS of digested proteins in the samples could be used for correctly sorting them in exact categories. The endogenous proteins were analyzed using a "bottom-up" proteomic approach that starts by a tryptic digestion of all the proteins in the samples and then analyzing the digests through LC and online electrospray ionization of the analytes into a quadrupole/hexapole FT-ICR MS. After isolation in the quadrupole, collision-induced dissociation (CID) of selected peptides can be performed in the hexapole collision cell, and using the acquired fragment spectra, proteins can be identified by database searches. For comparison and further validation, selected peptides were also fragmented and analyzed using liquid-chromatography-matrix-assisted laser desorption-ionization time-of-flight/time-of-flight mass spectrometry (LC-MALDI-TOF/TOF MS). With the combination of these techniques, the measurement of peptides from LC-separated samples results in precise mass information on the ppm-level, in very high resolution and in high sensitivity that in itself has been successfully used to explore the protein content of body fluids, for example, plasma and cerebrospinal fluid (CSF) [10–13]. In addition, sequence information essential for significant protein identification in minute tissue samples, for example, laser-dissected human spinal cord tissue can be obtained [14].

## 2. Materials and Methods

As a general overview for the method used, a flow chart is shown in Figure 1 depicting the sample preparation, mass spectrometric methods and measurements, as well as the data evaluation.

*2.1. Sample Selection and Data.* In 20 patients undergoing cardiac surgery with extracorporeal circulation, right atrial appendages were subject to removal and discarding for venous cannulation that were collected intraoperatively, after approval by the local ethical committee. The group of patients consisted of 10 individuals with hemodynamically significant, isolated AVD and 10 with isolated symptomatic CHD. Patient selection was made in such a way that all patients with CHD selected for this study were "healthy" in terms of the AVD and all selected AVD patients were "healthy" in terms of the coronary disease. The median age was 62 years (45–81) in the AVD group, and 66 years (37–83) in the CHD group. All antithrombotic agents were suspended 10–14 days prior to surgery. In all studied cases, patients stopped taking aspirin one week before the operation. Cardiac samples (each of them in the range of $1 \text{ cm}^3$) were immediately washed in Krebs-Henseleit solution (118 mM NaCl, 4.7 mM KCl, 1.2 mM $MgSO_4$, 1.25 mM $CaCl_2$, 1.2 mM $KH_2PO_4$, 25 mM $NaHCO_3$, 11 mM glucose) and fixed on wax plates at room temperature. After separation from the epicardium, the trabecular tissue was shock-frozen in liquid nitrogen and stored at $-75°C$.

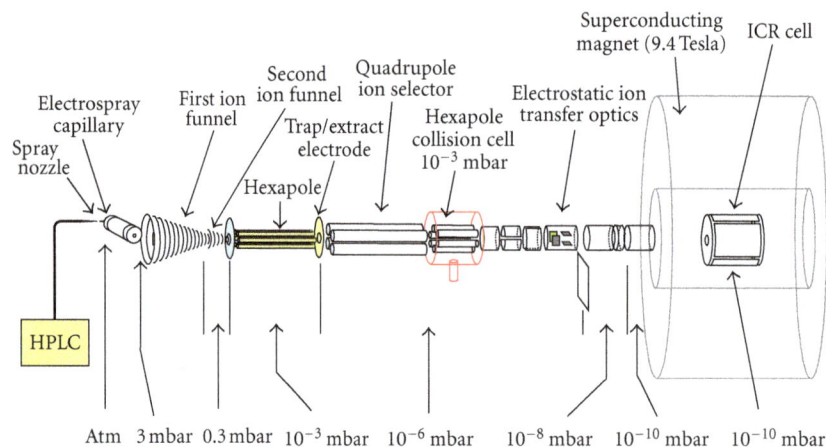

Figure 2: Non-scaled schematic view of an LC-FT-ICR MS system with a quadrupole mass selector and a hexapole collision cell. In the electrospray ion source the formed ions are captured after they pass the electrospray capillary in an ion funnel, which increases the sensitivity of the system by roughly up to an order of magnitude. Ions are transferred through two ion funnels into a hexapole ion guide, where they can also be trapped. Ions are selected in the quadrupole mass selector and can undergo collision induced dissociation in the hexapole collision chamber which is at a relatively high pressure. The ICR cell is in the magnetic center of a 9.4T superconducting magnet. The vacuum system, not shown in the figure, consists of pumping stages down to the range of $10^{-10}$ mbar in the ultra high vacuum chamber of the ICR cell. The numbers shown are approximate pressures in different pumping stages.

## 2.2. Sample Preparation.

Sample selection and preparation of the tissue specimens have been described in detail in our earlier study [9]. Briefly, the tissues were homogenized in 8 M urea, 0.4 M $NH_4HCO_3$ followed by reduction of the disulfide bridges, carbamidomethylation of cysteins, and tryptic digestion according to the manufacture's protocol (Modified Trypsin, sequencing grade, Promega GmbH). The digests were SPE extracted (Spec Plus $C_{18}$ AR, Ansys Diagnostics Inc, CA), vacuum-centrifuged to dryness, and then redissolved in 0.1% formic acid before injection into an LC system. Acetonitrile and acetic acid, formic acid and triflouro acetic acid were purchased from Merck (Darmstadt, Germany). Urea, ammonium carbonate, and iodoacetamide were obtained from Sigma Chemical Co. (St. Louis, MO, USA) and were used without further purification. Dithiothreitol was purchased from Amersham Biosciences (Uppsala, Sweden). All used chemicals were of analytical grade. Trypsin, sequence grade from bovine pancreas (1418475), was obtained from Roche Diagnostics (Mannheim, Germany). Water was purified with a Milli-Q purification system (Millipore, Bedford, MA, USA). All fused-silica capillaries were obtained from Polymicro Technologies (Phoenix, AZ, USA).

## 2.3. LC/FT-ICR MS/MS for Protein Identification.

The total amount of each digested sample was dissolved in 250 μL 0.1% trifluoroacetic acid (TFA), split into 5 vials each of 50 μL sample solution and Speed Vacced to dryness at medium temperature. The dried sample of one vial of each sample was dissolved in 50 μL 0.1% TFA and diluted 1 : 10 with 0.1% formic acid. An aliquot of 5 μL of the sample solution was injected on a Nano LC column (see below) for FT-ICR-NanoLC/MS/MS analysis.

Mass spectra were acquired with an APEX Qe FT-ICR (Bruker Daltonics Inc., Billerica, MA, USA) equipped with a 9.4 Tesla superconducting magnet (Bruker BioSpin, Wissembourg, France). A schematic description of the LC-FT-ICR MS system with a quadrupole mass selector and a hexapole collision cell is shown in Figure 2. Samples were ionized in an "Apollo II" electrospray ion source (Bruker Daltonics, Billerica, MA, USA) with ion funnel technology for efficient capture of the generated ions. Ions were detected in a cylindrical ion cyclotron resonance cell with segmented trapping plates (infinity cell) [15]. Nano LC-MS/MS measurements were performed with a Nano LC system Ultimate 3000 (Dionex, Sunnyvale, CA, US) fitted with 2 10-port valves. An online-nanospray ESI source (Bruker Daltonics, Billerica, MA, USA) was coupled to the NanoLC system. This ion source was equipped with an angled off-axis spraying system, that used a PicoTip adapter (New Objective, MA, USA) to connect the tubing of the NanoLC column to a distal-coated fused silica needle that had a 10 μm inner diameter (New Objective, MA, USA). A spring in the needle holder provided the connection from the needle to ground potential. The tip of the needle was placed a few mm in front of the orifice of the glass capillary at an angle of approximately 70°. The voltage applied to the metal-coated capillary entrance of the electrospray source was set to −1500 V. The NanoLC column PepMap C-18 (75 μm inner diameter, 15 cm length, 5 μm particle size, Dionex, Sunnyvale, CA, USA) was used for compound separation with a flow of 200 nL/min. The peptides were first captured on a C-18 precolumn for desalting of the sample and eluted from this column to the PepMap column with a 185 min gradient for peptide separation of highly complex sample (solvent A: water with 0.1% formic acid, solvent B: acetonitrile with 0.1% formic acid, 0 min 2% solvent B, 5 min 2% solvent B, 175 min 40% solvent B, 190 min 50% solvent B, 191 min 90% solvent B, 205 min 90% solvent B, 206 min 2% solvent B, 250 min 2% solvent B). After storing

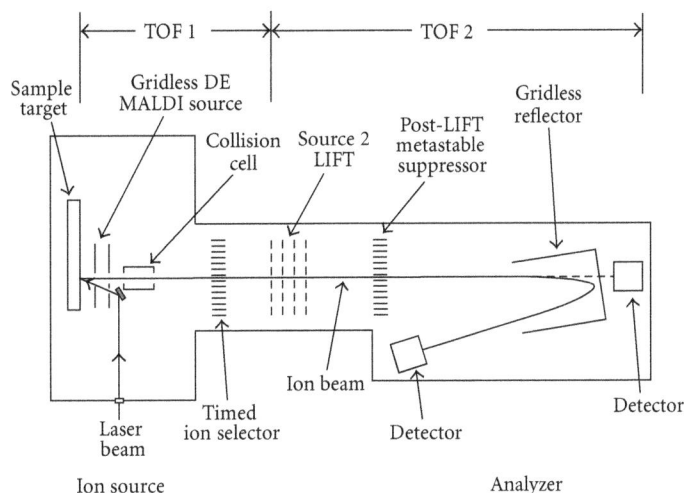

FIGURE 3: Non-scaled schematic view of a MALDI-TOF/TOF mass spectrometer.

the sample on the precolumn, this column was washed for 5 minutes with 0.1% TFA for desalting using a 30 $\mu$L/min flow and than switched to gradient flow. The pressure in the collision cell (located between the quadrupole filter and the ICR analyzer cell) for CID was set by increasing the read-out pressure of the source ion-gauge to about $4.5 \times 10^{-6}$ mbar (approx., a factor of 10 higher than without pressure in the collision cell). The default parameter setting was used for collision energy calculation for CID fragmentation. The FT-ICR mass spectra were acquired with the acquisition software Apex Control 1.0 (Bruker Daltonics, Billerica, MA, USA) using data-dependent MS/MS acquisition. The ion accumulation time was set to 0.5 s for the MS scan and 2 s for the MS/MS. The mass range was set to m/z 246 to m/z 2000 for the MS and MS/MS spectra using 512 k data points. The mass spectra were processed with 512 k data points and sine apodization. Further processing of the data including mass deconvolution and mgf-file generation for database search was performed with Data Analysis 3.3 (Bruker Daltonics, Billerica, MA, USA).

The mass spectra were processed using the protein analysis software Biotools 3.0 (Bruker Daltonics, Billerica, MA, USA) for database searches and results interpretation. The MASCOT search engine (Matrix Science, London, UK) [16, 17] was used to search the SwissProt database. MOWSE (molecular weight search) scores, assigned by MASCOT, were calculated based on the algorithm described by Pappin et al. [18]. Search criteria for all experiments included one missed cleavage, taxonomy human, fixed modification carbamidomethyl, and variable modifications oxidation of methionine. Mass tolerances for the MS/MS search were set to 4 ppm for the parent mass and 0.01 Da for the fragment masses. In a second search, 2 missed cleavages were allowed, and in a third search, also unspecific cleavages were accepted. Mass spectra were externally calibrated with a siloxane mixture of the chemical background signals using a two parameter linear calibration [19].

*2.4. LC/MALDI TOF/TOF MS for Protein Identification.* $\alpha$-cyano-hydroxy cinnamic acid (CHCA) and the profiling

kit Magnetic Beads based on Weak Cation Exchange Chromatography (MB-WCX) were obtained from Bruker Daltonik, Bremen, Germany. HPLC gradient grade and ammonium citrate, dibasic, ACS grade, were obtained from Sigma-Aldrich (Steinheim, Germany).

The capillary high-performance liquid chromatography (HPLC) separations were performed using an Agilent 1100 Cap-LC system equipped with a 15 cm $\times$ 180 $\mu$m I.D. PepMap column with a 3 $\mu$m $C_{18}$ stationary phase (LC Packings, Amsterdam, The Netherlands). For all separations eluent A consisted of 0.1% TFA/water and 2% $CH_3CN$ and eluent B was 0.05% TFA in 100% $CH_3CN$. An isocratic flow of eluent A, 2 $\mu$L/min for 5 min, was followed by a gradient from 2% to 40% $CH_3CN$ in 90 minutes. All separations were performed at ambient temperature and the injection volume was 8 $\mu$L. After each 15 s fractions of 0.5 $\mu$L were dispensed onto a spot on the MALDI target.

The MALDI-TOF-MS instrument used was an Ultraflex II TOF/TOF (Bruker Daltonik GmbH, Bremen, Germany) [20]. It was equipped with a Smart Beam laser [21]. The focus diameter was approximately 60 $\mu$m. All spectra were acquired using the reflectron mode of the instrument (Figure 3) at 25 kV acceleration voltage. While the MS data was recorded in a fully automated fashion the MS/MS data of peptides was acquired in a manual operation mode of the instrument. In order to acquire MS/MS spectra, precursor ions were accelerated out of ion source (S1) by means of time delayed extraction. Fragmentation was performed either by laser-induced or collision-induced dissociation or a combination of both. The collision cell was located right after S1. Argon was used as a collision gas. In the first field free region of the TOF1, the precursor ions including the already created related fragment ions are passed straight through an ion selection unit, while all the rest is deflected. The resolution of this unit is about 750 FWHH (single mass resolution at mass 750 Da). The selected precursor ions and their corresponding fragment ions then enter a second ion source (S2). Here, they are accelerated again, leading to a separation of their mass in the second field free drift region (TOF2). All fragment ions are separated in time and focused in energy according to their

$m/z$ ($z = 1$) by a combination of time-delayed extraction out of S2 and time focusing in the reflectron of TOF2. Depending on their mass, typical fragment ion resolution is in the range of 2000–5000.

Despite a 90 min separation of the peptides in the HPLC and a spread of fractions onto 384 spots, the acquired MS spectra were still very complex. On one side, complexity increased suppression effects during desorption/ionization, on the other side, selection of individual peptides for MS/MS often was ambiguous. In order to simplify MS spectra and avoid mixing of fragments from different precursor ions prior to the LC separation, selected samples were prefractionated using Magnetic Beads based on Weak Cation Exchange Chromatography (Profiling Kit 100 MB-WCX, Bruker Daltonik GmbH, Bremen, Germany). The freeze-dried samples were dissolved in $10\,\mu L$ 0.1% TFA. After a further 1 : 10 dilution, $5\,\mu L$ was used to bind the peptides to the beads. $8\,\mu L$ of the supernatant was used in the first run of the LC separation. After washing of the beads with washing buffer, the bound peptides were eluted using 3 × $3\,\mu L$ of elution buffer. In order to remove the acetonitrile from the eluent, the volume was reduced in a speedvac by a factor of two. The remaining volume of $4\,\mu L$ was then used for a second run of the LC separation.

The target of choice for the LC-MALDI approach was a prespotted Anchor Chip target (Bruker Daltonik GmbH, Bremen, Germany). Within a 90 min LC run, all 384 sample spots on the target were prepared. After evaporation of the solvent, the target was washed by dipping it for a few seconds into a solution of 10 mM ammonium citrate/0.1% TFA. The sample plate was then introduced into the Ultraflex II MALDI-TOF mass spectrometer, and mass spectra were recorded from each prepared sample spot. The calibration used was an external near-neighbor calibration by means of additional 96 prespotted calibration spots. The sample used for calibration was a mixture of peptides covering the mass range from 700 Da–3500 Da already prepared onto the disposable target. The acquisition process was controlled by the WARP-LC software (Bruker Daltonik GmbH, Bremen, Germany) and a compound list was created. From these peptide masses, scored according to their signal-to-noise and the complexity of the MS spectrum, the candidates of possible biomarkers were selected manually and their MS/MS spectrum was recorded.

The spectra were processed by means of Flex Analysis and the peak lists were sent to BioTools for the database search using Mascot (Matrix Science, London, UK, [17]). The following search parameters were chosen for the database search. Taxonomy: human; database: SwissProt, variable modification: carbamidomethylation (C) and oxidation (M): MS/MS tolerance: 0.5 Da, Partials: 1. In a second search, 2 missed cleavages were allowed, and in a third search also unspecific cleavages were accepted.

*2.5. Pattern Recognition and Extraction of Classification Features.* The pattern recognition and classification strategy have been described in detail by Ramström et al. [10] and in Baykut et al. [9]. Briefly, a numerical code was developed

for the analysis. The first part of the code normalizes the spectrogram intensity, removes the noise, and calibrates the individual HPLC/FT-ICR mass chromatograms in time to a common "table" sample. The second part of the code concerns the calibrated sample classification and extraction of the classification features. In case of a binary classification, the features are represented by those pattern peaks, which are more abundant in the majority of the samples of one class. These peaks are defined as the characteristic peaks. Thus, two lists of characteristic peaks were generated for each classification. To extract the classification features, the selection of best individual features was applied and the "nearest mean classifier" was used for the classification of test samples and the sample projection onto classification patterns.

## 3. Results

Examples of two-dimensional data obtained from HPLC-FT-ICR mass spectrometry are shown in Figure 4 for an aortic (a) and a coronary (b) sample.

*3.1. Comparison and Classification.* As described in detail in our recent publication [9], the classification using 240 characteristic peaks in the CHD group and 90 characteristic peaks in the AVD group analyzed using LC-FT-ICR MS (Figure 2) resulted in a clear difference in distribution patterns. In Figure 5, the symbols represent both cardiac disorders accumulated on either side of the diagonal line with increased distance from the diagonal related to the specification for the particular type of disease. Above the diagonal, samples from myocardial tissue with CHD and below the diagonal samples from myocardial tissue with AVD are located. Each filled symbol represents an individual supervised classified training sample while each open symbol shows an individual unsupervised classified test sample. The algorithm led to two ambiguous classifications of the unknown test samples. When the mass chromatograms for these ambiguous samples were manually inspected, large gaps in the total ion chromatograms due to unstable electrospray and sudden changes in elution rates could be observed, resulting in difficulties for the algorithm to correctly calibrate and align the datasets. There were intense and stable representative peaks found in both classes. Having a clear difference in proteomic pattern between both myocardial disorders, no interference between aortic and coronary samples was registered, even if the distribution of open spots on either sides of the diagonal (unsupervised classified test samples) displayed a closer distance to the diagonal. The AVD samples showed a closer accumulation pattern of the spots, indicating a more specific classification of the proteins compared to the coronary samples.

*3.2. Protein Identification Results.* Using the digested proteins from chosen samples from patients with CHD as well as with AVD, selected classifier masses were fragmented by collision-induced dissociation in the hexapole collision chamber of the FT-ICR system. Fragmentation spectra of these peptides are used for the identification of the proteins by database search

TABLE 1: Proteins identified as classifiers in coronary samples by LC-FT-ICR MS/MS. On-line measurements with electrospray ionization. Proteins identified as classifiers in aortic samples by LC-FT-ICR MS/MS. On-line measurements with electrospray ionization.

(a) FT-ICR MS/MS database search identification of classifier proteins from CHD patients

| Protein ID in coronary samples | Peptide sequence | Sequence tag | m/z (charge state) | Mascot score |
|---|---|---|---|---|
| (P13533) Myosin heavy chain, cardiac muscle alpha isoform (MyHC-alpha) MYH6_HUMAN | KLAEKDEEMEQAK NLQEEISDLTEQLGEGGKNVHELEKVR | 1577–1589 1506–1532 | 774.884 (2+), 516.925 (3+) 766.896 (4+) | 65 (92), 28 113 |
| (P12883) Myosin heavy chain, cardiac muscle beta isoform (MyHC-beta) MYH7_HUMAN | KLAEKDEEMEQAK | 1575–1587 | 774.884 (2+), 516.925 (3+) | 65, 28 |
| (P00568) Adenylate kinase isoenzyme 1 (EC 2.7.4.3) (ATP-AMP transphosphorylase) (AK1) (Myokinase) KAD1_HUMAN | GQIVPLETVLDMLR | 64–77 | 792.447 (2+) | 105 |
| (P45379) Troponin T, cardiac muscle (TnTc) (Cardiac muscle troponin T) (cTnT) TNNT2_HUMAN | VLAIDHLNEDQLR | 227–239 | 768.415 (2+) | 98 (86) |
| (P62736) Actin, aortic smooth muscle (Alpha-actin-2) ACTA_HUMAN | MQKEITALAPSTMK | 315–328 | 774.912 (2+) | 18 |

(b) FT-ICR MS/MS database search identification of classifier proteins in AVD patients

| Protein ID in aortic samples | Peptide sequence | Sequence tag | m/z (charge state) | Mascot score |
|---|---|---|---|---|
| (P45379) Troponin T, cardiac muscle (TnTc) (Cardiac muscle troponin T) (cTnT) TNNT2_HUMAN | DLNELQALIEAHFENR | 107–122 | 956.482 (2+) | 93 |
| (P17661) Desmin DESM_HUMAN | FLEQQNAALAAEVNR | 127–141 | 837.426 (2+) | 90 |
| (P02144) Myoglobin MYG_HUMAN | HPGDFGADAQGAMNK | 119–133 | 505.894 (3+) | 71 (20) |
| (P60709) Actin, cytoplasmic 1 (Beta-actin) ACTB_HUMAN or | IWHHTFYNELR | 85–95 | 505.922 (3+) | 70 (53) (55) |
| (P68032) Actin, alpha cardiac (Alpha-cardiac actin) ACTC_HUMAN | IWHHTFYNELR | 87–97 | 758.379 (2+), 505.922 (3+) | 53, 61 |
| (P51884) Lumican precursor (Keratan sulfate proteoglycan lumican) (KSPG lumican) LUM_HUMAN | ILGPLSYSK LKEDAVSAAFK | 297–305 171–181 | 489.288 (2+) 589.825 (2+) | 29 60 |
| (P12111) Collagen alpha-3 (VI) chain precursor CO6A3_HUMAN | VAVVQYSDR | 1067–1075 | 518.776 (2+) | 59 |
| (P12883) Myosin heavy chain, cardiac muscle beta isoform (MyHC-beta) MYH7_HUMAN | RKLEGDLK (also in MYH6_Human) AQLEFNQIK GSSFQTVSALHR | 1053–1060 1561–1569 641–652 | 479.788 (2+) 545.799 (2+) 645.334 (2+) | 21 25 25 (35) (62) |

(b) Continued.

| Protein ID in aortic samples | Peptide sequence | Sequence tag | m/z (charge state) | Mascot score |
|---|---|---|---|---|
| (P13533) Myosin heavy chain, cardiac muscle alpha isoform (MyHC-alpha) MYH6_HUMAN | RKLEGDLK AQLEFNQIK GSSFQTVSALHR GKLSYTQQMEDLKR | 1055–1062 1563–1571 643–654 1306–1319 | 479.788 (2+) 545.799 (2+) 645.334 (2+), 430.559 (3+) 566.295 (3+) | 21 25 (26) 25 (17), 47 37 |
| (P09493) Tropomyosin 1 alpha chain (Alpha-tropomyosin) TPM1_HUMAN | MEIQEIQLK | 141–149 | 566.309 (2+) | 36 |
| (P02768) Serum albumin precursor ALBU_HUMAN | KYLYEIAR | 161–168 | 528.299 (2+) | 29 |
| (P02511) Alpha crystallin B chain (Alpha(B)-crystallin) (Rosenthal fiber component) (Heat-shock) pro-CRYAB_HUMAN | HFSPEELK | 83–90 | 493.751 (2+) | 28 |
| (P09669) Cytochrome c oxidase polypeptide VIc precursor (EC 1.9.3.1) COX6C_HUMAN | KAGIFQSVK | 67–75 | 489.293 (2+) | 28 |
| (P12235) ADP/ATP translocase 1 (Adenine nucleotide translocator 1) (ANT 1) ADP, ATP carrier protein (ADT1_HUMAN) | TAVAPIER | 23–30 | 856.487 (1+) | 24 (22) |
| (P35555) Fibrillin-1 precursor FBN1_HUMAN | TICIETIK | 843–850 | 489.271 (2+) | 23 |
| (P02768) Serum albumin precursor ALBU_HUMAN | KYLYEIAR | 161–168 | 528.298 (2+) | 22 |
| (P19429) Troponin I, cardiac muscle (Cardiac troponin 1) TNNI3_HUMAN | AKESLDLR | 162–169 | 466.264 (2+) | 19 |
| (P06576) ATP synthase beta chain, mitochondrial precursor (EC 3.6.3.14) ATPB_HUMAN | FLSQPFQVAEVFTGHMGK | 463–480 | 675.008 (3+) | 16 |

TABLE 2: Proteins identified as classifiers in coronary samples by LC-MALDI-TOF/TOF mass spectrometry. Samples separated by LC deposited on plates, which are then analyzed by MALDI TOF/TOF MS. Proteins identified as classifiers in aortic samples by LC-MALDI-TOF/TOF mass spectrometry. Samples separated by LC deposited on plates, which are then analyzed by MALDI TOF/TOF MS.

(a) HPLC/MALDI TOF/TOF database search identification of classifier proteins in CHD patients

| Protein ID of coronary samples | Peptide sequence | Sequence tag | $m/z$ (charge state) | Mascot score | Protein summary score |
|---|---|---|---|---|---|
| Collagen HSCOLL NID:<br>-Homo sapiens CAA23761 or<br>AF004877 NID:<br>-Homo sapiens AAB93981 | GYPGNIGPVGAAGAPGPHGPVGPAGK<br>3: Hydroxyl(P) 15: Hydroxyl(P)<br>GYPGNIGPVGAAGAPGPHGPVGPAGK<br>3: Hydroxyl(P) 15: Hydroxyl(P) | 327–352<br>949–974 | 2284.147 | 185 (91)<br>91 | 88 |
| (P12883) Myosin heavy chain, cardiac muscle beta isoform (MyHC-beta) MYH7_HUMAN (Displayed; Variant CMH1-VAR_019864) | VIQYFAVIAAIGDR | 191–204 | 1535.858. | 72 | 88 (41.20) |
| (Q05639) Elongation factor 1-alpha 2 (EF-1-alpha-2) (Elongation factor 1 A-2) (eEF1A-2) (Statin S1) EF1A2_HUMAN | VETGILRPGMVVTFAPVNITTEVK | 267–280 | 2571.421 | 55 | 74.50 |
| Hypothetical protein DKFZp686P07163. -Homo sapiens (Human). Q5HYB7_HUMAN | SSSLLIPPLETALANFSSGPEGGVMQPVR | 19–47 | 2954.529 | 80 (66) | 105 (94.90) |
| AX885183 NID:<br>-Homo sapiens CAE99297 or<br>AX885189 NID:<br>-Homo sapiens CAE99303 or<br>AX885185 NID:<br>-Homo sapiens CAE99299 or<br>alpha-crystallin chain B (validated) -human CYHUAB or<br>AF007162 NID:<br>-Homo sapiens AAC19161 or<br>AX888028 NID:<br>-Homo sapiens CAE93953 | LFDQFFGEHLLESDLFPTSTSLSPFYLRPPSFLR<br>LFDQFFGEHLLESDXFPTSTSLSPFYLRPPSFLR | 23–56<br>18–51<br>98–131<br>23–56<br>23–56 | 4004.027 | 106 (34) | 133 (56.90) |
| (P02511) Alpha crystallin B chain (Alpha(B)-crystallin) (Rosenthal fiber component) (Heat-shock) pro CRYAB_HUMAN | LFDQFFGEHLLESDLFPTSTSL | 23–44 | 2543.234 | 128 | 131 |
| Crystallin, alpha B (Homo sapiens) gii13937813 | LFDQFFGEHLLESDLFPTSTSL | 23–44 | 2543.234 | 110 | 111 |
| Crystallin, alpha B (Homo sapiens) gii4503057 | LFDQFFGEHLLESDLFPTSTSL | 23–44 | 2543.234 | 95 | 95.3 |
| Adenylate kinase (EC 2.7.4.3) 1-human (tentative sequence) KIHUA or<br>AK1 protein (Adenylate kinase 1). -Homo sapiens, Q6FGX9_HUMAN or<br>BC001116 NID: Homo sapiens AAH01116 or<br>Adenylate kinase 1. -Homo sapiens (Human). Q5T9B7_HUMAN | GQIVPLETVLDMLR | 64–77<br>64–77<br>64–77<br>79–93 | 1583.882 | 64 (49) | 86.10 (69.80) |

a) Continued.

| Protein ID of coronary samples | Peptide sequence | Sequence tag | m/z (charge state) | Mascot score | Protein summary score |
|---|---|---|---|---|---|
| (MLRA_HUMAN) Myosin regulatory light chain 2, atrial isoform (Myosin light chain 2a) (MLC-2a) (MLC2a) (Myosin regulatory light chain 7) Myosin regulatory light Q01449 | QLLLTQADKFSPAEVEQMFALTPMDLAGNIDYK | 129–161 | 3697.849 | 106 | 131 |
| (P45379) Troponin T, cardiac muscle (TnTc) (Cardiac muscle troponin T) (cTnT) TNNT2_HUMAN | VLAIDHLNEDQLR | 227–239 | 1535.81 | 58 | 85 |

(b) HPLC/MALDI TOF/TOF database search identification of classifier proteins in AVD patients

| Protein ID of aortic samples | Peptide sequence | Sequence tag | m/z | Mascot score | Protein summary score |
|---|---|---|---|---|---|
| (P17661) Desmin DESM_HUMAN | FLEQQNAALAAEVNR | 127–141 | 1673.860 | 115 | 133.00 |
| mutant desmin (Homo sapiens) gi\|21358854 | FLEQQNAALAAEVNR | 128–142 | 1673.860 | 99 (65) | 117 (84.30) |
| (P60709) Actin, cytoplasmic 1 (Beta-actin) ACTB_HUMAN or | IWHHTFYNELR | 85–95 | | | 100 |
| (P63261) Actin, cytoplasmic 2 (Gamma-actin) ACTG_HUMAN or | IWHHTFYNELR | 85–95 | 1515.749 | 82 | |
| (P68133) Actin, alpha skeletal muscle (Alpha-actin 1) ACTS_HUMAN or | IWHHTFYNELR | 85–95 | | | |
| (P68032) Actin, alpha cardiac (Alpha-cardiac actin) ACTC_HUMAN | IWHHTFYNELR | 87–97 | | | |
| (P12883) Myosin heavy chain, cardiac muscle beta isoform (MyHC-beta) MYH7_HUMAN or | GSSFQTVSALHR | 641–652 | 1289.660 | 79 | 87.30 |
| (P13533) Myosin heavy chain, cardiac muscle alpha isoform (MyHC-alpha) MYH6_HUMAN | GSSFQTVSALHR | 643–654 | 1289.660 | 79 | 85.70 |
| MSTP161 (Homo sapiens) gi\|33338222 | SFPNLAFIR | 108–116 | 1064.589 | 74 | 97.40 |
| Myosin light chain 2a (Homo sapiens) gi\|10864037 | SLCYITHGDEKEE 3: Carbamidomethyl (C) | 162–175 | 1693.774 | 66 | 122 |
| actin-like protein (Homo sapiens) gi\|6421180 | IWHHTFYNELR | 2–12 | 1515.749 | 64 | 88.40 |
| Myosin heavy chain alpha subunit gi\|386971 | AQLEFNQIK | 13–21 | 1090.589 | 60 | 81 |
| alpha-1 type III collagen gi\|180413 or | | 7–15 | | | 87 |
| unnamed protein product (Homo sapiens) gi\|1340174 or | GDKGETGER | 28–36 | 948.438 | 57 | 80.1 |
| alpha1 (III) collagen (Homo sapiens) gi\|30054 or | | 141–153 | | | 74.50 |
| alpha-1 (III) collagen (Homo sapiens) gi\|930045 or | | 945–953 | | | 71.40 |
| III preprocollagen alpha 1 chain (Homo sapiens) gi\|16197601 | | 1092–1100 | | | 70.10 |
| (P12235) ADP, ATP carrier protein, heart/skeletal muscle isoform T1 (ADP/ATP translocase 1) ADT1_HUMAN or | TAVAPIER | 23–30 | 856.489 | 49 | 65.10 |
| (P05141) ADP, ATP carrier protein, fibroblast isoform (ADP/ATP translocase 2) ADT2_HUMAN or | TAVAPIER | 23–30 | 856.489 | 49 | 65.10 |

(b) Continued.

| Protein ID of aortic samples | Peptide sequence | Sequence tag | m/z | Mascot score | Protein summary score |
|---|---|---|---|---|---|
| (P12236) ADP, ATP carrier protein, liver isoform T2 (ADP/ATP translocase 3) ADT3_HUMAN | TAVAPIER | 23–30 | 856.489 | 49 | 65.10 |
| Cytochrome c oxidase subunit Va, (COX5A protein). -Homo sapiens (Human). Q8TB65_HUMAN | RLNDFASTVR | 98–107 | 1178.628 | 49 | 70.40 |
| (P12883) Myosin heavy chain, cardiac muscle beta isoform (MyHC-beta) MYH7_HUMAN or | ILYGDFR | 713–719 | 883.467 | 46 | 56.20 |
| (P13533) Myosin heavy chain, cardiac muscle alpha isoform (MyHC-alpha) MYH6_HUMAN | ILYGDFR | 715–721 | 883.467 | 46 | 56.20 |
| (P12883) Myosin heavy chain, cardiac muscle beta isoform (MyHC-beta) MYH7_HUMAN or (P13533) Myosin heavy chain, cardiac muscle alpha isoform (MyHC-alpha) MYH6_HUMAN | AVVEQTER | 1689–1697 1692–1699 | 931.484 | 38 | 49.20 |
| unnamed protein product (Homo sapiens) gi|34533821 | FLLVGQTMSTLLDEDLTK | 495–512 | 2024.062 | 29 | 47.90 |
| Myosin, heavy polypeptide 7, cardiac muscle, beta variant (Homo sapiens) gi|62088996 or cardiac beta myosin heavy chain (Homo sapiens) gi|29727 | AGLLGLLEEMRDER 10: Oxidation (M) | 406–419 | 1617.826 | 29 | 42 |
| (O14958) Calsequestrin, cardiac muscle isoform precursor (Calsequestrin 2) CASQ2_HUMAN | EHQRPTLR | 243–250 | 1036.565 | 24 | 41.60 |
| alpha integrin interacting protein 63 (Homo sapiens) gi|4468915 | ESVSSFVR | 27–35 | 910.463 | 25 | 39.60 |
| Myosin, heavy polypeptide 7, cardiac muscle, beta variant (Homo sapiens) gi|62088996 | GSSFQTVSALHR | 271–291 | 1289.660 | 47 | 63.60 |

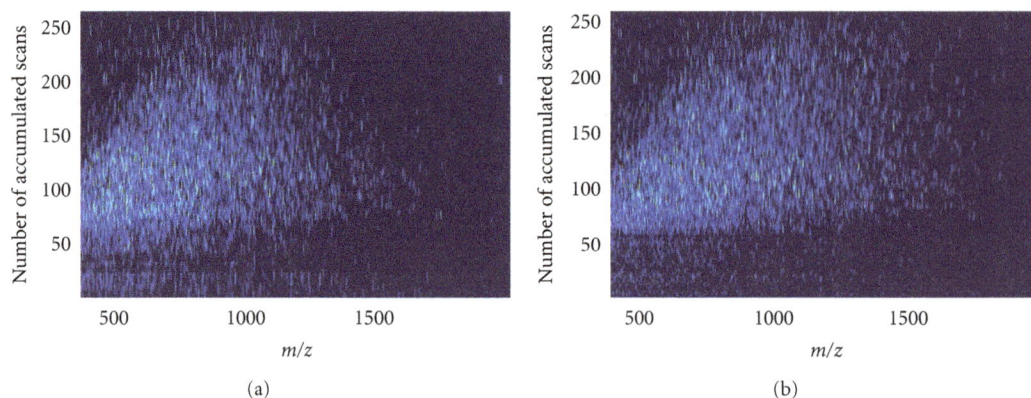

FIGURE 4: Examples of data (m/z versus number of accumulated scans (corresponds to LC retention time), 10 seconds each) obtained from HPLC-FT-ICR mass spectrometry of aortic (a) and coronary (b) samples. The light spots in the diagrams correspond to individual mass spectral peaks. In the original diagrams the peak intensities are color coded for easy recognition. (Figure modified from reference [9] with kind permission from the publisher).

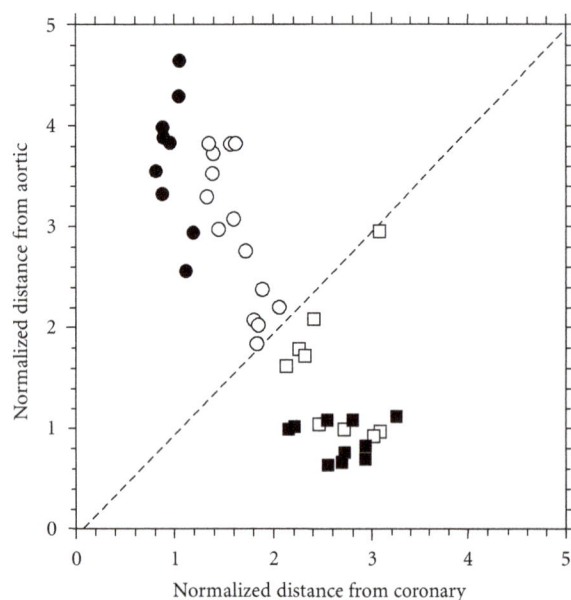

FIGURE 5: Pattern for the classification of coronary (circular symbols) versus AVD disease (square symbols) samples. Each point represent an individual sample where filled symbols (• and ■) represent supervised classified training samples while open symbols (○ and □) represent unsupervised classified test samples. All samples with exception of two were unambiguously correctly classified. (Figure modified from reference [9] with kind permission from the publisher).

(MASCOT search, [17]). Tables 1(a) and 1(b) show the results obtained from HPLC-FT-ICR MS/MS experiments for CHD and AVD samples, respectively. In samples from patients with CHD, four major proteins were identified from classifiers with a relatively high MASCOT score these are myosin heavy chain alpha isoform, myosin heavy chain beta isoform, adenylate kinase, and troponin T (Table 1(a)). Aortic smooth muscle actin was also identified, with a

relatively low MASCOT score, however, still in the useful range when MS/MS of single masses are performed. A larger number of proteins were identified from classifiers in samples from patients with AVD: these were mainly troponin T, desmin, myoglobin, alpha cardiac actin, collagen alpha chain precursor, lumican precursor, tropomyosin I alpha chain, alpha crystallin beta chain, cytochrome C oxydase, ADP/ATP translocase, and serum albumin precursor (Table 1(b)).

Tables 2(a) and 2(b) reveal the protein identification results from database searches with HPLC-MALDI-TOF/TOF spectra of samples from CHD and AVD groups, respectively. In samples from patients with CHD, seven main proteins are found: these were myosin heavy chain beta isoform, alpha crystallin B chain, collagen, elongation factor, hypothetical protein DKFZp686P07163, adenylate kinase, and myosin light chain 2a (Table 2(a)). Like in the FT-ICR results, samples from patients with AVD revealed a larger number of proteins: these were desmin, alpha cardiac actin, myosin heavy chain alpha isoform, myosin heavy chain beta isoform, myosin light chain 2a, ADP/ATP carrier protein, cytochrome C oxydase, alpha-1 (III) collagen, alpha integrin, and an unnamed protein product (Table 2(b)).

Examples from MS/MS results acquired with FT-ICR and MALDI-TOF mass spectrometry as comparative cases are shown in Figures 6 to 9. Figure 6(a) shows the Electrospray Ionisation-FT-ICR MS/MS spectrum obtained from one of the AVD samples leading to the identification of the protein Desmin (DESM_HUMAN), while Figure 6(b) shows the MS/MS spectrum from this sample by MALDI-TOF/TOF mass spectrometry. In both cases, sufficient sequence information is obtained to identify Human Desmin. In the ESI-FT-ICR MS/MS spectrum, the intensities of the y fragment peaks are higher than the b fragments. Yet, the sequence information from both y and b fragments were usable. Similarly, Figures 7(a) and 7(b) show the ESI-FT-ICR MS/MS spectra, and MALDI-TOF/TOF spectra, respectively, obtained from a CHD sample. Adenylate kinase (KAD1_HUMAN) could be identified from any of these two

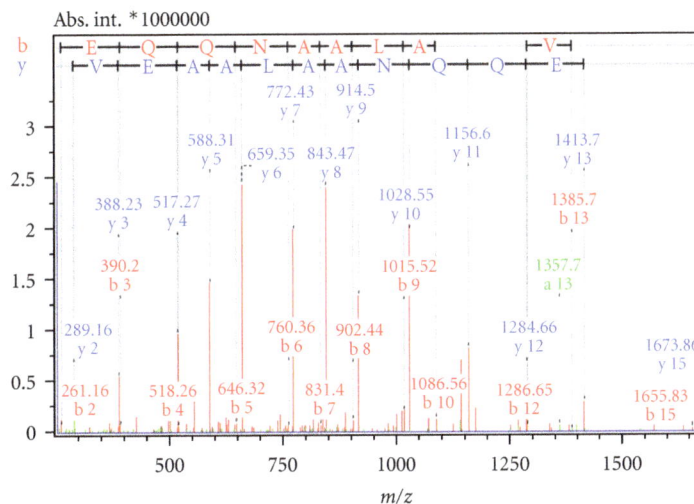

(a) Aortic Sample LC-ESI-FTMS/MS, Desmin DESM_HUMAN

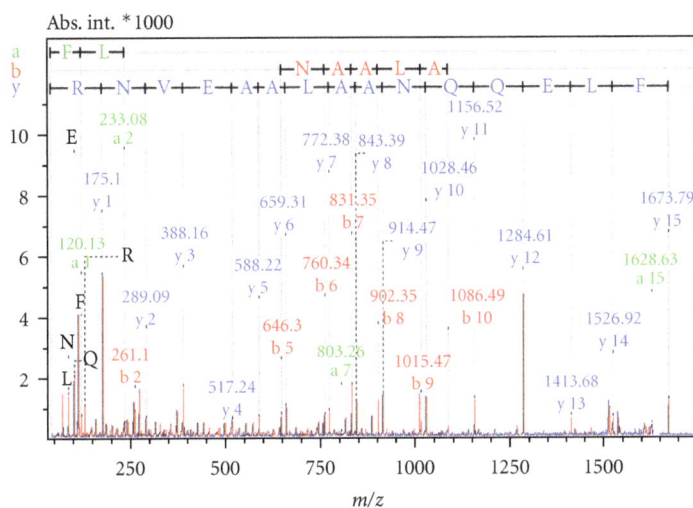

(b) Aortic Sample LC-MALDI-TOF/TOF, Desmin DESM_HUMAN

FIGURE 6: MS/MS spectra of the classifier peptide FLEQQNAALAAEVNR from the sample of an aortic patient. The protein is identified as Desmin upon database search. Spectrum (a) is obtained by LC-ESI-FTMS/MS, spectrum (b) by LC-MALDI-TOF/TOF.

spectra by database search. Here, the ESI-FT-ICR MS/MS spectrum mainly shows y fragments, while in the MALDI-TOF/TOF spectrum y, b, and a fragments are visible. Figures 8(a) and 8(b) show the ESI-FT-ICR MS/MS and MALDI-TOF/TOF mass spectra from an AVD sample, respectively, from any of which the protein myosin heavy chain beta isoform MY7_HUMAN could be identified due to the sequence information. Again here, the ESI-FT-ICR MS/MS spectrum shows mainly the y fragments while MALDI-TOF/TOF spectrum shows b, y, and a fragmentation. Figure 9 is the identification of beta actin ACTB_HUMAN with ESI-FT-ICR MS/MS (a) and MALDI TOF/TOF (b) again from an AVD sample.

## 4. Discussion

In this work, the sample collection was made in such a way that all patients with CHD selected for this study were

free from AVD, and all selected AVD patients were free from CHD. All comparisons presented in this work can, therefore, virtually be considered as to be "diseased" versus "healthy" case against each other. Differences in the mass chromatograms are determined as classifier masses. Thus, proteins identified from classifiers by this *differential mass spectrometry* method are biomarkers in the corresponding disease case (Figure 10).

Some of the identified proteins appear both in the list of CHD samples as well as in the list of the AVD samples. As an example, myosin heavy chain alpha and beta isoforms have been found both in CHD and AVD cases (Tables 1 and 2). However, classifier peptides leading to apparently the same protein by database search were different in CHD samples than in AVD samples. In samples from CHD group, myosin heavy chain alpha isoform was found by database search from MS/MS and identification of the two classifier peptides (Table 1(a)). In samples from AVD group, entirely

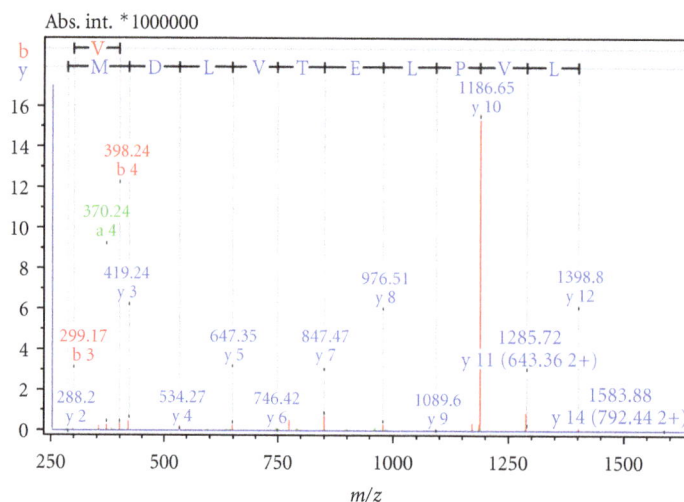

(a) Coronary Sample LC-ESI-FTMS/MS, Adenylate Kinase KAD1_HUMAN

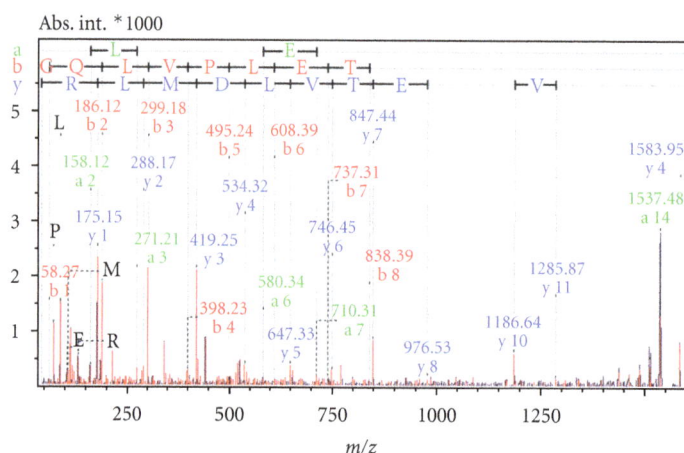

(b) Coronary Sample LC-MALDI-TOF/TOF, Adenylate Kinase KAD1_HUMAN

FIGURE 7: MS/MS spectra of the classifier peptide GQLVPLETVLDMLR from the sample of an aortic patient. The protein is identified as Adenylate Kinase upon database search. Spectrum (a) is obtained by LC-ESI-FTMS/MS, spectrum (b) by LC-MALDI-TOF/TOF.

different classifier peptides led to myosin heavy chain alpha by database search (Table 1(b)). These are displayed in a simplified table (Table 3). A targeted analysis of these potential biomarker peptides could thus result in a simplified diagnostic assay for different etiologies in cardiovascular diseases. The explanation of this observation is suggested as follows. The approach in the present study is the analysis of a peptide mixture resulting from a tryptic digestion of the initial protein mixture. A comparison of the LC-MS data leads to classifier masses for samples with AVD indicating that these particular masses (peptides) are here significantly more abundant than in the samples with CHD, and vice versa. However, since these peptides are the result of a tryptic digestion of the initial proteins, any factor influencing the digestion by trypsin can very well suppress the appearance of some peptides in the digest. If, for example, in CHD samples, some of the tryptic cleavages are suppressed, the comparison will show a higher abundance of these particular digestion products in the aortic samples that they become classifiers for AVD.

Similarly, if certain tryptic cleavages are suppressed in samples from AVD patients, the resulting peptides appear more abundant in samples from CHD group, and the comparison will indicate them as classifiers for CHD samples. One of the major factors altering the digestion specificity of trypsin is a posttranslational modification near the cleavage site of the protein. Thus, a tryptic cleavage at this particular site can sometimes be not successful. Heavy groups like large glycans may hinder a cleavage, or even if a cleavage occurs, digested peptide with the modification is too heavy and may be off the detected mass range. It is known that glycosylation can even completely block the tryptic digestion of a protein [22].

Based on the thoughts above, a possible explanation for the different classifier peptides in CHD and AVDs leading to the same myosin may be differences in tryptic digestion patterns caused by posttranslational modifications at different positions. An investigation of the correlation to these differences is the topic of our ongoing work. For this particular study, fragmentations by electron capture

(a) Aortic Sample LC-ESI-FTMS/MS, Myosin Heavy Chain Beta Isoform MYH7_HUMAN

(b) Aortic Sample LC-MALDI-TOF/TOF, Myosin Heavy Chain Beta Isoform MYH7_HUMAN

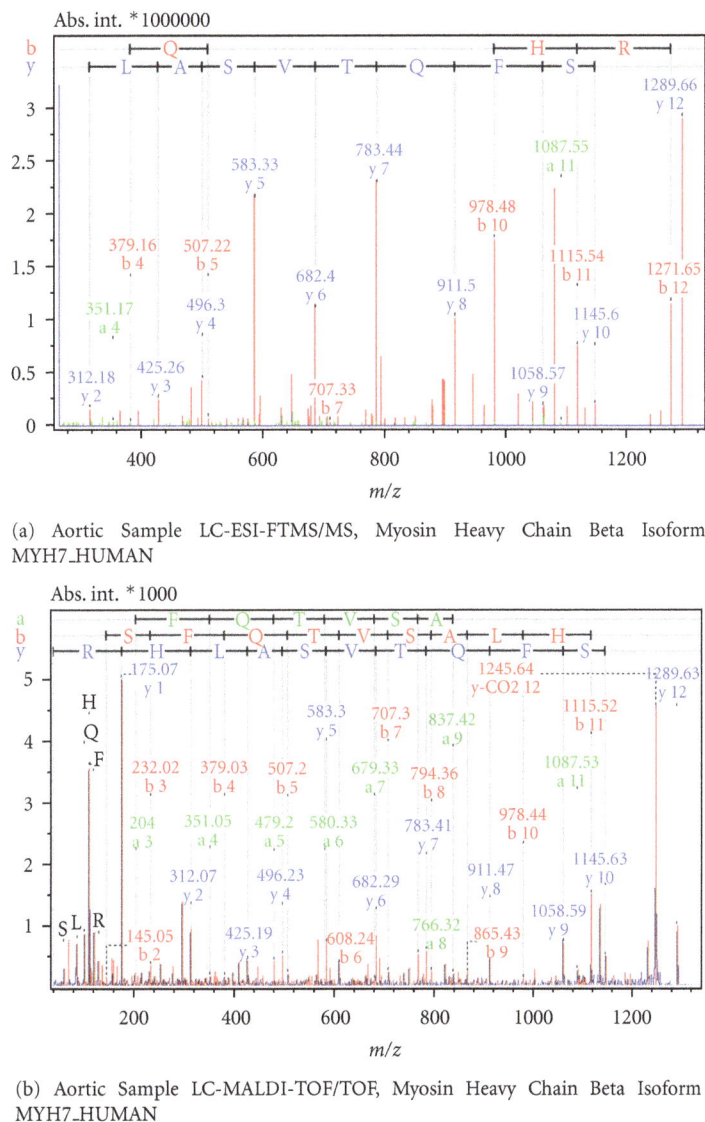

FIGURE 8: MS/MS spectra of the classifier peptide GSSFQTVSALHR from the sample of an aortic patient. The protein is identified as Myosin Heavy Chain Beta Isoform upon database search. Spectrum (a) is obtained by LC-ESI-FTMS/MS, spectrum (b) by LC-MALDI-TOF/TOF.

dissociation (ECD) or electron transfer dissociation (ETD) are more suitable than collision-induced dissociation (CID), since ECD and ETD protect posttranslational modifications while breaking the peptide backbone bonds.

Another possible reason for altered digestion specificity of trypsin could be the folding geometry. If the protein is not completely unfolded during the digestion process, the tryptic cleavage at particular sites can be sterically hindered. This would then lead to missed cleavages.

A number of proteins in ESI-FT-ICR MS/MS were also identified with the method MALDI-TOF/TOF mass spectrometry with the same peptide sequences. In samples from the CHD group, commonly identified proteins were adenylate kinase, myosin heavy chain beta isoform. In samples from the AVD group, the proteins commonly identified in FT-ICR MS/MS and TOF/TOF were desmin, alpha cardiac actin, myosin heavy chain alpha isoform, myosin heavy chain beta isoform, and ADP/ATP translocase [23].

As described previously, FT-ICR and MALDI-TOF mass spectrometry used different methods of ionization. For the FT-ICR MS, ions were separated by liquid-chromatography and online ionized by electrospray ionization and transferred into mass spectrometer—either directly or after collision-induced dissociation—for detection. The MALDI-TOF mass spectrometric method uses samples in deposited solid phase. Thus, the components in the samples were LC-separated first and fractions were deposited on a MALDI sample plate precoated with a alpha-cyano-4-hydroxy cinnamic acid as a matrix (Pre Spotted Anchor Chip target. Bruker Daltonik, Bremen, Germany). This plate was inserted into the ion source of the TOF mass spectrometer, irradiated with the laser beam, and the ions were generated by matrix assisted laser desorption/ionization and detected. These ions are in general singly charged. Although compounds in both MALDI and electrospray ionization become mildly ionized, due to the different ionization techniques in MALDI-TOF

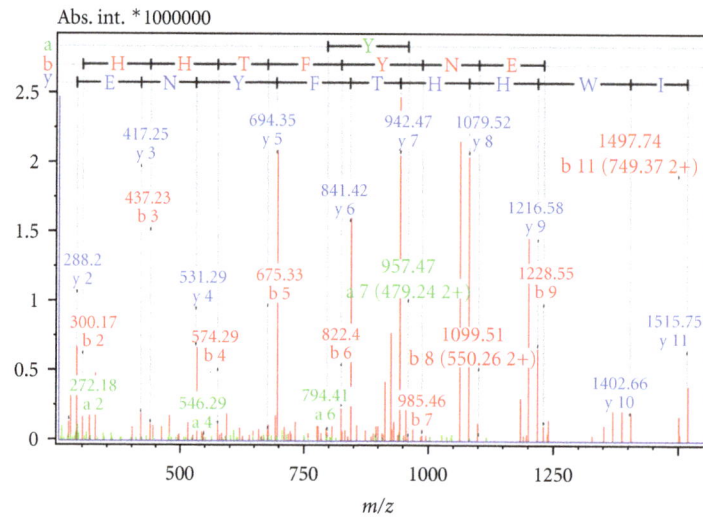

(a) Aortic Sample LC-ESI-FTMS/MS, Beta Actin ACTB_HUMAN

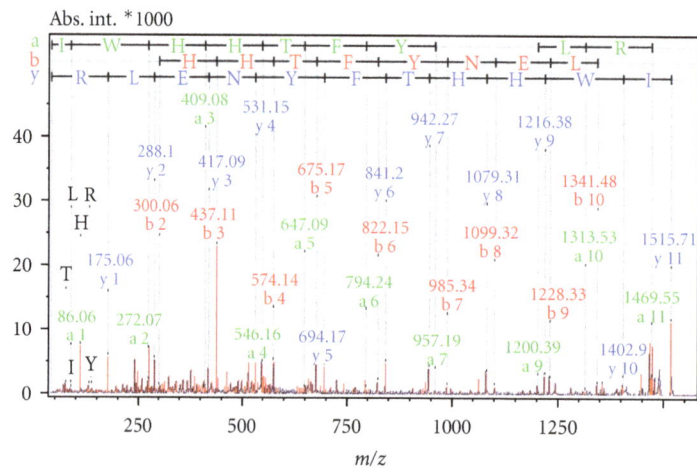

(b) Aortic Sample LC-MALDI-TOF/TOF, Beta Actin ACTB_HUMAN

FIGURE 9: MS/MS spectra of the classifier peptide IWHHTFYNELR from the sample of an aortic patient. The protein is identified as Beta Actin upon database search. Spectrum (a) is obtained by LC-ESI-FTMS/MS, spectrum (b) by LC-MALDI-TOF/TOF.

FIGURE 10: Proteomic profiles of myocardial tissue in two different etiologies of heart failure were investigated using right atrial appendages samples representative for the aortic valve disease and coronary heart disease using a quadrupole/hexapole FT-ICR MS that allowed collision induced dissociation (CID) of selected classifier masses. For comparison and further validation, classifier masses were also fragmented and analyzed using HPLC/Matrix assisted laser desorption ionization (MALDI) time-of-flight/time-of-flight (TOF/TOF) mass spectrometry.

TABLE 3: Classifier peptides in aortic and coronary diseases identifying myosin heavy chain alpha and beta isoforms as potential biomarkers.

| | Classifier peptides | |
| --- | --- | --- |
| Identified protein | CHD disease | AVD disease |
| Myosin heavy chain alpha isoform | KLAEKDEEMEQAK, <br><br> NLQEEISDLTEQLGEGGKNVHELEKVR | RKLEGDLK, <br> AQLEFNQIK, <br> GSSFQTVSALHR, <br> GKLSYTQQMEDLKR |
| Myosin heavy chain beta isoform | KLAEKDEEMEQAK | RKLEGDLK, <br> AQLEFNQIK, <br> GSSFQTVSALHR |

and FT-ICR instruments, and to the non-online method in the LC-MALDI-TOF, some of the ions may have different abundances than in the on-line system. Thus, some of the FT-ICR MS/MS or MALDI-TOF/TOF dissociations of classifiers could not be performed as the abundance in the corresponding system was low. One of the examples is the alpha crystallin B chain [24–27] in the CHD samples which is intensively found in the MALDI-TOF spectra, fragmented (TOF/TOF), detected, and identified, which was not sufficiently abundant in the experiments with the FT-ICR MS. Thus, no results regarding detection of alpha crystallin B chain have been displayed in the CHD samples by FT-ICR MS.

As an overview, the bottom-up proteomic approach was applied to proteins in human cardiac muscle tissue samples from two groups of patients in this study. The selection of patients enabled the examination of two clearly separated case etiologies. Proteins in the tissue samples were digested by trypsin, and the digest containing a mixture of peptides was analyzed by mass spectrometry after liquid chromatographic separation. The use of high-resolution mass spectrometry (in this case FT-ICR MS) allowed to resolve and display the mass spectrometric peaks in this complex picture of the LC-MS results and to compare both separate disease forms. By comparison of LC-MS diagrams, the classifier masses could be clearly identified. These were selected in the subsequent experiments and fragmented (LC-FT-ICR MS/MS), in order to identify the proteins which had generated the classifier masses. The comparison clearly separated both disease forms while the analysis and identification of the proteins which led to the classifiers helped to study the biomarkers related to CHD and AVD. An additional work for comparing LC-MALDI-TOF mass spectrometry and LC-MALDI TOF/TOF for the MS/MS fragmentation has also be performed.

The unique patient selection in this study, combined with the bottom-up proteomic approach using liquid chromatography and high resolution mass spectrometry, seems to be highly efficient in determination of the differences between selected disease groups. The *differential high-resolution mass spectrometry* performed subsequently to characterize the related proteins did require the MS/MS of the classifier masses only.

This study of the heart muscle tissue samples helped establish a first picture of the proteomic appearance of two virtually independent etiologies of heart disease. As our main target is to diagnose cardiac disease less invasively and directly, we are currently investigating blood plasma samples from CHD and AVD patients in order to identify the differences with the same technique using differential high-resolution mass spectrometry.

## Acknowledgments

The authors thank Sören Deininger and Arndt Asperger for their help with liquid-chromatography and sample deposition processes prior to MALDI-TOF and MALDI-TOF/TOF MS experiments. Financial support from the Swedish Research Council (Grant 621–2008-3562, 621-2011-4423, 342-2004-3944 (JB)) is gratefully acknowledged.

## References

[1] G. K. Hansson, "Mechanisms of disease: inflammation, atherosclerosis, and coronary artery disease," *New England Journal of Medicine*, vol. 352, no. 16, pp. 1685–1626, 2005.

[2] B. J. Maron, V. J. Ferrans, and W. C. Roberts, "Myocardial ultrastructure in patients with chronic aortic valve disease," *American Journal of Cardiology*, vol. 35, no. 5, pp. 725–739, 1975.

[3] F. De La Cuesta, G. Alvarez-Llamas, F. Gil-Dones et al., "Tissue proteomics in atherosclerosis: elucidating the molecular mechanisms of cardiovascular diseases," *Expert Review of Proteomics*, vol. 6, no. 4, pp. 395–409, 2009.

[4] F. De La Cuesta, G. Alvarez-Llamas, A. S. Maroto et al., "A proteomic focus on the alterations occurring at the human atherosclerotic coronary intima," *Molecular and Cellular Proteomics*, vol. 10, no. 4, 2011.

[5] J. Zhang, M. J. Guy, H. S. Norman et al., "Top-down quantitative proteomics identified phosphorylation of cardiac troponin I as a candidate biomarker for chronic heart failure," *Journal of Proteome Research*, vol. 10, pp. 4054–4065, 2011.

[6] R. E. Gerszten, A. Asnani, and S. A. Carr, "Status and prospects for discovery and verification of new biomarkers of cardiovascular disease by proteomics," *Circulation Research*, vol. 109, pp. 463–474, 2011.

[7] W. J. Huang, R. Zhou, X. R. Zeng et al., "Comparative proteomic analysis of atrial appendages from rheumatic heart disease patients with sinus rhythm and atrial fibrillation," *Molecular Medicine Reports*, vol. 4, no. 4, pp. 655–661, 2011.

[8] E. Dubois, M. Fertin, J. Burdese, P. Amouyel, C. Bauters, and F. Pinet, "Cardiovascular proteomics: translational studies to

develop novel biomarkers in heart failure and left ventricular remodeling," *Proteomics*, vol. 5, no. 1-2, pp. 57–66, 2011.

[9] D. Baykut, M. Grapow, M. Bergquist et al., "Molecular differentiation of ischemic and valvular heart disease by liquid chromatography/fourier transform ion cyclotron resonance mass spectrometry," *European Journal of Medical Research*, vol. 11, no. 6, pp. 221–226, 2006.

[10] M. Ramström, I. Ivonin, A. Johansson et al., "Cerebrospinal fluid protein patterns in neurodegenerative disease revealed by liquid chromatography-Fourier transform ion cyclotron resonance mass spectrometry," *Proteomics*, vol. 4, no. 12, pp. 4010–4018, 2004.

[11] J. Bergquist, M. Palmblad, M. Wetterhall, P. Håkansson, and K. E. Markides, "Peptide mapping of proteins in human body fluids using electrospray ionization fourier transform ion cyclotron resonance mass spectrometry," *Mass Spectrometry Reviews*, vol. 21, no. 1, pp. 2–15, 2002.

[12] S. L. Wu, G. Choudhary, M. Ramström, J. Bergquist, and W. S. Hancock, "Evaluation of shotgun sequencing for proteomic analysis of human plasma using HPLC coupled with either ion trap or Fourier transform mass spectrometry," *Journal of Proteome Research*, vol. 2, no. 4, pp. 383–393, 2003.

[13] J. Bergquist, "FTICR mass spectrometry in proteomics," *Current Opinion in Molecular Therapeutics*, vol. 5, no. 3, pp. 310–314, 2003.

[14] T. Ekegren, J. Hanrieder, S. M. Aquilonius, and J. Bergquist, "Focused proteomics in post-mortem human spinal cord," *Journal of Proteome Research*, vol. 5, no. 9, pp. 2364–2371, 2006.

[15] P. Caravatti and M. Allemann, "'The infinity cell': a new trapped-ion cell with radiofrequency covered trapping electrodes for Fourier transform ion cyclotron resonance mass spectrometry," *Organic Mass Spectrometry*, vol. 26, pp. 514–518, 1991.

[16] D. J. C. Pappin, D. Rahman, H. F. Hansen, M. Bartlet-Jones, W. Jeffery, and A. J. Bleasby, "Chemistry Mass Spectrometry and Peptide-Mass Databases: evolution of methods for the rapid identification and mapping of cellular proteins," in *Mass Spectrometry in the Biological Sciences*, A. L. Burlingame and S. A. Carr, Eds., pp. 135–150, Humana, Totowa, NJ, USA, 1996.

[17] MASCOT, http://www.matrixscience.com/.

[18] D. J. C. Pappin, P. Hojrup, and A. J. Bleasby, "Rapid identification of proteins by peptide-mass fingerprinting," *Current Biology*, vol. 3, no. 6, pp. 327–332, 1993.

[19] S. D. H. Shi, J. J. Drader, M. A. Freitas, C. L. Hendrickson, and A. G. Marshall, "Comparison and interconversion of the two most common frequency-to-mass calibration functions for Fourier transform ion cyclotron resonance mass spectrometry," *International Journal of Mass Spectrometry*, vol. 195-196, pp. 591–598, 2000.

[20] D. Suckau, A. Resemann, M. Schuerenberg, P. Hufnagel, J. Franzen, and A. Holle, "A novel MALDI LIFT-TOF/TOF mass spectrometer for proteomics," *Analytical and Bioanalytical Chemistry*, vol. 376, no. 7, pp. 952–965, 2003.

[21] A. Holle, A. Haase, M. Kayser, and J. Höhndorf, "Optimizing UV laser focus profiles for improved MALDI performance," *Journal of Mass Spectrometry*, vol. 41, no. 6, pp. 705–716, 2006.

[22] S. O. Deininger, L. Rajendran, F. Lottspeich et al., "Identification of teleost Thy-1 and association with the microdomain/lipid raft reggie proteins in regenerating CNS axons," *Molecular and Cellular Neuroscience*, vol. 22, no. 4, pp. 544–554, 2003.

[23] M. A. Portman, "Adenine nucleotide translocator in heart," *Molecular Genetics and Metabolism*, vol. 71, no. 1-2, pp. 445–450, 2000.

[24] X. Wang, R. Klevitsky, W. Huang, J. Glasford, F. Li, and J. Robbins, "αB-Crystallin Modulates Protein Aggregation of Abnormal Desmin," *Circulation Research*, vol. 93, no. 10, pp. 998–1005, 2003.

[25] S. P. Bhat, J. Horwitz, A. Srinivasan, and L. Ding, "αB-crystallin exists as an independent protein in the heart and in the lens," *European Journal of Biochemistry*, vol. 202, no. 3, pp. 775–781, 1991.

[26] G. Lutsch, R. Vetter, U. Offhauss et al., "Abundance and location of the small heat shock proteins HSP25 and αB-crystallin in rat and human heart," *Circulation*, vol. 96, no. 10, pp. 3466–3476, 1997.

[27] J. Horwitz, "α-Crystallin can function as a molecular chaperone," *Proceedings of the National Academy of Sciences of the United States of America*, vol. 89, no. 21, pp. 10449–10453, 1992.

# 15

# Plasma Fractionation Enriches Post-Myocardial Infarction Samples Prior to Proteomics Analysis

**Lisandra E. de Castro Brás,**[1, 2] **Kristine Y. DeLeon,**[1, 2] **Yonggang Ma,**[1, 2] **Qiuxia Dai,**[1, 2] **Kevin Hakala,**[1, 3] **Susan T. Weintraub,**[1, 3] **and Merry L. Lindsey**[1, 2]

[1] *San Antonio Cardiovascular Proteomics Center, The University of Texas Health Science Center at San Antonio, San Antonio, TX 78245, USA*
[2] *Division of Geriatrics, Gerontology & Palliative Medicine, Department of Medicine, UTHSCSA, San Antonio, TX 78245, USA*
[3] *Department of Biochemistry, UTHSCSA, San Antonio, TX 78245, USA*

Correspondence should be addressed to Lisandra E. de Castro Brás, decastrobras@uthscsa.edu
and Merry L. Lindsey, LindseyM@uthscsa.edu

Academic Editor: William C. S. Cho

Following myocardial infarction (MI), matrix metalloproteinase-9 (MMP-9) levels increase, and MMP-9 deletion improves post-MI remodeling of the left ventricle (LV). We provide here a technical report on plasma-analysis from wild type (WT) and MMP-9 null mice using fractionation and mass-spectrometry-based proteomics. MI was induced by coronary artery ligation in male WT and MMP-9 null mice (4–8 months old; $n = 3$/genotype). Plasma was collected on days 0 (pre-) and 1 post-MI. Plasma proteins were fractionated and proteins in the lowest (fraction 1) and highest (fraction 12) molecular weight fractions were separated by 1-D SDS-PAGE, digested in-gel with trypsin and analyzed by HPLC-ESI-MS/MS on an Orbitrap Velos. We tried five different fractionation protocols, before reaching an optimized protocol that allowed us to identify over 100 proteins. Serum amyloid A substantially increased post-MI in both genotypes, while alpha-2 macroglobulin increased only in the null samples. In fraction 12, extracellular matrix proteins were observed only post-MI. Interestingly, fibronectin-1, a substrate of MMP-9, was identified at both day 0 and day 1 post-MI in the MMP-9 null mice but was only identified post-MI in the WT mice. In conclusion, plasma fractionation offers an improved depletion-free method to evaluate plasma changes following MI.

## 1. Introduction

Acute myocardial infarction (MI) remains a leading cause of morbidity and mortality worldwide. According to the latest report of the American Heart Association, every 25 seconds, an American will have a coronary event, and approximately every minute, someone will die of a coronary event [1]. In 2010, 785,000 Americans experienced an MI, and approximately 470,000 had a recurring MI [1]. Heart failure can result from adverse remodeling of the collagenous scar that replaces the damaged myocardium in the left ventricle (LV) after MI. LV remodeling is mediated by cell survival, inflammation, angiogenesis, and turnover of the extracellular matrix (ECM). Markers of LV remodeling can be either determined in the circulation (e.g., serum or

plasma) or detected in the heart by imaging technologies or biopsy. Post-MI, levels of specific matrix metalloproteinases (MMPs) increase and mediate left ventricular remodeling. MMP-9 has been reported as a prognostic indicator of cardiac dysfunction in MI patients [2, 3]. MMP-9 deletion has also been shown to improve remodeling of the LV in mice [4, 5]. We hypothesized that the analysis of plasma proteins post-MI in wild-type (WT) and MMP-9 null mice will identify prospective markers of early MI that are MMP-9 dependent.

Termed as the most complex proteome, plasma is an intricate body fluid, containing a wide diversity of proteins [6]. Plasma has been investigated using targeted evaluations, to measure markers that detect MI or predict outcomes following MI. For examples, the muscle form of creatine kinase (CK-Mb), troponins, and C-reactive protein are used clinically to determine both presence of MI and extent of myocardial damage [7, 8]. MMP-9, galectin-3, and brain natriuretic peptide have been used to evaluate LV responses to MI [9–11]. Plasma has also been investigated using proteomic approaches, but this has been fraught with technical issues, primarily because the range of protein levels in the plasma is $10^{10}$, and the ten most abundant proteins account for 90% of the total protein concentration [12, 13]. Serum albumin is a high abundant protein in plasma, and it is the leading candidate for selective removal prior to proteomics analysis of less abundant proteins in plasma. Several albumin-depletion methods are commercially available, mainly based in immunoaffinity columns. Albumin can also be removed by ligand chromatography [14, 15], and isoelectric trapping [16]. Nonetheless, the use of depletion methods may also result in specific removal of low abundant cytokines, lipoproteins, and peptide hormones of interest [17].

Accordingly, we hypothesized that using a fractionation protocol for the analysis of plasma proteins post-MI in wild-type (WT) and MMP-9 null mice would identify prospective markers of early MI that are MMP-9 dependent. In our study, we performed protein fractionation prior to protein separation by 1D-PAGE and MS analysis. By doing so, we avoided using depletion methods and concomitantly reduced the presence of albumin and enriched for lower abundance proteins.

## 2. Materials and Methods

*2.1. Animals and Surgery.* All animal procedures were conducted according to the "Guide for the Care and Use of Laboratory Animals" (NIH Notice Number: NOT-OD-12-020) and were approved by the Institutional Animal Care and Use Committee at the University of Texas at San Antonio. Male 4–8 months old C57BL6/J wild-type (WT) ($n$ = 3) and MMP-9 null mice ($n$ = 3) were used in this study. Animals were housed at constant temperature (22 ± 2°C) on a 12 h light/dark cycle. They were fed *ad libitum* on standard laboratory mice chow and had free access to tap water. MI was made by permanent ligation of the left anterior descending coronary artery as described previously [18]. Animals without MI (day 0) were used as controls ($n$ = 3/genotype). At one day post-MI, mice were anesthetized with 5% isoflurane, plasma was collected, the coronary vasculature was flushed with 0.9 M saline, and the hearts were excised. The hearts were separated between right and left ventricles and were stained with 1% 2,3,5-triphenyltetrazolium chloride (Sigma) and photographed for measurement of infarct area.

*2.2. Plasma Fractionation.* Plasma was collected at days 0 and 1 post-MI, snap frozen and stored at −80°C. At sacrifice, heparin (100 μL of 1000 USP Units/mL) was injected intraperitoneally, and 5 min after heparin injection, blood was collected from the carotid artery of the mouse. Total protein quantification was determined using Quick Start Bradford Protein Assay (Biorad). Plasma was fractionated using the GellFree 8100 Fractionation System (Protein Discovery, Inc.). Five hundred micrograms of total protein were reduced for 10 min at 50°C with 1x acetate sample buffer (Protein Discovery, Inc.) and 0.053 M dithiothreitol (DTT). After samples being cooled down to room temperature, 15 mM iodoacetamide was added, and samples were alkylated in the dark for 10 min.

For protocol optimization, we used six different protocols where samples were either run in an 8%, 10%, or 12% Tris-acetate cartridge combined with one of three fractionation programs. We tested three different fractionation programs shown in Figures 2, 3, and 4. For all of the programs, MES was used as the running buffer (0.05 M MES, 0.05 M Tris, 0.1% SDS pH 7.9). For each of the six protocols tested, twelve fractions (150 μL/fraction) were collected per sample and proteins were visualized on a 12% Bis-Tris gel by SDS-PAGE.

*2.3. Mass Spectrometry.* The proteins in fraction 1 were separated in a 10–20% Tricine/peptide gel. The gel lane for each replicate was divided into six slices. The gel region containing visually detectable proteins from the lane for fraction 12 (the highest molecular weight fraction) on the Bis-Tris gel was excised into three slices. Each slice was separately destained and dehydrated and the proteins digested *in situ* with trypsin (Promega). The digests were analyzed by capillary HPLC-electrospray ionization tandem mass spectrometry (HPLC-ESI-MS/MS) on a Thermo Fisher LTQ Orbitrap Velos mass spectrometer fitted with a New Objective Digital PicoView 550 NanoESI source. Online HPLC separation of the digests was accomplished with an Eksigent/AB Sciex NanoLC-Ultra 2-D HPLC system: column, PicoFrit (New Objective; 75 μm i.d.) packed to 15 cm with C18 adsorbent (Vydac; 218MS 5 μm, 300 Å). Precursor ions were acquired in the Orbitrap in profile mode at 60,000 resolution ($m/z$ 400); data-dependent collision-induced dissociation (CID) spectra of the six most intense ions in the precursor scan above a set threshold were acquired at the same time in the linear trap. Mascot (versions 2.3.02; Matrix Science) was used to search the uninterpreted CID spectra against a combination of the mouse subset of the NCBInr database (Mus. (145,083 sequences)) and a database of common contaminants (179 sequences). Methionine oxidation was considered as a variable modification; trypsin was specified as the proteolytic enzyme, with one missed cleavage allowed. A secondary search of the CID spectra using X! Tandem, cross-correlation of the X! Tandem and Mascot results, and determination of protein and peptide identity probabilities were accomplished by Scaffold (version 3; Proteome Software). The thresholds for acceptance of peptide and protein assignments in Scaffold were 95% and 99.9%, respectively. The results for the individual slices were combined for presentation purposes.

*2.4. Immunoblotting.* Proteins of interest were further analyzed by immunoblotting. Total proteins (10 μg) were loaded onto either 4–12% Bis-Tris gels (proteins >50 kDa) or 10–20% tricine gels (proteins <50 kDa) and run by SDS-PAGE. Proteins were transferred to a nitrocellulose membrane

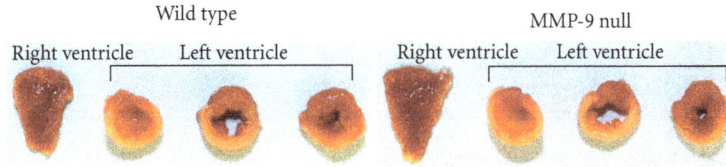

FIGURE 1: Infarct area was measured in the left ventricle. Infarct areas were similar between WT ($52 \pm 8\%$) and MMP-9 null ($54 \pm 2\%$) mice ($P = 0.85$).

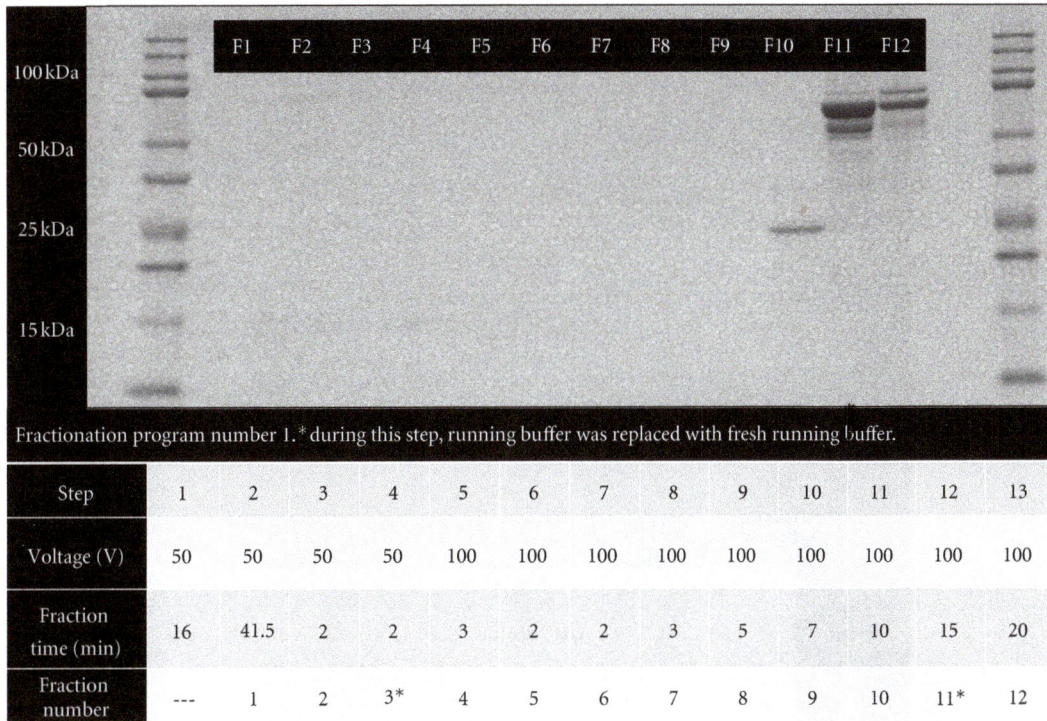

Fractionation program number 1.* during this step, running buffer was replaced with fresh running buffer.

| Step | 1 | 2 | 3 | 4 | 5 | 6 | 7 | 8 | 9 | 10 | 11 | 12 | 13 |
|---|---|---|---|---|---|---|---|---|---|---|---|---|---|
| Voltage (V) | 50 | 50 | 50 | 50 | 100 | 100 | 100 | 100 | 100 | 100 | 100 | 100 | 100 |
| Fraction time (min) | 16 | 41.5 | 2 | 2 | 3 | 2 | 2 | 3 | 5 | 7 | 10 | 15 | 20 |
| Fraction number | --- | 1 | 2 | 3* | 4 | 5 | 6 | 7 | 8 | 9 | 10 | 11* | 12 |

FIGURE 2: Plasma fractionation using an 8% acetate cartridge and program number 1; samples were run on 12% Bis-Tris gel. This fractionation scheme was not optimal because all of the proteins were observed in the last three fractions, rather than being evenly spread across fractions.

which was hybridized overnight at 4°C with primary antibody. Primary antibodies used were antiserum amyloid A1 (number AF2948, R&D), anti-alpha-2 macroglobulin (number ab52651, Abcam) and antineutrophil-associated gelatinase lipocalin (NGAL, aka lipocalin 2; number ab63929, Abcam) (number ab63929, Abcam). Protein quantification was determined by densitometry analysis using ImageJ.

*2.5. Statistical Analysis.* Data are reported as mean ± SEM. Immunoblot intensities (arbitrary units) were assessed using a one-way ANOVA with Newman-Keuls multiple comparison test. A $P < 0.05$ was considered significant.

## 3. Results

Infarct areas were similar between WT and MMP-9 null mice ($P = 0.85$; Figure 1), indicating that both groups received a similar injury stimulus. We tested several methods to optimize the plasma fractionation prior to MS analysis. The different fractionation programs are shown in the figures. When using fractionation program number 1 and an 8% acetate cartridge, proteins were only observed in fractions 10 to 12 (Figure 2). By changing voltage intensities and step duration, we were able to visualize proteins in all 12 fractions. The protein' profiles differed depending of the type of cartridge used (Figure 3). Since serum albumin is approximately 66 kDa, we focused on protocols that provided fractions with reduced albumin content. The combination of program number 3 with the 8% acetate cartridge yielded fractions with these characteristics, where most of the albumin was seen in fractions 2 to 11 (Figure 4). These conditions were considered optimal for our examination, and fractions 1 and 12, which showed reduced levels of albumin, were further analyzed by HPLC-ESI-MS/MS on an Orbitrap Velos.

Supplemental Tables 1 and 2 list the proteins identified in both fractions (see Supplementary Material available online at doi:10.1155/2012/397103), per genotype and time point. Of the 145 proteins identified in the WT mice, 12

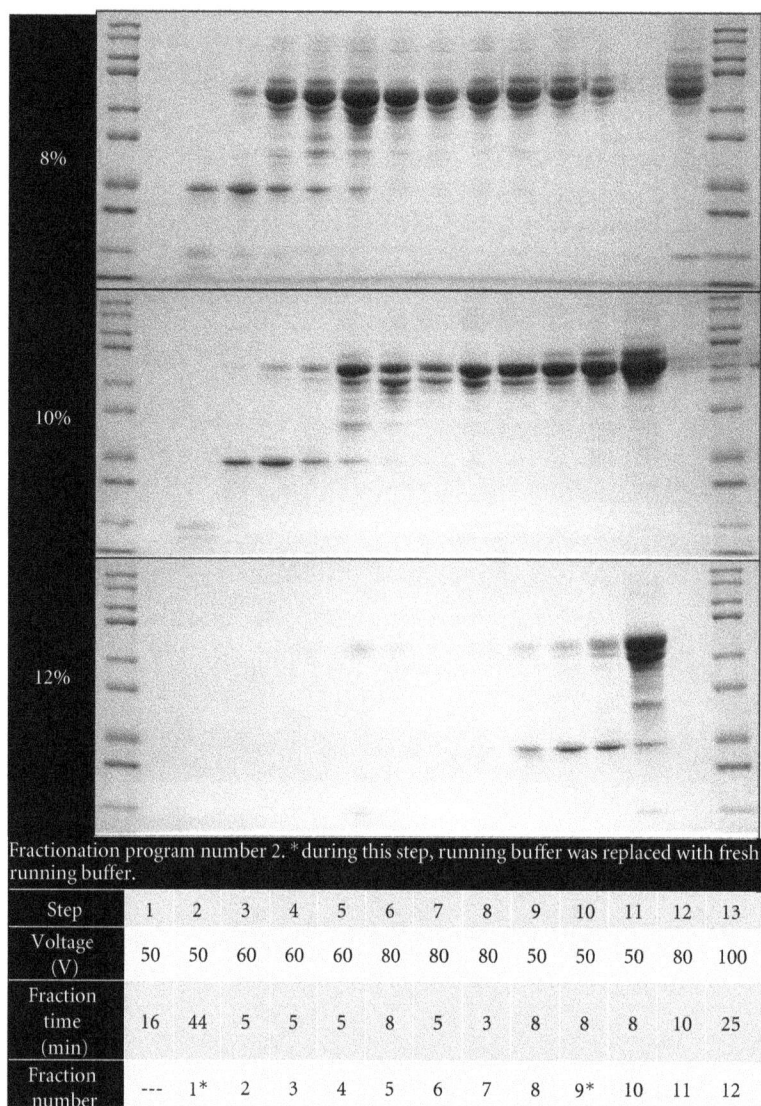

| Fractionation program number 2. * during this step, running buffer was replaced with fresh running buffer. | | | | | | | | | | | | | |
|---|---|---|---|---|---|---|---|---|---|---|---|---|---|
| Step | 1 | 2 | 3 | 4 | 5 | 6 | 7 | 8 | 9 | 10 | 11 | 12 | 13 |
| Voltage (V) | 50 | 50 | 60 | 60 | 60 | 80 | 80 | 80 | 50 | 50 | 50 | 80 | 100 |
| Fraction time (min) | 16 | 44 | 5 | 5 | 5 | 8 | 5 | 3 | 8 | 8 | 8 | 10 | 25 |
| Fraction number | --- | 1* | 2 | 3 | 4 | 5 | 6 | 7 | 8 | 9* | 10 | 11 | 12 |

FIGURE 3: Three cartridges with different acetate percentages were used with the same fractionation program to study fraction protein profile. Program number 2 on 8%, 10% and 12% acetate cartridges and 12% Bis-Tris gels gave interesting results, in that the samples were spread out across fractions, but fraction 12 still showed high albumin abundance.

proteins were present only at day 0, and 45 proteins were just observed 1 day post-MI (Figure 5(a)). In the MMP-9 null mice, 195 proteins were identified; of which 19 were unique to day 0 and 61 proteins were observed only post-MI (Figure 5(b)). The molecular weight of proteins observed in fraction 1 ranged from 7 kDa to 69 kDa, although fragments of higher molecular weight proteins (e.g., C-terminus of alpha-2 macroglobulin) were also present. The majority of proteins observed in fraction 12 had molecular weights ranging from 45 kDa to 263 kDa; nevertheless, lower molecular weight proteins such as transthyretin (16 kDa) were also observed. The UniProt protein database was used to classify proteins by biological function. The unweighted spectrum counts were used to provide measure of relative abundance (Figure 6). Two percent of the proteins identified in WT animals were

ECM proteins, while ECM proteins accounted for 3% of the total identified proteins in MMP-9 null mice.

We used immunoblotting as a secondary method to examine the proteins identified by MS. Serum amyloid A (SAA), a marker of inflammation, was observed at both time points. SAA was identified in fraction 1 of both genotypes and levels at day 0 were significantly different ($P < 0.001$) from levels at day 1 post-MI (Figure 7(a)). Eighty nine proteins were identified only post-MI, including NGAL. NGAL levels post-MI were significantly higher than at day 0 ($P < 0.05$) but no differences were observed between genotypes (Figure 7(b)). Alpha-2 macroglobulin, a generic MMP inhibitor and an MMP-9 substrate, was observed only post-MI in the WT group. Nevertheless, alpha-2 macroglobulin precursor was observed in both groups at days 0 and

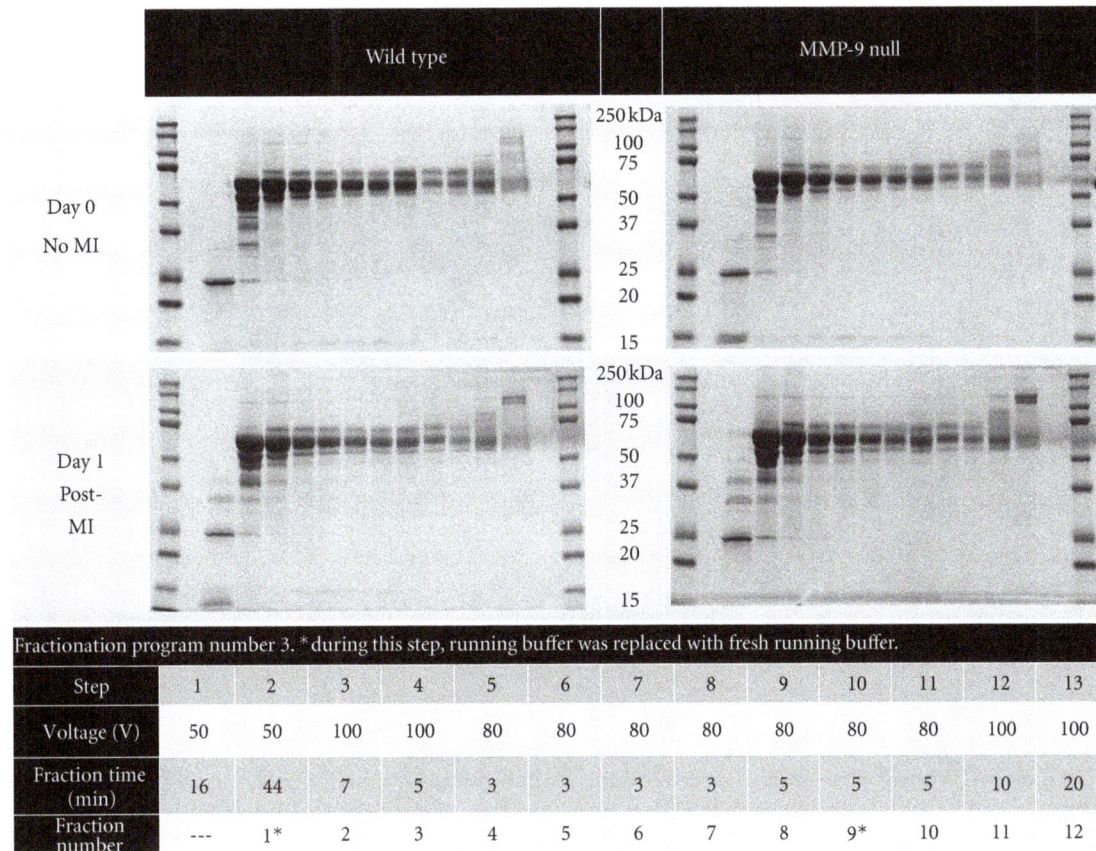

| Step | 1 | 2 | 3 | 4 | 5 | 6 | 7 | 8 | 9 | 10 | 11 | 12 | 13 |
|---|---|---|---|---|---|---|---|---|---|---|---|---|---|
| Voltage (V) | 50 | 50 | 100 | 100 | 80 | 80 | 80 | 80 | 80 | 80 | 80 | 100 | 100 |
| Fraction time (min) | 16 | 44 | 7 | 5 | 3 | 3 | 3 | 3 | 5 | 5 | 5 | 10 | 20 |
| Fraction number | --- | 1* | 2 | 3 | 4 | 5 | 6 | 7 | 8 | 9* | 10 | 11 | 12 |

FIGURE 4: Each plasma sample was separated by electrophoretic mobility into 12 fractions, using fractionation program number 3 on an 8% acetate cartridge and were run on 12% Bis-Tris gels. The figure shows a representative gel from each group ($n = 3$ for each group).

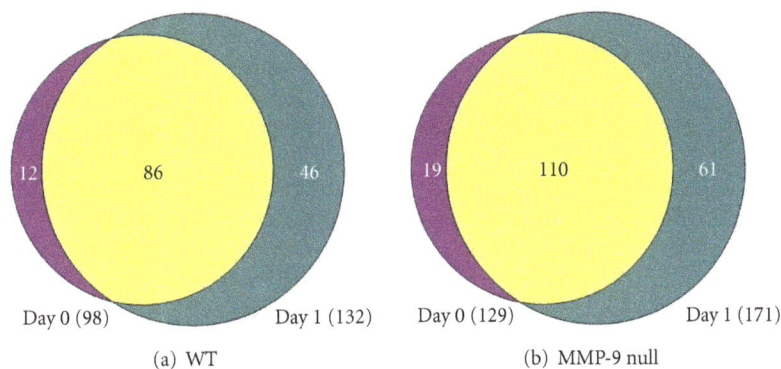

FIGURE 5: Venn diagram representing the number of proteins identified in combined fractions 1 and 12 of the plasma from each group. The purple is the number of proteins identified only in day 0. The green is the number of proteins identified at both time points. The Venn diagram was made using Venn Diagram Plotter software.

1 post-MI. Quantification of alpha-2 macroglobulin showed a significant difference between genotypes 1 day post-MI (Figure 7(c)).

## 4. Discussion

The discovery of plasma markers remains challenging due to the complexity of the samples and the wide range of protein concentrations. In addition, the analysis of proteomics data is a complex multistep process [17]. Therefore, to overcome these problems, effective sample preparation is of outmost importance. Efficient sample preparation will reduce component complexity and enrich for lower abundance proteins while depleting or reducing the most abundant ones. We used a novel fractionation technique to interrogate plasma from WT and MMP-9 null mice at day 1 post-MI. This novel technique provides all of the advantages of 1-D gel electrophoresis, with the additional

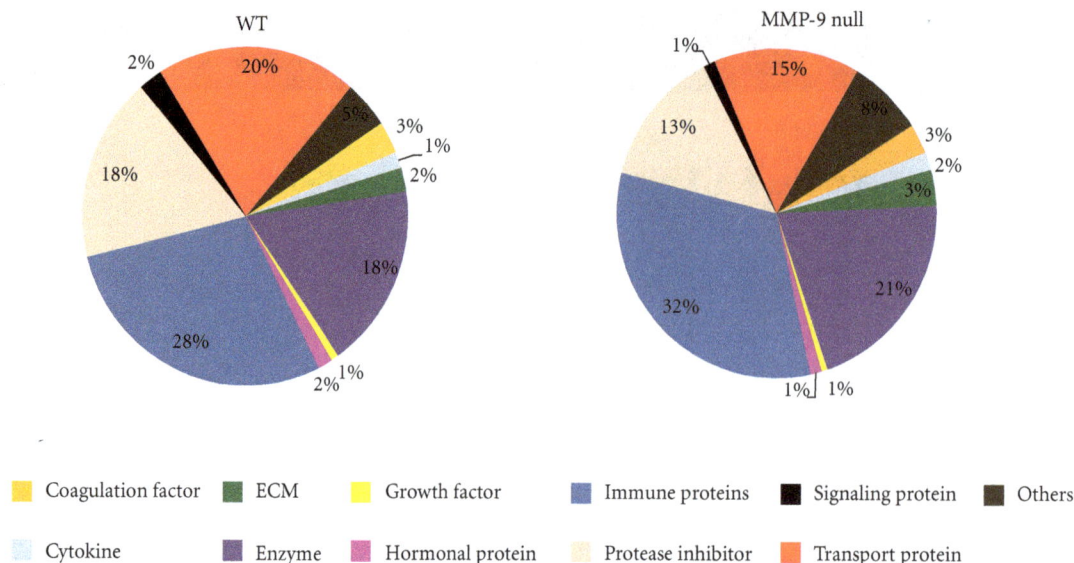

FIGURE 6: Proteins classification by biological function. The graph was created using the number of unweighted spectrum counts as a measure of relative abundance.

benefits of increased loading capacity and high yield liquid phase recovery. The most significant findings of this study were that (1) multiple proteins were identified in post-MI plasma, compared with day 0 control plasma; (2) serum amyloid A is good marker of early MI but it is not MMP-9 dependent; (3) alpha-2 macroglobulin may be an MMP-9 dependent marker. These results combined indicate that a fractionation followed by 1-D gel/LC/MS analysis strategy is effective to isolate and identify plasma proteins changes in response to MI.

We identified a total of 145 unique proteins in the WT samples and 195 unique proteins in the MMP-9 null samples. Known markers of inflammation, such as haptoglobin and SAA, were among the proteins identified. Studies from the Malmö Preventive Study, Sweden, have shown that elevated plasma levels of haptoglobin are a risk factor for MI [19]. Recently, Devaux's group identified haptoglobin as a potential biomarker of prognosis of heart failure in patients with acute MI [20]. Interestingly, they state that low levels of haptoglobin early post-MI favor heart failure. We identified haptoglobin as a potentially increased post-MI marker; nevertheless, studies with longer time points post-MI will have to be developed to confirm the role of haptoglobin in progression to heart failure. SAA is a known marker of inflammation, and SAA levels inversely correlate with cardiac function [21]. One-day post-MI the levels of SAA were significantly higher, in both genotypes, confirming an association with MI. Our future work involves a temporal study of plasma biomarkers post-MI. We plan to investigate if changes in plasma SAA are correlated with progression to heart failure post-MI.

From the proteins present only post-MI, we performed immunoblots against NGAL and alpha-2 macroglobulin. NGAL is a marker of kidney injury [22], as well as matrix degradation and inflammation [23]. This protein has previously been reported to be associated with MI and heart

failure [24, 25]. A recent paper by Akcay and colleagues shows that 1-year mortality rates were significantly higher in patients with high levels of NGAL [26]. Our results were in accordance with the previous reports, showing a robust increase in NGAL levels post-MI. Alpha-2 macroglobulin is a generic proteinase inhibitor with broad specificity [27] and an MMP-9 substrate [28]. Although the association between MMP-9 and alpha-2 macroglobulin has been previously reported, this is the first time that alpha-2 macroglobulin is associated with MMP-9 in the myocardial infarction setting. The higher levels of alpha-2 macroglobulin observed in the MMP-9 mice post-MI suggest that this protein may be an MI biomarker that is MMP-9 dependent.

The protocols developed in this study can be used for other biological samples besides plasma. We are currently developing a fractionation method to investigate secreted proteins in cell culture media. Like in plasma, albumin is highly abundant in the commercially available serums used to supplement culture media. In vitro, the levels of secreted proteins are very low compared to the values observed in culture serum, making it very difficult to identify and quantify the proteins produced by the cells. Fractionation of samples is an easy and reproducible technique that can be used in a variety of models and biological samples.

Mouse models of MI are very useful and important given the unique ability to genetically manipulate these animals [29]. However, it is important to remember that the MI mouse model does not fully mimic the human disease. Thus, postinfarct remodeling of the LV likely has differences between the mouse and human that will need to be taken into account before full translation can occur. Acute MI remains a leading cause of morbidity and mortality worldwide. Thus, the discovery and development of biomarkers has high potential for providing a real benefit for screening, diagnosis, prognosis, prediction of recurrence, and therapeutic monitoring of MI patients.

(a) SAA

(b) NGAL

(c) $\alpha$2M

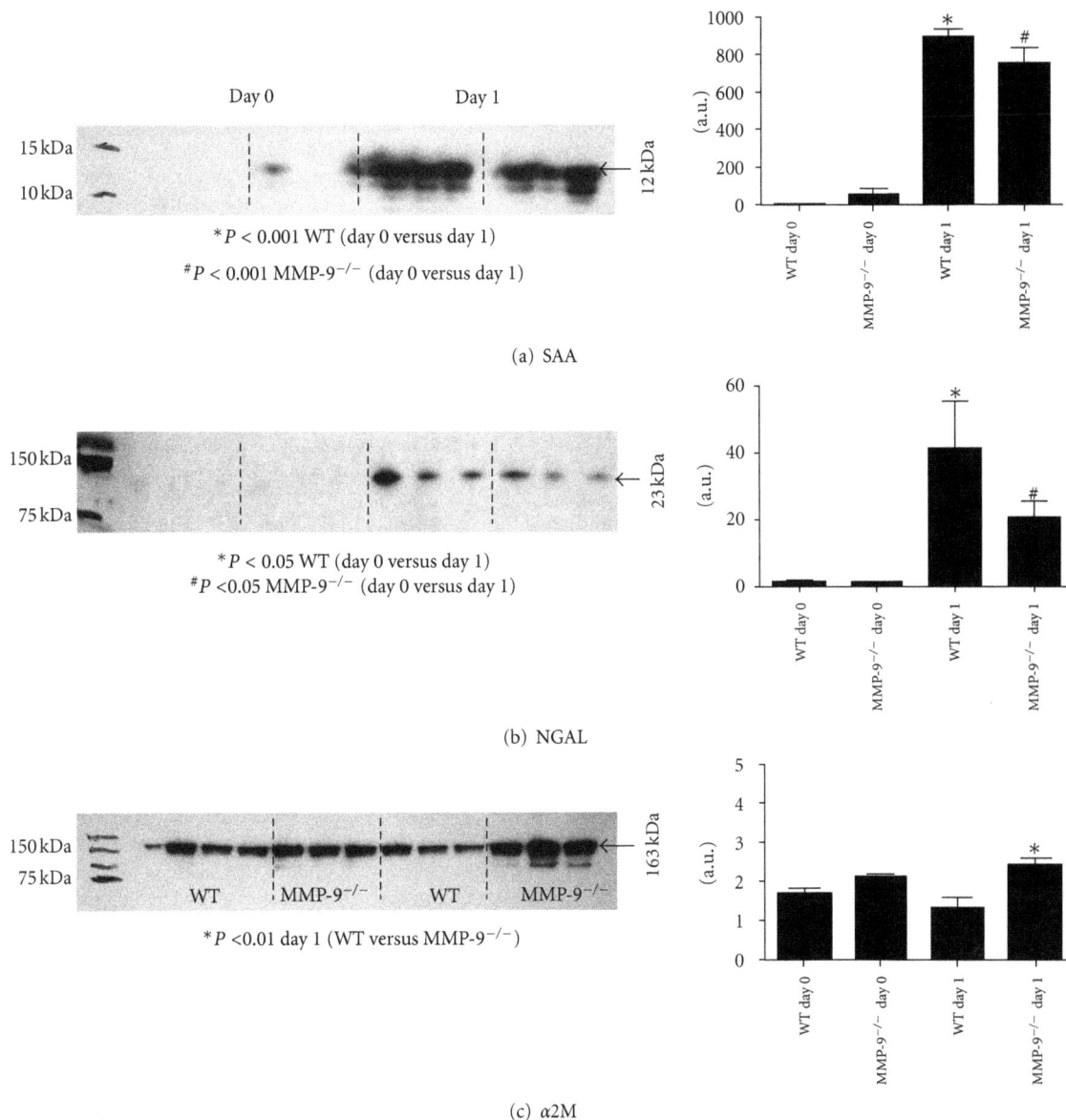

FIGURE 7: Immunoblots for: (a) Serum amyloid A. Protein levels increased significantly post-MI for both genotypes ($P < 0.001$). (b) NGAL. NGAL was observed only post-MI in both genotypes ($P < 0.05$). (c) Alpha-2 macroglobulin ($\alpha$2M). MMP-9 null mice showed higher levels of $\alpha$2M at 1 day post-MI compared to the WT mice counterpart. Densitometry, measured as arbitrary units (a.u.), was used to quantify protein levels in all immunoblots. $n = 3$/group.

In conclusion, by performing plasma fractionation prior to proteomics analysis, we were able to reduce the presence of high abundant proteins, such as albumin, and enrich samples for the detection of lower abundance proteins. We compared plasma samples from wild-type and MMP-9 null mice post-MI, and identified alpha-2 macroglobulin as a prospective MI marker which may be MMP-9 dependent. This technical report revealed that a fractionation approach is a useful technique to evaluate plasma samples.

## Acknowledgments

The authors acknowledge support from NIH NHLBI T32 HL07446 to KYD and from NHLBI HHSN 268201000036C (N01-HV-00244) for the UTHSCSA Cardiovascular Proteomics Center and R01 HL075360, the Max and Minnie Tomerlin Voelcker Fund, and the Veteran's Administration (Merit) to M. L. Lindsey. They acknowledge support from the University of Texas Health Science Center at San Antonio for the Institutional Mass Spectrometry Laboratory.

## References

[1] V. L. Roger, A. S. Go, D. M. Lloyd-Jones, E. J. Benjamin, and J. D. Berry, "Heart disease and stroke statistics—2012 update: a report from the American Heart Association," *Circulation*, vol. 125, pp. e2–e220, 2012.

[2] A. T. Yan, R. T. Yan, F. G. Spinale et al., "Plasma matrix metalloproteinase-9 level is correlated with left ventricular volumes

and ejection fraction in patients with heart failure," *Journal of Cardiac Failure*, vol. 12, no. 7, pp. 514–519, 2006.

[3]  D. Kelly, G. Cockerill, L. L. Ng et al., "Plasma matrix metalloproteinase-9 and left ventricular remodelling after acute myocardial infarction in man: a prospective cohort study," *European Heart Journal*, vol. 28, no. 6, pp. 711–718, 2007.

[4]  M. L. Lindsey, J. Gannon, M. Aikawa et al., "Selective matrix metalloproteinase inhibition reduces left ventricular remodeling but does not inhibit angiogenesis after myocardial infarction," *Circulation*, vol. 105, no. 6, pp. 753–758, 2002.

[5]  A. Ducharme, S. Frantz, M. Aikawa et al., "Targeted deletion of matrix metalloproteinase-9 attenuates left ventricular enlargement and collagen accumulation after experimental myocardial infarction," *Journal of Clinical Investigation*, vol. 106, no. 1, pp. 55–62, 2000.

[6]  N. L. Anderson, M. Polanski, R. Pieper et al., "The human plasma proteome," *Molecular and Cellular Proteomics*, vol. 3, no. 4, pp. 311–326, 2004.

[7]  L. Karpinski, R. Płaksej, R. Derzhko, A. Orda, and M. Witkowska, "Serum levels of interleukin-6, interleukin-10 and C-reactive protein in patients with myocardial infarction treated with primary angioplasty during a 6-month follow-up," *Polskie Archiwum Medycyny Wewnetrznej*, vol. 119, no. 3, pp. 115–121, 2009.

[8]  N. D. Brunetti, D. Quagliara, and M. Di Biase, "Troponin ratio and risk stratification in subjects with acute coronary syndrome undergoing percutaneous coronary intervention," *European Journal of Internal Medicine*, vol. 19, no. 6, pp. 435–442, 2008.

[9]  M. Szulik, J. Stabryla-Deska, J. Boidol, R. Lenarczyk, and Z. Kalarus, "Echocardiography-based qualification and response assessment to cardiac resynchronisation therapy in patients with chronic heart failure. The matrix metalloproteinase-9 substudy," *Kardiologia Polska*, vol. 69, pp. 1043–1051, 2011.

[10]  K. Sakata, K. Iida, N. Mochiduki, and Y. Nakaya, "Brain natriuretic peptide (BNP) level is closely related to the extent of left ventricular sympathetic overactivity in chronic ischemic heart failure," *Internal Medicine*, vol. 48, no. 6, pp. 393–400, 2009.

[11]  P. A. McCullough, A. Olobatoke, and T. E. Vanhecke, "Galectin-3: a novel blood test for the evaluation and management of patients with heart failure," *Reviews in Cardiovascular Medicine*, vol. 12, pp. 200–210, 2011.

[12]  P. G. Righetti, A. Castagna, F. Antonucci et al., "Proteome analysis in the clinical chemistry laboratory: myth or reality?" *Clinica Chimica Acta*, vol. 357, no. 2, pp. 123–139, 2005.

[13]  M. Nissum, S. Kuhfuss, M. Hauptmann et al., "Two-dimensional separation of human plasma proteins using iterative free-flow electrophoresis," *Proteomics*, vol. 7, no. 23, pp. 4218–4227, 2007.

[14]  J. Travis, J. Bowen, and D. Tewksbury, "Isolation of albumin from whole human plasma and fractionation of albumin depleted plasma," *Biochemical Journal*, vol. 157, no. 2, pp. 301–306, 1976.

[15]  A. K. Sato, D. J. Sexton, L. A. Morganelli et al., "Development of mammalian serum albumin affinity purification media by peptide phage display," *Biotechnology Progress*, vol. 18, no. 2, pp. 182–192, 2002.

[16]  D. L. Rothemund, V. L. Locke, A. Liew, T. M. Thomas, V. Wasinger, and D. B. Rylatt, "Depletion of the highly abundant protein albumin from human plasma using the Gradiflow," *Proteomics*, vol. 3, no. 3, pp. 279–287, 2003.

[17]  K. Chandramouli and P. Y. Qian, "Proteomics: challenges, techniques and possibilities to overcome biological sample 6 complexity," *Human Genomics and Proteomics*, vol. 2009, Article ID 239204, 2009.

[18]  M. L. Lindsey, G. P. Escobar, L. W. Dobrucki et al., "Matrix metalloproteinase-9 gene deletion facilitates angiogenesis after myocardial infarction," *American Journal of Physiology*, vol. 290, no. 1, pp. H232–H239, 2006.

[19]  S. Adamsson Eryd, J. G. Smith, O. Melander, B. Hedblad, and G. Engström, "Inflammation-sensitive proteins and risk of atrial fibrillation: a population-based cohort study," *European Journal of Epidemiology*, vol. 26, no. 6, pp. 449–455, 2011.

[20]  B. Haas, T. Serchi, D. R. Wagner et al., "Proteomic analysis of plasma samples from patients with acute myocardial infarction identifies haptoglobin as a potential prognostic biomarker," *Journal of Proteomics*, vol. 75, pp. 229–236, 2011.

[21]  R. Di Stefano, V. Di Bello, M. C. Barsotti et al., "Inflammatory markers and cardiac function in acute coronary syndrome: difference in ST-segment elevation myocardial infarction (STEMI) and in non-STEMI models," *Biomedicine and Pharmacotherapy*, vol. 63, no. 10, pp. 773–780, 2009.

[22]  D. Bolignano, V. Donato, G. Coppolino et al., "Neutrophil Gelatinase-Associated Lipocalin (NGAL) as a marker of kidney damage," *American Journal of Kidney Diseases*, vol. 52, no. 3, pp. 595–605, 2008.

[23]  S. H. Nymo, T. Ueland, E. T. Askevold, T. H. Flo, and J. Kjekshus, "The association between neutrophil gelatinase-associated lipocalin and clinical outcome in chronic heart failure: results from CORONA," *Journal of Internal Medicine*, vol. 271, no. 5, pp. 436–443, 2012.

[24]  A. Şahinarslan, S. A. Kocaman, D. Bas et al., "Plasma neutrophil gelatinase-associated lipocalin levels in acute myocardial infarction and stable coronary artery disease," *Coronary Artery Disease*, vol. 22, no. 5, pp. 333–338, 2011.

[25]  A. Yndestad, L. Landrø, T. Ueland et al., "Increased systemic and myocardial expression of neutrophil gelatinase-associated lipocalin in clinical and experimental heart failure," *European Heart Journal*, vol. 30, no. 10, pp. 1229–1236, 2009.

[26]  A. B. Akcay, M. F. Ozlu, N. Sen, S. Cay, and O. H. Ozturk, "Prognostic significance of neutrophil gelatinase-associated lipocalin in ST-segment elevation myocardial infarction," *Journal of Investigative Medicine*, vol. 60, pp. 508–513, 2012.

[27]  L. C. Cáceres, G. R. Bonacci, M. C. Sánchez, and G. A. Chiabrando, "Activated $\alpha 2$ macroglobulin induces matrix metalloproteinase 9 expression by low-density lipoprotein receptor-related protein 1 through MAPK-ERK1/2 and NF-$\kappa$B activation in macrophage-derived cell lines," *Journal of Cellular Biochemistry*, vol. 111, no. 3, pp. 607–617, 2010.

[28]  L. F. Arbeláez, U. Bergmann, A. Tuuttila, V. P. Shanbhag, and T. Stigbrand, "Interaction of matrix metalloproteinases-2 and -9 with pregnancy zone protein and $\alpha 2$-macroglobulin," *Archives of Biochemistry and Biophysics*, vol. 347, no. 1, pp. 62–68, 1997.

[29]  N. A. Trueblood, Z. Xie, C. Communal et al., "Exaggerated left ventricular dilation and reduced collagen deposition after myocardial infarction in mice lacking osteopontin," *Circulation Research*, vol. 88, no. 10, pp. 1080–1087, 2001.

# 16

# Characterization of the Phosphoproteome in Human Bronchoalveolar Lavage Fluid

**Francesco Giorgianni,[1] Valentina Mileo,[2] Dominic M. Desiderio,[3,4] Silvia Catinella,[2] and Sarka Beranova-Giorgianni[1]**

[1] *Department of Pharmaceutical Sciences, The University of Tennessee Health Science Center, Memphis, TN 38163, USA*
[2] *Corporate Preclinical R&D, Analytics and Early Formulations Department, Chiesi Farmaceutici S.p.A., 43122 Parma, Italy*
[3] *Department of Neurology, The University of Tennessee Health Science Center, Memphis, 38163 TN, USA*
[4] *Charles B. Stout Neuroscience Mass Spectrometry Laboratory, The University of Tennessee Health Science Center, Memphis, 38163 TN, USA*

Correspondence should be addressed to Sarka Beranova-Giorgianni, sberanova@uthsc.edu

Academic Editor: Visith Thongboonkerd

Global-scale examination of protein phosphorylation in human biological fluids by phosphoproteomics approaches is an emerging area of research with potential for significant contributions towards discovery of novel biomarkers. In this pilot work, we analyzed the phosphoproteome in human bronchoalveolar lavage fluid (BAL) from nondiseased subjects. The main objectives were to assess the feasibility to probe phosphorylated proteins in human BAL and to obtain the initial catalog of BAL phosphoproteins, including protein identities and exact description of their phosphorylation sites. We used a gel-free bioanalytical workflow that included whole-proteome digestion of depleted BAL proteins, enrichment of phosphopeptides by immobilized metal ion affinity chromatography (IMAC), LC-MS/MS analyses with a linear ion trap mass spectrometer, and searches of a protein sequence database to generate a panel of BAL phosphoproteins and their sites of phosphorylation. Based on sequence-diagnostic MS/MS fragmentation patterns, we identified a collection of 36 phosphopeptides that contained 26 different phosphorylation sites. These phosphopeptides mapped to 21 phosphoproteins including, for example, vimentin, plastin-2, ferritin heavy chain, kininogen-1, and others. The characterized phosphoproteins have diverse characteristics in terms of cellular origin and biological function. To the best of our knowledge, results of this study represent the first description of the human BAL phosphoproteome.

## 1. Introduction

Posttranslational modification of proteins by phosphorylation plays a complex and critical role in the regulation of numerous biological processes. In recent years, large efforts have been devoted to global-scale analysis of protein phosphorylation sites using various phosphoproteomics methodologies [1, 2]. These phosphoproteomics studies have focused chiefly on large-scale characterization of the phosphoproteomes in cultured cells or tissues. In contrast, investigation of the phosphoproteomes in biological fluids is an emerging area, and studies of this type are relatively scarce. Characterization of protein phosphorylation in biological fluids presents a major challenge. Phosphoproteins released into the fluid are diluted and mostly of low abundance, and they are present in a highly complex mixture, that is composed predominantly of nonphosphorylated proteins. The issue is often compounded by overabundance of certain proteins such as albumin and immunoglobulins. Highly advanced bioanalytical strategies that have been developed and applied successfully in the context of cell and tissue phosphoproteomics are now being tailored for biological fluid phosphoproteomics.

Recent studies of phosphoproteomics of biological fluids include serum and plasma [3–5], CSF [6, 7], saliva [8], and urine [9]. In particular, examination of phosphoproteomes in biological fluids obtained from sites proximal to specific organs represents a potential route to important mechanistic information as well as to biomarker discovery.

Human bronchoalveolar lavage fluid (BAL) is a proximal fluid commonly used for diagnosis of lung diseases including chronic obstructive pulmonary disease (COPD) and lung cancer. Procurement of clinical BAL specimens involves washing of the epithelial lining of the lung with saline using a fiberoptic bronchoscope. Molecular composition of BAL reflects the status of the respiratory tract, and analysis of human BAL composition at the molecular level therefore provides an attractive way towards improved understanding of disease mechanisms or discovery of biomarker signatures that are directly relevant to specific lung diseases. The proteome of human BAL has been studied numerous times in the context of various lung diseases [10–14]. In contrast, the phosphoproteome of human BAL has not been characterized yet.

In this study, we undertook a pilot interrogation of the human BAL phosphoproteome. Our ongoing research program focuses on proteomics of human BAL [15], and we aim to expand this program to encompass studies of posttranslational modifications. Initially, we set out to determine if phosphorylated proteins can be characterized in human BAL using a mass spectrometry-based analytical platform, and to obtain a first description of the BAL phosphoproteome, including assignments of the sites of phosphorylation.

## 2. Methods

*2.1. Characteristics of BAL Specimens.* The human BAL specimens were provided by Chiesi Farmaceutici, Parma Italy; the project was approved by the IRB at The University of Tennessee Health Science Center. The human BAL samples were obtained from subjects without clinical diagnosis of COPD or lung cancer. Information on the characteristics of the BAL specimen donors is listed in Table 1. The lavage was performed with four aliquots of 50 mL of saline delivered via a fiberoptic bronchoscope. After centrifugation, the liquid component of BAL was aliquoted and stored at −80°C until analysis. To provide sufficient amount of protein, pooled BAL samples were used. Two separate pools of 3 (Pool 1) and 7 samples (Pool 2), respectively, were analyzed in two independent experiments.

*2.2. Sample Desalting and Protein Depletion.* Prior to analysis, the BAL samples were centrifuged to remove cell debris. Processing of each sample included removal of salts and depletion of overabundant contaminant proteins. Desalting was performed by ultrafiltration with spin concentrators (MW cutoff of 5,000 Da). The samples in the concentrators were centrifuged (25 min; 5,000 g; 4°C) to produce *ca.* 100–200 μL of retentate. After the first concentration step, water (4 mL) was added to the retentate and the concentration step was repeated for a total of three times. The final retentates (*ca.* 100 μL) were dried in a vacuum centrifuge.

Table 1: Characteristics of BAL specimen donors.

| Donor | Disease status | Gender | Age |
|---|---|---|---|
| 1 | control | F | 48 |
| 2 | control | F | 68 |
| 3 | control | F | 58 |
| 4 | control | F | 75 |
| 5 | control | F | 58 |
| 6 | control | F | 64 |
| 7 | control | F | 63 |
| 8 | control | F | 60 |
| 9 | control | F | 65 |
| 10 | control | F | 73 |

Albumin and five other high-abundance proteins were removed with the Hu-6 Multiple Affinity Removal System (MARS) spin cartridge (Agilent) following procedure provided by the manufacturer. After MARS depletion, the samples were desalted by ultrafiltration as described above. Protein concentration before and after MARS depletion was determined with the micro BCA assay (Pierce). After pooling, the final protein content was 450 μg (Pool 1) and 900 μg (Pool 2).

*2.3. Whole Proteome Digestion and IMAC Enrichment.* The proteins were digested with trypsin using an in-solution digestion procedure. Briefly, the dried proteins in each pooled sample were redissolved in 45 μL of 400 mM ammonium bicarbonate buffer containing 8 M urea (pH 8). Prior to digestion, the proteins were reduced with DTT (5 μL of 50 mM solution, incubation for 1 h at 56°C) followed by alkylation with iodoacetamide (5 μL of 200 mM solution, incubation for 45 min at room temperature in the dark). The sample was diluted with water to 2 M final urea concentration, and 20 μg of sequencing-grade trypsin (Promega) were added. The mixture was incubated overnight at 37°C.

After digestion, the mixture was acidified with TFA and subjected to solid phase extraction using a home-made SPE minicolumn packed with C18 stationary phase. After elution from the minicolumn, the sample was dried and the redissolved in 90% water/10% acetic acid, as required for immobilized metal ion affinity chromatography (IMAC).

The IMAC procedure, which serves to enrich the proteolytic digests for phosphopeptides, was performed with the Phosphopeptide Isolation Kit (gallium/IDA, Pierce). Each BAL peptide digest was applied to the column, and the phosphopeptides were bound by incubation at room temperature for 1 h. The column was washed with the following solutions: 40 μL of 0.1% acetic acid (2 washes), 40 μL of 0.1% acetic acid/10% ACN (2 washes), and 40 μL of water (2 washes). The phosphopeptides were eluted from the IMAC column with two 40 μL-aliquots of 200 mM sodium phosphate (pH 8.4), followed by a single elution with 40 μL of 100 mM sodium phosphate/50% ACN. The eluates were combined, the resulting sample was acidified, and its volume was reduced to ca 25 μL in a vacuum centrifuge. Prior to LC-MS/MS analysis, the IMAC-enriched phosphopeptides were desalted with ZipTipC18 (Millipore, Billerica, MA, USA), using the procedure provided by the manufacturer. The phosphopeptides bound to the ZipTipC18 column were eluted with $4\,\mu L$ of 50% ACN/0.1% formic acid and diluted with $6\,\mu L$ of 0.5% formic acid; aliquots of these samples were injected onto the LC-MS/MS instrument.

*2.4. LC-MS/MS and Phosphoprotein Identification.* The LC-MS/MS analyses were performed with an LTQ linear ion trap mass spectrometer (Thermo Electron) that was interfaced with a nano-LC system (Dionex). The IMAC-enriched peptide digests were loaded onto a fused-silica microcapillary column/spray needle (Picofrit, 15 cm length, $75\,\mu m$ I.D.; New Objective) packed in-house with C18 stationary phase (Michrom Bioresources). The peptides were separated using a 90-min linear gradient from 0% to 90% mobile phase B. Mobile phase B was 10% water/90% methanol/0.05% formic acid; mobile phase A was 98% water/2% methanol/0.05% formic acid. The LC-MS/MS data were acquired in the data-dependent mode. Each of the pooled samples (Pool 1 and Pool 2) was analyzed in triplicate.

The LC-MS/MS datasets were used to search the UniProt database (subset of human proteins) using TurboSEQUEST search engine that was part of Bioworks 3.2 (Thermo Electron). The following parameters were used in the searches: full-trypsin specificity, dynamic modifications of phosphorylated S, T, and Y (+80.0), and dynamic modifications of oxidized M (+16.0). The search results were filtered to include peptides retrieved XCorr values $\geq 2.00$, and 3.50 for doubly and triply charged precursor ions, respectively. All MS/MS spectra for the individual phosphopeptides that passed this initial filtering were inspected manually. This manual validation checked for the presence of a product ion that corresponds to the neutral-loss of phosphoric acid ($[M+2H-98]^{2+}$ for doubly charged ions or $[M+3H-98]^{3+}$ for triply charged ions); and for coverage of the phosphopeptide sequence by the b- and/or y product-ion series. Assignments of the sites of phosphorylation were verified by inspecting the b- and/or y-product ions that flanked the phosphorylation site assigned by the search engine. Data from analyses of Pool 1 and Pool 2 were combined to produce the final phosphoprotein panel. Additional information about the phosphorylation sites/phosphoproteins was obtained from the UniProt annotations, the Phosphosite knowledgebase (http://www.phosphosite.org/), the Human Protein Atlas knowledgebase (http://www.proteinatlas.org/), the Ingenuity Pathway Analysis tool (IPA), and from searches of primary literature.

# 3. Results and Discussion

For this pilot study, a simple gel-free bioanalytical strategy was employed. The general outline of the bioanalytical workflow is shown in Figure 1. Specific characteristics of human BAL have to be taken into account for sample processing and protein extraction. First, proteins in BAL are diluted in saline, and therefore sample concentration and desalting are needed. Second, high background created by overabundant plasma proteins would interfere with analysis of low-level phosphoproteins, and removal of these proteins must be accomplished. In our study, to process the BAL samples for phosphoproteome analysis, salts were removed by ultrafiltration, and overabundant plasma proteins were depleted using immunoaffinity capture. Proteins in the depleted BAL samples were digested with trypsin, and the digests were subjected to immobilized metal ion affinity chromatography (IMAC) enrichment for phosphopeptides. The enriched digests were analyzed by LC-MS/MS on an LTQ ion trap mass spectrometer to obtain MS/MS data that indicate the phosphopeptide sequences and phosphosite locations in these peptides. The phosphopeptides and phosphoproteins were identified through searches of the UniProt protein sequence database. Manual inspection of all phosphopeptide search results and of the corresponding MS/MS data was performed to confirm the validity of the phosphopeptide matches. Of diagnostic value in the context of MS/MS fragmentation was the neutral loss of the elements of phosphoric acid from the phosphopeptide molecular ions. This fragmentation pathway, which is prominent in the ion trap mass spectrometer, leads to the appearance of a characteristic product ion in the MS/MS spectrum of a phosphopeptide [16]. This well-known scenario is illustrated in Figure 2, which shows the MS/MS spectrum for the phosphopeptide IEDVGpSDEEDDSGKDK. This spectrum displays a prominent product ion at $m/z$ 860.7, which corresponds to the loss of the elements of phosphoric acid from the doubly charged precursor ion. In addition, a number of product ions from the b- and y-series are present that determine the amino acid sequence of the phosphopeptide. Peaks at $m/z$ 682 ($b_6$) and $m/z$ 1137 ($y_{10}$) indicate the exact location of the phosphorylation site on Ser 255 of human heat shock HSP 90-beta.

Each of the two pooled BAL samples that were analyzed produced a set of 13 phosphoproteins. Five of these phosphoproteins were common to both samples; in addition, each sample yielded a unique group of phosphoproteins. This is not unexpected given the large biological variability associated with clinical specimens, and variable phosphoprotein signatures have been also observed for other clinical samples [17]. The results of our BAL phosphoproteome analyses are summarized in Table 2. Overall, interrogation of the IMAC-enriched digests of depleted BAL samples with LC-MS/MS resulted in the characterization of 36 unique phosphopeptides that contained a total of 26 phosphorylation sites and mapped to 21 proteins. Our results demonstrate that characterization of BAL phosphoproteome is feasible, and the phosphoprotein panel represents new findings that expand our knowledge of the molecular characteristics of BAL proteins.

Since the focus of our pilot study reported here was on first description of the human BAL phosphoproteome, the scope of the study was limited to qualitative analyses of a small number of specimens from female donors only. This initial examination was not intended to characterize phosphoprotein biomarkers associated with a specific lung disease but to initiate the building of a detailed phosphoprotein/phosphosites catalog as a starting point for future

TABLE 2: Phosphopeptides and phosphoproteins characterized in human BAL.

| Database accession code | Entry name | Protein name<br>Phosphopeptide characterized[a] | Site[b] |
|---|---|---|---|
| (1) P08670 | VIME_HUMAN | Vimentin | |
| | | QVQS*LTCEVDALK | S325 |
| (2) P02794 | FRIH_HUMAN | Ferritin heavy chain | |
| | | KM#GAPESGLAEYLFDKHTLGDS*DNES | S179 |
| | | KMGAPESGLAEYLFDKHTLGDS*DNES | S179 |
| | | MGAPESGLAEYLFDKHTLGDS*DNES | S179 |
| | | HTLGDS*DNES | S179 |
| | | KMGAPESGLAEYLFDKHTLGDSDNES* | (S183) |
| | | MGAPESGLAEYLFDKHTLGDSDNES* | S183 |
| (3) P30086 | PEBP1_HUMAN | Phosphatidylethanolamine-binding protein 1 | |
| | | NRPTS*ISWDGLDSGK | S52 |
| (4) P13796 | PLSL_HUMAN | Plastin-2 | |
| | | GS*VSDEEM#M#ELR | S5 |
| | | GS*VSDEEMM#ELR | S5 |
| | | GS*VSDEEMMELR | S5 |
| | | EGES*LEDLMK | S257 |
| (5) Q9H3Z4 | DNJC5_HUMAN | DnaJ homolog subfamily C member 5 | |
| | | S*LSTSGESLYHVLGLDK | (S8) |
| (6) P27816 | MAP4_HUMAN | Microtubule-associated protein 4 | |
| | | DVT*PPPETEVVLIK | T521 |
| (7) P21333 | FLNA_HUMAN | Filamin-A | |
| | | RAPS*VANVGSHCDLSLK | S2152 |
| | | CSGPGLS*PGMVR | S1459 |
| (8) P08575 | PTPRC_HUMAN | Receptor-type tyrosine-protein phosphatase C | |
| | | NRNS*NVIPYDYNR | S973 |
| (9) P01042 | KNG1_HUMAN | Kininogen-1 | |
| | | ETTCSKES*NEELTESCETK | S332 |
| (10) P02765 | FETUA_HUMAN | Alpha-2-HS-glycoprotein | |
| | | CDSSPDS*AEDVRK | S138 |
| | | CDSSPDS*AEDVR | S138 |
| (11) Q15637 | SF01_HUMAN | Splicing factor 1 | |
| | | TGDLGIPPNPEDRS*PS*PEPIYNSEGK | S80; S82 |
| (12) Q9UK76 | HN1_HUMAN | Hematological and neurological expressed 1 protein | |
| | | RNS*SEASSGDFLDLK | (S87) |
| (13) Q7Z3D4 | LYSM3_HUMAN | LysM and putative peptidoglycan-binding domain-containing protein 3 | |
| | | S*TSRDRLDDIIVLTK | (S53) |
| (14) P02671 | FIBA_HUMAN | Fibrinogen alpha chain | |
| | | PGSTGTWNPGS*SER | S364 |
| (15) P09651 | ROA1_HUMAN | Heterogeneous nuclear ribonucleoprotein A1 | |
| | | SES*PKEPEQLR | (S6) |
| (16) P51858 | HDGF_HUMAN | Hepatoma-derived growth factor | |
| | | AGDLLEDS*PKRPK | S165 |
| | | RAGDLLEDS*PK | S165 |
| | | AGDLLEDS*PK | S165 |
| | | GNAEGSS*DEEGKLVIDEPAK | (S133) |
| (17) Q9H2C0 | GAN_HUMAN | Gigaxonin | |
| | | FGAVACGVAMELY*VFGGVR | Y471 |
| (18) Q13637 | RAB32_HUMAN | Ras-related protein Rab-32 | |
| | | DSS*QSPSQVDQFCK | (S152) |

| Database accession code | Entry name | Protein name<br>Phosphopeptide characterized[a] | Site[b] |
|---|---|---|---|
| (19) P35579 | MYH9_HUMAN | Myosin-9 | |
| | | KGAGDGS*DEEVDGK | S1943 |
| (20) P07900 | HS90A_HUMAN | Heat shock protein HSP 90-alpha | |
| | | DKEVS*DDEAEEK | S231 |
| (21) P08238 | HS90B_HUMAN | Heat shock protein HSP 90-beta | |
| | | IEDVGS*DEEDDSGKDKK | S255 |
| | | IEDVGS*DEEDDSGKDK | S255 |
| | | IEDVGS*DEEDDSGK | S255 |

[a]STY* denotes phosphorylated amino acid. M# denotes oxidized methionine.
[b]Phosphorylation sites were assigned based on MS/MS product ions. Parentheses indicate cases where an alternative site is possible.

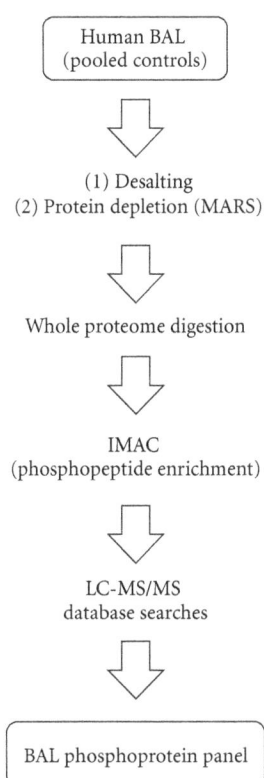

FIGURE 1: Flowchart depicting the bioanalytical workflow used for BAL phosphoproteome mapping. Abbreviations: Multiple Affinity Removal System—MARS; immobilized metal ion affinity chromatography—IMAC.

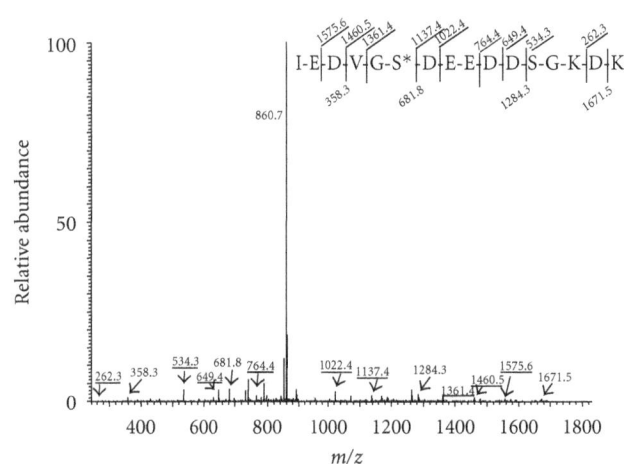

FIGURE 2: Representative MS/MS spectrum obtained in analyses of IMAC-enriched digests of depleted BAL proteomes. The spectrum displays a prominent product ion at $m/z$ 860.7 that corresponds to loss of $H_3PO_4$ from the molecular ion ([M+2H]$^{2+}$ at $m/z$ 909.9). Furthermore, y- and b-ions are present that define phosphopeptide sequence and site of phosphorylation. The peptide IEDVGpS-DEEDDSGKDK belongs to HSP 90-beta.

differential phosphoproteomics efforts. In terms of the size of our initial BAL phosphoprotein panel, our results are comparable, for example, to a CSF phosphoproteome study that revealed 44 phosphoproteins [6], or to a recently published catalog of the urine phosphoproteome that included 45 phosphopeptides from 31 proteins [9]. Our initial exploration of the BAL phosphoproteome was not expected to yield a complete description of all BAL phosphoproteins, and it is possible that some phosphoproteins escaped detection due to their low abundance, unfavorable properties of the corresponding phosphopeptides influencing their behavior during analyses, and other issues. Clearly, these pilot results can be expanded in future efforts to enhance the BAL phosphoproteome coverage through modifications of the bioanalytical workflow such as incorporation of additional separation/enrichment dimensions.

Bronchoalveolar lavage samples components of the epithelial lining fluid, and proteins that are found in BAL are of diverse origin [14]. They may be released by different types of resident and/or infiltrating cells; many plasma proteins are also identified in BAL. To supplement our experimental findings on the phosphorylation status of BAL proteins, we compiled additional information on protein localization from several protein knowledgebases, including Ingenuity and the Human Protein Atlas, HPA (see Table 3 and Figure 3). Regarding tissue-specific protein expression, inspection of protein expression profiles in HPA showed that the majority of proteins from our dataset are expressed in the lung and

TABLE 3: Tissue expression and subcellular location for proteins from our panel. This information was compiled from Human Protein Atlas (HPA) and from Ingenuity Pathway Analysis.

| No. | Database code | Entry name | Protein name | Tissue expression in the lung (from HPA; March 2012) | | | Subcellular location (from Ingenuity; March 2012) |
|---|---|---|---|---|---|---|---|
| | | | | Pneumocytes | Macrophages | Other tissues | |
| 1 | P08670 | VIME_HUMAN | Vimentin | Y (strong) | Y (strong) | Y | Cytoplasm |
| 2 | P02794 | FRIH_HUMAN | Ferritin heavy chain | N | Y (strong) | Y | Cytoplasm |
| 3 | P30086 | PEBP1_HUMAN | Phosphatidylethanolamine-binding protein 1 | Y (weak) | Y (weak) | Y | Cytoplasm |
| 4 | P13796 | PLSL_HUMAN | Plastin-2 | N | Y (strong) | Y (limited) | Cytoplasm |
| 5 | Q9H3Z4 | DNJC5_HUMAN | DnaJ homolog subfamily C member 5 | Y (moderate) | Y (moderate) | Y | Plasma membrane |
| 6 | P27816 | MAP4_HUMAN | Microtubule-associated protein 4 (MAP 4) | Y (moderate) | Y (weak) | Y | Cytoplasm |
| 7 | P21333 | FLNA_HUMAN | Filamin-A | Y (weak) | Y (strong) | Y | Plasma membrane |
| 8 | P08575 | PTPRC_HUMAN | Receptor-type tyrosine-protein phosphatase C | N | Y (weak) | Y (limited) | Plasma membrane |
| 9 | P01042 | KNG1_HUMAN | Kininogen-1 | N | N | Y (distinct in plasma; other limited) | Extracellular space |
| 10 | P02765 | FETUA_HUMAN | Alpha-2-HS-glycoprotein | N | Y (moderate) | Y (plasma; other limited) | Extracellular space |
| 11 | Q15637 | SF01_HUMAN | Splicing factor 1 | Y (moderate) | Y (moderate) | Y | Nucleus |
| 12 | Q9UK76 | HN1_HUMAN | Hematological and neurological expressed 1 protein | | Y (moderate) | Y | Nucleus |
| 13 | Q7Z3D4 | LYSM3_HUMAN | LysM and putative peptidoglycan-binding domain-containing protein 3 | Y (moderate) | Y (strong) | Y | unknown |
| 14 | P02671 | FIBA_HUMAN | Fibrinogen alpha chain | N | Y (moderate) | Y (limited) | Extracellular space |
| 15 | P09651 | ROA1_HUMAN | Heterogeneous nuclear ribonucleoprotein A1 | Y (strong) | Y (strong) | Y | Nucleus |
| 16 | P51858 | HDGF_HUMAN | Hepatoma-derived growth factor; | Y (strong) | Y (moderate) | Y | Extracellular space |
| 17 | Q9H2C0 | GAN_HUMAN | Gigaxonin | N | Y (moderate) | Y | Cytoplasm |
| 18 | Q13637 | RAB32_HUMAN | Ras-related protein Rab-32 | Y (moderate) | Y (weak) | Y | Cytoplasm |
| 19 | P35579 | MYH9_HUMAN | Myosin-9 | Y (strong) | Y (moderate) | Y | Cytoplasm |
| 20 | P07900 | HS90A_HUMAN | Heat shock protein HSP 90-alpha | N | N | Y (limited) | Cytoplasm |
| 21 | P08238 | HS90B_HUMAN | Heat shock protein HSP 90-beta | Y (moderate) | Y (moderate) | Y | Cytoplasm |

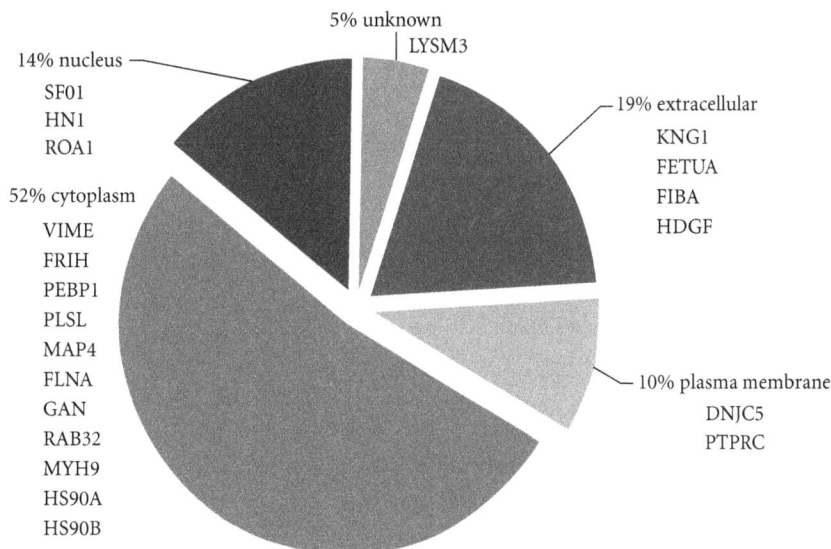

FIGURE 3: Subcellular location distribution of the characterized proteins; compiled from Ingenuity.

in other tissues/organs. Analysis of subcellular compartment categories for proteins from our panel showed a strong representation of cytoplasmic proteins (52%); four proteins (19%) were classified as extracellular and include plasma proteins kininogen-1 and alpha-2-HS-glycoprotein whose phosphorylated counterparts have also been characterized in human plasma/serum phosphoproteome [3, 5].

The BAL phosphoprotein panel (Table 2) includes proteins with diverse functional characteristics, including structural proteins (vimentin, plastin-2), transcriptional regulators (Splicing factor 1, hepatoma-derived growth factor), chaperones (heats shock protein HSP 90-alpha and -beta), and others. Several of the proteins whose phosphorylation was characterized here have known connection to lung function and perturbations due to environmental stresses including smoking, and/or to lung disease.

For example, ferritin is an important mediator of iron homeostasis, and increased levels of ferritin have been found in the lavage of smokers [18]. The rationale for this increase is, at least is part, that smoke particles cause iron accumulation in the respiratory tract, and increased expression of ferritin is part of the host response, aimed to sequester the iron.

Another phosphoprotein found in the present study is the actin-bundling protein plastin-2. Phosphorylation of plastin-2 modulates its function in the assembly of actin networks, and it is associated with leukocyte activation in response to various stimuli [19, 20]. Recently, plastin-2 has been identified in human BAL proteome as a component of a pulmonary disease marker profile [13].

In conclusion, this study presents novel findings towards description of the human BAL phosphoproteome. Since aberrant protein phosphorylation associated with specific lung diseases could potentially be reflected as alterations in BAL phosphoproteins, this study lays an important foundation for future differential phosphoprotein profiling for biomarker discovery.

## Acknowledgments

This study has been funded by Chiesi Farmaceutici. Funds for the LTQ mass spectrometer have been provided in part by the NIH Shared Instrumentation Grant S10RR16679 (to D.M.D).

## References

[1] P. A. Grimsrud, D. L. Swaney, C. D. Wenger, N. A. Beauchene, and J. J. Coon, "Phosphoproteomics for the masses," *ACS Chemical Biology*, vol. 5, no. 1, pp. 105–119, 2010.

[2] B. Eyrich, A. Sickmann, and R. P. Zahedi, "Catch me if you can: mass spectrometry-based phosphoproteomics and quantification strategies," *Proteomics*, vol. 11, no. 4, pp. 554–570, 2011.

[3] W. Zhou, M. M. Ross, A. Tessitore et al., "An initial characterization of the serum phosphoproteome," *Journal of Proteome Research*, vol. 8, no. 12, pp. 5523–5531, 2009.

[4] S. D. Garbis, T. I. Roumeliotis, S. I. Tyritzis, K. M. Zorpas, K. Pavlakis, and C. A. Constantinides, "A novel multidimensional protein identification technology approach combining protein size exclusion prefractionation, peptide zwitterion-ion hydrophilic interaction chromatography, and nano-ultraperformance RP chromatography/nESI-MS2 for the in-depth analysis of the serum proteome and phosphoproteome: application to clinical sera derived from humans with benign prostate hyperplasia," *Analytical Chemistry*, vol. 83, no. 3, pp. 708–718, 2011.

[5] M. Carrascal, M. Gay, D. Ovelleiro, V. Casas, E. Gelpí, and J. Abian, "Characterization of the human plasma phosphoproteome using linear ion trap mass spectrometry and multiple search engines," *Journal of Proteome Research*, vol. 9, no. 2, pp. 876–884, 2010.

[6] J. M. C. Bahl, S. S. Jensen, M. R. Larsen, and N. H. H. Heegaard, "Characterization of the human cerebrospinal fluid phosphoproteome by titanium dioxide affinity chromatography and mass spectrometry," *Analytical Chemistry*, vol. 80, no. 16, pp. 6308–6316, 2008.

[7] X. Yuan and D. M. Desiderio, "Proteomics analysis of phosphotyrosyl-proteins in human lumbar cerebrospinal fluid," *Journal of Proteome Research*, vol. 2, no. 5, pp. 476–487, 2003.

[8] M. D. Stone, X. Chen, T. McGowan et al., "Large-scale phosphoproteomics analysis of whole saliva reveals a distinct phosphorylation pattern," *Journal of Proteome Research*, vol. 10, no. 4, pp. 1728–1736, 2011.

[9] Q. R. Li, K. X. Fan, R. X. Li et al., "A comprehensive and non-prefractionation on the protein level approach for the human urinary proteome: touching phosphorylation in urine," *Rapid Communications in Mass Spectrometry*, vol. 24, no. 6, pp. 823–832, 2010.

[10] H. Chen, D. Wang, C. Bai, and X. Wang, "Proteomics-based biomarkers in chronic obstructive pulmonary disease," *Journal of Proteome Research*, vol. 9, no. 6, pp. 2798–2808, 2010.

[11] T. Oumeraci, B. Schmidt, T. Wolf et al., "Bronchoalveolar lavage fluid of lung cancer patients: mapping the uncharted waters using proteomics technology," *Lung Cancer*, vol. 72, no. 1, pp. 136–138, 2011.

[12] A. Plymoth, C. G. Löfdahl, A. Ekberg-Jansson et al., "Protein expression patterns associated with progression of chronic obstructive pulmonary disease in bronchoalveolar lavage of smokers," *Clinical Chemistry*, vol. 53, no. 4, pp. 636–644, 2007.

[13] C. Landi, E. Bargagli, B. Magi et al., "Proteome analysis of bronchoalveolar lavage in pulmonary langerhans cell histiocytosis," *Journal of Clinical Bioinformatics*, vol. 1, article 31, 2011.

[14] R. Wattiez and P. Falmagne, "Proteomics of bronchoalveolar lavage fluid," *Journal of Chromatography B*, vol. 815, no. 1-2, pp. 169–178, 2005.

[15] F. Giorgianni, V. Mileo, L. Chen, D. M. Desiderio, and S. Beranova-Giorgianni, "Proteomics of human bronchoalveolar lavage fluid: discovery of biomarkers of chronic obstructive pulmonary disease (COPD) with difference gel electrophoresis (DIGE) and mass spectrometry (MS)," in *Modern Methods in Protein Biochemistry*, H. Tschesche, Ed., pp. 219–234, Walter De Gruyter, Berlin, Germany, 2012.

[16] P. J. Boersema, S. Mohammed, and A. J. R. Heck, "Phospho-peptide fragmentation and analysis by mass spectrometry," *Journal of Mass Spectrometry*, vol. 44, no. 6, pp. 861–878, 2009.

[17] L. Chen, B. Fang, F. Giorgianni, J. R. Gingrich, and S. Beranova-Giorgianni, "Investigation of phosphoprotein signatures of archived prostate cancer tissue specimens via proteomic analysis," *Electrophoresis*, vol. 32, no. 15, pp. 1984–1991, 2011.

[18] A. J. Ghio, E. D. Hilborn, J. G. Stonehuerner et al., "Particulate matter in cigarette smoke alters iron homeostasis to produce a biological effect," *American Journal of Respiratory and Critical Care Medicine*, vol. 178, no. 11, pp. 1130–1138, 2008.

[19] B. Janji, A. Giganti, V. de Corte et al., "Phosphorylation on Ser5 increases the F-actin-binding activity of L-plastin and promotes its targeting to sites of actin assembly in cells," *Journal of Cell Science*, vol. 119, part 9, pp. 1947–1960, 2006.

[20] H. Shinomiya, "Plastin family of actin-bundling proteins: its functions in leukocytes, neurons, intestines, and cancer," *International Journal of Cell Biology*, vol. 2012, Article ID 213492, 2012.

# 17

# Rapid Screening of the Epidermal Growth Factor Receptor Phosphosignaling Pathway via Microplate-Based Dot Blot Assays

**Amedeo Cappione III, Janet Smith, Masaharu Mabuchi, and Timothy Nadler**

*EMD Millipore, Merck KGaA, 17 Cherry Hill Drive, Danvers, MA 01923, USA*

Correspondence should be addressed to Amedeo Cappione III, amedeo.cappione@merckgroup.com

Academic Editor: Gary B. Smejkal

Expression profiling on a large scale, as is the case in drug discovery, is often accomplished through use of sophisticated solid-phase protein microarrays or multiplex bead technologies. While offering both high-throughput and high-content analysis, these platforms are often too cost prohibitive or technically challenging for many research settings. Capitalizing on the favorable attributes of the standard ELISA and slot blotting techniques, we developed a modified dot blot assay that provides a simple cost-effective alternative for semiquantitative expression analysis of multiple proteins across multiple samples. Similar in protocol to an ELISA, but based in a membrane bound 96-well microplate, the assay takes advantage of vacuum filtration to expedite the tedious process of washing in between binding steps. We report on the optimization of the assay and demonstrate its use in profiling temporal changes in phosphorylation events in the well-characterized EGF-induced signaling cascade of A431 cells.

## 1. Introduction

Signaling through receptor tyrosine kinases (RTKs) is a highly conserved cellular mechanism, controlling fate determination, proliferation, survival, and migration [1, 2]. In most instances, ligand binding initiates conformational changes in the externally facing receptor molecule leading to autophosphorylation on the internal portion of the receptor. A subsequent chain of phosphorylation events propagates the signal to the nucleus culminating in the transcription of genes required to direct changes in cell function (the EGFR cascade is outlined in Figure 1). Given the dynamic interplay of cells with their surrounding microenvironment and owing to the presence of a myriad of other simultaneously activated paths, this process must be tightly regulated to ensure proper responses occur. The broad importance of RTK signaling is highlighted by the well-documented role of pathway dysregulation in human disease, most notably cancer. RTK mutations have been implicated in a variety of cancers, specifically, members of the epidermal growth factor receptor (EGFR) family in brain, lung, and breast cancer. In fact,

thirty percent of all solid tumors possess Ras or Raf mutations, including almost 90% of pancreatic adenocarcinomas [3, 4].

Due to the inherent complexity of the global signaling network and the involvement of their constituents in malignancy, these pathways have been extensively studied by researchers looking for insight into the mechanisms underlying both normal and aberrant growth. Detecting alterations in phosphorylation patterns within signaling profiles is a technique commonly employed to map the dose dependence and specificity of small-molecule inhibitors targeting upstream components. Traditionally, compound screening was performed in cell-free assays using purified enzymes as the target. More recently, a greater significance has been placed on the use of cell-based approaches where multiple components in a single pathway and multiple signaling cascades can be monitored simultaneously. Such analyses require complex and expensive detection platforms such as flow cytometers or high-content imaging systems. While well suited for the high throughput needs of large-scale screens at the industrial level, such platforms may not fit the workflow

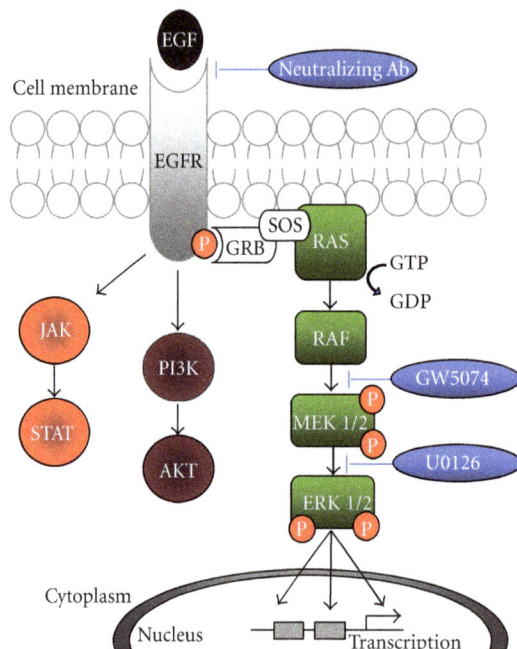

FIGURE 1: The EGFR signaling cascade. The binding of EGF to the EGF receptor (EGFR) results in receptor dimerization and conformational changes triggering autophosphorylation. Under proper conditions, phosphorylated EGFR activates any number of three downstream signaling pathways through Ras, PI3K, and JAK, respectively. The work performed in this study focuses on the Ras cascade (highlighted in green). The three inhibitors (blue) are shown acting at their specific sites of signal interruption.

or demand of smaller research groups. Alternative detection systems offering semiquantitative measurement of multiple proteins in parallel include the enzyme-linked immunosorbent assay (ELISA), multiplex bead arrays, western blots, and slot blots. Although plate based, a fact that simplifies setup and signal detection, ELISAs require a pair of protein-specific antibodies which identify unique epitopes and are also quite time consuming due to multiple binding steps and extensive washing. Slot blotting apparatuses have been developed that offer increased throughput over standard western blots yet retain the same overall labor-intensive protocol that is not amenable to automation. In addition, signal quantitation for both blotting techniques is limited by the dynamic range of the developing film and method of densitometric analysis.

In this paper, we present a modified dot blotting technique for protein detection where purified proteins or cell lysates are applied directly to membrane-based 96-well microplates. This dot blot assay combines the plate-based ease of handling offered by ELISAs with vacuum filtration to greatly expedite the process of semiquantitative analysis of protein expression. Following the addition of protein sample, a two-step antibody binding process using an HRP-conjugated secondary detection antibody is performed. In the final step, the conversion of chemiluminescent substrate provides the signal in each well that can be quantified using a standard plate reader. For a 96-well plate, the entire process, from sample addition to data acquisition, requires less than 90 minutes. Our assay was validated using lysates derived from the extensively studied EGFR signaling cascade of the A431 epidermal carcinoma cell line. Temporal changes in

phosphoactivation of three proteins (EGFR, MEK1/2, and ERK1/2) were measured at five-minute intervals across a twenty-minute time course of EGF stimulation. Pathway mapping was further interrogated using a set of three site-specific inhibitors of the EGFR cascade.

## 2. Materials and Methods

*2.1. Cell Culture.* A431 (CRL-1555, ATCC, Manassas, VA), a human skin carcinoma cell line, was maintained in complete media (DMEM + 10% FBS) and passaged routinely by trypsinization (TrypLESelect, GIBCO/Life Technologies, Grand Island, NY) to ensure log phase growth. For induction experiments, 100 K live cells were seeded per well in 6-well plates, cultured for 2 days, and then serum-starved for 20 HR. Following synchronization, cells were exposed to 100 ng/mL human EGF (Cell Signaling Technology, Danvers, MA) for 5–20 minutes. For inhibitor studies, cultures were pre-incubated with $10 \mu m$ U0126 (Cell Signaling Technology), $5 \mu m$ GW5074 (Sigma-Aldrich, St. Louis, MO), or 10 μg/mL anti-EGFR neutralizing Ab (EMD Millipore), for 2 hrs prior to EGF exposure. Following induction, total cell lysates were prepared as described below (for the EGFR pathway schematic, see Figure 1).

*2.2. Cell Counting and Viability.* $10 \mu L$ sample was mixed with $190 \mu L$ guava ViaCount reagent and incubated for 5 minutes at RT. Sample data was acquired on a guava easyCyte HT instrument and analyzed using guava ViaCount software (all EMD Millipore).

*2.3. Cell Lysis.* Lysis was performed using two buffers: (1) Cytobuster Protein Extraction Reagent (EMD Millipore) and a modified RIPA buffer (25 mM Tris-HCl pH 7.6, 150 mM NaCl, 1% NP-40, 1% sodium deoxycholate, and 0.1% SDS). All buffers were supplemented with protease inhibitors and phosphatase inhibitors (EMD Millipore). All buffers were chilled on ice prior to use. Following induction, cells were washed twice with ice-cold PBS. 400 μL of lysis buffer was added to each well. Samples were incubated on ice for 5 minutes with occasional swirling. To pellet cellular debris, resulting extracts were centrifuged at 16000 g × 15 min at 4°C. Cell lysates were removed, aliquoted, and stored at −20°C until assayed.

*2.4. IR-Based Protein Quantitation.* Proteins were quantified using the Direct Detect assay-free sample card and Direct Detect Spectrometer (EMD Millipore). Each card contained four hydrophilic polytetrafluoroethylene (PTFE) membrane positions, each surrounded by a hydrophobic ring to retain analyzed sample within the device's IR beam. All measurements were performed using 2 μL of sample per membrane position. A "buffer only" sample was also analyzed as a reference blank. Sample concentration was determined in reference to a calibration method. For all experiments, the system was initially calibrated using National-Institute-of-Standards-&-Technology- (NIST-) certified BSA SRM927d in phosphate-buffered saline (PBS). A series of ten concentration points (0.125–5 mg/mL) was used to generate the instrument calibration curve.

*2.5. Antibody Validation.* One of the limiting factors in biochemistry is the availability and quality of antibodies. Prior to use in the dot blot assay, each candidate antibody was subjected to a stringent validation procedure [5]. As part of the initial screening process, we reviewed the certificate of analysis for each of the four antibodies: anti-phospho-EGFR (TYR1069, clone 9H2), anti-phospho-Mek1/2 (SER218/SER222, clone E237), anti-phospho-Erk 1/2 (THR202/TYR204, clone 12D4), and anti-GAPDH (clone 6C5) employed in this study (all Abs are from EMD Millipore). All four were validated for use in western blotting analysis as part of the standard quality control testing by EMD Millipore. Staining with each of the four antibodies resulted in detection of a single prominent band at the approximate molecular weight and the lack of any nonspecific binding. In addition, all four were validated using lysates derived from A431 cells, the only cell line employed in this study. Antibodies against phosphorylated epitopes had to demonstrate specificity to stimulated (ex. EGF) or inhibited (anti-EGF neutralizing Ab) to yield phosphorylated (signal) or nonphosphorylated forms (no signal) of the protein, respectively.

*2.6. Dot Blot Protocol (see Figure 2).* Prewet PVDF membrane Multiscreen plates (EMD Millipore) with 100 uL 70% Ethanol for 15 seconds then immediately wash 2X with 100 μL Milli-Q H$_2$0 by vacuum filtration using the Multiscreen$_{HTS}$ Vacuum Manifold (EMD Millipore) with pressure

set to 4″Hg. In all wash steps use vacuum filtration. Add diluted lysate (50 μL/well) and incubate for 30 minutes at RT on a plate shaker at low speed. Wash plates 2X with Tris-Buffered Saline (TBS). Block sample wells in 0.5% nonfat dried milk (in TBS) for 5 minutes on the shaker then remove blocking agent by vacuum filtration. Add 50 μL/well diluted primary antibody and incubate on a shaker for 10 minutes. Each antibody was previously titrated to optimize performance. The antibodies included anti-phospho-EGFR (TYR1069, clone 9H2), anti-phospho-Mek1/2 (SER218/SER222, clone E237), anti-phospho-Erk1/2 (THR202/TYR204, clone 12D4), and anti-GAPDH (clone 6C5). Wash plates 3X with TBS + 0.1% Tween-20 (TBST). Add 50 μL/well diluted goat anti-rabbit IgG HRP (EMD Millipore) and incubate on a shaker for 10 minutes. Wash plates 3X as above. Add 100 μL/well of Luminata Forte Western HRP Reagent (EMD Millipore) and incubate for 5 minutes on a shaker. Read signal using BioTek Synergy microplate reader (BioTek, Winooski, VT). For each well, chemiluminescent signal was measured and presented as counts per second (CPS).

## 3. Results

Initial feasibility studies and optimization of the dot blot assay were performed using purified Glyceraldehyde-3-Phosphate Dehydrogenase (GAPDH) protein in PBS buffer. Representative results from a serial dilution of GAPDH are presented in Figure 3. From the graph, the assay was able to detect down to ~4 ng and was linear in response up to 100 ng protein loaded. The assay shows a robust signal:noise ratio and does not appear to be limited by binding capacity of the membrane; the theoretical binding capacity of a single well (0.3 cm$^2$) of membrane is approximately 90 μg [6]. The assay was optimized as follows for each step: 30 min protein binding, 10 min primary Ab binding, and 10 min secondary Ab binding. All binding steps were performed with low-speed agitation at room temperature. Overall assay time is 90 minutes. It is possible that protein range could be increased with greater Ab input or increased binding reaction times although this may lead to elevations in nonspecific binding. Due to differences in binding kinetics, optimization of reaction conditions may be required for each protein analyzed.

For the dot blot assay to have broader application, it must perform in the context of total cell lysates where relative protein concentration, buffering conditions, and lysate clarity may impact not only membrane binding but also antibody detection characteristics. To assess such issues, we first extracted lysates from A431 cells using two distinct lysis buffers, a modified RIPA buffer and a nondetergent-based commercial extraction reagent. Following extraction, protein samples were quantified using the Direct Detect IR-based quantitation system. On average, the RIPA buffer liberated five times greater total protein than the nondetergent buffer (data not shown); this may be due to the presence of harsher detergent in the RIPA buffer resulting in greater protein solubilization.

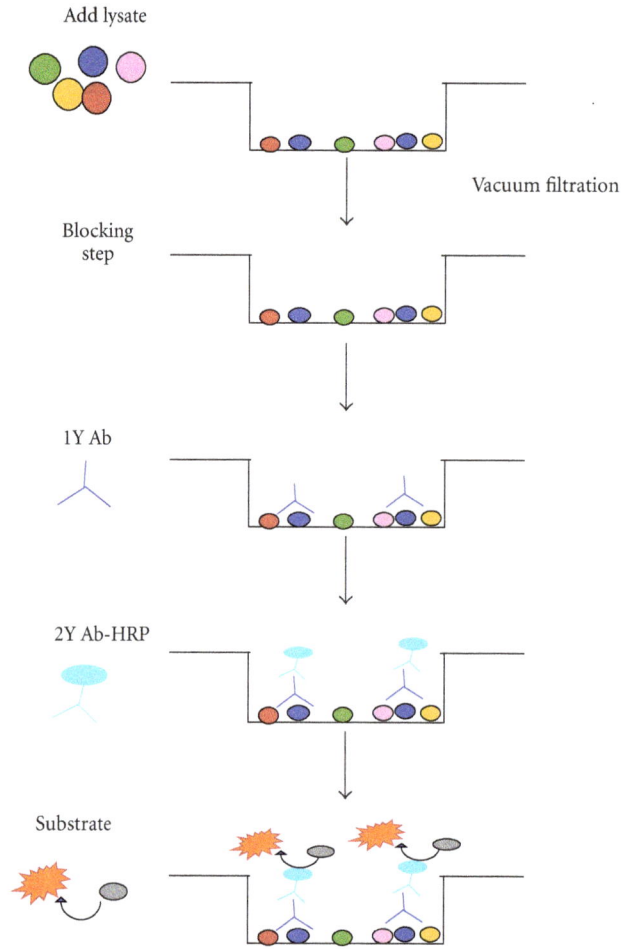

FIGURE 2: The dot blot protocol. Depicted in the diagram is the assay for one representative well from a 96-well microplate. After protein binding, a blocking step is performed to reduce nonspecific Ab binding to any unoccupied region of membrane. The remaining portion of the assay is similar to ELISA except that all wash steps are performed via vacuum filtration.

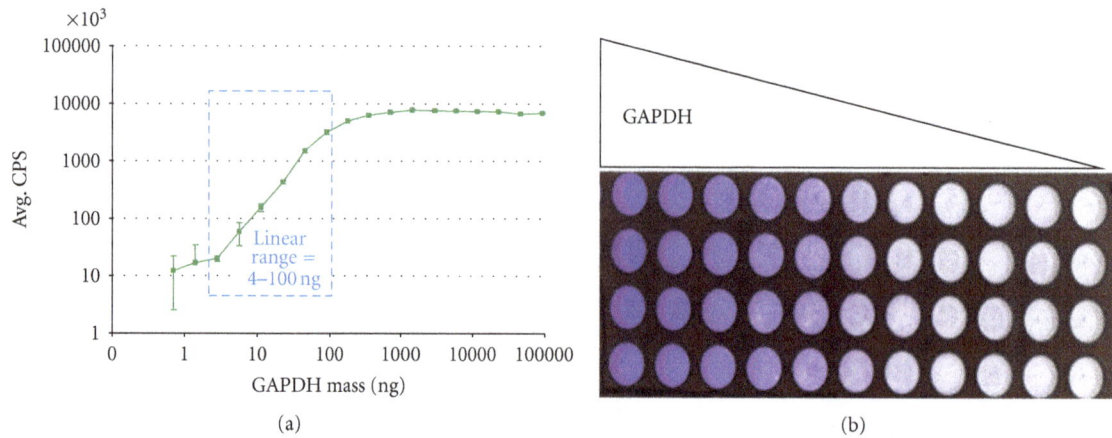

FIGURE 3: Dot blot assay feasibility—titration of pure GAPDH. (a) The titration curve demonstrates a range 4–100 ng input protein whereby linear signal could be detected. Each point represents the mean of 4 replicates. (b) Coomassie staining of membrane wells after protein loading. Four replicates are shown for each concentration point.

(a)

(b)

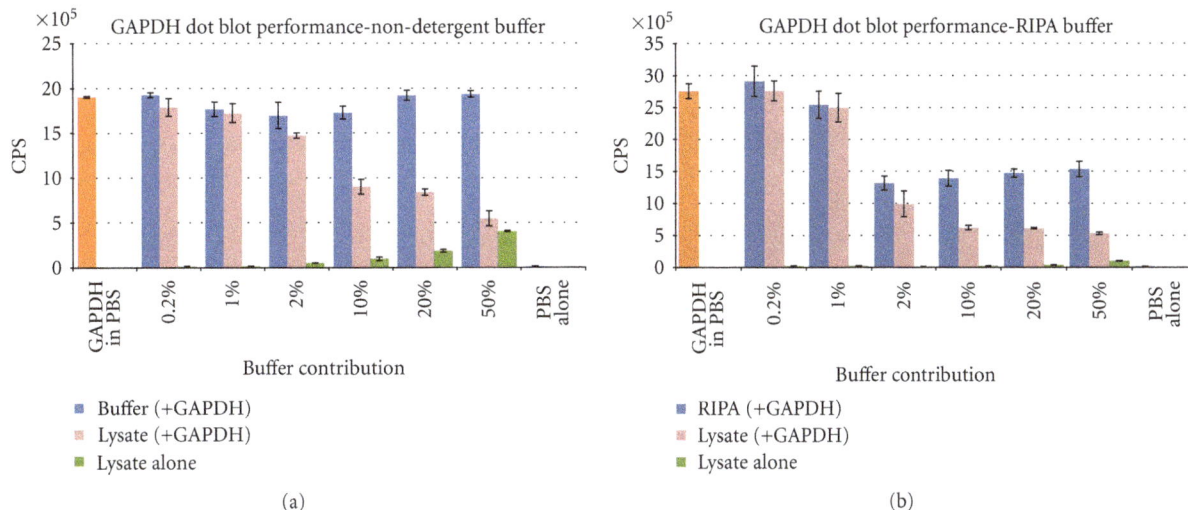

FIGURE 4: Impact of lysis buffer components on dot blot performance—Detergents. Lysates (pink bars) or buffers alone (blue) were diluted in PBS, spiked with 100 ng GAPDH and assayed for changes in GAPDH detection. The green bars show detection of the native GAPDH present in A431 lysates (no GAPDH was spiked into these samples). The CPS signal for GAPDH in PBS is displayed by the orange Bar. Each bar represents the average of 3 individual replicates. Protein concentrations for the lysates used were determined to be (a) nondetergent buffer, 494 ng/$\mu$L, and (b) RIPA buffer, 2443 ng/$\mu$L. The RIPA buffer was diluted to 494 ng/$\mu$L prior to setup. Total protein lysate loaded was as follows: 0.2% = 49 ng; 1% = 295 ng; 2% = 494 ng; 10% = 2950; 20% = 4940 ng; 50% = 12350 ng.

Total lysates from each extraction condition were used to assess the potential effects of protein concentration and buffer components on dot blot assay performance. The results of this experiment are outlined in Figure 4. Briefly, lysates, or buffers alone, were diluted to varying degrees with PBS. Samples were spiked with 100 ng purified GAPDH and loaded onto microplates. A standard dot blot assay was then performed. For the RIPA buffer, any contribution greater than 1% (0.5 $\mu$L in 50 $\mu$L reaction volume) caused a significant decrease in GAPDH signal; this is most likely due to detergents interfering with membrane binding and limiting protein-protein interactions. By contrast, the nondetergent-based buffer alone had little or no effect on GAPDH signal even at 50% sample dilution; this result may be important for situations where either total protein concentrations are low or the protein of interest is expressed at relatively low levels. We also found a reduction in GAPDH signal in both buffer types when lysate load/well was increased. Signal reduction was slightly greater in the homebrew samples due to the contributing detergent effect. Signal loss is most likely due to protein crowding and/or competition for membrane binding. More importantly, native GAPDH was easily detected in nondetergent-derived samples with signal >100X over background for 12.5 $\mu$g lysate loaded.

We sought to determine the assay's linear range and define the optimal concentration of primary antibody required for protein detection. Given the wide variability in relative protein expression within a cell and between different cells, such optimization may be required for each protein (and each cell type's lysate) to be measured. The graphs in Figure 5 depict the results for titration curves performed on EGF-stimulated A431 lysates using antibodies specific for the house-keeping GAPDH protein and phosphorylated

ERK1/2. The dot blot assay demonstrated a linear range of detection for 400–6000 ng/well. Irrespective of primary Ab concentration, little to no signal could be detected below 400 ng lysate. The assay also failed to detect any greater signal levels in wells with $\geq$6000 ng lysate. In fact, at higher sample loads, counts per second (CPS) values tended to decrease; this was most likely due to the higher complexity of the lysate solution as well as increased competition for membrane binding. In both cases, a 1 : 500 dilution of primary antibody (2 $\mu$g/mL GAPDH Ab, 1 $\mu$g/mL pERK1/2 Ab) provided optimal detection.

We next applied the dot blot assay in a proof-of-concept study to track changes in protein phosphorylation for the EGFR signaling cascade of A431 cells cultured with EGF. Ligand binding activates a chain of signaling events, which includes successive phosphorylation of the EGFR, MEK, and ERK proteins. Given the temporal nature of phosphoactivation within the cascade, synchronized A431 cells were exposed to EGF and harvested at 5 minute intervals for a total of 20 minutes. The time course was performed with EGF stimulus alone and in the presence of three pathway inhibitors. In the absence of inhibition, a clear temporal order of phosphorylation events is seen (Figure 6). EGF stimulation resulted in an almost immediate increase in the presence of phosphorylated EGFR, which was maintained at high levels for the entire 20 minutes. Among the four proteins measured, the phospho-EGFR signal was by far the highest level detected; this finding is not unexpected given that the A431 cell line expresses abnormally high levels of EGFR [7]. Phosphorylated Mek1/2 was first detected at 10 minutes followed by ERK1/2 at 15 minutes. For the latter two proteins, the phosphorylated state was far more transient appearing to decrease soon after initial appearance.

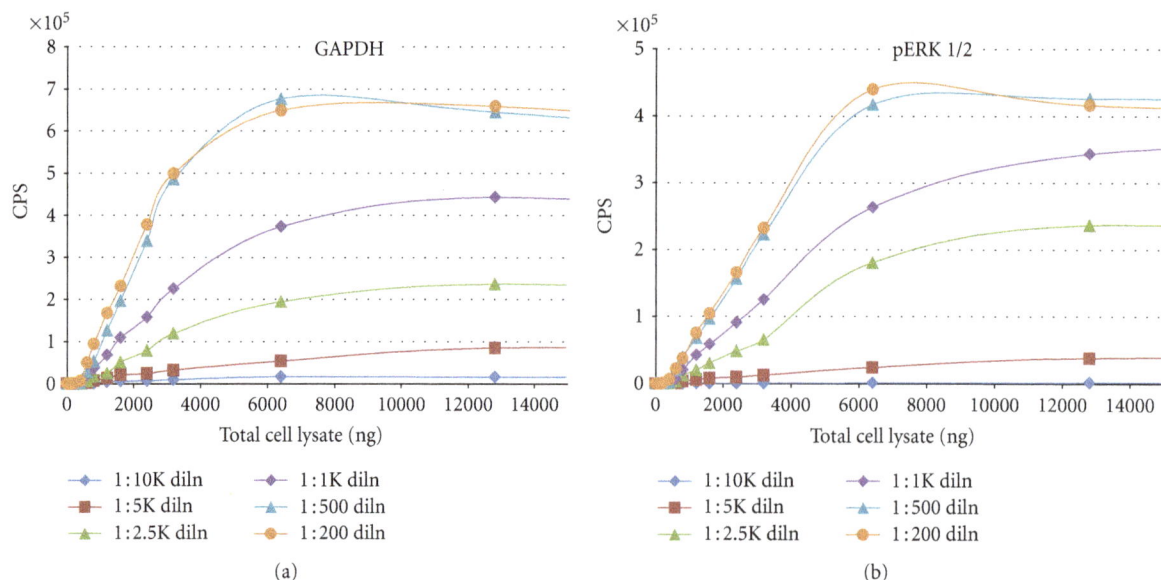

FIGURE 5: Native protein detection—primary antibody optimization. Serum-starved A431cells were stimulated with EGF for 20 min and harvested and lysates prepared as described previously. Dot blot assays were performed to optimize detection by (a) GAPDH and (b) phospho-ERK1/2 antibodies. A series of seven lysate concentrations (100–10000 ng/well) was used to assess six different dilutions of each primary antibody. All points were run in triplicate.

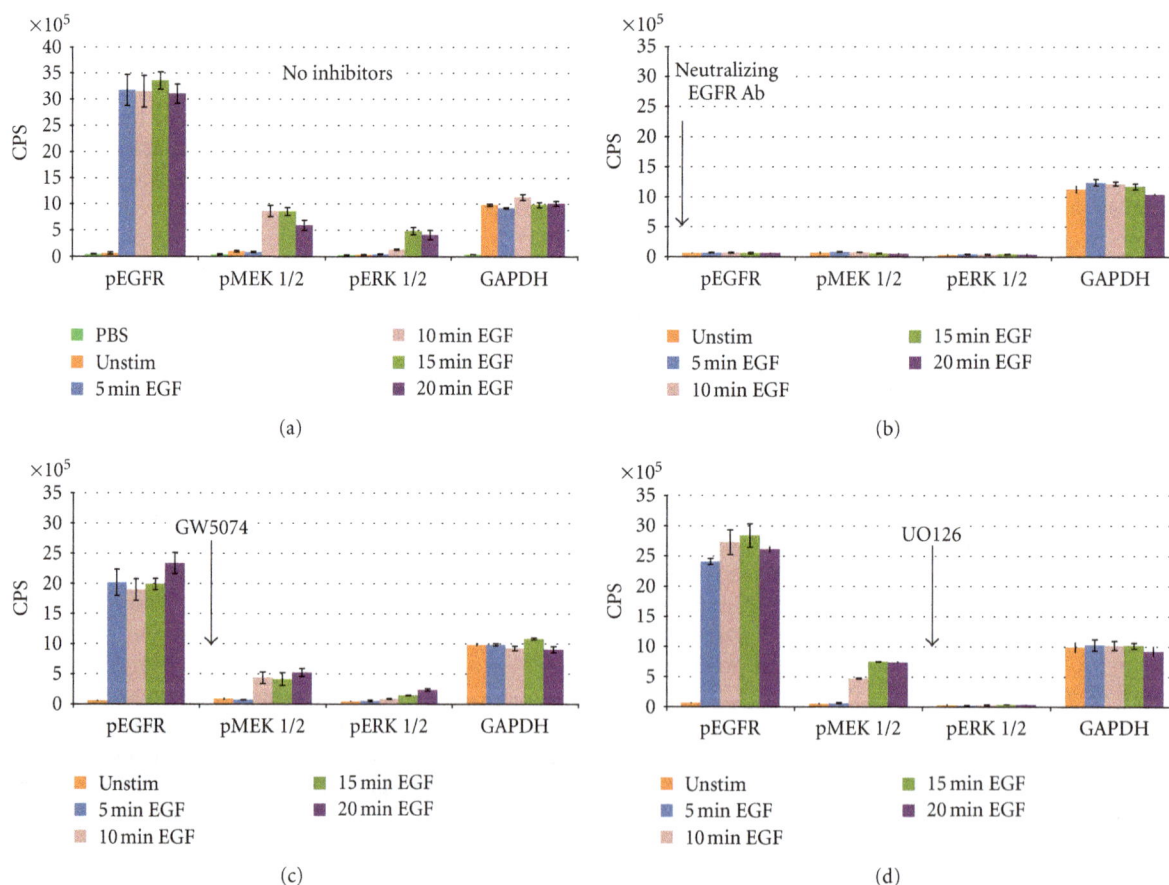

FIGURE 6: Temporal mapping of the EGFR cascade using site-specific inhibitor molecules. The four bar graphs represent results for the following: (a) EGF stimulation, no inhibitor, and EGF stimulation in the presence of (b) neutralizing EGFR Ab, (c) GW5074—an inhibitor of Raf kinase activity, and (d) U0126—a highly selective inhibitor of MEK1/2. For each inhibitor, the arrow indicates the point of pathway inhibition. Each graph contains the time course (0–20 minutes) of expression profiles for each of the four proteins analyzed. Each bar value represents the average of three replicates.

GAPDH detection was included in each data set as a loading control to permit cross-sample comparisons. Overall, for 60 GAPDH wells analyzed across 20 experimental conditions (3 replicates for each), the mean CPS value was $1,026,671 \pm 87,405$ with a coefficient of variation of 8.5%. Pre-incubation with the EGFR blocking antibody completely abolished all downstream phosphorylation events. U0126 treatment prevented phosphorylation of ERK1/2 without affecting either upstream event. At the concentration applied, GW5074 caused only partial inhibition of MEK1/2 and ERK1/2. Interestingly, GW5074 also caused a reduction in EGFR phosphorylation suggesting a potential positive feedback loop involving intermediary signaling proteins. A more extensive study involving challenges with various concentrations of GW5074 may offer greater insight into this phenomenon.

## 4. Discussion

Standard dot blotting is a method of protein detection similar to the western blot technique but differing in that protein samples are not initially separated electrophoretically on polyacrylamide gels but simply applied directly onto the membrane's surface. Once applied, proteins are driven to bind the membrane through either active pressure (vacuum) or gentle agitation and passive absorption. Dot and slot blotting techniques have been used extensively by molecular biology researchers to affix proteins and nucleic acids on membranes for the purpose of quantitation, DNA homology assessment, protein-DNA/RNA interactions, enzymatic activity, and the study of ligand-receptor binding. The standard device is comprised of three main parts: the upper block with an array of slots for sample loading, a middle component that holds the inserted membrane, and a bottom block with a connector permitting vacuum filtration. A set of screws clamps the assembled device in place thereby minimizing sample bleed-over. Given this format, slot blots offer greater throughput capacity than the standard western blot and are therefore ideal for screening applications. However, the device's main value is in sample loading; all subsequent steps are performed in the same labor-intensive manner as a western blot. To expedite the process without sacrificing throughput, our modified dot blot takes advantage of 96-well microplates equipped with PVDF membrane. The plate-based format minimizes sample bleed-over and permits easy reagent loading at each step. Since the plate is membrane based, all wash steps can be performed via vacuum filtration. The plate format and simple reaction steps are also well suited for automation and expanded screening needs. A final benefit is being able to use a standard plate reader for chemiluminescent signal detection; this format offers greater dynamic range than film densitometry enhancing quantitative capacity of the assay.

The work presented here clearly demonstrates the feasibility of the plate-based dot blot application for semiquantitative detection or comparative analyses of multiple proteins and/or multiple samples in parallel. The assay performed well on pure protein samples but more importantly worked for total cell lysates although the linear ranges of detection were considerably different. The assay is, however, quite sensitive to detergent interference, an important consideration when choosing extraction reagents. As well, samples with high viscosity or large amounts of debris had a tendency to cause a reduction in filtration flow rate and, in more severe cases, complete clogging of the membrane. Dilution of viscous samples or precentrifugation to clear particulates ameliorated clogging issues. In summary, the dot blot assay offers a cost-effective protein expression screening tool for researchers with moderate throughput needs.

## Acknowledgment

We would like to thank Dr. Elene Chemokalskaya for manuscript review, advice, and assistance with the submission process.

## References

[1] J. Schlessinger, "Cell signaling by receptor tyrosine kinases," *Cell*, vol. 103, no. 2, pp. 211–225, 2000.

[2] I. Rebay, "Keeping the receptor tyrosine kinase signaling pathway in check: lessons from *Drosophila*," *Developmental Biology*, vol. 251, no. 1, pp. 1–17, 2002.

[3] G. Pearson, F. Robinson, T. B. Gibson et al., "Mitogen-activated protein (MAP) kinase pathways: regulation and physiological functions," *Endocrine Reviews*, vol. 22, no. 2, pp. 153–183, 2001.

[4] G. L. Johnson and R. Lapadat, "Mitogen-activated protein kinase pathways mediated by ERK, JNK, and p38 protein kinases," *Science*, vol. 298, no. 5600, pp. 1911–1912, 2002.

[5] R. Tibes, Y. H. Qiu, Y. Lu et al., "Reverse phase protein array: validation of a novel proteomic technology and utility for analysis of primary leukemia specimens and hematopoietic stem cells," *Molecular Cancer Therapeutics*, vol. 5, no. 10, pp. 2512–2521, 2006.

[6] A. Weiss, "Overview of membranes and membrane plates used in research and diagnostic ELISPOT assays," *Methods in Molecular Biology*, vol. 792, pp. 243–256, 2012.

[7] R. Fabricant, J. Delarco, and G. Todaro, "Nerve growth factors on human melanoma cells in culture," *Proceedings of the National Academy of Sciences of the United States of America*, vol. 74, pp. 565–569, 1977.

# 18

# Seven-Signal Proteomic Signature for Detection of Operable Pancreatic Ductal Adenocarcinoma and Their Discrimination from Autoimmune Pancreatitis

**Kiyoshi Yanagisawa,[1,2] Shuta Tomida,[2] Keitaro Matsuo,[3] Chinatsu Arima,[2] Miyoko Kusumegi,[4] Yukihiro Yokoyama,[5] Shigeru B. H. Ko,[6] Nobumasa Mizuno,[7] Takeo Kawahara,[5] Yoko Kuroyanagi,[4] Toshiyuki Takeuchi,[4] Hidemi Goto,[6] Kenji Yamao,[7] Masato Nagino,[5] Kazuo Tajima,[3] and Takashi Takahashi[2]**

[1] Institute for Advanced Research, Nagoya University, Furo-cho, Chikusa-ku, Nagoya 464-8601, Japan
[2] Division of Molecular Carcinogenesis, Center for Neurological Diseases and Cancer, Nagoya University Graduate School of Medicine, Nagoya 466-8550, Japan
[3] Division of Epidemiology and Prevention, Aichi Cancer Center, Nagoya 464-8681, Japan
[4] Division of Research and Development, Oncomics Co., Ltd., Nagoya 464-0858, Japan
[5] Division of Surgical Oncology, Department of Surgery, Nagoya University Hospital, Nagoya 466-8550, Japan
[6] Department of Gastroenterology, Nagoya University Hospital, Nagoya 466-8550, Japan
[7] Department of Gastroenterology, Aichi Cancer Center, Nagoya 464-8681, Japan

Correspondence should be addressed to Kiyoshi Yanagisawa, kyana@med.nagoya-u.ac.jp

Academic Editor: Visith Thongboonkerd

There is urgent need for biomarkers that provide early detection of pancreatic ductal adenocarcinoma (PDAC) as well as discrimination of autoimmune pancreatitis, as current clinical approaches are not suitably accurate for precise diagnosis. We used mass spectrometry to analyze protein profiles of more than 300 plasma specimens obtained from PDAC, noncancerous pancreatic diseases including autoimmune pancreatitis patients and healthy subjects. We obtained 1063 proteomic signals from 160 plasma samples in the training cohort. A proteomic signature consisting of 7 mass spectrometry signals was used for construction of a proteomic model for detection of PDAC patients. Using the test cohort, we confirmed that this proteomic model had discrimination power equal to that observed with the training cohort. The overall sensitivity and specificity for detection of cancer patients were 82.6% and 90.9%, respectively. Notably, 62.5% of the stage I and II cases were detected by our proteomic model. We also found that 100% of autoimmune pancreatitis patients were correctly assigned as noncancerous individuals. In the present paper, we developed a proteomic model that was shown able to detect early-stage PDAC patients. In addition, our model appeared capable of discriminating patients with autoimmune pancreatitis from those with PDAC.

## 1. Introduction

Pancreatic ductal adenocarcinoma (PDAC) is the fifth leading cause of cancer death in Japan with more than 24,000 deaths annually [1], while 35,000 deaths each year in the United States are caused by the disease [2]. Long-term survival for PDAC patients remains unsatisfactory, with only 3–5% surviving for more than 5 years after surgical resection,

with the remainder succumbing to widespread metastasis or massive local recurrence. Since surgical resection is the only reliable curative treatment, early detection is essential to improve the outcomes of affected individuals. However, the clinical symptoms of PDAC are often unremarkable until advanced stages of the disease, and the anatomic location of the pancreas deep in the abdomen makes physical detection and imaging approaches difficult. Thus, less than

Seven-Signal Proteomic Signature for Detection of Operable Pancreatic Ductal Adenocarcinoma and Their Discrimination from Autoimmune Pancreatitis

165

10% of patients diagnosed with PDAC are eligible for surgical resection [3]. Although serum markers for PDAC including carcinoembryonic antigen (CEA) and carbohydrate antigen 19-9 (CA19-9) play important roles in current clinical practice for monitoring progression and treatment response, as well as surveillance for recurrence, these markers are not ideal for cancer screening due to their low specificity and/or sensitivity in early stages of the disease [4–6].

The concept of autoimmune pancreatitis (AIP) is supported by recent advances in elucidating its pathogenesis as a unique systemic disease. AIP has several characteristic features, such as infiltration of CD4-positive T cells and IgG4-positive plasmacytes, irregular narrowing of the pancreatic duct, and diffuse enlargement of the pancreas [7–9]. Although intensive investigations into the pathogenesis of AIP have been conducted, its underlying molecular mechanism remains unclear. The most important and difficult step in diagnosing AIP is to distinguish it from PDAC. Clinical symptoms such as obstructive jaundice are not helpful for discrimination, while IgG4, the most accurate serum marker for AIP, is not adequately specific to exclude the existence of cancer. Furthermore, AIP is sometimes accompanied by PDAC; thus percutaneous or endoscopic biopsy findings are often needed for final diagnosis. Unfortunately, those examinations are invasive for the patient and may fail to detect small regions of cancer cells. As a result, unnecessary surgery because of misdiagnosis performed for AIP patients without cancer or those undergoing treatment for existing cancer is a critical issue in clinical practice. Accordingly, there is urgent need for elucidation of novel biomarker(s) and noninvasive diagnostic strategies useful for early detection of PDAC, as well as discrimination of patients with AIP to improve clinical management and prognosis.

Comprehensive analysis of protein expression patterns in biological materials might improve understanding of the molecular complexities of human diseases [10] and could be useful to detect diagnostic or predictive protein expression patterns that reflect clinical features. Matrix-assisted laser desorption/ionization mass spectrometry (MALDI MS) can profile proteins up to 50 kDa in size in serum, tissues, and other various clinical specimens. Protein profiles obtained may contain thousands of data points and provide proteomic signatures that allow detection of patients with various diseases [11, 12]. We previously employed MALDI MS for expression profiling of proteins in human lung cancer specimens and found that the resultant proteomic patterns could predict various clinical features, as well as the potential of recurrence in stage I lung cancer patients [13, 14].

In the present study, protein expression profiling with MALDI MS was conducted to identify proteomic patterns in plasma samples for discrimination of PDAC from AIP as well as chronic pancreatitis (CP) using 3 independent datasets. We found that a proteomic model consisting of 7 mass spectrometry signals constructed by use of the training cohort could detect 82.6% (38 of 46, 95% CI 68.6–92.2) of known PDAC cases, including 62.5% (5 of 8, 95% CI 24.5–91.5) of the stage I and II cases in the independent test cohort, which successfully confirmed its discrimination power. We further applied our model for discrimination of AIP as well

as CP from PDAC and found that it correctly assigned 100% of the AIP and CP patients (19 of 19, 95% CI 82.4–100 and 11 of 11, 95% CI 71.5–100, resp.) as noncancerous. These results indicate that our 7-signal proteomic model may contribute to accurate decisions regarding the therapeutic plan for patients with chronic pancreatic diseases, especially PDAC and AIP.

## 2. Methods

*2.1. Patients and Specimens.* Plasma specimens from 96 PDAC patients were obtained from the Department of Epidemiology and Prevention, Aichi Cancer Center Research Institute, Nagoya, Japan, collected from January 2001 and November 2005. Of those, 80 were randomly assigned to the training set and 16 to the test set. An additional 30 plasma specimens from PDAC patients were obtained from the Department of Surgery, Nagoya University Hospital, Nagoya, Japan, collected from May 2004 to July 2006, and assigned to the test set. Plasma specimens from 147 healthy control subjects were also obtained from the Department of Epidemiology and Prevention, Aichi Cancer Center Research Institute, and used. Of those, 80 were randomly assigned to the training set and 67 to the test set. Plasma specimens from 2 acute pancreatitis, 11 chronic pancreatitis, and 3 autoimmune pancreatitis patients were obtained from the Department of Gastroenterology, Nagoya University Hospital, collected from April 2005 and November 2007, and assigned to the test set. In addition, 16 plasma specimens from autoimmune pancreatitis were obtained from the Department of Gastroenterology, Nagoya University Hospital, collected from September 2003 and August 2009, and assigned to the confirmation set. More detailed information is available in Supplementary Material available on line at doi: 10.1155/2012/510397. The characteristics of the patients and healthy subjects in the training, test, and confirmation cohorts are summarized in Supplementary Table S1, which shows that there were no statistically significant differences in regard to clinicopathologic features among the cohorts. All specimens were processed in the same manner and stored at $-80°C$ within 180 minutes after being collected from the patients and healthy subjects, and not thawed until analysis. Requisite approval from our institutional review boards and written informed consent from all subjects were obtained. One plasma specimen per patient or healthy subject was analyzed, and the training, test, and confirmation datasets were independently analyzed as different batches. Further details are available in supplementary Material.

*2.2. Proteomic Analysis.* Five microliters of nonpre-treated plasma was mixed with 5 nL drops of an energy absorbing matrix solution (saturated Sinapinic acid in water/aceto-nitrile/trifluoroacetic acid (500:500:1, by volume), which allows molecules to be protonated and desorbed from tissue surfaces). Then, 1 $\mu$L mixtures were deposited into individual wells of MALDI MS sample plates (PE Biosystems, Foster City, CA) and dried at room temperature for 5 minutes. Six spots were generated for each plasma-matrix mixture sample and spectra were acquired from all 6 using a 4800 Instrument (Applied Biosystems, Foster City, CA), essentially as described previously [13, 14]. Further details are available in Supplementary Material.

*2.3. Statistical Methods.* Protein profiles obtained by MALDI MS were analyzed using 3 distinct statistical methods, Fisher's exact test, the Kruskal-Wallis test, and a significance analysis of microarray (SAM) test [15], to investigate MS signals that appeared to differentiate PDAC patients from healthy individuals in the training set. MS signals that met at least 1 of the 3 selection criteria were further analyzed.

To construct a generally applicable proteomic classifier without specifically overfitting it to the training cohort, we used a weighted voting algorithm, a well-established technique for supervised classification, in which each weight value was calculated as the signal-to-noise ratio and a leave-one-out cross-validation strategy was utilized [16].

It is possible that unintended biased resubstitution or partial cross-validation can result in underestimation of the error rate after cross-validation; thus the performance of any class prediction rule is best assessed by applying the rule created by use of 1 dataset (the training set) to an independent dataset (the validation or test set) [17]. In the present study, the proteomic classifier constructed with the training dataset of 160 individuals was validated using a completely independent validation set composed of 145 individuals.

An agglomerative hierarchical clustering algorithm was applied to investigate the pattern among the statistically significant discriminator proteins as well as the biological status with Eisen's software [18].

*2.4. Identification of Individual Proteins in the Proteomic Signature.* $40\,\mu$ of serum samples was pretreated with high abundant protein depletion column (Agilent, Palo Alto, CA) according to manufacturer's instruction. The pretreated serum samples were separated over a polymeric column (Toso, Tokyo, Japan) with a high-performance liquid chromatography (HPLC) pump (Shimadzu, Osaka, Japan) and HPLC fractions were collected every minute for 80 minutes. Each fraction was lyophilized, reconstituted with a 50% acetonitrile in water containing 0.1% trifluoroacetic acid, and analyzed by MALDI mass spectrometry to identify the HPLC fractions that contained proteins corresponding to the peaks in the signature with molecular weights selected by bioinfomatic analysis as candidate molecular markers for the PDAC. The selected fractions were lyophilized and reconstituted with a mixture of $10\,\mu$L of 0.4 M ammonium hydrogen carbonate and $5\,\mu$L of 45 mM dithiothreitol, and then $10\,\mu$L of 100 mM iodoacetamide was added. This mixture was incubated for 4 hours at 37°C with $5\,\mu$L of 200 nM mass-grade trypsin (Promega, Madison, WI) to obtain peptides. The peptides were separated and sequenced by a microcapillary reverse-phase column (KYA technologies, Tokyo, Japan) with an HPLC pump (KYA) and MALDI mass spectrometer (Applied Biosystems). These spectra were compared with those in the human databases of the National Center for Biotechnology Information (nonredundant) by use of Mascot version 2.1.0 (Matrix Science Inc., Boston, MA). A minimum of two peptide matches and a positive association between the m/z values detected with MALDI mass spectrometry and the molecular weight of the intact protein (including posttranslational modifications) were required for protein identification.

## 3. Results

*3.1. Protein Expression Profiling in the Training Cohort.* We obtained protein expression profiles for the 160 human plasma specimens obtained from 80 PDAC patients and 80 healthy subjects at Aichi Cancer Center (Figure 1(a)) and Supplementary Table S1) using MALDI MS. Spectra were obtained from 6 replicates of single plasma specimens. MarkerView (Applied Biosystems) and custom software were used to bin the peaks across the spectra obtained from 960 samples, and then we calculate the average intensity of each signal individually among the 160 cases. As a result, we obtained expression profiles containing 1063 distinct proteomic signals. To extract a proteomic signature able to discriminate PDAC patients from healthy individuals, we compared MS signals from the 80 healthy subjects and 80 PDAC patients using our statistical selection criteria (signals met at least 2 of the following criteria: $P$ value corrected with Bonferroni was less than 0.05 in Fisher's exact test and Kruskal-Wallis test, and FDR < 0.1% for SAM). As a result, 134 MS signals were found to be differentially expressed. Agglomerative hierarchical clustering analysis using the identified proteomic signature showed a clear separation of plasma specimens from PDAC patients as compared to those from healthy individuals (Figure 1(b)), which confirmed that the selected MS signals were informative for discrimination of PDAC cases from healthy individuals. The left branch mostly consisted of PDAC cases (81.3%, 65 of 80 cases, 95% CI 71.0–89.1), whereas the right branch consisted of healthy subjects (78.8%, 63 of 80 cases, 95% CI 68.2–87.1). Next, we investigated whether our proteomic prediction model could best distinguish noncancerous individuals from cancer patients. For this purpose, the 134 selected MS signals, which were informative for discrimination, were further ranked according to the SAM and weighted-voting proteomic discriminatory models were constructed using increasing numbers of the differentially expressed proteomic signals (up to 134), for which learning errors were calculated by leave-one-out cross-validation (Figure 2(a)). This cross-validation analysis showed that the use of 7 MS signals gave the lowest number of misclassifications, while 7 MS signals (8562.3, 8684.4, 8765.1, 9423.5, 13761.5, 14145.2, and 17250.8 m/z) were extracted as the most shared ones. Using this proteomic model, plasma samples from both PDAC patients and healthy subjects were classified as either positive or negative for cancer, which showed that the sensitivity for prediction was 76.3% (61 of 80 of the cancer patients, 95% CI 65.4–85.1) and for specificity was 91.3% (73 of 80 of the healthy subjects, 95% CI 82.8–96.4, Table 1), for an overall classification accuracy of 83.8% (134 of 160, 95% CI 77.1–89.1). We also calculated positive and negative predictive

(a)

(b)

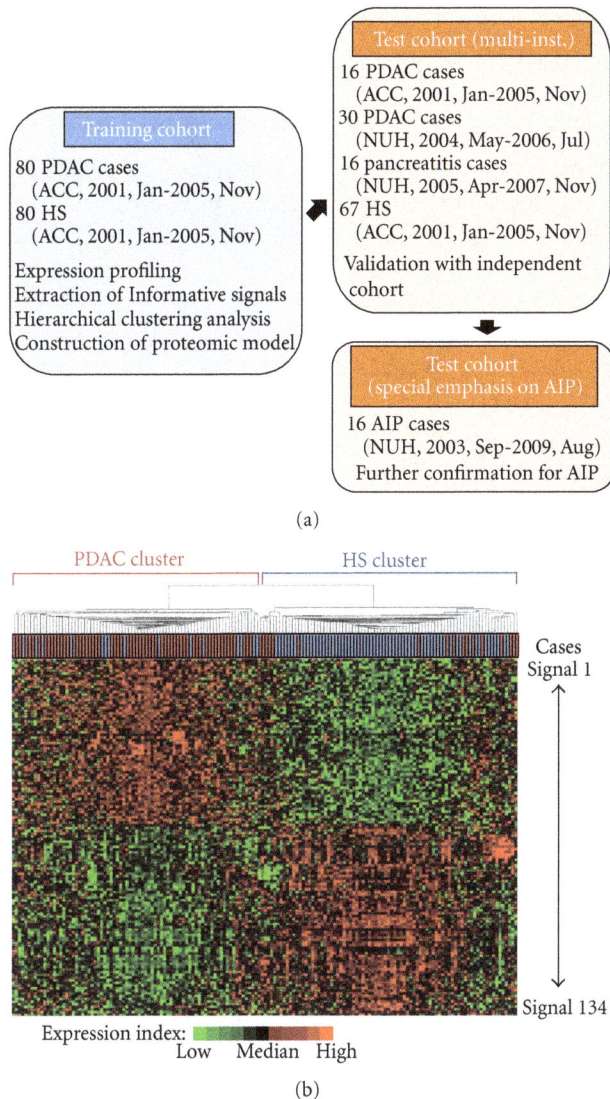

FIGURE 1: MALDI MS analysis of plasma specimens from human PDAC patients and healthy subjects in the training cohort. (a) Independent training-validation-confirmation datasets of 160 training cases, 129 validation cases, and 16 confirmation cases. (b) Unsupervised hierarchical clustering analysis of 80 human PDAC patients and 80 healthy subjects in the training cohort according to the protein expression patterns of 134 MS signals. Each row represents an individual proteomic signal and each column an individual sample. The dendrogram at the top shows the similarities in protein expression profiles among the samples. Substantially elevated (red) expression of the proteins was observed in individual plasma samples. HS: healthy subjects; PDAC: pancreatic ductal adenocarcinoma. Red box case: PDAC: blue box case: healthy subject.

values (PPV and NPV, resp.) to confirm the diagnostic power of our model, which were 89.8% and 79.3%, respectively. We observed no significant difference for detection of PDAC patients related to lymph node positivity and prognosis. Furthermore, we analyzed the relationship between the age of PDAC patients (≤60 or >60 years old) and detection

power of the 7 MS signals. Those results showed that the sensitivity for prediction was 69.8% (30 of 43, 95% CI 53.9–82.8) and 83.8% (31 of 37, 95% CI 68.0–93.8) in the younger and older groups, respectively (Table 1), with no significance in discrimination found ($P = 0.142$, Fisher's exact test). Representative spectra that comprised the 7-signal proteomic model for the healthy subjects and PDAC patients are shown in Figure 2(b). It is of note that our model was able to correctly distinguish 72.7% (8 of 11 cases, 95% CI 39.0–94.0) of the stage I and II cases from the healthy subjects, while it also correctly classified 78.8% (26 of 33, 95% CI 61.1–91.0) of the PDAC patients eligible for surgical resection as positive for cancer (Table 1).

*3.2. Protein Expression Profiling in the Test Cohort.* It has been well reported that the robustness, including accuracy, of a prediction model should be assessed using an independent validation cohort, even when cross-validation methods, such as LOOCV or n-fold CV, were properly used for developing the prediction model [19]. To examine the robustness of the 7-signal proteomic model constructed with data from MALDI-MS analysis of the training cohort, we applied it to an independent test dataset obtained from plasma samples collected at two different institutions. We also determined whether the identified proteomic model could discriminate between acute and chronic pancreatitis patients, as well as autoimmune pancreatitis, as the discovery of biomarkers applicable for differential diagnosis between PDAC and noncancerous pancreatic diseases has great potential for clinical practice. For the test cohort, plasma samples were obtained from 46 PDAC patients (16 and 30 cases of ACC and NUH, resp.) and 67 healthy subjects from the ACC group, while 16 pancreatitis samples obtained from Nagoya University hospital (NUH) consisted of 2 acute pancreatitis, 11 chronic pancreatitis, and 3 autoimmune pancreatitis cases (Figure 1(a), Supplementary Tables S1 and S2 for additional clinical information for AIP patients). With the 7-signal proteomic model, 82.6% (38 of 46, 95% CI 68.6–92.2) of the cancer cases were classified into the positive group, while 89.2% (74 of 83, 95% CI 80.4–94.9) of the noncancerous subjects were assigned to the group negative for cancer (Figure 3 and Table 2). We calculated PPV and NPV, which were 80.9% and 90.2%, respectively, and the overall accuracy of the classification with the test cohort was 86.8% (112 of 129, 95% CI 79.7–92.1). We also evaluated the relationship between blood vessel invasion (surgery with or without mesenteric venous tract resection) and detection power of the 7 MS signals. Our results showed that the sensitivity for prediction was 88.8% (8 of 9, 95% CI 51.8–99.7) for PDAC patients who underwent mesenteric venous tract resection and 78.6% (11 of 14, 95% CI 49.2–95.3) for those who did not, with no significant difference found ($P = 0.524$, Fisher's exact test). Future studies with a larger number of PDAC patients treated with surgery are warranted to validate the clinical usefulness of our 7-signal proteomic signature. It is of note that our model was able to correctly distinguish 62.5% (5 of 8 cases, 95% CI 24.5–91.5) of the stage I and II cases from the healthy subjects and also classified 78.9% (30 of 38,

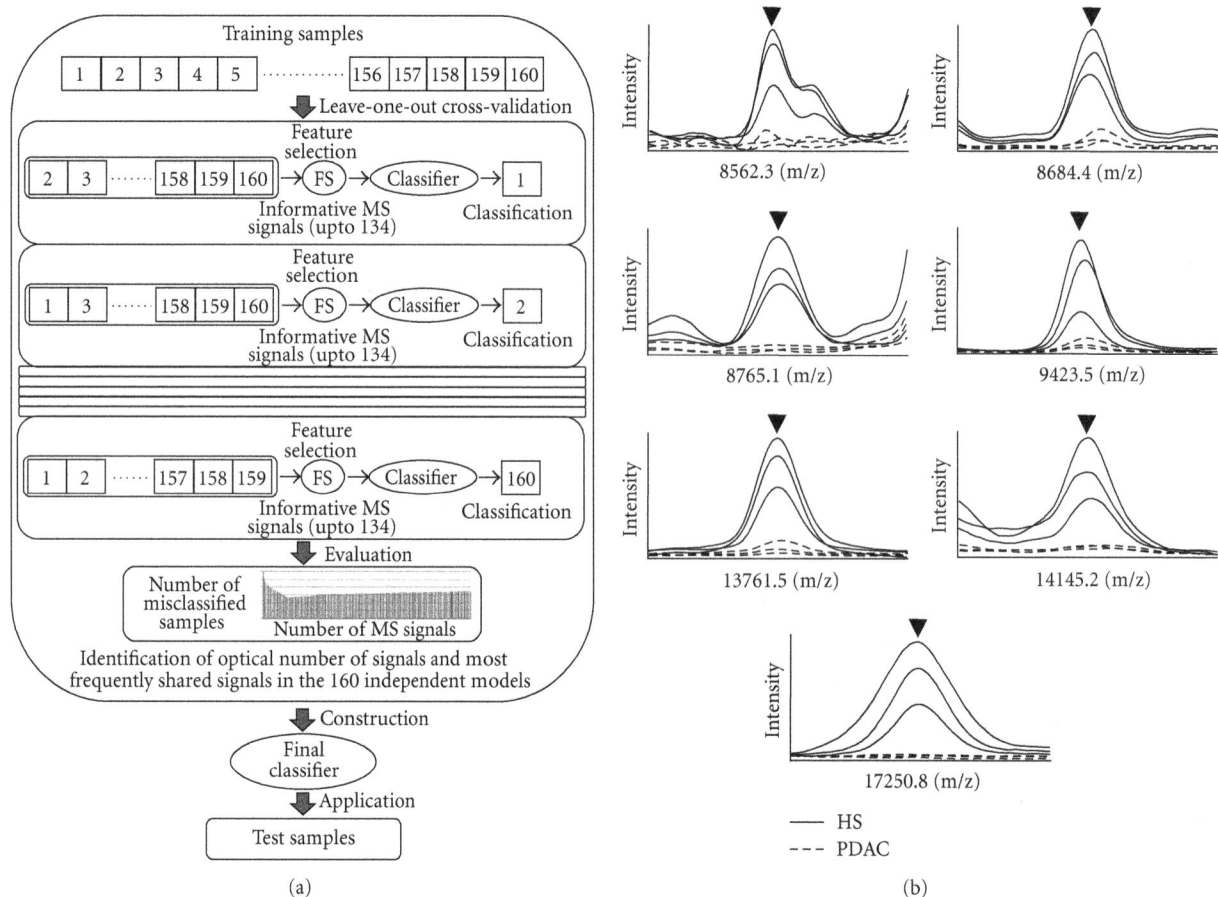

(a)

(b)

FIGURE 2: Construction of proteomic model for discrimination of PDAC cases from healthy subjects. (a) Schematic diagram of construction of proteomic discrimination model. (b) Representative mass spectra comprising 7-signal proteomic signature. Arrowheads show informative peaks for discrimination between healthy subjects and PDAC patients. Blue lines show representative spectra from healthy subjects and red lines show representative spectra from PDAC patients.

TABLE 1: Discrimination of samples in the training cohort according to 7-signal proteomic model.

|  | Number of cases analyzed | Number of correctly assigned cases (%) | 95% C.I.* (%) |
|---|---|---|---|
| All samples | 160 | 134 (83.8) | 77.1–89.1 |
| Pancreatic ductal adenocarcinoma | 80 | 61 (76.3) | 65.4–85.1 |
| Healthy subjects | 80 | 73 (91.3) | 82.8–96.4 |
| age |  |  |  |
| ≤60 | 43 | 30 (69.8) | 53.9–82.8 |
| >60 | 37 | 31 (83.8) | 68.0–93.8 |
| Clinical stage of pancreatic ductal adenocarcinoma patients |  |  |  |
| 0/I | 3 | 3 (100) | 29.2–100 |
| II | 8 | 5 (62.5) | 24.5–91.5 |
| III | 8 | 8 (100) | 63.1–100 |
| IVa | 14 | 10 (71.4) | 41.9–91.6 |
| IVb | 47 | 35 (74.5) | 59.7–86.1 |

* 95% confidence interval.

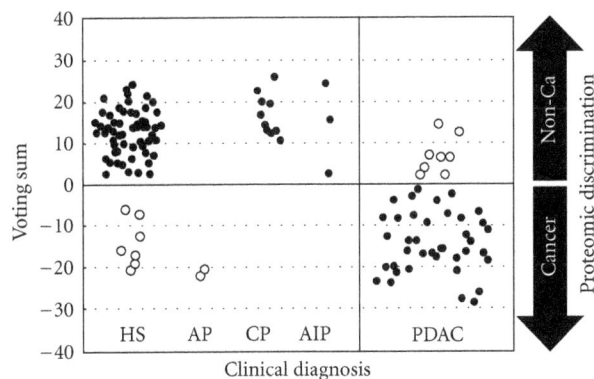

FIGURE 3: Assessment of 7-signal proteomic model with the validation cohort using weighted voting algorithm. The results of proteomic analyses of the training cohort are shown. Each circle represents a voting sum for a single patient. Solid circles: specimens whose prediction with proteomic model matched clinical diagnosis; open circles: specimens whose prediction with proteomic model did not match clinical diagnosis; HS: healthy subjects; AP: acute pancreatitis; CP: chronic pancreatitis; AIP: autoimmune pancreatitis; PDAC: pancreatic ductal adenocarcinoma.

95% CI 62.7–90.5) of the PDAC patients eligible for surgical resection as positive for cancer. It is also noteworthy that the identified proteomic model distinguished 100% of the patients with chronic pancreatitis (11 of 11, 95% CI 71.5–100) and AIP (3 of 3, 95% CI 29.2–100) from cancer cases (Figure 3 and Table 2).

*3.3. Discrimination of Autoimmune Pancreatitis from PDAC Using 7-Signal Proteomic Model.* Autoimmune pancreatitis is a systemic inflammatory disease of the pancreas and several diagnostic criteria have been proposed. However, their usefulness is under debate and accurate differential diagnosis remains difficult. Moreover, an important step in diagnosing AIP is to discriminate it from PDAC. In the present study, all (3 of 3) of the AIP patients were correctly discriminated from those with PDAC in the analysis with the test dataset; thus we next performed a confirmatory analysis using plasma samples collected from 16 AIP patients treated at NUH (Figure 1(a) and Supplementary Table S2). For this, we employed our 7-signal proteomic model to investigate whether it would classify the AIP patients as noncancerous and found that it correctly assigned those patients as negative for cancer with 100% accuracy (16 of 16 cases, 95% CI 79.4–100). Therefore, the high potential for discrimination of AIP from PDAC was validated with an independent confirmatory dataset used in a blinded manner. The serum level of CA19-9 was elevated in 4 (21.1%, 95% CI 7.3–52.4) of the AIP cases in our cohort, while IgG4 levels have been reported to be elevated in 10–30% of PDAC cases [7, 20]. Thus, our proteomic model may be applicable as a novel serological test to discriminate AIP from PDAC in clinical practice. Representative spectra obtained from the AIP and PDAC cases are shown in Figure 4.

*3.4. Combination of MALDI Proteomic Signature and CA19-9 for Cancer Screening.* Our 7-signal proteomic model was able to detect 82.6% (38 of 46, 95% CI 68.6–92.2) of the PDAC patients in the test cohort (Table 2). Moreover, it assigned 78.9% (30 of 38, 95% CI 62.7–90.5) of the patients eligible for an operation to the cancerous group, while 62.5% (5 of 8 and 95% CI 24.5–91.5) of the stage I and II cases were also detected with the identified model. Since it is possible that our 7-signal proteomic model and CA19-9 level are complementary, we investigated whether their combined use would improve the detection rate of patients who may benefit from surgery. The overall sensitivity of CA19-9 (cutoff value, 37 units/mL) alone for stage 0–IVa patients was 71.1% (27 of 38, 95% CI 54.1–84.6), while a combination of our 7-signal proteomic model and CA19-9 level detected 89.5% (34 of 38, 95% CI 75.2–97.1) of operable cases. Notably, for detection of stage I and II PDAC patients, CA19-9 assigned only 50.0% (4 of 8, 95% CI 15.7–84.3) of the cases to the positive group and no additional discrimination power of that marker was observed when combined with our proteomic model. Accordingly, we consider that our 7-signal proteomic model might be more sensitive for detection of early stage PDAC patients than CA19-9, which would improve clinical outcomes following surgical treatment.

*3.5. Identification of Individual Proteins in the Proteomic Signature.* As an initial step toward elucidating the biologic mechanism of the association between the proteomic signature and carcinogenesis, we identified a couple of proteins that correspond to the mass spectrometry signals in the proteomic signature obtained from serum. Extracts from two serum samples of healthy individual were fractionated by reverse phase-HPLC and analyzed by MALDI MS to identify the HPLC fractions that contained proteins corresponding to peaks in the proteomic signature. These selected fractions were subjected to sequence analysis of tryptic peptides by use of MALDI MS. Accordingly, we identified the following proteins as part of the proteomic signature: apolipoprotein A-I ([M + H]$^+$ = 17,250.8 m/z) and C-III ([M + H]$^+$ = 8765.1), and transthyretin ([M + H]$^+$ = 13761.5).

## 4. Discussion

In the present study, we analyzed the protein expression profiles of plasma specimens obtained from patients with PDAC, as well as acute and chronic pancreatitis cases, and autoimmune pancreatitis (AIP) patients with MALDI MS. Using bioinformatic analysis, we derived 7 MS signals that allowed us to produce a proteomic model for discrimination of PDAC from noncancerous individuals. When we used our proteomic model with both independent test cohort and confirmation group, 62.5% (5 of 8, 95% CI 24.5–91.5) of stage 0–II cases were correctly assigned to the cancerous group, while all AIP patients (19 of 19, 95% CI 82.4–100) were correctly assigned to the noncancerous group. Discrimination of AIP from cancer is obviously important; however it is currently problematic in clinical practice. Although previous reports have shown discrimination power

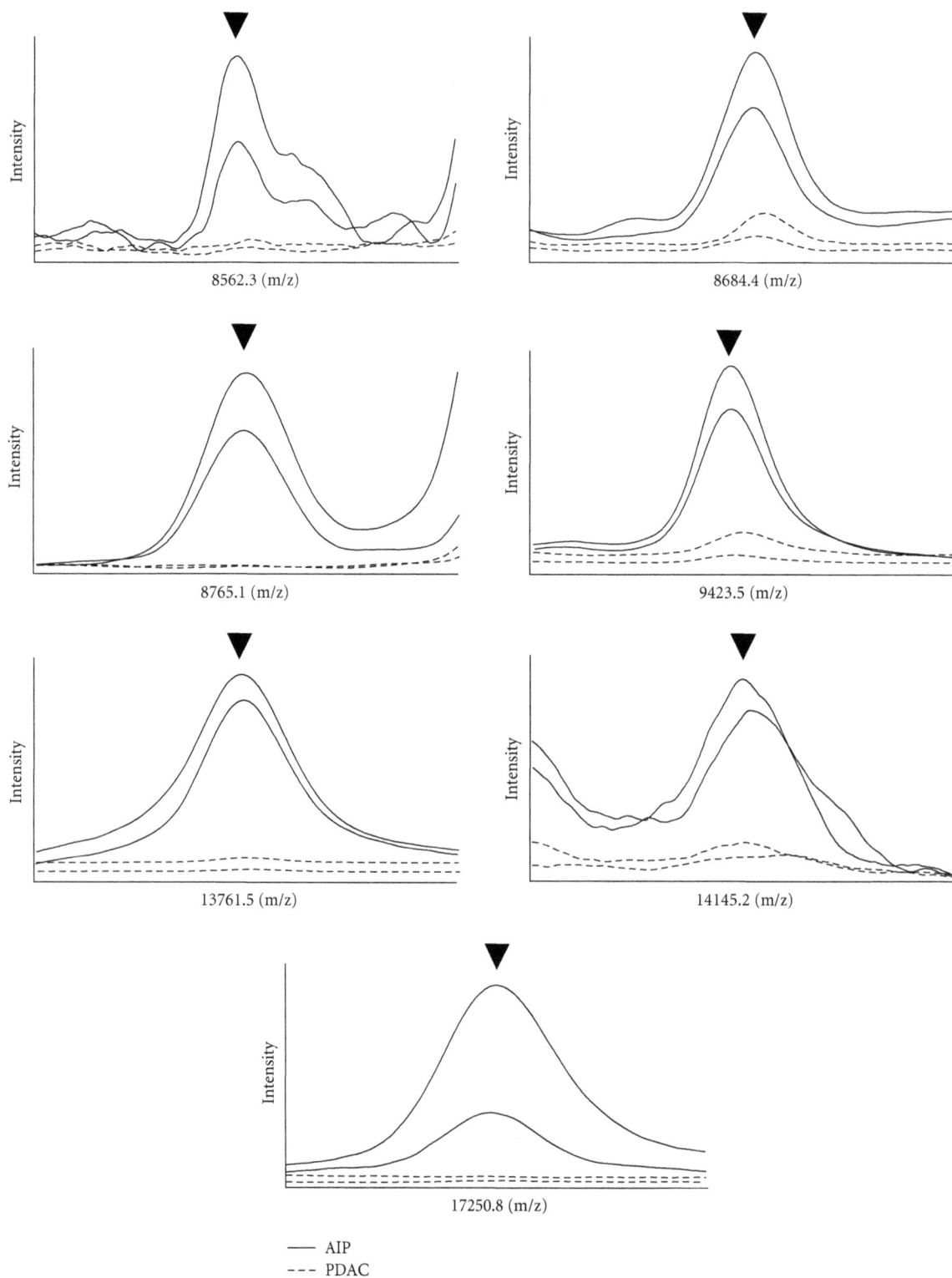

FIGURE 4: Representative mass spectra comprising 7-signal proteomic signature in autoimmune pancreatitis patients and PDAC patients. Arrowheads show informative peaks for discrimination between autoimmune pancreatitis patients and patients with pancreatic cancer. Blue solid and dotted lines show representative spectra from autoimmune pancreatitis patients, and red solid and dotted lines show representative spectra from pancreatic cancer patients. AIP: autoimmune pancreatitis; PDAC: pancreatic ductal adenocarcinoma.

Seven-Signal Proteomic Signature for Detection of Operable Pancreatic Ductal Adenocarcinoma and Their Discrimination from
Autoimmune Pancreatitis

171

TABLE 2: Discrimination of samples in the test cohort according to 7-signal proteomic model.

| | Number of cases analyzed | Number of correctly assigned cases (%) | 95% C.I.* (%) |
|---|---|---|---|
| All samples | 129 | 112 (86.8) | 79.7–92.1 |
| Healthy subjects | 67 | 60 (89.6) | 79.7–95.7 |
| Pancreatic ductal adenocarcinoma (ACCH) | 16 | 13 (81.3) | 54.4–96.0 |
| Pancreatic ductal adenocarcinoma (NUH) | 30 | 25 (83.3) | 65.3–94.4 |
| Acute pancreatitis (NUH) | 2 | 0 (0) | 0–84.2 |
| Chronic pancreatitis (NUH) | 11 | 11 (100) | 71.5–100 |
| Autoimmune pancreatitis (NUH) | 3 | 3 (100) | 29.2–100 |
| Clinical stage of pancreatic ductal adenocarcinoma patients at ACCH | | | |
| 0/I | 0 | NA | NA |
| II | 1 | 0 (0) | 0–97.5 |
| III | 3 | 3 (100) | 29.2–100 |
| IVa | 4 | 2 (50) | 6.8–93.2 |
| IVb | 8 | 8 (100) | 63.1–100 |
| Clinical stage of pancreatic ductal adenocarcinoma patients at NUH | | | |
| 0/I | 1 | 0 (0) | 0–97.5 |
| II | 6 | 5 (83.3) | 35.9–99.6 |
| III | 13 | 11 (84.6) | 54.6–98.1 |
| IVa | 10 | 9 (90) | 55.5–99.7 |
| IVb | 0 | NA | NA |

*95% confidence interval
NA: not available.

of proteomic signature between PDAC patients and control subjects [21–24], to the best of our knowledge, the present 7-signal proteomic model is the first system of proteomic prediction based upon mass spectrometry found capable to both detect early-stage PDAC cases and discriminate AIP patients.

Early detection is essential for improving the outcomes of PDAC patients. However, those in stages 0–II are difficult to detect with current diagnostic approaches, including computerized tomography scanning, positron emission tomography scanning, and tissue-based diagnostic tests. CA19-9 is a tumor marker widely used for evaluations of therapeutic effects and detection of PDAC recurrence, though it is not considered to be applicable for mass screening when used alone [4, 6, 25, 26]. Recent advances in molecular biology have also revealed that clinical features cannot be adequately characterized or predicted by a single marker. Thus, microarray analysis has been employed to simultaneously investigate the expression levels of thousands of genes and identify mRNA patterns associated with various human diseases including PDAC [27–29]. However, mRNA expression does not always indicate which of the corresponding proteins are expressed or provide information regarding their posttranslational regulation. Moreover, blood and body fluids, such as pancreatic juice and urine, do not contain mRNA. Thus, proteome analysis of such specimens is considered to better reflect the underlying clinical characteristics of human diseases as compared to gene expression profiling, while proteomic technologies including MS have been employed

to analyze proteomes in clinical specimens [10–14, 30–32]. Previous proteomics studies of PDAC with healthy controls have shown promising results in distinguishing PDAC, with a sensitivity ranging from 78 to 91% and specificity from 75 to 100% [21–24, 33, 34]. These discrimination power results are better than those obtained with the current CA19-9 marker, while improved diagnostic performance has been observed when serum MS markers were combined with CA19-9 [21, 22, 24]. In the present study, we found that the combination of our 7-signal proteomic model and CA19-9 level improved the positive rate of detection of PDAC patients eligible for surgical resection to 89.5% (34 of 38, 95% CI 75.2–97.1). It is noteworthy that detection of stage I-II cases was also attainable at a sensitivity of 62.5% (5 of 8, 95% CI 24.591.5) without further improvement by adding CA19-9. These results support the usefulness of our 7-signal proteomic model for detection of early stage cases. Since we constructed the present 7-signal model independent from CA19-9, further optimization of selection of a proteomic signature with focus on early detection possibly along with adjustment of the CA19-9 cutoff value is warranted to obtain increased sensitivity. The present 7-signal proteomic model showed high potential to assign inflammatory pancreatic disease patients to the noncancerous group (93.8%; 30 of 32, 95% CI 79.2–99.2). Interestingly, 2 of the misclassified patients suffered from acute pancreatitis; however, all of the patients of chronic pancreatitis and AIP (11 of 11, 95% CI 71.5–100; and 19 of 19, 95% CI 82.4–100) were correctly assigned to the noncancerous group by our proteomic model. Discrimination of AIP from PDAC is difficult in clinical practice, as symptoms such as obstructive jaundice or space occupying lesions in the pancreas are commonly observed in both cases. Actually, most of the AIP patients in this study showed at least one of these symptoms. Our proteomic model distinguished between AIP patients and those with PDAC with high accuracy; thus it is considered to be effective in future clinical applications, especially for selecting those who are eligible for invasive diagnostic procedures followed by inevitably invasive surgical treatment for PDAC. During the course of our study, Frulloni et al. reported that autoantigens against the plasminogen binding protein of helicobacter pylori and ubiquitin-protein ligase E3 component n-recognin 2 were detected in most of the AIP patients tested, as well as a small number of PDAC cases [35]. It would be interesting to combine our proteomic model with testing for those autoantigens for diagnosis of chronic pancreatic diseases.

In this study, 2 acute pancreatitis patients and 14 healthy subjects were assigned to the cancerous group by our 7-signal proteomic model in the training (7 healthy subjects) and test (2 acute pancreatitis patients and 7 healthy subjects) cohorts. Since that time, we have carefully followed their clinical courses of these healthy subjects and found that 5 suffered from cancerous disease within 3 years, including 2 with rectal cancer, 1 with prostate cancer, 1 with hepatocellular carcinoma, and 1 with a metastatic bone tumor from an unknown primary site. In addition, another false positive healthy subject later developed polyposis in the colon. These observations suggest potential relation of our proteomic model with these malignancies, although further in-depth investigations are apparently required to draw definitive conclusions.

Mass spectrometry profiles obtained from complex protein mixtures can contain thousands of data points derived from real protein signatures. However, they can also be contaminated by electronic and chemical noise, variability in instrumentation, and variable crystallization of the matrix, necessitating careful analytical techniques [11, 13, 14]. In the present study, we employed multiple statistical methods and leave-one-out cross-validation to combine differentially expressed proteins with the clinical variables and found that a minimal set of 7 low-molecular weight proteins was sufficient to distinguish between healthy subjects and PDAC patients. The discriminating power of the extracted proteomic signature was further validated using independent test datasets obtained from plasma specimens collected at 2 different institutions. With this protocol, we carefully eliminated accidental identification of overly optimistic and nonbiological/mathematical multivariate signatures within a closed cohort by overfitting.

The primary goal of this study was development of a bioassay applicable to clinical practice for detection of PDAC and discrimination from AIP, as attempts to identify proteins that comprise a proteomic model have not been fully successful to date. However, the high reproducibility of MALDI MS indicates that direct application of its findings would be successful. In the previous study, Koomen et al. reported that a set of 4 peaks could be used to detect

PDAC, of which one MS signal was downregulated in PDAC patients and found to be derived from apolipoprotein A-I [23], while Yan et al. found that transthyretin levels were independently associated with PDAC likelihood when obstructive jaundice was considered [36]. Accordingly, our identification of apolipoprotein A-I and transthyretin, which is a constituent of our proteomic model and downregulated in PDAC patients in this study, is in accord with previous reports from different institutes. We also identified the downregulation of apolipoprotein C-III in serum samples obtained from PDAC patients [37, 38]. Further investigations are warranted to identify discriminating proteins for ascertainment of their functional significance. Notably, 2 downregulated peaks (8765 and 13762 m/z), which were previously extracted as proteomic serum markers for lung cancer [39], were also identified as downregulated proteomic signals in PDAC patients in the present study.

Prospective multi-institutional studies with a larger number of patients including those with early-stage PDAC, AIP, and other pancreatic diseases are apparently warranted to validate further significance of our 7-signal proteomic signature for clinical application. Given that it has potential for early detection of PDAC as well as accurate discrimination of AIP, our 7-signal proteomic model may ultimately lead to a reduction in the large number of deaths caused by devastating cancer and also provide better management for chronic inflammatory disease of pancreas.

## Ethical Approval

Requisite approval from the institutional review boards and written informed consent from all subjects were obtained.

## Acknowledgments

This work was supported in part by a Grant-in-Aid for Exploratory Research and Program for Improvement of Research Environment for Young Researchers from Special Coordination Funds for Promoting Science and Technology commissioned and a Grant-in-Aid for Scientific Research on Priority Areas, a Grant-in-Aid for Scientific Research (C) by the Ministry of Education, Culture, Sports, Science and Technology of Japan. The sponsors of the study had no role in its design, data collection, analysis, and interpretation of data, the decision to submit the manuscript for publication, or writing the manuscript.

## References

[1] http://ganjoho.ncc.go.jp/public/statistics/backnumber/2009 _en.html

[2] http://www.cancer.gov/cancertopics/types/pancreatic.

[3] M. Yamamoto, O. Ohashi, and Y. Saitoh, "Japan pancreatic cancer registry: current status," *Pancreas*, vol. 16, no. 3, pp. 238–242, 1998.

[4] M. Goggins, M. Canto, and R. Hruban, "Can we screen high-risk individuals to detect early pancreatic carcinoma?" *World Journal of Surgical Oncology*, vol. 74, no. 4, pp. 243–248, 2000.

[5] R. A. Abrams, L. B. Grochow, A. Chakravarthy et al., "Intensified adjuvant therapy for pancreatic and periampullary adenocarcinoma: survival results and observations regarding patterns of failure, radiotherapy dose and CA 19-9 levels," *International Journal of Radiation Oncology Biology Physics*, vol. 44, no. 5, pp. 1039–1046, 1999.

[6] R. E. Ritts and H. A. Pitt, "CA 19-9 in pancreatic cancer," *Surgical Oncology Clinics of North America*, vol. 7, no. 1, pp. 93–101, 1998.

[7] H. Hamano, S. Kawa, A. Horiuchi et al., "High serum IgG4 concentrations in patients with sclerosing pancreatitis," *The New England Journal of Medicine*, vol. 344, no. 10, pp. 732–738, 2001.

[8] D. L. Finkelberg, D. Sahani, V. Deshpande, and W. R. Brugge, "Autoimmune pancreatitis," *The New England Journal of Medicine*, vol. 355, no. 25, pp. 2670–2676, 2006.

[9] T. Pickartz, J. Mayerle, and M. M. Lerch, "Autoimmune pancreatitis," *Nature Clinical Practice Gastroenterology & Hepatology*, vol. 4, no. 6, pp. 314–323, 2007.

[10] F. Taguchi, B. Solomon, V. Gregorc et al., "Mass spectrometry to classify non-small-cell lung cancer patients for clinical outcome after treatment with epidermal growth factor receptor tyrosine kinase inhibitors: a multicohort cross-institutional study," *Journal of the National Cancer Institute*, vol. 99, no. 11, pp. 838–846, 2007.

[11] E. F. Petricoin, A. M. Ardekani, B. A. Hitt et al., "Use of proteomic patterns in serum to identify ovarian cancer," *The Lancet*, vol. 359, no. 9306, pp. 572–577, 2002.

[12] B. L. Adam, Y. Qu, J. W. Davis et al., "Serum protein fingerprinting coupled with a pattern-matching algorithm distinguishes prostate cancer from benign prostate hyperplasia and healthy men," *Cancer Research*, vol. 62, no. 13, pp. 3609–3614, 2002.

[13] K. Yanagisawa, Y. Shyr, B. J. Xu et al., "Proteomic patterns of tumour subsets in non-small-cell lung cancer," *The Lancet*, vol. 362, no. 9382, pp. 433–439, 2003.

[14] K. Yanagisawa, S. Tomida, Y. Shimada, Y. Yatabe, T. Mitsudomi, and T. Takahashi, "A 25-signal proteomic signature and outcome for patients with resected non-small-cell lung cancer," *Journal of the National Cancer Institute*, vol. 99, no. 11, pp. 858–867, 2007.

[15] V. G. Tusher, R. Tibshirani, and G. Chu, "Significance analysis of microarrays applied to the ionizing radiation response," *Proceedings of the National Academy of Sciences of the United States of America*, vol. 98, no. 9, pp. 5116–5121, 2001.

[16] T. R. Golub, D. K. Slonim, P. Tamayo et al., "Molecular classification of cancer: class discovery and class prediction by gene expression monitoring," *Science*, vol. 286, no. 5439, pp. 531–527, 1999.

[17] R. Simon, M. D. Radmacher, K. Dobbin, and L. M. McShane, "Pitfalls in the use of DNA microarray data for diagnostic and prognostic classification," *Journal of the National Cancer Institute*, vol. 95, no. 1, pp. 14–18, 2003.

[18] M. B. Eisen, P. T. Spellman, P. O. Brown, and D. Botstein, "Cluster analysis and display of genome-wide expression patterns," *Proceedings of the National Academy of Sciences of the United States of America*, vol. 95, no. 25, pp. 14863–14868, 1998.

[19] A. Dupuy and R. M. Simon, "Critical review of published microarray studies for cancer outcome and guidelines on statistical analysis and reporting," *Journal of the National Cancer Institute*, vol. 99, no. 2, pp. 147–157, 2007.

[20] A. Ghazale, S. T. Chari, T. C. Smyrk et al., "Value of serum IgG4 in the diagnosis of autoimmune pancreatitis and in distinguishing it from pancreatic cancer," *American Journal of Gastroenterology*, vol. 102, no. 8, pp. 1646–1653, 2007.

[21] G. M. Fiedler, A. B. Leichtle, J. Kase et al., "Serum peptidome profiling revealed platelet factor 4 as a potential discriminating peptide associated with pancreatic cancer," *Clinical Cancer Research*, vol. 15, no. 11, pp. 3812–3819, 2009.

[22] K. Honda, Y. Hayashida, T. Umaki et al., "Possible detection of pancreatic cancer by plasma protein profiling," *Cancer Research*, vol. 65, no. 22, pp. 10613–10622, 2005.

[23] J. M. Koomen, L. N. Shih, K. R. Coombes et al., "Plasma protein profiling for diagnosis of pancreatic cancer reveals the presence of host response proteins," *Clinical Cancer Research*, vol. 11, no. 3, pp. 1110–1118, 2005.

[24] J. Koopmann, Z. Zhang, N. White et al., "Serum diagnosis of pancreatic adenocarcinoma using surface-enhanced laser desorption and ionization mass spectrometry," *Clinical Cancer Research*, vol. 10, no. 3, pp. 860–868, 2004.

[25] F. Safi, W. Schlosser, G. Kolb, and H. G. Beger, "Diagnostic value of CA 19-9 in patients with pancreatic cancer and non-specific gastrointestinal symptoms," *Journal of Gastrointestinal Surgery*, vol. 1, no. 2, pp. 106–112, 1997.

[26] H. Narimatsu, H. Iwasaki, F. Nakayama et al., "Lewis and secretor gene dosages affect CA19-9 and DU-PAN-2 serum levels in normal individuals and colorectal cancer patients," *Cancer Research*, vol. 58, no. 3, pp. 512–518, 1998.

[27] C. A. Iacobuzio-Donahue, A. Maitra, M. Olsen et al., "Exploration of global gene expression patterns in pancreatic adenocarcinoma using cDNA microarrays," *American Journal of Pathology*, vol. 162, no. 4, pp. 1151–1162, 2003.

[28] H. Han, D. J. Bearss, L. W. Browne, R. Calaluce, R. B. Nagle, and D. D. Von Hoff, "Identification of differentially expressed genes in pancreatic cancer cells using cDNA microarray," *Cancer Research*, vol. 62, no. 10, pp. 2890–2896, 2002.

[29] B. Ryu, J. Jones, N. J. Blades et al., "Relationships and differentially expressed genes among pancreatic cancers examined by large-scale serial analysis of gene expression," *Cancer Research*, vol. 62, no. 3, pp. 819–826, 2002.

[30] R. M. Caprioli, T. B. Farmer, and J. Gile, "Molecular imaging of biological samples: localization of peptides and proteins using MALDI-TOF MS," *Analytical Chemistry*, vol. 69, no. 23, pp. 4751–4760, 1997.

[31] S. A. Schwartz, R. J. Weil, R. C. Thompson et al., "Proteomic-based prognosis of brain tumor patients using direct-tissue matrix-assisted laser desorption ionization mass spectrometry," *Cancer Research*, vol. 65, no. 17, pp. 7674–7681, 2005.

[32] M. Stoeckli, P. Chaurand, D. E. Hallahan, and R. M. Caprioli, "Imaging mass spectrometry: a new technology for the analysis of protein expression in mammalian tissues," *Nature Medicine*, vol. 7, no. 4, pp. 493–496, 2001.

[33] M. Ehmann, K. Felix, D. Hartmann et al., "Identification of potential markers for the detection of pancreatic cancer through comparative serum protein expression profiling," *Pancreas*, vol. 34, no. 2, pp. 205–214, 2007.

[34] J. Guo, W. Wang, P. Liao et al., "Identification of serum biomarkers for pancreatic adenocarcinoma by proteomic analysis," *Cancer Science*, vol. 100, no. 12, pp. 2292–2301, 2009.

[35] L. Frulloni, C. Lunardi, R. Simone et al., "Identification of a novel antibody associated with autoimmune pancreatitis," *The New England Journal of Medicine*, vol. 361, no. 22, pp. 2135–2142, 2009.

[36] L. Yan, S. Tonack, R. Smith et al., "Confounding effect of obstructive jaundice in the interpretation of proteomic plasma profiling data for pancreatic cancer," *Journal of Proteome Research*, vol. 8, no. 1, pp. 142–148, 2009.

[37] H. L. Huang, T. Stasyk, S. Morandell et al., "Biomarker discovery in breast cancer serum using 2-D differential gel electrophoresis/MALDI-TOF/TOF and data validation by routine clinical assays," *Electrophoresis*, vol. 27, no. 8, pp. 1641–1650, 2006.

[38] R. D. Oleschuk, M. E. McComb, A. Chow et al., "Characterization of plasma proteins adsorbed onto biomaterials by MALDI- TOFMS," *Biomaterials*, vol. 21, no. 16, pp. 1701–1710, 2000.

[39] P. B. Yildiz, Y. Shyr, J. S. M. Rahman et al., "Diagnostic accuracy of MALDI mass spectrometric analysis of unfractionated serum in lung cancer," *Journal of Thoracic Oncology*, vol. 2, no. 10, pp. 893–901, 2007.

# 19

# Application of iTRAQ Reagents to Relatively Quantify the Reversible Redox State of Cysteine Residues

**Brian McDonagh,**[1] **Pablo Martínez-Acedo,**[2] **Jesús Vázquez,**[2] **C. Alicia Padilla,**[1] **David Sheehan,**[3] **and José Antonio Bárcena**[1]

[1] *Department of Biochemistry and Molecular Biology, University of Córdoba and IMIBIC, 14071 Córdoba, Spain*
[2] *Cardiovascular Proteomics Laboratory, National Center for Cardiovascular Research, 28026 Madrid, Spain*
[3] *Department of Biochemistry, University College Cork, Cork, Ireland*

Correspondence should be addressed to José Antonio Bárcena, ja.barcena@uco.es

Academic Editor: Qiangwei Xia

Cysteines are one of the most rarely used amino acids, but when conserved in proteins they often play critical roles in structure, function, or regulation. Reversible cysteine modifications allow for potential redox regulation of proteins. Traditional measurement of the relative absolute quantity of a protein between two samples is not always necessarily proportional to the activity of the protein. We propose application of iTRAQ reagents in combination with a previous thiol selection method to relatively quantify the redox state of cysteines both within and between samples in a single analysis. Our method allows for the identification of the proteins, identification of redox-sensitive cysteines within proteins, and quantification of the redox status of individual cysteine-containing peptides. As a proof of principle, we applied this technique to yeast alcohol dehydrogenase-1 exposed in vitro to $H_2O_2$ and also in vivo to the complex proteome of the Gram-negative bacterium *Bacillus subtilis*.

## 1. Introduction

The dynamic nature of the proteome ensures that the cell is able to respond to perturbation of environmental, genetic, biochemical, and pathological conditions. How the proteome responds to these stimuli is of considerable interest as it can relate to the cell's stress response and can take the form of posttranslational modifications and interprotein interactions with subsequent effects on translation and transcription. Improvements in mass spectrometry has led to the development of a number of techniques to quantify the relative protein abundance within a given sample. These include isotope-coded affinity tags (ICATs) [1], stable isotope labeling of amino acids in cell culture (SILAC) [2], and isobaric tags for relative and absolute quantification (iTRAQ) [3]. However, measuring the relative quantity of a protein between two samples does not tell us anything about the activity of the protein itself. This is especially important in reference to redox proteins that contain thiol switches susceptible to activation or inactivation.

Cysteine is the most important redox-responsive amino acid within proteins largely due to the wide range of oxidation states that sulfur can occupy—so called, "sulfur switches" [4]. Indeed, it has been demonstrated that cysteines are characterized by the most extreme conservation pattern, being highly conserved in functional positions of proteins but poorly conserved otherwise [5]. Within an individual protein there may be a number of cysteines which could allow for multiple thiol modifications. Cysteines often form part of active sites, allowing for the protein to be switched on or off depending on redox state. One of the best-known examples of this is glyceraldehyde 3-phosphate dehydrogenase [6]. In proteins where cysteine is not within the active site, activity can be modulated by changing conformation or by influencing its regulatory role, for example, iron sulfur complexes (ISCs) in aconitase possess cysteines required for its activity [7]. Interactions with other proteins or molecules are another feature of cysteines that can affect protein activity. Allosterically regulated proteins that require an activator are sometimes based on a thiol exchange interaction involving cysteines, for example, pyruvate kinase uses fructose bisphosphate (FBPs) as a heterotrophic activator and it contains a cysteine in its FBP binding site [8]. Reversible modification of cysteines such as disulfide bond formation, glutathionylation, and nitrosylation may also be a means of protection from further, generally irreversible, modifications to sulfinic ($-SO_2H$) or sulfonic ($-SO_3H$) acids [9]. Thus, reversible cysteine modifications can influence protein activity and the relative quantification of the status of the thiol can potentially provide valuable insights into protein activity where the protein exists in a range of redox states. Redox proteomics has taken advantage of the thiol specificity of ICAT reagents not only to identify targets of ROS but also to quantify oxidative thiol modifications in individual proteins. The first applications of this technology involved exposing purified proteins to either OS or normal condition before labeling with either heavy or light ICAT reagents, respectively. This facilitated study of the activity of p21ras GTPase, a redox protein essential for cellular proliferation and differentiation which contains cysteines targeted for reversible glutathionylation and nitrosylation [10, 11]. The versatility of ICAT reagents has been further exploited in using the same technique (termed OxICAT) to determine the oxidation state of an individual protein thiol in a complex protein mixture [12].

iTRAQ has become a popular choice for researchers as it allows up to eight samples to be analyzed simultaneously. In this technique, digested peptides are labeled with amine-specific isobaric reagents to label primary amines of peptides from up to eight different biological samples [3]. We propose a novel method that exploits the accuracy and flexibility of iTRAQ together with a previous thiol selection method [13, 14] to quantify the redox state of cysteines both within and between samples in a single analysis (outlined in Figure 1). This technique allows the identification of the protein, identification of redox sensitive cysteines within the protein, and quantification of its redox state. We used yeast alcohol dehydrogenase-1 (ADH-1) as a model redox protein for proof of principle of the technique. The activity and number of free thiols in this protein decrease in a concentration-dependent manner upon exposure to $H_2O_2$. In addition, we applied the technique to a complex proteome of a Gram-negative bacterium exposed to $H_2O_2$.

## 2. Materials and Methods

All chemicals and reagents were from either Sigma or GE Healthcare unless stated and were of AnalR grade or above.

### 2.1. Alcohol Dehydrogenase.
Yeast ADH-1 (100 μg) in 100 mM HEPES pH 8.0 was exposed to different concentrations of $H_2O_2$ for 5 minutes and the reaction terminated by the addition of excess catalase. Enzyme activity was measured according to [15] by the formation of NADH in the

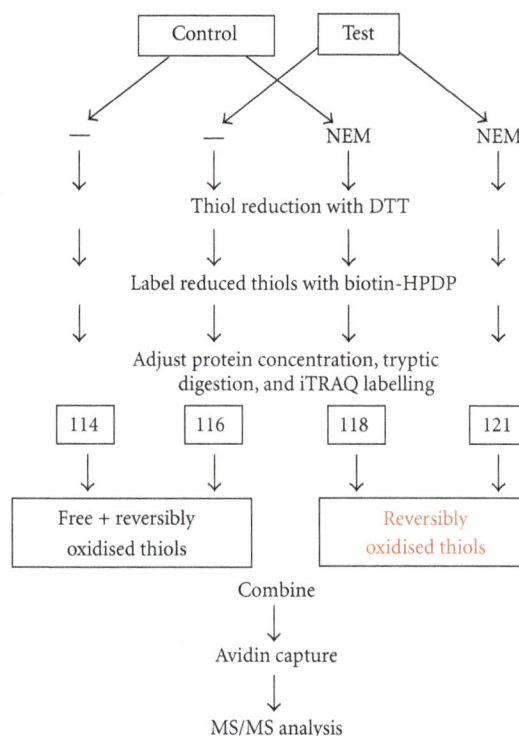

FIGURE 1: Schematic diagram for the relative quantification of the redox state of cysteine-containing peptides between two samples. Each sample (control and test) is split in two. One set has its free thiols initially blocked with the alkylating reagent NEM. Once excess NEM is removed, all samples have their reversibly oxidized thiols reduced with DTT. Free thiols in all samples are then labeled with thiol-specific biotin-HPDP, and protein concentration is measured so all samples have equivalent protein content. Proteins are tryptic digested and peptides labeled with iTRAQ reporter tags according to the scheme outlined. Labeled peptides are combined. Biotinylated cysteine-containing peptides are purified using streptavidin, and purified peptides are analysed and quantified by MS/MS.

first 5 minutes. Free thiol content in alcohol dehydrogenase was measured using Ellman's reagent (5,5′-dithiobis-2-nitrobenzoic acid, DTNB) at 412 nm in denaturing conditions. All activities and measurements were performed in triplicate and with $N = 3$.

### 2.2. Protein Preparation of iTRAQ.
ADH-1 was prepared for analysis adapted from a method described previously [13] and outlined in Figure 1, the major difference being that Tris-HCl was replaced with HEPES due to the reactivity of iTRAQ reagents with amines. Briefly, after each treatment, the protein sample was split in two, one with a population of cysteines with free thiols blocked with NEM and the other with free thiols (without NEM). From this point on, all samples were treated identically. The protein was precipitated and washed to remove any free NEM, dissolved in 180 μL denaturing buffer (8 M Urea, 4% CHAPS and 100 mM HEPES, pH 8.0) with 20 μL of 200 mM DTT, and

incubated for 45 min on a rotator. Protein was precipitated and washed with acetone to remove excess DTT and redissolved in denaturing buffer containing 0.5 mM biotin-HPDP (Pierce Biotechnology). Excess biotin-HPDP was removed using zebra spin trap columns (Pierce) and buffer exchanged for 100 mM HEPES, pH 8.0, using repeated cycles with microcon 3 filters. Protein concentration was measured using Bradford reagent (BioRad) [16] with BSA as a standard.

ADH-1 (10 $\mu$g) from control or either 1 mM or 5 mM $H_2O_2$ exposure was tryptic digested (Promega) at a ratio of 1 : 20 trypsin : protein and incubated at 37$^\circ$C for 3 hours. Peptides were labeled with iTRAQ isobaric tags (ABSciex) according to the manufactures' instructions in the following order: control (without NEM – total thiols) reporter 114, control (plus NEM – reversibly oxidized thiols) reporter 118, test 1 or 5 mM $H_2O_2$ (without NEM – total thiols) reporter 116, and 1 or 5 mM $H_2O_2$ (plus NEM – reversibly oxidized thiols) reporter 121. Replicate peptides (see Supplementary information available online at doi:10.1155/2012/514847) were labeled in the same order with 113, 115, 117, and 119 iTRAQ reagents. After labeling, the four distinct isobaric-labeled peptides were combined and incubated with Streptavidin-Sepharose resin. This was prepared by washing twice in binding buffer (4 M urea, 2% CHAPS, 50 mM NaCl and 50 mM HEPES, pH 8.0), and 100 $\mu$L of this slurry was incubated with peptides overnight at 4$^\circ$C on a rotator. Following overnight incubation, the resin was washed once with binding buffer, twice with wash buffer A (8 M urea, 4% CHAPS, 1 M NaCl and 50 mM HEPES, pH 8.0) and three times with wash buffer B (8 M urea, 4% CHAPS, and 50 mM HEPES. In order to remove urea, the resin was washed four times with wash buffer C (5 mM HEPES/20% acetonitrile). Biotinylated peptides were eluted from the resin by adding 30 $\mu$L of wash buffer C containing 20 mM DTT and incubated for 30 mins. Peptides were collected by centrifugation and stored at −70$^\circ$C until analysis by MS/MS.

### 2.3. Bacterial Culture.
A Gram-negative bacterial *Bacillus subtilis* strain available in our laboratory was used to assess the potential of this technique to analyze complex proteomes. Exponentially growing cells (O.D.$_{600}$ = 1–1.5) grown in standard media [17] were exposed to 1 mM $H_2O_2$ and harvested for analysis. Cell cultures were split in two for analysis, one for lysis in a buffer containing 100 mM HEPES, 8 M urea, 2 mM EDTA and 0.1% Triton and the other in the same buffer but also containing 50 mM NEM. All analyses were performed on two independent cultures. Cell lysis and protein preparation were carried out as previously described [13]. The same protocol was used for complex protein samples as with ADH-1 except 100 $\mu$g of protein sample was tryptic digested and labeled with each iTRAQ reagent.

### 2.4. Sample Analysis by nLC-MALDI MS/MS.
Labeled peptides were separated by reverse phase nano HPLC using the integrated autosampler Famos, switch pump, and micropump Ultimate (LC Packings). Solvent A was 10 mM $Na_2HPO_4$ in 0.1% TFA (v/v) and solvent B, 10 mM $Na_2HPO_4$ in 70% acetonitrile (ACN) and 0.1% TFA (v/v). Labeled peptides were desalted and concentrated in a reverse phase C18 PepMap column (0.3–5 mm, 5 mm, 100 Å LC Packings) for 15 min. The peptides were separated manually in a reverse phase C18 analytical column (0.075–0.1 mm, Thermo C18Aq, 5 mm, 100 Å Thermo) using a 60 min linear 6–60% gradient followed by 20 min linear increase 60–100% solvent B with a flow rate 300 $\mu$L/min. Eluted fractions were collected at 12 s intervals and directly spotted onto MALDI plate OptiTOF (ABSciex) using the Suncollect system. The eluent spotted was 60 nl and mixed with 200 nL matrix $\alpha$-cyano-4-hydroxycinnamic acid (CHCA), 7 mg/mL (w/v) in 70% ACN (v/v), and 0.1% TFA (v/v). Eluent deposition time was dependent on chromatography separation time.

nLC-MALDI fractions were analyzed using an Applied Biosystems 4800 MALDI TOF-TOF Analyzer (ABSciex) in positive ion reflector mode with a mass range of 800–4000 Da controlled by analysis programme 4000 Explorer Series v3.5 (ABSciex). A rate of 2500 laser spots per mass spectrum was used with a uniform standard. In each mass spectrum, the 20 most abundant peaks were selected for MS/MS using the ion exclusion method for ions with an S/N greater than 50, leaving out identical peaks from adjacent spots and selecting for only the highest precursor ions. Weaker precursor ions with a lower S/N ratio were acquired first to obtain a stronger signal for less abundant peptides. The peptide angiotensin was used for internal calibration of MS spectra. To obtain fragmentation MS/MS spectra, 1 kV collision energy was used. A window of 250 (total average mass width) relative to precursor ion and using CID activated collision allowed suppression of metastable ions. MS/MS spectra selected were obtained using a fixed laser shot range 1000–3000 and 50 for subspectra. The minimum criteria were set at 100 S/N in more than 7 peaks after a minimum of 1000 shots.

### 2.5. Data Analysis.
The peptide data obtained by MALDI-TOF/TOF were analyzed with ProteinPilot 1.0 software using the Paragon protein database search algorithm (ABSciex). Using this software, peptide analysis data obtained with the iTRAQ system were converted into the differential analysis data for peptide matching identification and relative quantification. The parameters for the analysis were set as follows: sample type: iTRAQ 8-plex (peptide labeled); Cys alkylation: NEM and including all biological modifications; digestion: trypsin; instrument: MALDI TOF/TOF. MS/MS data were searched against all entries in the UniProt nonredundant database (517,802 sequences; 161,091,005 residues). Crude data were limited to peptide confidence (minimum 95%), the peak area of reporter ion, error of peak area of reporter ion, accession number, taxonomy, peptide sequence, assigned peptide, and the relative quantification of peptides. Rates of false positive identifications were estimated using a concatenated reversed sequence database. Only peptides with a confidence of at least 95% were used to quantify the relative abundance of each peptide determined by ProteinPilot using

the peak areas of signature ions from the iTRAQ-labeled peptides.

## 3. Results

*3.1. Alcohol Dehydrogenase.* To test the performance of the method, we used pure commercial ADH-1. Yeast ADH-1 is a tetrameric protein composed of identical 36 kDa subunits and containing two zinc ions co-coordinated to cysteine residues [15]. Of the eight cysteine residues within ADH-1, three are contained in tryptic peptides that are amenable to MS/MS analysis (Figure 2(a)). Cys[44] contained within peptide 40–60 has been reported to coordinate to a zinc ion forming part of the catalytic centre, and oxidation plays a major role in $H_2O_2$ induced deactivation [15]. Cys[277&278] are contained in peptide 277–287 and have been identified as forming disulfide bonds after $H_2O_2$ oxidation [15]. Exposure of ADH to $H_2O_2$ resulted in a concentration-dependent reduction in activity and free thiols. Enzyme activity decreased to about 40% after 5 minutes exposure to 5 mM $H_2O_2$ (Figure 2(b)) and free thiols as measured using Ellman's reagent also decreased and correlated with the decrease in catalytic activity (Figure 2(b)). There was also an increase in irreversible protein carbonylation at this concentration (Figure 2(c)). Once the redox behavior of the enzyme was determined, we checked whether the iTRAQ methodology could provide parallel consistent results.

*3.2. iTRAQ Relative Quantification.* A schematic outline of our approach in applying iTRAQ reagents to relatively quantify individual cysteine-containing peptides after exposure to $H_2O_2$ is outlined in Figure 1. Samples are divided in two, one group has its free thiols blocked with NEM. Reversibly oxidized thiols are then reduced in all groups with dithiothreitol (DTT) and free thiols subsequently labeled with biotin-HPDP. After tryptic digestion, iTRAQ labeling and mixing of samples, labeled peptides are selected, analyzed, and quantified by MS/MS. Peaks are quantified relative to the control cysteine-containing peptides (labeled with 114-total thiols) which include both reduced or free thiols and reversibly oxidized thiols. The second peak (116) is the corresponding value after treatment with $H_2O_2$ (1 or 5 mM). The third peak (118) is the proportion of the peptide with reversibly oxidized thiols in controls only and the last peak (121) is the proportion of reversibly oxidized thiols after exposure to $H_2O_2$. Table 1 lists the relative proportion of free thiols and reversibly oxidized thiols in the amenable ADH peptides. Further analysis of the results for peptide 40–60 after treatment with 5 mM $H_2O_2$ is presented in Table 2. If we take the reporter 114 from control as total detectable thiols to be 100%, then we can calculate both the proportion of that reversibly oxidized cysteine (118/114) and that in a reduced state (1 − (118/114)). Similarly, after peroxide exposure, we can calculate the proportion of the thiol remaining reversibly oxidized (121/114) and reduced (116/114) − (121/114). The remaining proportion (1 − (116/114)) is presumably over-oxidized. Inspection of the results indicates that under control conditions, approximately half of these thiols were

reversibly oxidized (47%) and half were in a reduced state (53%). After exposure to 5 mM $H_2O_2$, the proportion of reversibly oxidized thiols decreased to 26%, free thiols decreased to 22%, and the overoxidized proportion was 52%. This cysteine forms part of the active site and these results correlated well with the decrease in ADH activity (∼50%), loss of free thiols, and increase in carbonylation at this concentration (Figure 2). This suggests that cys[44] is redox sensitive and subject to oxidation. Figure 3, shows fragmentation of the precursor ion 3028 *m/z* that corresponds to peptide [40]YSGVCHTDLHAWHGDWPLPTK[60] in ADH-1 in control and after exposure to 1 mM (Figure 3(a)) or 5 mM $H_2O_2$ (Figure 3(b)). The reporter tags can be seen in the inset and it is clear that, after exposure to 5 mM $H_2O_2$, there is a significant decrease in iTRAQ reporter ion 121 (inset Figure 3(b)) corresponding to the relative proportion of reversibly oxidized after peroxide exposure. Exposure to 1 mM $H_2O_2$ had little effect on reversibly oxidized cysteines, coincident with lack of significant change in either enzyme activity, or in free thiols at this peroxide concentration (Figure 2(b)).

Analysis of ADH-1 peptide 277–287 is more complex due to the presence of two cysteine residues that have previously been reported to be involved in a disulfide bond [15]. The potential oxidation of either or both cysteine residues as well as thiol exchange and oxidation (especially under higher oxidative conditions) make the relative quantification complex for this technique. In-depth analysis of this peptide after differential alkylation of cysteines by selective MS/MS ion monitoring (SMIM) [19] indicated that the two cysteines can exist alternatively in both reduced and reversibly oxidized forms. Application of SMIM indicated the peptide exists in at least twelve distinct oxidation states and even with both cysteines in a $-SO_3H$ form after 5 mM $H_2O_2$ (Supplementary information Figures 1 and 2). This is further supported by our results after application of iTRAQ in which we have seen both alternative cysteine residues irreversibly oxidized to $-SO_2H$ forms and a consistent relative increase in the peptide signal after exposure to 5 mM $H_2O_2$. Taken together with the fact that at least one of the thiols needs to be either in a reduced state or reversibly oxidized to be able to capture the cysteine-containing peptide, analysis of the redox state of individual cysteines in such peptides is complex.

Application of this technique to the redox proteome of *B. subtilis* resulted in identification and relative quantification of the redox status of 23 cysteine-containing peptides from 18 known redox-sensitive proteins (Supplementary Table 1). A number of these proteins known to be sensitive to redox changes and have been well characterized, for example, thiol peroxidase, elongation factors, and ribosomal proteins. Application of the same criteria used in Table 2 for a selection of these cysteine-containing peptides, is presented in Table 3. In general, results are as would be expected with a large number of proteins having a decreased value for total detectable thiols (116 : 114) ratio after exposure to 1 mM $H_2O_2$. We also have an estimation of the proportion of the total thiols that are reversibly oxidized in both controls (118 : 114 ratio) and after peroxide exposure (121 : 116 ratio). The advantage of this technique can clearly be seen

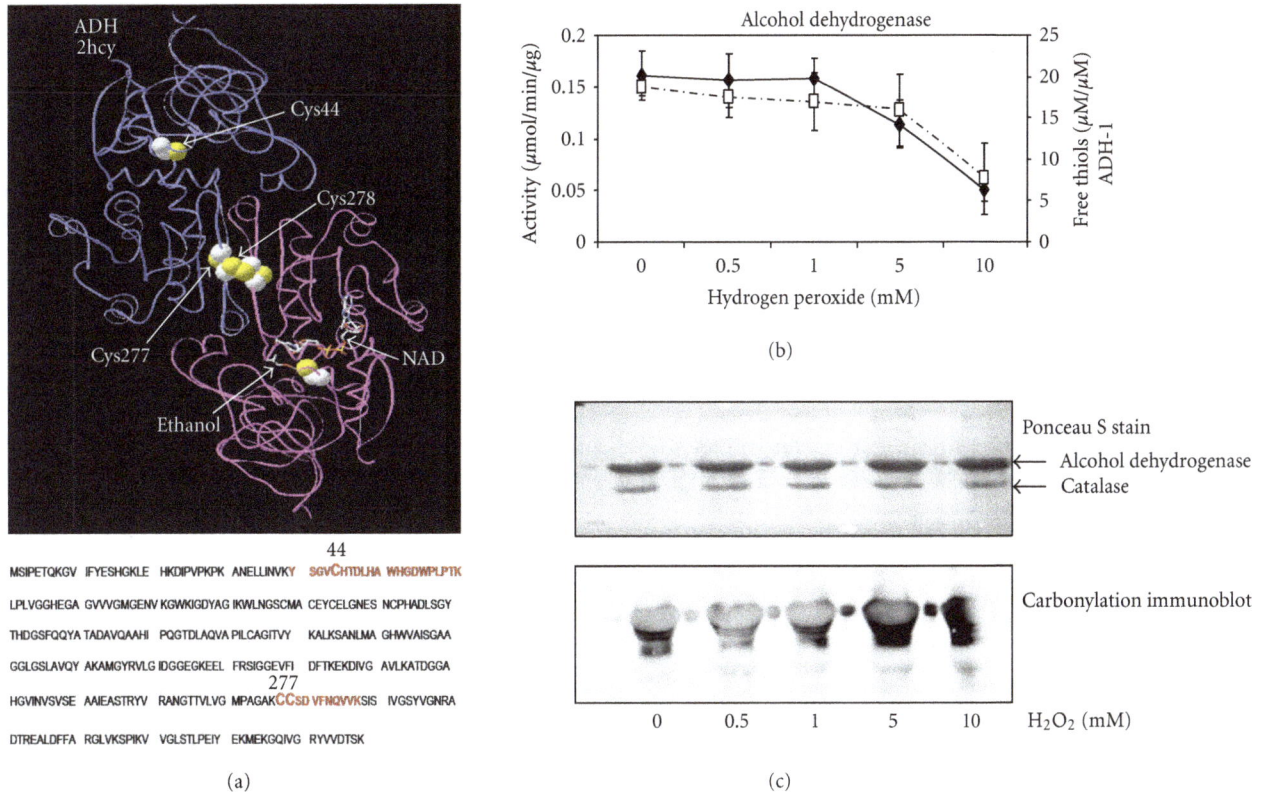

FIGURE 2: (a) ADH-1 homodimer is represented with substrate ethanol and coenzyme A at the active site. Coordinates were downloaded from the Protein Data Bank as a PDB file 2HCY and manipulated with the DeepView free software [18]. Cys$^{44}$ and Cys$^{277,278}$ are highlighted, and analysis of ADH-1 amino acid sequence indicates Cys-containing tryptic peptides in red that are amenable to analysis by MS/MS. Cys$^{44}$ forms part of the catalytic centre, and Cys$^{277,278}$ is involved in a disulfide. (b) Activity ($\blacklozenge - \blacklozenge$) and free thiols ($\square$- -$\square$) present in ADH-1 after exposure to increasing concentrations of H$_2$O$_2$. (c) Ponceau S stain and carbonylation immunoblot of ADH after H$_2$O$_2$ exposure; there is equivalent protein loading, but an increase in irreversible carbonylation after exposure to 5 mM H$_2$O$_2$ is evident.

TABLE 1: Relative quantification of the redox state of Cys-containing tryptic peptides from yeast ADH-1 after exposure to either 1 or 5 mM H$_2$O$_2$. The ratio of free and reversibly oxidized thiols are compared to control levels (taken as 1.0*). 116:114 are the relative amounts of total thiols after H$_2$O$_2$ exposure. Shaded boxes are the relative amounts of reversible oxidized thiols only, referred to total thiols in control; thus, 118:114 and 121:114 are the relative amounts of reversibly oxidized thiols in controls and after exposure, respectively.

| Amenable Cys tryptic peptides from yeast ADH-1 | Free + reversibly oxidized thiols* | | | Reversibly oxidized thiols | | |
|---|---|---|---|---|---|---|
| | Control (114:114) | 1 mM H$_2$O$_2$ (116:114) | 5 mM H$_2$O$_2$ (116:114) | Control (118:114) | 1 mM H$_2$O$_2$ (121:114) | 5 mM H$_2$O$_2$ (121:114) |
| $^{40}$YSGV**C**HTDLHAWHGDWPLPTK$^{60}$ | 1.00 | 0.341 ± 0.029 | 0.478 ± 0.143 | 0.467 ± 0.18 | 0.548 ± 0.221 | 0.258 ± 0.195 |
| $^{277}$**CC**SDVFNQVVK$^{287}$ | 1.00 | 0.288 ± 0.021 | 2.293 ± 1.041 | 0.397 ± 0.15 | 0.549 ± 0.139 | 0.346 ± 0.293 |

TABLE 2: Relative quantification of the redox status of cys$^{44}$ in the ADH-1 peptide (40–60) in controls and after exposure to 5 mM H$_2$O$_2$. There is a decrease in both reversibly oxidized (47% to 26%) and reduced thiols (53% to 22%) and an increase in over oxidized thiols (52%) at this peroxide concentration.

| Protein/peptide | Total detectable thiols (%) | Sample | Reversibly oxidized thiols (%) | Free thiols (%) | Overoxidized thiols (%) |
|---|---|---|---|---|---|
| Example | 100 | Control | (118/114) | 1 − (118/114) | Not detectable |
| | | Test | (121/114) | (116/114) − (121/114) | 1 − (116/114) |
| ADH-1 (40–60) | 100 | Control | 47 | 53 | ND |
| | | Test (5 mM H$_2$O$_2$) | 26 | 22 | 52 |

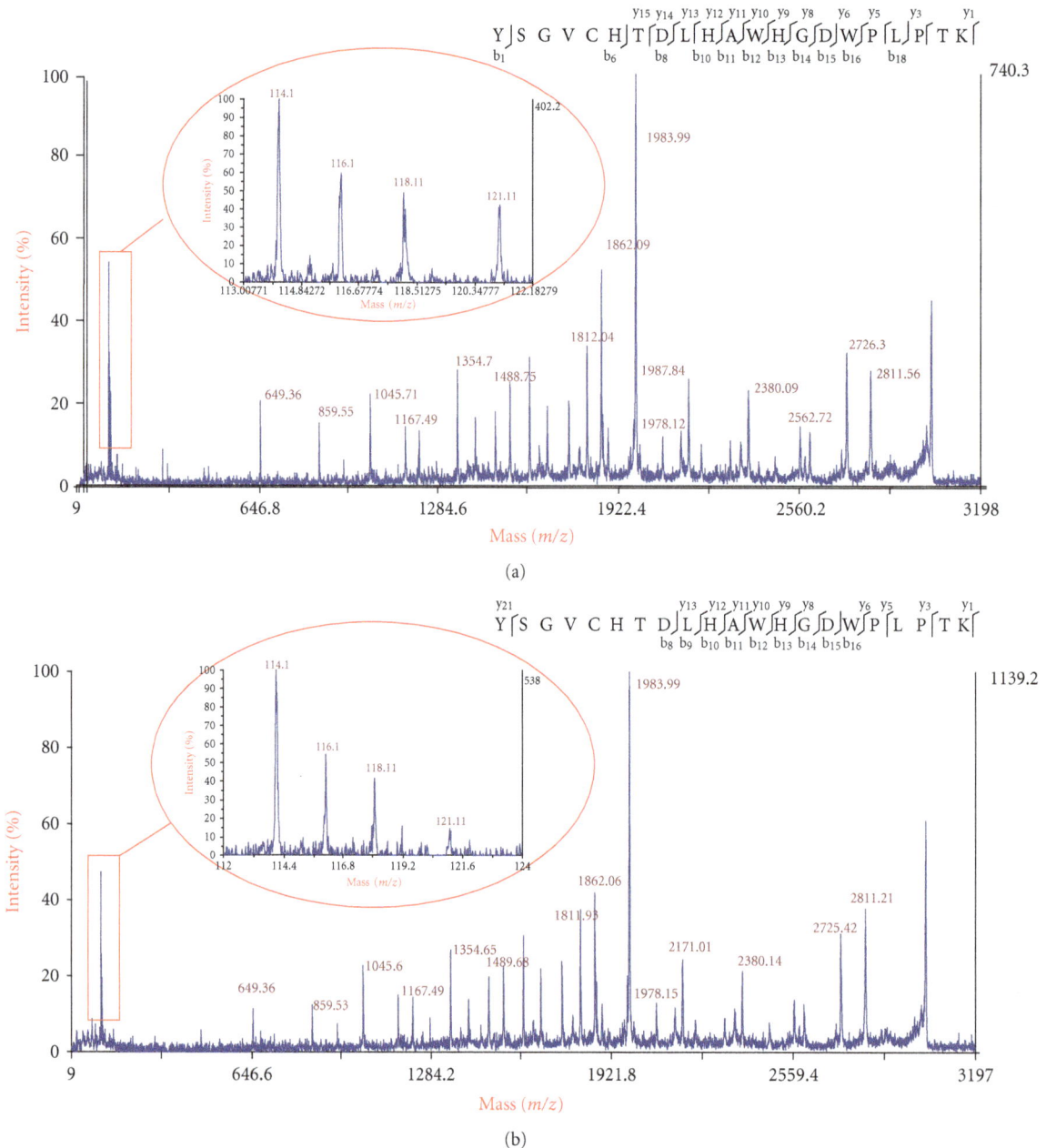

FIGURE 3: Fragmentation spectrum of peptide $^{40}$YSGV<u>C</u>HTDLHAWHGDWPLPTK$^{60}$ with iTRAQ reporter ions magnified. (a) Reporter ions 114 and 118 are for controls and indicate approximately half of this Cys population is in a free thiol state. After exposure to 1 mM $H_2O_2$, there is a decrease in reporter ion 116 for total free thiols and reporter 121 indicates that it is predominantly reversibly oxidized and not present as a free thiol. (b) Reporter ions 114 and 118 are again controls and are equivalent to the control results in (a), that is, approximately half of the Cys in the peptide are in the free thiol form. After exposure to 5 mM $H_2O_2$, there is a dramatic reduction in reversibly oxidized thiols (reporter 121) indicating, at this concentration, that the Cys residue is susceptible to irreversible oxidation.

when we examine the peptides for elongation factor G (Q8CQ82) protein 5 with two cysteine-containing peptides detected. Relative quantification of the cysteines within the two peptides indicates that under control conditions, the majority of the thiols are reversibly oxidized (85% and 95%, resp.) In the first peptide $^{595}$CNPVILEPISK$^{605}$ the proportion of thiols reversibly oxidized did not change dramatically after exposure 75% (121 : 114) and approximately 25% of

the thiols were over-oxidized. However, quantification of the second peptide $^{381}$DTTTGDTLCDEK$^{392}$ indicates that after exposure, the proportion of reversibly oxidized thiols decreased from 95% (118 : 114) to 30% (121 : 114) while the proportion irreversibly oxidized ($-SO_2H$ or $-SO_3H$) increased to 75%. Elongation factor G is redox sensitive and known to be inactivated by sulfhydryl reagents in other species [20, 21]. Yet this technique allowed us to identify

TABLE 3: A selection of peptides identified from the Gram-negative bacteria, *B. subtilis* with relative quantification of the redox state of identified cysteine peptides. Total detectable thiols refer to both reversibly oxidized and reduced thiols and quantification is relative to control values. As overoxidized thiols are not amenable to selection they are not detected in controls (N.D.).

| Protein (accession number) | Cys tryptic peptide | Total detectable thiols (%) | Sample | Reversibly oxidized thiols (%) | Free thiols (%) | Overoxidized thiols (%) |
|---|---|---|---|---|---|---|
| Triose phosphate isomerase (Q65ENO) | [85]DLGVEY**C**VIGHSER[98] | 100 | Control | 41 | 59 | N.D. |
| | | | Test | 35 | 32 | 33 |
| | [179]SSTSEDANEM**C**AHVR[193] | 100 | Control | 51 | 49 | N.D. |
| | | | Test | 52 | 32 | 17 |
| Elongation factor Ts (Q65JJ8) | [15]TGAGMMD**C**K[23] | 100 | Control | 53 | 47 | N.D. |
| | | | Test | 52 | 8 | 40 |
| | [234]YFEEI**C**LLDQAFVK[247] | 100 | Control | 66 | 34 | N.D. |
| | | | Test | 44 | 4 | 52 |
| Elongation factor G (Q65PB0) | [595]**C**NPVILEPISK[605] | 100 | Control | 85 | 15 | N.D. |
| | | | Test | 75 | −10 | 25 |
| | [381]DTTTGDTL**C**DEK[392] | 100 | Control | 95 | 5 | N.D. |
| | | | Test | 30 | −6 | 75 |
| Elongation factor Tu (Q5P334) | [138]**C**DMVDDEELLELVEMEVR[155] | 100 | Control | 117 | (−17) | N.D. |
| | | | Test | 91 | −1 | 8 |
| | [76]HYAHVD**C**PGHADYVK[90] | 100 | Control | 173 | (−73) | N.D. |
| | | | Test | 150 | (−41) | −50 |
| Adenylate kinase (P35140) | [75]ND**C**DGGFLLDGFPR[88] | 100 | Control | 94 | 4 | N.D. |
| | | | Test | 68 | 1 | 31 |
| Purine nucleoside phosphorylase (Q65IE9) | [27]YIADTYLENVE**C**YNEVR[43] | 100 | Control | 91 | 9 | N.D. |
| | | | Test | 50 | 9 | 41 |
| Transition state regulatory protein AbrB (P08874) | [50]YKPNMT**C**QVTGEVSDDNLK[68] | 100 | Control | 39 | 61 | N.D. |
| | | | Test | 46 | 23 | 31 |

the redox sensitive cysteine within the protein, which would not be detected by relative quantification alone. Elongation factor Tu is also known to be redox sensitive, and indeed both elongation factors G and Tu have previously been purified using covalent chromatography with thiol sepharose beads [22, 23] indicating that they possess free thiols and are redox dependent. It is known that EfTu cys[81] and cys[137] are associated with aminoacyl-tRNA and guanosine nucleotide binding, respectively, in *Escherichia coli* [24] equivalent to the peptides containing cys[82] and cys[138] detected here. Interestingly, cys[81] has been reported as the site for nucleotide binding in *E. coli* and the equivalent cys[82] increases in relative abundance in both control and treated samples even after initial alkylation with NEM, which is probably due to overoxidation of sensitive thiol groups during the relatively harsh conditions used for cell lysis, resulting in an under estimation for the reference "total detectable thiols" and hence an artificially higher value for the proportion of thiols reversibly oxidized.

## 4. Discussion

Cysteines are one of the most rarely used amino acids in proteins [25]. Therefore, when conserved, they usually play critical roles in structure, function, or regulation of the protein. The average $pK_a$ value of cysteines has been calculated as 6.8 ± 2.7, indicating that at physiological pH, they may exist in both charged thiolate form and uncharged form depending on a number of factors [26, 27]. The location and sequence of surrounding amino acids strongly influence the $pK_a$ and hence, reactivity of a particular cysteine residue. In unstressed mammalian cells, it has been demonstrated that proteins disulfides (PSSP) account for 6% and 9.5% of protein sulfhydryls in HEK and HeLa cells, respectively. After treatment with the thiol-specific oxidant diamide, this increased to 24% and 25%. The steady state level of glutathione-protein mixed disulfides (PSSG) was less than 1% but this increased to 15% after prooxidant treatment [28]. Protein thiols therefore represent an important and significant redox buffer within the cell so application of a relative quantification method is now especially timely. iTRAQ is a flexible and multiplexed quantitative method and we had successfully developed a high throughput method for oxidized cysteine selection. A combination of both techniques could in principle be appropriate for quantitatively analyze the redox proteome. Here we demonstrate that the combined approach is feasible and provides useful information, despite some limitations.

Key goals in identifying redox-regulated proteins involve determining which proteins are involved, which cysteines within those proteins are redox sensitive, and identifying thiol modifications within particular cysteines [29]. Although the technique described herein cannot distinguish the type of reversible modifications of cysteines, it does allow for quantification of the proportion of the cysteine that is reversibly modified (and also free thiols) in both control and test conditions. Each cysteine-containing peptide is monitored independently so it is applicable to proteins that contain various cysteines reacting at different rates or which are involved in different protein functions. The relative merits and drawbacks regarding precision and accuracy of iTRAQ reagents have been extensively studied elsewhere [30, 31]. This paper aims to present the results of a novel application of these reagents in redox proteomics. Our results indicate that, when this technique is applied to study the redox state of purified proteins (in this case ADH-1), quantification of the catalytic cys$^{44}$ with iTRAQ correlates with observed decrease in enzyme activity and loss of free thiols. When applied to a complex proteome, it can identify and relatively quantify the redox state of amenable cysteines within abundant proteins. Abundant proteins are both predominantly identified and quantified because iTRAQ labeling is optimized for a maximum of 100 $\mu$g protein and we are dealing with a small percentage of amenable peptides form the total proteome. Disulfides in proteins have been classified as forming subproteomes, redox responsive, or the more resistant structural disulfides [32]. One of the advantages of the technique employed in this analysis is that redox-responsive cysteines can be distinguished from structural cysteines by change in relative abundance not only after initial blocking but also after exposure to OS. For instance, elongation factor G has two very distinct cysteine peptides in terms of their sensitivity to OS; cys$^{389}$ is more sensitive to oxidation by OS than cys$^{595}$. This is also an important aspect when proteins have an altered function dependent on their redox state. For instance, it is known that the peroxiredoxins may act as peroxidases, redox sensors, or chaperones depending on oligomerization, which is, in turn, dependent on the redox state [33].

Our approach also provides meaningful information regarding both the sensitivity and oxidation states of individual cysteine residues and may provide clues to regulation and catalytic centres when there is no structural information available for a given protein. When applying this technique to quantification of sensitive cysteines in complex mixtures, care must be taken to minimize oxidation during cell lysis. One shortcoming of the technique is that, when there are two or more cysteine residues within a peptide it cannot distinguish the cysteine involved and so quantification of the redox state is not possible. This was demonstrated with a two-cysteine-containing peptide from ADH-1 that existed in up to twelve distinct states after differential oxidation. Nevertheless, this technique provides both an informative and powerful tool in the study of redox proteomics with all the advantages of the iTRAQ reagents and protocols regarding precision, accuracy, multiplexing, and availability in conventional Proteomics facilities.

## Acknowledgments

Mass spectrometry of labeled samples was performed at the Proteomics Facility, SCAI, University of Córdoba, node 6 of the ProteoRed Consortium financed by ISCIII. This work was supported by Grants P06-CVI-01611 from the Andalusian Government, BFU2006-02990 and BFU2009-08004 from the Spanish Government to J. A. Bárcena, and by Grants BIO2006-10085, GR/SAL/0141/2004 (CAM), and CAM BIO/0194/2006 from the Fondo de Investigaciones Sanitarias (Ministerio de Sanidad y Consumo, Instituto Salud Carlos III, RECAVA) and by an institutional grant by Fundación Ramón Areces to CBMSO. P. Martínez-Acedo is recipient of a fellowship from the Comunidad Autónoma de Madrid (supported by the European Social Fund).

## References

[1] S. P. Gygi, B. Rist, S. A. Gerber, F. Turecek, M. H. Gelb, and R. Aebersold, "Quantitative analysis of complex protein mixtures using isotope-coded affinity tags," *Nature Biotechnology*, vol. 17, no. 10, pp. 994–999, 1999.

[2] S. E. Ong, B. Blagoev, I. Kratchmarova et al., "Stable isotope labeling by amino acids in cell culture, SILAC, as a simple and accurate approach to expression proteomics," *Molecular & Cellular Proteomics*, vol. 1, no. 5, pp. 376–386, 2002.

[3] P. L. Ross, Y. N. Huang, J. N. Marchese et al., "Multiplexed protein quantitation in *Saccharomyces cerevisiae* using amine-reactive isobaric tagging reagents," *Molecular and Cellular Proteomics*, vol. 3, no. 12, pp. 1154–1169, 2004.

[4] F. Q. Schafer and G. R. Buettner, "Redox environment of the cell as viewed through the redox state of the glutathione disulfide/glutathione couple," *Free Radical Biology and Medicine*, vol. 30, no. 11, pp. 1191–1212, 2001.

[5] S. M. Marino and V. N. Gladyshev, "Cysteine function governs its conservation and degeneration and restricts its utilization on protein surfaces," *Journal of Molecular Biology*, vol. 404, no. 5, pp. 902–916, 2010.

[6] S. Mohr, H. Hallak, A. de Boitte, E. G. Lapetina, and B. Brüne, "Nitric oxide-induced S-glutathionylation and inactivation of glyceraldehyde-3-phosphate dehydrogenase," *Journal of Biological Chemistry*, vol. 274, no. 14, pp. 9427–9430, 1999.

[7] X. J. Chen, X. Wang, and R. A. Butow, "Yeast aconitase binds and provides metabolically coupled protection to mitochondrial DNA," *Proceedings of the National Academy of Sciences of the United States of America*, vol. 104, no. 34, pp. 13738–13743, 2007.

[8] D. Susan-Resiga and T. Nowak, "Proton donor in yeast pyruvate kinase: chemical and kinetic properties of the active site Thr 298 to cys mutant," *Biochemistry*, vol. 43, no. 48, pp. 15230–15245, 2004.

[9] D. Sheehan, B. McDonagh, and J. A. Brcena, "Redox proteomics," *Expert Review of Proteomics*, vol. 7, no. 1, pp. 1–4, 2010.

[10] M. Sethuraman, M. E. McComb, H. Huang et al., "Isotope-coded affinity tag (ICAT) approach to redox proteomics: identification and quantitation of oxidant-sensitive cysteine thiols in complex protein mixtures," *Journal of Proteome Research*, vol. 3, no. 6, pp. 1228–1233, 2004.

[11] M. Sethuraman, N. Clavreul, H. Huang, M. E. McComb, C. E. Costello, and R. A. Cohen, "Quantification of oxidative

posttranslational modifications of cysteine thiols of p21ras associated with redox modulation of activity using isotope-coded affinity tags and mass spectrometry," *Free Radical Biology and Medicine*, vol. 42, no. 6, pp. 823–829, 2007.

[12] L. I. Leichert, F. Gehrke, H. V. Gudiseva et al., "Quantifying changes in the thiol redox proteome upon oxidative stress in vivo," *Proceedings of the National Academy of Sciences of the United States of America*, vol. 105, no. 24, pp. 8197–8202, 2008.

[13] B. McDonagh, S. Ogueta, G. Lasarte, C. A. Padilla, and J. A. Bárcena, "Shotgun redox proteomics identifies specifically modified cysteines in key metabolic enzymes under oxidative stress in *Saccharomyces cerevisiae*," *Journal of Proteomics*, vol. 72, no. 4, pp. 677–689, 2009.

[14] B. McDonagh, C. A. Padilla, J. R. Pedrajas, and J. A. Barcena, "Biosynthetic and iron metabolism is regulated by thiol proteome changes dependent on glutaredoxin-2 and mitochondrial peroxiredoxin-1 in *Saccharomyces cerevisiae*," *Journal of Biological Chemistry*, vol. 286, no. 17, pp. 15565–15576, 2011.

[15] L. Men and Y. Wang, "The oxidation of yeast alcohol dehydrogenase-1 by hydrogen peroxide in vitro," *Journal of Proteome Research*, vol. 6, no. 1, pp. 216–225, 2007.

[16] M. M. Bradford, "A rapid and sensitive method for the quantitation of microgram quantities of protein utilizing the principle of protein dye binding," *Analytical Biochemistry*, vol. 72, no. 1-2, pp. 248–254, 1976.

[17] J. Thaniyavarn, N. Roongsawang, T. Kameyama et al., "Production and characterization of biosurfactants from Bacillus licheniformis F2.2," *Bioscience, Biotechnology and Biochemistry*, vol. 67, no. 6, pp. 1239–1244, 2003.

[18] N. Guex and M. C. Peitsch, "SWISS-MODEL and the Swiss-PdbViewer: an environment for comparative protein modeling," *Electrophoresis*, vol. 18, no. 15, pp. 2714–2723, 1997.

[19] I. Jorge, E. M. Casas, M. Villar et al., "High-sensitivity analysis of specific peptides in complex samples by selected MS/MS ion monitoring and linear ion trap mass spectrometry: application to biological studies," *Journal of Mass Spectrometry*, vol. 42, no. 11, pp. 1391–1403, 2007.

[20] K. Kojima, K. Motohashi, T. Morota et al., "Regulation of translation by the Redox State of Elongation factor G in the cyanobacterium *Synechocystis* sp. PCC 6803," *Journal of Biological Chemistry*, vol. 284, no. 28, pp. 18685–18691, 2009.

[21] D. R. Southworth, J. L. Brunelle, and R. Green, "EFG-independent translocation of the mRNA:tRNA complex is promoted by modification of the ribosome with thiol-specific reagents," *Journal of Molecular Biology*, vol. 324, no. 4, pp. 611–623, 2002.

[22] T. D. Caldas, A. El Yaagoubi, M. Kohiyama, and G. Richarme, "Purification of elongation factors EF-Tu and EF-G from *Escherichia coli* by covalent chromatography on thiol-sepharose," *Protein Expression and Purification*, vol. 14, no. 1, pp. 65–70, 1998.

[23] W. Hu, S. Tedesco, B. McDonagh, J. A. Bárcena, C. Keane, and D. Sheehan, "Selection of thiol- and disulfide-containing proteins of *Escherichia coli* on activated thiol-Sepharose," *Analytical Biochemistry*, vol. 398, no. 2, pp. 245–253, 2010.

[24] P. H. Anborgh, A. Parmeggiani, and J. Jonak, "Site-directed mutagenesis of elongation factor Tu. The functional and structural role of residue Cys81," *European Journal of Biochemistry*, vol. 208, no. 2, pp. 251–257, 1992.

[25] I. Pe'er, C. E. Felder, O. Man, I. Silman, J. L. Sussman, and J. S. Beckmann, "Proteomic signatures: amino acid and oligopeptide compositions differentiate among phyla," *Proteins*, vol. 54, no. 1, pp. 20–40, 2004.

[26] G. R. Grimsley, J. M. Scholtz, and C. N. Pace, "A summary of the measured pK values of the ionizable groups in folded proteins," *Protein Science*, vol. 18, no. 1, pp. 247–251, 2009.

[27] C. N. Pace, G. R. Grimsley, and J. M. Scholtz, "Protein ionizable groups: pK values and their contribution to protein stability and solubility," *Journal of Biological Chemistry*, vol. 284, no. 20, pp. 13285–13289, 2009.

[28] R. E. Hansen, D. Roth, and J. R. Winther, "Quantifying the global cellular thiol-disulfide status," *Proceedings of the National Academy of Sciences of the United States of America*, vol. 106, no. 2, pp. 422–427, 2009.

[29] J. Ying, N. Clavreul, M. Sethuraman, T. Adachi, and R. A. Cohen, "Thiol oxidation in signaling and response to stress: detection and quantification of physiological and pathophysiological thiol modifications," *Free Radical Biology and Medicine*, vol. 43, no. 8, pp. 1099–1108, 2007.

[30] Y. O. Saw, M. Salim, J. Noirel, C. Evans, I. Rehman, and P. C. Wright, "ITRAQ underestimation in simple and complex mixtures: "the good, the bad and the ugly"," *Journal of Proteome Research*, vol. 8, no. 11, pp. 5347–5355, 2009.

[31] N. A. Karp, W. Huber, P. G. Sadowski, P. D. Charles, S. V. Hester, and K. S. Lilley, "Addressing accuracy and precision issues in iTRAQ quantitation," *Molecular and Cellular Proteomics*, vol. 9, no. 9, pp. 1885–1897, 2010.

[32] S. W. Fan, R. A. George, N. L. Haworth, L. L. Feng, J. Y. Liu, and M. A. Wouters, "Conformational changes in redox pairs of protein structures," *Protein Science*, vol. 18, no. 8, pp. 1745–1765, 2009.

[33] S. Barranco-Medina, J. J. Lázaro, and K. J. Dietz, "The oligomeric conformation of peroxiredoxins links redox state to function," *FEBS Letters*, vol. 583, no. 12, pp. 1809–1816, 2009.

# 20

# Functional Proteomic Profiling of Phosphodiesterases Using SeraFILE Separations Platform

**Amita R. Oka, Matthew P. Kuruc, Ketan M. Gujarathi, and Swapan Roy**

*ProFACT Proteomics Inc., 1 Deer Park Drive, Suite M, Monmouth Junction, NJ 08852, USA*

Correspondence should be addressed to Matthew P. Kuruc, mkuruc@profactproteomics.com

Academic Editor: Winston Patrick Kuo

Functional proteomic profiling can help identify targets for disease diagnosis and therapy. Available methods are limited by the inability to profile many functional properties measured by enzymes kinetics. The functional proteomic profiling approach proposed here seeks to overcome such limitations. It begins with surface-based proteome separations of tissue/cell-line extracts, using SeraFILE, a proprietary protein separations platform. Enzyme kinetic properties of resulting subproteomes are then characterized, and the data integrated into proteomic profiles. As a model, SeraFILE-derived subproteomes of cyclic nucleotide-hydrolyzing phosphodiesterases (PDEs) from bovine brain homogenate (BBH) and rat brain homogenate (RBH) were characterized for cAMP hydrolysis activity in the presence (challenge condition) and absence of cGMP. Functional profiles of RBH and BBH were compiled from the enzyme activity response to the challenge condition in each of the respective subproteomes. Intersample analysis showed that comparable profiles differed in only a few data points, and that distinctive subproteomes can be generated from comparable tissue samples from different animals. These results demonstrate that the proposed methods provide a means to simplify intersample differences, and to localize proteins attributable to sample-specific responses. It can be potentially applied for disease and nondisease sample comparison in biomarker discovery and drug discovery profiling.

## 1. Introduction

Proteomic profiling based on enzyme activity is assuming significance in drug discovery as it becomes possible to profile selectivity of drugs and their mechanism of action [1]. Such an approach focuses on protein function, an aspect which has been missing from expression proteomics [1]. A functional proteomic profiling approach has the potential not only to help identify targets for diagnosis and therapy [2], specifically in personal medicine [3, 4], but also to reveal the underlying mechanisms of action of disease-sustaining proteins [5].

Methods for global analysis of protein expression and function, including liquid chromatography with mass spectrometry (MS) for shotgun analysis [6, 7], yeast two-hybrid methods [8], and protein microarrays [9], have been crucial in developing the field of proteomics, but they do not provide an accurate assessment of functional states of proteins in cells and tissues [10]. Activity-based protein profiling (ABPP) was first demonstrated for serine hydrolyses [11] and has

now been applied to other enzyme classes such as kinases, phosphatases, and histone deacetylates [10, 12]. ABPP typically uses active site-directed covalent probes to interrogate specific subsets (families) of enzymes in complex proteomes to provide a quantitative assessment of the functional state of individual enzymes in the family [10]. The probe-bound enzymes can be visualized with SDS-PAGE or purified using affinity tools for peptide or labeling site identification with MS [10]. Although this approach is promising, it is limited by the availability of suitable synthetic probes. Also, while ABPP categorizes the active site in enzymes, it does not measure the functional kinetics of enzymes and therefore can be considered only as an indirect measure of protein function.

This article proposes a novel approach for localization of a functional enzyme. It forms the central component of the workflow strategy, which has the potential to identify functional biomarkers from natural cellular sources. The proposed method would fill an unmet need for research in drug response and biomarker discovery for investigations in natural cellular source environments. The physiological relevance of working with natural cellular sources is especially significant for discovery, which targets proteins whose function may be altered by post-translational modification, noncovalent regulatory factors or splice variants. Such an approach may help to reconcile data from high-throughput screening of recombinant proteins to natural cellular sources. It is anticipated that select subproteomes will be subjected to downstream characterization by liquid chromatography mass spectrometry (LC-MS), and other suitable identification methods in common use, so as to annotate sequence and structure to function.

While the term functional proteomics encompasses a variety of phenotypic descriptions of known or measurable functional consequences including cellular response to stimuli [13] and binding interactions [14, 15], etc., the model approach reported herein is limited to characterizing enzyme kinetic properties.

The proposed profiling strategy starts with subfractionation of complex proteomes using SeraFILE [16] (USPTO 20040106131, ProFACT Proteomics, Monmouth Junctions, NJ, USA). This proprietary protein separations platform is configured as a surface library with associated interrogation methods designed to retain bioactivity of the samples. As a result, subproteome pools obtained after SeraFILE separations can be characterized for their enzyme activity properties (e.g., enzyme activity with and without inhibitors, activators, or cosubstrates). Then, a collective functional profile of the original proteome is generated as an integrated profile of the functional properties of the characterized subproteomes. This approach provides multiple data points to characterize and compare samples, thereby increasing the robustness and reliability of analysis. It also allows localization of proteins responsible for sample-specific responses. This profiling method can compare one proteome sample to another (intersample analysis, e,g., tissue type versus tissue type, or normal versus diseased tissue) and can compare different subproteomes of the same complex proteome (intrasample analysis).

The cyclic nucleotide phosphodiesterases (PDEs) enzyme family has been used in this study as a model class of proteins to demonstrate the proposed strategy. PDEs are enzymes that hydrolyze the second messenger adenosine 3′, 5′-cyclic monophosphate (cAMP) or guanosine 3′, 5′-cyclic monophosphate (cGMP), or both. These small molecules along with other nucleotides, lipids, and ions function as secondary messengers [17]. The second messenger cAMP mediates a wide variety of actions of hormones and neurotransmitters and influences cell growth, differentiation, survival, and inflammatory processes [18]. Class I PDEs (found in protozoa and metazoa) are cAMP specific (PDE4, 7 and 8) or cGMP specific (PDE5, 6 and 9) or can hydrolyze both cAMP and cGMP (PDE1, 2, 3, 10, and 11) [17, 19, 20]. A comprehensive review of PDEs can be found in [17, 19, 21].

PDEs are widely acknowledged and explored as drug targets in pulmonary, neurodegenerative, and vascular diseases, and in diabetes, osteoporosis, cancer, rheumatoid arthritis, and depression [22]. Inhibitors of PDE5 and PDE3

are already in clinical use [23], but numerous other PDE inhibitors have not been used for therapeutic purposes due to side effects such as nausea and emesis [24]. The proposed approach to proteomic profiling is guided by the principle that, by discriminating and characterizing PDE variants in natural sources, greater disease-specific therapeutic inhibition/activation can be achieved along with a better understanding of disease pathway dynamics.

This research article demonstrates functional proteomic profiling of cAMP-hydrolyzing phosphodiesterases from bovine and rat brains. Although earlier studies have documented the presence of different types of PDEs in rat and bovine brains, a comprehensive comparative profile of PDE proteomes based on function and content/identity has not been established. It is known that bovine brain exhibits calmodulin-activated PDE activity (PDE1), as well as PDE2, and PDE4 activity [25, 26]. The cAMP hydrolysis activity of PDEs in bovine brain can be stimulated [27] or inhibited [28] by cGMP. Studies on rat brain have identified calmodulin-stimulated PDEs [29–31] (PDE1), as well as PDE4 isoforms [32–34].

SeraFILE was first applied for fractionation of each brain homogenate (sample proteome) into subproteomes, in order to reduce the complexity of PDEs in the sample proteome. Then, these subproteomes were interrogated for cAMP hydrolysis activity in the presence and absence of cGMP. cGMP is another substrate of PDEs and is used as a challenge condition in these experiments. The results were compiled into a signature profile of cAMP hydrolysis characteristics of each sample proteome, defined as an integrated profile of characteristics of SeraFILE-generated subproteomes. The hypothesis was that SeraFILE and associated interrogation methods would generate distinct profiles of enzyme catalyzed cAMP hydrolysis-activities from bovine brain and rat brain homogenates because these are different mammalian species.

## 2. Materials and Methods

*2.1. SeraFILE Surfaces.* The SeraFILE inventions [16] encompass the surface characteristics and protocols suitable for differential proteomic fractionation. Each surface architecture was designed to have moderate binding capacity and was prepared with Nugel Epoxy (Biotech Support Group Inc., Monmouth Junction, NJ, USA). The epoxy-coated silica was modified by reacting it with different ligands to generate unique surfaces selectivities and was based on the premise that important ligand protein interactions include hydrogen bonds, ionic interactions (salt bridges), hydrophobic interactions and ring structures. Table 1 illustrates the differences in the properties of the surfaces in the library; however for proprietary protection, details of the surface chemistries remain undisclosed.

An initial screen of 13 surfaces from the library (Table 1) and one underivatized control was performed. Further study was limited to a set of five surfaces (A, B, D, M, and N) from the surface library because the subproteomes obtained from these surfaces had the most distinguishing characteristics (enzyme activity and its response to

TABLE 1: Mixed-mode properties of SeraFILE surface structures[a]. Table shows potential numbers of hydrogen bond donor/acceptor groups, numbers of cationic/anionic groups, and number of ring structures in the surface ligands, along with relative hydrophobicity of the ligands.

| | Surface | Hydrogen bond | | Cationic groups | Anionic groups | Relative hydrophobicity[b] | Rings |
|---|---|---|---|---|---|---|---|
| | | Donor groups | Acceptor groups | | | | |
| Surfaces used in the study | A | 1 | 2 | | 1 | 2 | |
| | B | 1 | 6 | | 3 | 2 | 1 |
| | D | 1 | 4 | | 2 | 2 | 1 |
| | M | Multipolymer | | | | 1 | |
| | N | | | 1 | | 1 | |
| Surfaces initially screened, but not used in the study | PN | 3 | | 3 | | 3 | 1 |
| | E | | | Multipolymer | | 3 | |
| | AP | | | | Multipolymer | 1 | |
| | AM | 1 | 2 | 1 | | 1 | |
| | S | | | | Multipolymer | 5 | Multipolymer |
| | F | | | 1 | | 4 | |
| | C | 1 | 2 | | 1 | 5 | 1 |
| | PL | | | 1 | | 3 | |
| | PA | 1 | 1 | | 1 | 4 | 1 |
| | PC | | | 1 | | 5 | 1 |

[a] In cases of polymers, only predominant effect is considered.
[b] Scale 1–5: low-high.

rolipram/vinpocetine/calmodulin, protein concentrations, and SDS profile, data not shown).

*2.2. Preparation of Brain Homogenates.* Rat brain homogenate (RBH) and bovine brain homogenate (BBH) were supplied by Lampire Biologicals (Pipersville, PA, USA). Whole bovine or rat brain was homogenized in a prechilled blender using 100 mL of extraction buffer for every 50 g of brain tissue. Extraction buffer for BBH was 0.1 M Tris, 2 mM EDTA, and pH 7.5, and for RBH it was 1 mM EDTA, 10 mM HEPES, and pH 7.4. Each extraction buffer was made with protease inhibitor cocktail (Roche, Indianapolis, IN, U.S.A.). Homogenized brain-buffer mixtures were centrifuged at 4°C, and the supernatant was used for the experiments.

*2.3. Brain Homogenate Pretreatment (Clarification).* RBH and BBH samples were mixed with Cleanascite (Biotech Support Group, Monmouth Junction, NJ, U.S.A.) in a 1 : 16 ratio of Cleanascite-to-homogenate, to remove lipids and particulates. Clarified homogenates were obtained by following mixing and centrifugation steps as given in the manufacturer's protocol.

*2.4. Sample Separation.* The pretreated homogenates were each subjected to separation by five SeraFILE surfaces (A, B, D, M, and N) [35–38]. For separation of each homogenate, 50 mg of each surface contained in a Spin-X tube (Corning Inc., Corning, NY, U.S.A.) was equilibrated with binding buffer (0.05 M HEPES, 1 mM MgCl₂, and pH 6.5). Clarified BBH and RBH were diluted in the binding buffer, to pH 6.5-6.6, and 200 μL of each diluted homogenate (load, 1.16 mg of total protein) was added separately to each of the five surfaces, mixed for 10 mins, and then centrifuged. (Note that the total protein amounts used for SeraFILE separations were based on the sensitivity of the cAMP hydrolysis assay used in our experiments for downstream analysis. The SeraFILE methodology is nevertheless amenable to protocols that can use μg amounts of protein loads). The flowthrough was collected as the 1st SeraFILE fraction, represented as subproteomes A1, B1, D1, M1, and N1, from surfaces A, B, D, M, and N, respectively. The proteins bound on the surfaces were eluted with 200 μL of elution buffer (0.05 M HEPES, 1 mM MgCl₂, 0.5 M NaCl, and pH 8.0) using mixing and centrifugation steps as above. The flow-through collected in this process was the 2nd SeraFILE fraction, represented as subproteomes A2, B2, D2, M2, and N2, from surfaces A, B, D, M, and N, respectively. Mixing steps were performed using a MixMate (Eppendorf, Hauppauge, NY, U.S.A.) at 1150 rpm following an initial pulse of mixing on a vortex mixer. Centrifugation steps were performed using a tabletop centrifuge at 16873 rcf for 3 mins. Each brain homogenate, bovine and rat, was used for separations in triplicates.

*2.5. cAMP Hydrolysis Activity Assays and Protein Assays.* Activity of cAMP hydrolysis in each subproteome was measured using a real-time kinetic assay [39, 40]. This assay links cAMP hydrolysis to NADH oxidation using coupling enzymes (adenylate kinase, pyruvate kinase, and lactate dehydrogenase), and NADH loss can be measured at 340 nm. For each assay, a mixture of reaction buffer and coupling enzymes was equilibrated at room temperature for 16 mins (stage I). Then, subproteomes were each individually added to the reaction mix, and loss in absorbance was measured for 16 mins (stage II). Finally, substrate cAMP or a mix of cAMP and cGMP was added to the assay, and the loss in absorbance was measured as above (stage III). Final concentrations of assay components were as follows: 9 mM $MgCl_2$, 0.46 mM $CaCl_2$, 46 mM KCl, 46 mM HEPES, 1 mM phosphoenolpyruvate, 46 $\mu$M ATP, 0.4 $\mu$M NADH, 50 $\mu$M cAMP, 0.8 units pyruvate kinase, 4 units lactate dehydrogenase, and 0.06 units adenylate kinase (Sigma, St. Louis, MO, U.S.A.) with 6.25 $\mu$L of each enzyme sample. The cAMP hydrolysis activity of each sample was measured as the basal activity (in the absence of cGMP) and as challenged activity (in presence of 25 $\mu$M cGMP or 50 $\mu$M cGMP). Final volume of the assay was 0.1 mL. Volume-normalized enzyme assays were performed on each replicate subproteomes in a 96-well format using a Multiskan MMC346 plate reader (Thermo Scientific, Hudson, NH, U.S.A.).

To measure cAMP hydrolysis activity in the unfractionated brain homogenates, the clarified homogenates were diluted with binding buffer to obtain 5-6 dilutions of each homogenate, which were then used for the assays as described above.

Protein content of all RBH and BBH proteomes and subproteomes was measured using a BCA assay kit (Pierce, Rockford, IL, U.S.A.). Replicate subproteomes were pooled before protein analysis.

*2.6. Calculations.* (a) The cAMP hydrolysis activity of each sample was calculated as follows:

$$\text{Enzyme activity } \left(\text{nmoles mL}^{-1}\text{min}^{-1}\right)$$

$$= \frac{\text{Path length correction factor} \times \text{corrected PDE rate } \left(\text{min}^{-1}\right) \times \text{reaction volume (mL)} \times \text{dilution factor}}{\text{Molar absorption coeffcient } \left(\text{M}^{-1}\text{cm}^{-1}\right) \times \text{sample volume (mL)}}, \tag{1}$$

where corrected PDE rate is $\Delta A_{340\,nm}(\text{min}^{-1})$ of stage III $-\Delta A_{340\,nm}(\text{min}^{-1})$ stage II, molar absorption coefficient is $1.25 \times 10^4\,\text{M}^{-1}\,\text{cm}^{-1}$, dilution factor is 1, and path length correction (to 10 mm) is 3.16.

(b) The % change in cAMP hydrolysis activity of each sample was calculated as follows:

$$\% \text{ Change in cAMP hydrolysis activity} = \frac{(\text{Challenged enzyme activity} - \text{Basal enzyme activity}) \times 100}{\text{Basal enzyme activity}}, \tag{2}$$

where challenged enzyme activity is cAMP hydrolysis activity in presence of cGMP, and basal enzyme activity is cAMP hydrolysis activity in absence of cGMP.

## 3. Results and Discussion

*3.1. Sample Pretreatment.* Cleanascite [41–44], a solid-phase, nonionic adsorbent for lipid removal, significantly improved the clarity of brain homogenates and eliminated clogging of the surfaces during the SeraFILE process. A 1 : 16 ratio of Cleanascite to untreated BBH gave optimal results, with minimum loss of cAMP hydrolysis activity. Consequently, the same ratio of Cleanascite to brain homogenate was used for RBH clarification.

*3.2. cAMP Hydrolysis Activity in the Sample Proteomes.* Enzyme activity and protein analysis of the unfractionated brain homogenates showed that the mean specific activity of clarified RBH and BBH was comparable, between 8 and 8.8 units/mg, measured at 50 $\mu$M cAMP concentration.

*3.3. Effect of cGMP on cAMP Hydrolysis of Unfractionated Brain Homogenates.* A comparison of the basal and challenged cAMP hydrolysis activities in the dilutions of each homogenate is shown in Figure 1. As expected, increase in activity (basal and cGMP challenged) was observed with increasing concentration of clarified homogenates. The comparison also shows that at relatively lower concentrations of the homogenates, cGMP inhibited cAMP hydrolysis, while, at higher concentrations, cAMP hydrolysis activity increased. In addition, the change in cAMP hydrolysis activity was more pronounced in the presence of 50 $\mu$M cGMP than 25 $\mu$M cGMP. Specifically, at 50 $\mu$M cGMP (Figures 2(a) and 2(b)), the change from inhibition to activation of cAMP hydrolysis activity occurred above ∼1.5 mg/mL protein in RBH and above 4 mg/mL protein in BBH proteomes. Thus, it is a characteristic in the PDEs of RBH and BBH, that the effect of cGMP on cAMP hydrolysis is a function both of the concentration of the homogenate and of the concentration of cGMP.

*3.4. SeraFILE-Derived Subproteomes and Generation of Enzyme Activity Profiles.* Buffer-diluted, protein-normalized, and clarified RBH and BBH samples were used for separations. Each subproteome obtained was analyzed for protein content and cAMP hydrolysis activity under basal and cGMP-challenged conditions, and then the change in cAMP hydrolysis activity was calculated.

(a)

(b)

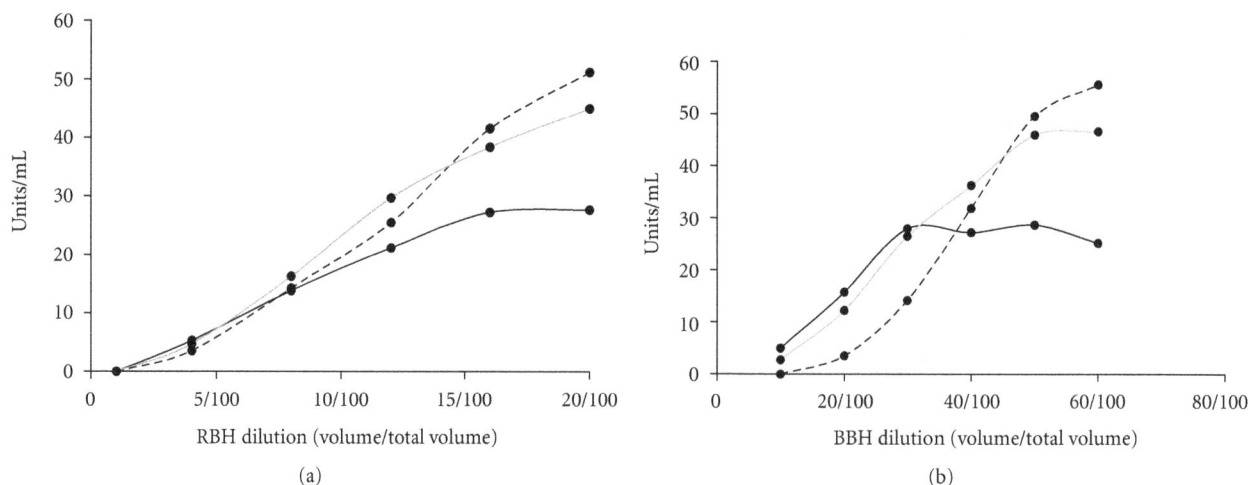

FIGURE 1: The cAMP hydrolysis activity in clarified rat brain homogenate (RBH), (a), and bovine brain homogenate (BBH), (b), proteomes. The cAMP hydrolysis activity was measured by using dilutions of the clarified homogenates in the absence of cGMP (solid black) or in presence of 25 $\mu$M cGMP (solid gray) or 50 $\mu$M cGMP (dashes).

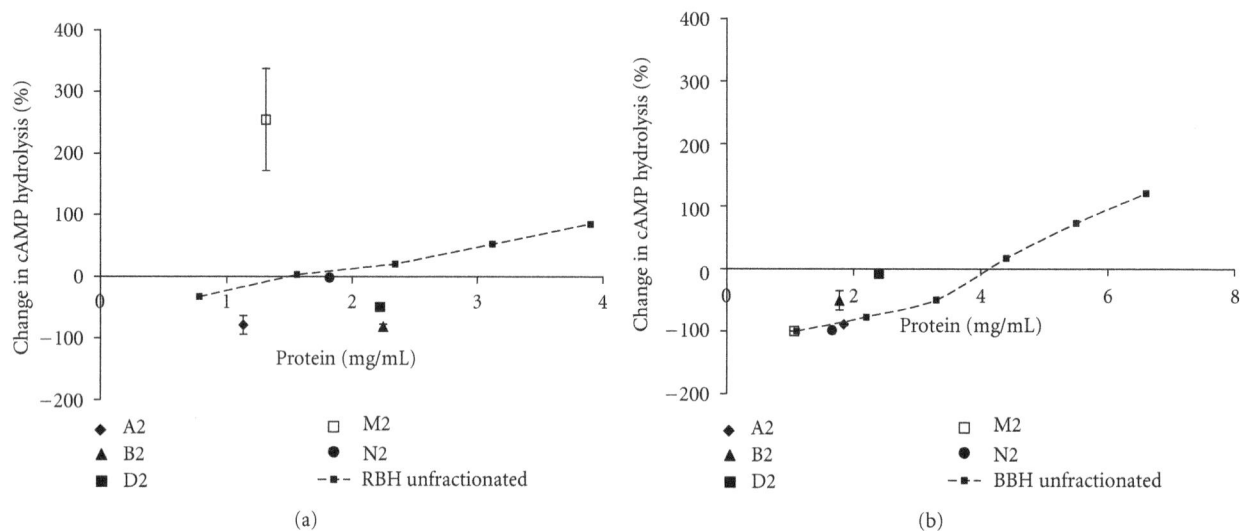

(a)

(b)

FIGURE 2: Relationship between change in cAMP hydrolysis activity and protein content of the unfractionated brain homogenates and SeraFILE-generated subproteomes. The figure shows change in cAMP hydrolysis of rat brain homogenate (RBH) and generated subproteomes, (a), and bovine brain homogenate (BBH) and generated subproteomes, (b). $X$-axis represents protein concentration. $Y$-axis represents percentage change in cAMP hydrolysis activity due to the challenge of 50 $\mu$M cGMP, as compared to basal cAMP hydrolysis activity. A2, B2, D2, M2, and N2 represent subproteomes from the homogenates.

To ensure that the observed properties of the subproteomes were not an effect of the dilution of the sample proteome, the properties of RBH and BBH proteomes and their respective SeraFILE subproteomes were compared (Figures 2(a) and 2(b)). The data in Figure 2(a) show that not all RBH subproteomes follow the activity versus protein content relationship of the RBH proteome. Similar observations were made with respect to BBH (Figure 2(b)). These outliers indicate that SeraFILE produces differential subproteomes. Data show that some subproteomes do share the activity versus protein content relationship of the sample proteome, likely indicating a comparable distribution of cAMP hydrolyzing PDEs to total proteins.

To generate an intersample functional proteomic profile of RBH and BBH proteomes, the change in cAMP hydrolysis activity of each subproteome due to cGMP challenge was calculated and plotted as shown in Figure 3. A functional proteomic profile of the brain homogenates in these experiments is defined by the collective response of individual SeraFILE subproteomes to cGMP challenge. A comparison between the functional profiles of the two homogenates (Figure 3) shows that, overall, these two profiles have a similar pattern (i.e., % change in cAMP hydrolysis is similar, positive or negative, in comparable fractions of the two homogenates). However, a major difference is found in subproteome M2 of RBH and BBH (refer to Figure 3(a) versus Figure 3(e),

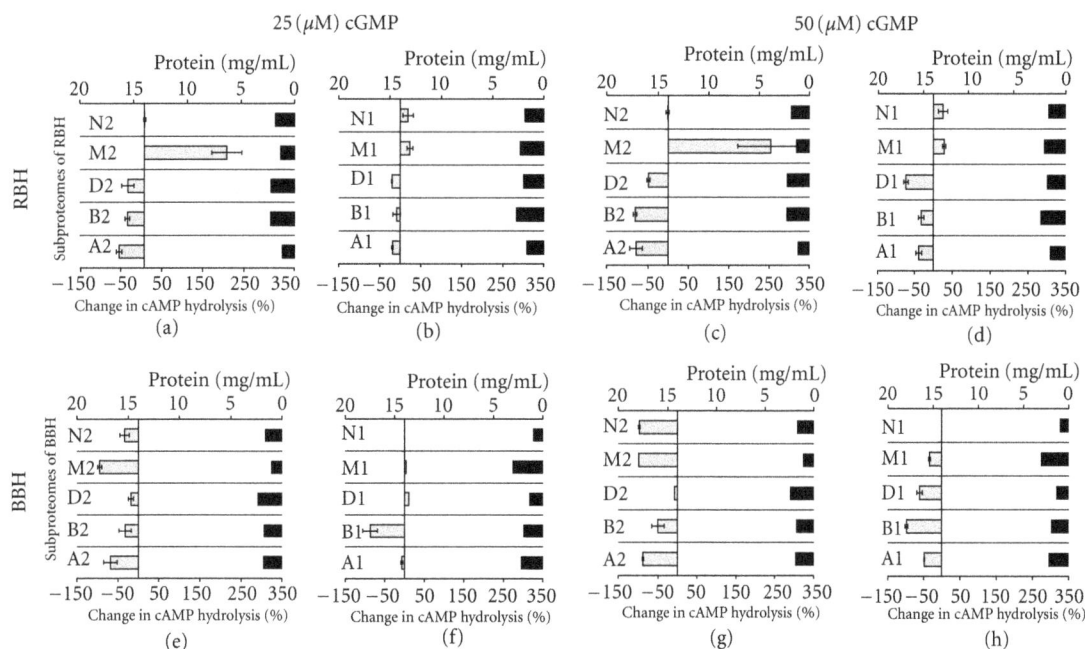

FIGURE 3: Comparison of functional profiles of rat brain homogenate (RBH) and bovine brain homogenate (BBH). Figure shows percentage change in cAMP hydrolysis activity in each subproteome of RBH ((a)–(d)) and BBH ((e)–(h)) due to cGMP challenge of $25\,\mu M$ or $50\,\mu M$. In each panel, the primary X-axis represents the change in cAMP hydrolysis due to cGMP, the secondary X-axis represents protein concentration, and the Y-axis represents subproteomes. The pair of subproteomes A1 and A2 (and similarly others) was derived sequentially from the same surface in the library as described in the protocol. Grey bars represent mean percent change ($n = 3$), in cAMP hydrolysis of each subproteome due to presence of cGMP. Error bars represent ($\pm 1$) std. Black bars represent protein concentration of each subproteome.

and Figure 3(c) versus Figure 3(g). Subproteome M2 of RBH shows over 190% increase in cAMP hydrolysis in the presence of cGMP (both $25\,\mu M$ and $50\,\mu M$), while subproteome M2 of BBH shows over 90% decrease in cAMP hydrolysis in the presence of cGMP (both $25\,\mu M$ and $50\,\mu M$). Thus, subproteome M2 is a differentiating feature of this intersample analysis and therefore can be used for further sample characterization.

These model data demonstrate that our proposed methods of protein separation generate subproteomes that are sufficiently differentiated for intersample functional analysis. As a result, these methods can be potentially applied to effectively differentiate functional properties of complex proteomes and can be used to localize subset of proteins attributable to sample-specific responses. The localized proteins can then be used for further analysis, characterization, and subsequent MS identification (gene sequence annotation/reconciliation).

SeraFILE separations use mild-to-moderate elution conditions with buffers like phosphate or HEPES that are commonly used in the laboratory. In addition, the solid-phase surface ($50\,\mu$ derivatized silica) can be easily removed by filtration. Thus SeraFILE separations methods do not introduce substances like urea or SDS that may restrict downstream compatibility with existing reporting assay and LC-MS detection methodologies [45]. Therefore, the proposed methodology is considered to have a broad scope of applicability within the pathway to identification and can be potentially applied to profile narrowly defined

therapeutically important classes of enzymes such as Kinases or cyclic nucleotide phosphodiesterases.

Another important characteristic of SeraFILE separations methodology is its reproducibility at different protein loads. In separate experiments, surface separation of 0.25 mg to 1 mg protein per 50 mg of surface was shown to have only 10% variation (data not shown, [35]). The reproducibility in sample separation can be significant for heterogeneous samples of clinical origin.

In addition to separations, SeraFILE can also be applied for enrichment of proteins. Incremental increase in pH was applied for enrichment of alkaline phosphatase (data not shown, [35]) with an enrichment factor up to 20X.

The applications of SeraFILE separations can be based on two basic types of sample and data analysis of (i) intersample analysis whereby samples such as tissues, cellular models, or biofluids are compared and contrasted and (ii) intrasample analyses, or differential analysis within a sample whereby the subproteomes are monitored with respect to a challenging modulation condition such as in drug response profiling. It is envisioned that these will complement one another for personalized medicine applications.

Inter-sample analysis of complex proteomes, as demonstrated, potentially applies to disease and nondisease comparisons, to identify differences in samples by compartmentalizing the most distinctive subproteomes associated with disease. Deeper characterization of these fractions with enrichment (e.g., with pH optimization of SeraFILE separations, or with conventional separations 2DE or HPLC), followed by LC-MS analyses, can help identify prospective biomarkers. It is important to recognize that any biomarker panel selected in this context would require more characterization, with larger sample sets and statistical validation.

Intra-sample analysis, on the other hand, can be used to catalogue or index the effects of functional modulation of the daughter subproteomes. This will be especially valuable for establishing localized panels of proteins that are responsive to modulation with drug compounds, with the same caveats as the aforementioned inter-sample analyses.

The two data analysis strategies, profiling between samples, and cataloging within samples, are complementary insofar as molecular profiles that characterize and compartmentalize drug-responsive proteins from complex mixtures, can potentially, through coincident iterations with disease profiling, create a bulls-eye effect for drug repurposing.

We envision that, for the drug development industry, the proposed methods for localizing proteins with known functional attributes offer new resources for biomarker discovery, complementing conventional methods of identification and sequence annotation. For drug compounds, a challenge/response method, as described, can help address the problems of drug promiscuity and discern the subtleties of protein attributes; when the same or similar underlying sequences, have multiple conformations and functions, and when different sequences sometimes perform the same or similar function.

As a way to begin sifting through these biological complexities, a more efficient method to characterize protein function and corresponding modulation is now possible. Starting with the enrichment of prospective functional biomarkers in localized subproteomes, we suggest that structural and sequence relationships can be determined. Such an approach has the potential to provide new and useful service to biomarker discovery and personalized medicine.

## Conflict of Interests

The authors declare that there is no conflict of interests.

## Acknowledgments

The authors would like to thank Dr. Miles Houslay, University of Glasgow, for his guidance in the initial development of this project. The authors are also thankful to Dr. Faribourz Payvandi, NeoloMed BioSciences, for providing facility for sample treatment and for his continuous guidance throughout the project. The authors would like to acknowledge work of Dr. Meghan Tierney on reproducibility of separations and enrichment of proteins with SeraFILE library. This work was funded in part by NJCST Post-Doctoral Fellowship Grant (10-2042-014-54) to A. R. Oka for work at ProFACT Proteomics Inc. (2009-2010), and in part by the IRS Qualifying Therapeutic Discovery Grant to ProFACT Proteomics Inc. (2009-2010).

## References

[1] P. Mallick and B. Kuster, "Proteomics: a pragmatic perspective," *Nature Biotechnology*, vol. 28, no. 7, pp. 695–709, 2010.

[2] S. Pitteri and S. Hanash, "A systems approach to the proteomic identification of novel cancer biomarkers," *Disease Markers*, vol. 28, no. 4, pp. 233–239, 2010.

[3] A. H. J. Danser, W. W. Batenburg, A. H. van den Meiracker, and S. M. Danilov, "ACE phenotyping as a first step toward personalized medicine for ACE inhibitors. Why does ACE genotyping not predict the therapeutic efficacy of ACE inhibition?" *Pharmacology and Therapeutics*, vol. 113, no. 3, pp. 607–618, 2007.

[4] S. M. Danilov, I. V. Balyasnikova, R. F. Albrecht, and O. A. Kost, "Simultaneous determination of ACE activity with 2 substrates provides information on the status of somatic ACE and allows detection of inhibitors in human blood," *Journal of Cardiovascular Pharmacology*, vol. 52, no. 1, pp. 90–103, 2008.

[5] M. Sanchez-Carbayo, "Antibody array-based technologies for cancer protein profiling and functional proteomic analyses using serum and tissue specimens," *Tumor Biology*, vol. 31, no. 2, pp. 103–112, 2010.

[6] S. P. Gygi, B. Rist, S. A. Gerber, F. Turecek, M. H. Gelb, and R. Aebersold, "Quantitative analysis of complex protein mixtures using isotope-coded affinity tags," *Nature Biotechnology*, vol. 17, no. 10, pp. 994–999, 1999.

[7] M. P. Washburn, D. Wolters, and J. R. Yates, "Large-scale analysis of the yeast proteome by multidimensional protein identification technology," *Nature Biotechnology*, vol. 19, no. 3, pp. 242–247, 2001.

[8] T. Ito, K. Ota, H. Kubota et al., "Roles for the two-hybrid system in exploration of the yeast protein interactome.," *Molecular & Cellular Proteomics*, vol. 1, no. 8, pp. 561–566, 2002.

[9] G. MacBeath, "Protein microarrays and proteomics," *Nature Genetics*, vol. 32, supplement 5, pp. 526–532, 2002.

[10] B. F. Cravatt, A. T. Wright, and J. W. Kozarich, "Activity-based protein profiling: from enzyme chemistry to proteomic chemistry," *Annual Review of Biochemistry*, vol. 77, pp. 383–414, 2008.

[11] Y. Liu, M. P. Patricelli, and B. F. Cravatt, "Activity-based protein profiling: the serine hydrolases," *Proceedings of the National Academy of Sciences of the United States of America*, vol. 96, no. 26, pp. 14694–14699, 1999.

[12] M. Bantscheff, D. Eberhard, Y. Abraham et al., "Quantitative chemical proteomics reveals mechanisms of action of clinical ABL kinase inhibitors," *Nature Biotechnology*, vol. 25, no. 9, pp. 1035–1044, 2007.

[13] D. Kültz, D. Fiol, N. Valkova, S. Gomez-Jimenez, S. Y. Chan, and J. Lee, "Functional genomics and proteomics of the cellular osmotic stress response in 'non-model' organisms," *Journal of Experimental Biology*, vol. 210, no. 9, pp. 1593–1601, 2007.

[14] R. D. Gietz, B. Triggs-Raine, A. Robbins, K. C. Graham, and R. A. Woods, "Identification of proteins that interact with a protein of interest: applications of the yeast two-hybrid system," *Molecular and Cellular Biochemistry*, vol. 172, no. 1-2, pp. 67–79, 1997.

[15] M. Fromont-Racine, J. C. Rain, and P. Legrain, "Toward a functional analysis of the yeast genome through exhaustive two- hybrid screens," *Nature Genetics*, vol. 16, no. 3, pp. 277–282, 1997.

[16] S. Roy, J. Krupey, and M. Kuruc, "Composition and methods for proteomic investigations," in *US Patent and Trade Mark*

*Office*, P. Proteomics, Ed., ProFACT Proteomics, Monmouth Junction, NJ, USA, 2003.

[17] M. Conti and J. Beavo, "Biochemistry and physiology of cyclic nucleotide phosphodiesterases: essential components in cyclic nucleotide signaling," *Annual Review of Biochemistry*, vol. 76, pp. 481–511, 2007.

[18] K. M. Torgersen, E. M. Aandahl, and K. Taskén, "Molecular architecture of signal complexes regulating immune cell function.," *Handbook of Experimental Pharmacology*, no. 186, pp. 327–363, 2008.

[19] J. A. Beavo, M. Conti, and R. J. Heaslip, "Multiple cyclic nucleotide phosphodiesterases," *Molecular Pharmacology*, vol. 46, no. 3, pp. 399–405, 1994.

[20] C. Lugnier, "Cyclic nucleotide phosphodiesterase (PDE) superfamily: a new target for the development of specific therapeutic agents," *Pharmacology and Therapeutics*, vol. 109, no. 3, pp. 366–398, 2006.

[21] J. Beavo, S. H. Francis, and M. D. Houslay, *Cyclic Nucleotide Phosphodiesterases in Health and Disease*, CRC Press, Boca Raton, Fla, USA, 2006.

[22] J. Beavo, S. H. Francis, and M. D. Houslay, *Cyclic Nucleotide Phosphodiesterases in Health and Disease*, CRC Press/Taylor & Francis, Boca Raton, Fla, USA, 2007.

[23] Y. H. Jeon, Y. S. Heo, C. M. Kim et al., "Phosphodiesterase: overview of protein structures, potential therapeutic applications and recent progress in drug development," *Cellular and Molecular Life Sciences*, vol. 62, no. 11, pp. 1198–1220, 2005.

[24] M. D. Houslay, P. Schafer, and K. Y. J. Zhang, "Keynote review: phosphodiesterase-4 as a therapeutic target," *Drug Discovery Today*, vol. 10, no. 22, pp. 1503–1519, 2005.

[25] R. K. Sharma, A. M. Adachi, K. Adachi, and J. H. Wang, "Demonstration of bovine brain calmodulin-dependent cyclic nucleotide phosphodiesterase isozymes by monoclonal antibodies," *Journal of Biological Chemistry*, vol. 259, no. 14, pp. 9248–9254, 1984.

[26] T. Kyoi, M. Oka, K. Noda, and Y. Ukai, "Phosphodiesterase inhibition by a gastroprotective agent irsogladine: preferential blockade of cAMP hydrolysis," *Life Sciences*, vol. 75, no. 15, pp. 1833–1842, 2004.

[27] S. Murashima, T. Tanaka, S. Hockman, and V. Manganiello, "Characterization of particulate cyclic nucleotide phosphodiesterases from bovine brain: purification of a distinct cGMP-stimulated isoenzyme," *Biochemistry*, vol. 29, no. 22, pp. 5285–5292, 1990.

[28] K. Sankaran, I. Hanbauer, and W. Lovenberg, "Heat-stable low molecular weight form of phosphodiesterases from bovine pineal gland," *Proceedings of the National Academy of Sciences of the United States of America*, vol. 75, no. 7, pp. 3188–3191, 1978.

[29] S. Kakiuchi, R. Yamazaki, Y. Teshima, and K. Uenishi, "Regulation of nucleoside cyclic 3':5' monophosphate phosphodiesterase activity from rat brain by a modulator and Ca$^{2+}$," *Proceedings of the National Academy of Sciences of the United States of America*, vol. 70, no. 12, 1973.

[30] J. A. Smoake, S. Y. Song, and W. Y. Cheung, "Cyclic 3',5' nucleotide phosphodiesterase. Distribution and developmental changes of the enzyme and its protein activator in mammalian tissues and cells," *Biochimica et Biophysica Acta*, vol. 341, no. 2, pp. 402–411, 1974.

[31] R. S. Hansen and J. A. Beavo, "Differential recognition of calmodulin-enzyme complexes by a conformation-specific anti-calmodulin monoclonal antibody," *Journal of Biological Chemistry*, vol. 261, no. 31, pp. 14636–14645, 1986.

[32] I. McPhee, L. Pooley, M. Lobban, G. Bolger, and M. D. Houslay, "Identification, characterization and regional distribution in brain of RPDE-6 (RNPDE4A5), a novel splice variant of the PDE4A cyclic AMP phosphodiesterase family," *Biochemical Journal*, vol. 310, no. 3, pp. 965–974, 1995.

[33] F. Ohsawa, M. Yamauchi, H. Nagaso, S. Murakami, J. Baba, and A. Sawa, "Inhibitory effects of rolipram on partially purified phosphodiesterase 4 from rat brains," *Japanese Journal of Pharmacology*, vol. 77, no. 2, pp. 147–154, 1998.

[34] M. Shepherd, T. McSorley, A. E. Olsen et al., "Molecular cloning and subcellular distribution of the novel PDE4B4 cAMP-specific phosphodiesterase isoform," *Biochemical Journal*, vol. 370, no. 2, pp. 429–438, 2003.

[35] ProFACT Proteomics, Reproducibility Of SeraFILE Derived Functional Proteomic Profiles, ProFACT Proteomics, http://www.profactproteomics.com/technical_notes.html, 2007.

[36] ProFACT Proteomics, Molecular profiling with SeraFILE, Nature Methods, Application Notes, ProFACT Proteomics, http://www.profactproteomics.com/technical_notes.html, 2008.

[37] ProFACT Proteomics, SeraFILE—A Biomarker and Drug Discovery Engine, ProFACT Proteomics, http://www.profactproteomics.com/serafile.html, 2004.

[38] ProFACT Proteomics, Functional Proteomic Signatures of the Ubiquitin/Proteasome Pathway, ProFACT Proteomics, http://www.profactproteomics.com/technical_notes.html, 2007.

[39] S. P. Chock and C. Y. Huang, "An optimized continuous assay for cAMP phosphodiesterase and calmodulin," *Analytical Biochemistry*, vol. 138, no. 1, pp. 34–43, 1984.

[40] A. B. Burgin, O. T. Magnusson, J. Singh et al., "Design of phosphodiesterase 4D (PDE4D) allosteric modulators for enhancing cognition with improved safety," *Nature Biotechnology*, vol. 28, no. 1, pp. 63–70, 2010.

[41] M. S. S. Alhamdani, C. Schröder, and J. D. Hoheisel, "Analysis conditions for proteomic profiling of mammalian tissue and cell extracts with antibody microarrays," *Proteomics*, vol. 10, no. 17, pp. 3203–3207, 2010.

[42] A. Farina, J. M. Dumonceau, J. L. Frossard, A. Hadengue, D. F. Hochstrasser, and P. Lescuyer, "Proteomic analysis of human bile from malignant biliary stenosis induced by pancreatic cancer," *Journal of Proteome Research*, vol. 8, no. 1, pp. 159–169, 2009.

[43] B. Chen, J. Q. Dong, Y. J. Chen et al., "Two-dimensional electrophoresis for comparative proteomic analysis of human bile," *Hepatobiliary and Pancreatic Diseases International*, vol. 6, no. 4, pp. 402–406, 2007.

[44] T. Z. Kristiansen, J. Bunkenborg, M. Gronborg et al., "A proteomic analysis of human bile," *Molecular and Cellular Proteomics*, vol. 3, no. 7, pp. 715–728, 2004.

[45] E. I. Chen, D. Cociorva, J. L. Norris, and J. R. Yates, "Optimization of mass spectrometry-compatible surfactants for shotgun proteomics," *Journal of Proteome Research*, vol. 6, no. 7, pp. 2529–2538, 2007.

# 21

# Optimization of an Efficient Protein Extraction Protocol Compatible with Two-Dimensional Electrophoresis and Mass Spectrometry from Recalcitrant Phenolic Rich Roots of Chickpea (*Cicer arietinum* L.)

**Moniya Chatterjee, Sumanti Gupta, Anirban Bhar, and Sampa Das**

*Division of Plant Biology, Bose Institute, Centenary Campus, P 1/12, CIT Scheme VII-M, Kankurgachi, West Bengal, Kolkata 700054, India*

Correspondence should be addressed to Sampa Das, sampa@bic.boseinst.ernet.in

Academic Editor: Paul P. Pevsner

Two-dimensional electrophoresis and mass spectrometry are undoubtedly two essential tools popularly used in proteomic analyses. Utilization of these techniques however largely depends on efficient and optimized sample preparation, regarded as one of the most crucial steps for recovering maximum amount of reliable information. The present study highlights the optimization of an effective and efficient protocol, capable of extraction of root proteins from recalcitrant phenolic rich tissues of chickpea. The widely applicable TCA-acetone and phenol-based methods have been comparatively evaluated, amongst which the latter appeared to be better suited for the sample. The phenol extraction-based method further complemented with sodium dodecyl sulphate (SDS) and pulsatory treatments proved to be the most suitable method represented by greatest spot number, good resolution, and spot intensities. All the randomly selected spots showed successful identification when subjected to further downstream MALDI-TOF and MS/MS analyses. Hence, the information obtained collectively proposes the present protein extraction protocol to be an effective one that could be applicable for recalcitrant leguminous root samples.

## 1. Introduction

Presence of intricate photosynthetic machinery, cell wall and other organelles, complex primary and secondary metabolic processes, and their cellular regulation adds to the complexity of functional biology of plants. In recent years, proteomics has become one of the most enthralling fields in molecular biology as it targets the molecular link in the information chain from protein to its coding sequence and its manifestation in the form of phenotype. In contrast to the relative ease of mRNA isolation, c-DNA synthesis and analysis, protein extraction presents numerous challenges due to its heterogeneous nature, structural complexity and instability. Such features dramatically complicate their extraction, solubilization, handling, separation, and ultimately identification. Moreover no technology currently exists that is

equivalent to PCR, which can amplify low abundance proteins [1].

The most critical step in any proteomic study is protein extraction and sample preparation. However, the difficulties involving plant protein extractions especially from roots are quite complicated as compared to other organisms. Root tissues are highly vacuolated with relatively low protein content. They are often rich in proteases, storage polysaccharides, lipids, phenolics and a broad array of secondary metabolites [2–4]. Such contaminants cause major obstacles for two-dimensional electrophoresis (2DE) resulting in horizontal and vertical streaking, smearing, and reduction in the number of distinctly resolved protein spots [5].

The present investigation deals with protein extraction from chickpea roots. Chickpea is the most important legume crop in India and ranks third in the world's list of important

Optimization of an Efficient Protein Extraction Protocol Compatible with Two-Dimensional Electrophoresis and Mass Spectrometry from Recalcitrant Phenolic Rich Roots of Chickpea (Cicer arietinum L.)

193

legumes. Its production is greatly hampered by different abiotic and biotic factors. Major yield loss is caused by root invading pathogens like *Sclerotium rolfsii* (collar rot), *Fusarium solani* (black root rot), *Thielaviopsis basicola* (black streak root rot), *Phytophthora* sp. (*Phytophthora* root rot), *Fusarium* sp. (*Fusarium* root rot), *Fusarium oxysporum* f.sp. ciceris (*Fusarium* wilt), and so forth. Hence, root proteins serve to be excellent target to study early signaling in plant-pathogen interaction involving root invading pathogens in particular.

Most common and basic protocols used for protein extraction from plant tissue are TCA-acetone and phenol-based extraction methods. TCA-acetone precipitation was initially developed by Damerval et al. [6]. This method increases the protein concentration and helps removing contaminants, although some polymeric contaminants are often coextracted. This appears as a problem with tissues that are rich in compounds such as soluble cell wall polysaccharides and polyphenols. Another method involves protein solubilization in phenol, with or without using SDS followed by precipitation with methanol and ammonium acetate and subsequent resolubilization in IEF (isoelectric focusing) sample buffer [5, 7, 8]. This method can efficiently generate protein extracts from resistant tissues such as wood [9], olive leaves [10], maize roots [11], and hemp roots [12], and so forth. Similar studies also suggested that phenol-based method reduces protein degradation during extraction and helps in solubilizing membrane proteins and glycoproteins [5, 13]. However, requirement of extensive time appears to be the major limitation of this method. Thus, these extraction protocols demand optimization for particular organisms, tissue or cell compartment.

In current study attempts were made to optimize the phenol SDS method along with sonication for protein extraction from small amount of recalcitrant chickpea roots. Evaluations of other different extraction methods were also done in comparison to the optimized phenol SDS sonication method and its compatibility with high throughput method like mass spectrometry analysed.

## 2. Materials and Methods

*2.1. Plant Material.* Experiments were performed using chickpea seeds (JG62) obtained from International Crops Research Institute for the Semi Arid Tropics (ICRISAT), Patancheru, Andhra Pradesh, India. Seeds sown in a mixture of sand and synthetic soil (1:1) were allowed to grow in natural green house conditions suited for the crop [14]. Roots of 15–20 days old seedlings were thoroughly washed, frozen in liquid nitrogen, and stored at −80°C prior to extraction of protein.

*2.2. Extraction Protocols*

*(A) TCA-Acetone Precipitation Method.* TCA-acetone precipitation was carried out according to Damerval et al. with some modifications [6]. One gram of root material was ground in a precooled mortar in the presence of liquid nitrogen. Approximately 100–150 mg of ground tissue powder

was precipitated overnight with freshly prepared 2 mL of 10% TCA, 0.07% β-mercaptoethanol in cold acetone. Following precipitation the set was centrifuged at 10,000 g for 15–20 min at 4°C and the supernatant discarded. The obtained pellet was rinsed twice in ice-cold acetone with 0.07% β-mercaptoethanol. An additional modification was introduced between the rinsing steps by incubating the sample for 60 min at −20°C [15]. The pellet was air dried, resuspended in 100 μL sample buffer (8 M Urea, 2% CHAPS, 50 mM DTT, 0.2% Biolyte 3/10 Ampholyte, 0.001% Bromophenol Blue) (Biorad), and vortexed for 1 hour at room temperature. The supernatant was used for downstream analyses (Figure 1).

*(B) Phenol Extraction Method.* Phenol extraction method was used both singly and in combinations of extraction buffer and SDS along with variations of with and without sonication (Figure 1).

*(B.1) Phenol-SDS Buffer Extraction with Sonication (PSWS).* Phenol extraction of proteins was carried out as described by Hurkman and Tanaka [7] in the presence of SDS buffer designated as phenol-SDS extraction by Wang et al. [10]. One gram of root tissue was ground in a mortar in the presence of liquid nitrogen and extracted with 3 mL of SDS buffer (30% sucrose, 2% SDS, 0.1 M Tris-Cl, 5% β-mercaptoethanol, and 1 mM phenyl methyl sulfonyl fluoride (PMSF), pH 8.0). The extract was sonicated 6 times for 15 seconds at 60 amps. Following sonication 3 mL of Tris buffered phenol was added to the mixture and vortexed for 10 mins at 4°C. The set was centrifuged at 8,000 g for 10 min at 4°C, phenolic phase collected and reextracted with 3 mL SDS buffer and shaken for 3–10 min. Centrifugation was further repeated using the same settings, phenolic phase collected and precipitated overnight with four volumes of 0.1 M ammonium acetate in methanol at −20°C. Precipitate obtained by centrifugation at 10,000 g for 30 min at 4°C was washed thrice with cold 0.1 M ammonium acetate and finally with cold 80% acetone. The pellet was dried and resuspended in 100 μL sample buffer (Biorad) and used for further analyses.

*(B.2) Phenol-SDS Buffer Extraction without Sonication (PSWOS).* This method was same as mentioned in case of PSWS only with the elimination of the sonication step.

*(B.3) Phenol-Extraction Buffer with Sonication (PEWS).* One gram of frozen root tissue was homogenized in liquid nitrogen and was extracted with ice-cold extraction buffer (500 mM Tris-Cl, 50 mM EDTA, 700 mM sucrose, 100 mM KCl, pH 8.0) at 4°C. The extract was sonicated 6 times at 60 amps for 15 sec and further extracted with Tris buffered phenol as described in PSWS.

*(B.4) Phenol-Extraction Buffer without Sonication (PEWOS).* Protein extraction was carried out in the same way as described in case of PEWS with elimination of the sonication step.

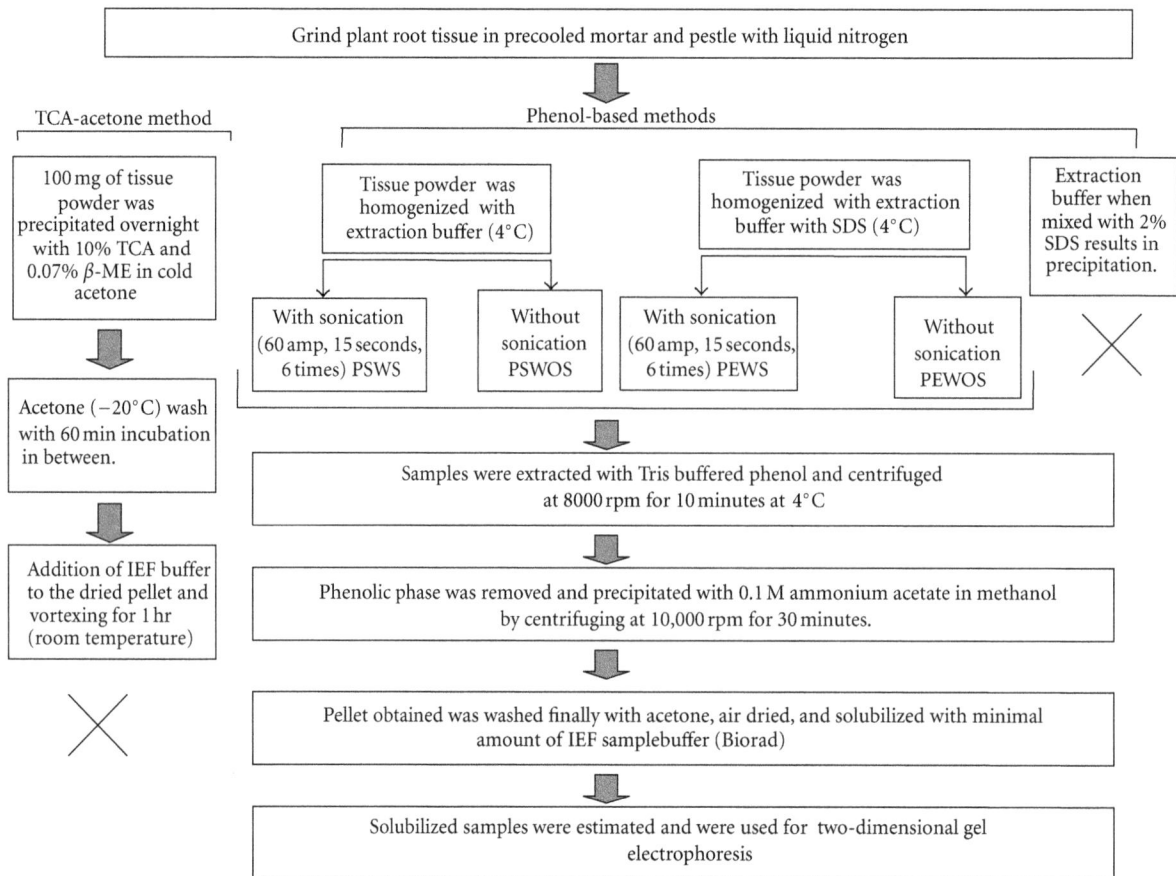

FIGURE 1: Schematic representation of extraction of protein from chickpea roots using TCA-acetone and phenol based extraction protocols.

*(B.5) Phenol-Extraction Buffer with SDS.* This protocol was similar to phenol extraction method. The buffer composition was the same as mentioned in PEWS pH 8.0 with 2% SDS as an additional component. However appearance of a white precipitate following SDS addition to the basal phenol extraction buffer prevented further processing of the sample using this buffer (Figure 1).

*2.3. Protein Quantification.* Protein concentrations were quantified using the Bradford protein assay method using BSA as a standard [16].

*2.4. Two-Dimensional Electrophoresis (2DE).* IPG strips (11 cm, 3–10 nonlinear, Readystrip, Biorad) were passively rehydrated overnight with rehydration sample buffer containing 250 $\mu$g of isolated protein. IEF was carried out on PROTEAN IEF Cell (Biorad) at field strength of 600 V/cm and 50 mA/IPG strip. The strips were focused at 250 V for 20 mins, 8000 V for 2 hours 30 mins with linear voltage amplification, and finally to 20,000 volt hour with rapid amplification. Following IEF, the strips were reduced with 135 mM DTT in 4 mL of equilibration buffer (20% (v/v) glycerol, 0.375 M Tris-Cl, 6 M urea, 2% (w/v) SDS, pH 8.8) for 15 mins and alkylated with 135 mM iodoacetamide in 4 mL equilibration buffer for 15 mins. The 2DE was

performed using 12% polyacrylamide gels (13.8 cm × 13.0 cm × 1 mm) in an AE-6200 Slab Electrophoresis Chamber (Atto Biosciences and Technology, China) at constant volt (200 V) for 3 hours 30 mins in Tris glycine-SDS running buffer. All 2DE gel separation was performed in triplicates for all the methods. The gels were stained with 0.1% (w/v) coomassie brilliant blue R-250 (Sigma) overnight, destained, and stored in 5% acetic acid at 4°C for further analysis.

*2.5. Image Analysis of 2D PAGE Gels.* Coomassie stained 2-D gels were visualized using Versa Doc (Model 4000) Imaging System (Biorad) and analyzed with PD Quest Advanced 2-D Analysis software (version 8.0.1, Biorad). Spots were detected automatically by the Spot Detection Parameter Wizard using the Gaussian model with standard parameters. Comparison between spot quantities across gels was performed accurately, and normalization was done using local regression model. Only spots present in each of the three replicate gels, with high and low intensity, were randomly chosen for subsequent analyses. Selected protein spots were subjected to in-gel digestion for identification by MALDI-TOF MS and MS/MS analyses.

*2.6. MALDI-TOF MS and MS/MS Analysis and Database Search.* Spots were excised from protein gels, and in-gel digestion was performed as described by Shevchenko et al.

TABLE 1: Protein yield/fresh weight of root tissue ($\mu$g/gm) using Bradford method.

| Methods | Protein yield ($\mu$g/gm) |
|---|---|
| PSWS | 603 ± 6.08 |
| PSWOS | 406 ± 5.77 |
| PEWS | 302 ± 5.51 |
| PEWOS | 408 ± 7.64 |
| TCA | 73 ± 2 |

TABLE 2: Total number of spots using different methods.

| Methods | Average number of spots |
|---|---|
| PSWS | 446 ± 9.07 |
| PSWOS | 287 ± 6.43 |
| PEWS | 338 ± 6.11 |
| PEWOS | 348 ± 1.53 |

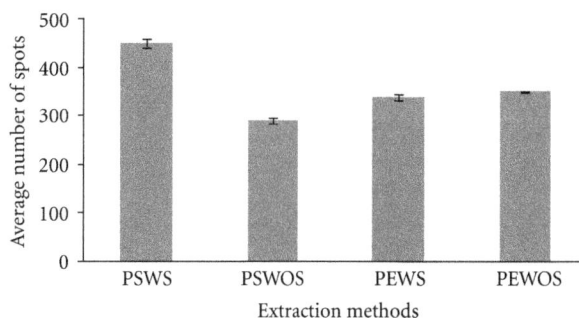

FIGURE 2: A comparative graphical representation showing the average number of protein spots detected in 2DE gels using PSWS, PSWOS, PEWS, and PEWOS protein extraction protocols.

with minor modifications [17]. Proteins were digested in gel using porcine trypsin (Promega) and were extracted using 25% acetonitrile and 1% trifluoroacetic acid. One microlitre of sample and matrix ($\alpha$-cyano-4-hydroxy cinnamic acid, HCCA) (Bruker, Daltonics) was loaded in a Anchor Chip MALDI Plate (Bruker, Daltonics).

Mass spectra were obtained on an Autoflex II MALDI TOF/TOF (Bruker, Daltonics, Germany) mass spectrometer equipped with a pulsed nitrogen laser ($\lambda$-337 nm, 50 Hz). Then the spectra were analysed with Flex Analysis Software (version 2.4, Bruker, Daltonics). The processed spectra were then searched using MS Biotools (version 3.0) program, against the taxonomy of Viridiplantae (green plants) in the MSDB database using MASCOT search engine (version 2.2). The peptide mass fingerprinting parameters included peptide mass tolerance ($\leq$100 ppm); proteolytic enzyme (trypsin); global modification (carbamidomethyl, Cys); variable modification (oxidation, Met); peptide charge state (1+) and maximum missed cleavage 1. The significance threshold was set to a minimum of 95% ($P \leq 0.05$). The criteria used to accept protein identification were based on molecular weight search (MOWSE) score, the percentage of the sequence coverage, and match with minimum five peptides. MS/MS was performed to confirm the identification with matched peptides, selected on the basis of suitability for fragmentation (signal strength and relative isolation).

## 3. Results

### 3.1. Protein Quantification

*3.1.1. TCA-Acetone Precipitation Method.* Protein yield using the classical TCA-acetone precipitation method was extremely low (data not shown). However a modification of incubating the sample at −20°C for 60 minutes in-between the rinsing step yielded a measurable amount of protein. Approximately seventy-three micrograms of protein were obtained from one gram of root tissue using this method (Table 1). However, when the obtained protein was subjected

to electrophoresis in SDS PAGE (polyacrylamide gel electrophoresis) gel, no banding profile was visualized (data not shown). Hence, this protocol was eliminated from further downstream analysis.

*3.1.2. Phenol-Based Methods.* In case of phenol-based methods, protein yields obtained from PSWS, PSWOS, PEWS, and PEWOS were 600 $\mu$g, 406 $\mu$g, 408 $\mu$g, and 300 $\mu$g, respectively, (Table 1). One gram of fresh chickpea roots yielded maximum amount of protein with PSWS method as compared to protein obtained by methods PSWOS, PEWS, and PEWS.

*3.2. Data Analysis of 2DE Gels.* The 2DE patterns of extracted protein when compared with equal amount of initial protein load revealed that protein extracted by PSWS method displayed a comparatively good resolution with lesser contamination, whereas proteins extracted with methods PSWOS, PEWS and PEWOS resolved fewer protein spots (Figure 2). Approximately 446 detectable spots (as estimated by PD Quest software) were obtained by PSWS method while 287 spots by PSWOS method, 338 by PEWS, and 348 by PEWOS method were detected (Table 2). The number of spots described in Table 2 is the average number of spots across the triplicates. In addition we also found that many spots were diffused or absent in these methods (PSWOS, PEWS, PEWOS) as indicated in the marked areas (Figures 3A, 3B, 3C, and 3D). Intensities of all the spots randomly selected for downstream MS and MS/MS were more in PSWS method as compared to other methods (Figures 4 and 5).

*3.3. MALDI-TOF MS and MS/MS Analysis for Protein Identification.* All the 9 spots selected for MALDI analysis (Figures 4 and 5), consisting of both less abundant (sp 36, 80, 212) and more abundant (sp 19, 55, 109, 165, 248, 267) proteins, were successfully identified and listed in Table 3 (Figure 6). Data listed in the table include assigned spot number, spot identity, protein identity (MSDB database), number of peptide matches, sequence coverage (%), MOWSE score, accession number, experimental and theoretical molecular weight and pI.

TABLE 3: Proteins identified by MALDI-TOF MS analyses.

| S no. | Spot ID. | Protein identity | Peptides matched | Sequence coverage (%) | MOWSE score | Accession number (NCBI) | Mr(kDa)/pI experimental (theoretical) | Plant species |
|---|---|---|---|---|---|---|---|---|
| 1 | sp 165 | NADP specific isocitrate dehydrogenase | 10 | 17% | 70 | Q9XGU7_ORYSA | 46.4/6.29 (46.0/6.0) | *Oryza sativa* |
| 2 | sp 212 | Glyceraldehyde 3 phosphate dehydrogenase | 9 | 24% | 86 | Q6K5G8_ORYSA | 36.716/7.68 (37/6.5) | *Oryza sativa* |
| 3 | sp 109 | Triose phosphate isomerase | 6 | 20% | 71 | Q38IW8_SOYBN | 27.4/5.87 (25/5.5) | *Glycine max* |
| 4 | sp 55 | Fructokinase-like protein | 9 | 40% | 94 | Q8LPE5_CICAR | 26.26/5.03 (35.5, 4.5) | *Cicer arietinum* |
| 5 | sp 36 | ATP synthase (subunit D chain) | 13 | 36% | 88 | ATPQ_ARATH | 19.4/5.09 (20/5.0) | *Arabidopsis thaliana* |
| 6 | sp 267 | Porin of Pea, channel protein | 2 | 11% | 134 | T12558 | 29.7/8.56 (30/9.5) | *Phaseolus coccineus* |
| 7 | sp 19 | Plasma membrane intrinsic polypeptide | 10 | 38% | 74 | Q9SMK5_CICAR | 23.3 /4.95 (24.5/5.0) | *Cicer arietinum* |
| 8 | sp 248 | Unidentified protein | 11 | 35% | 80 | CAA06491 | 22.12/9.91 (44.0,9.0) | *Cicer arietinum* |
| 9 | sp 80 | Putative pyruvate dehydrogenase E1 beta subunit isoform 1 protein | 2 | 6% | 55 | Q6Z1G7_ORYSA | 40.2/5.25 (38.5/5.3) | *Oryza sativa* |

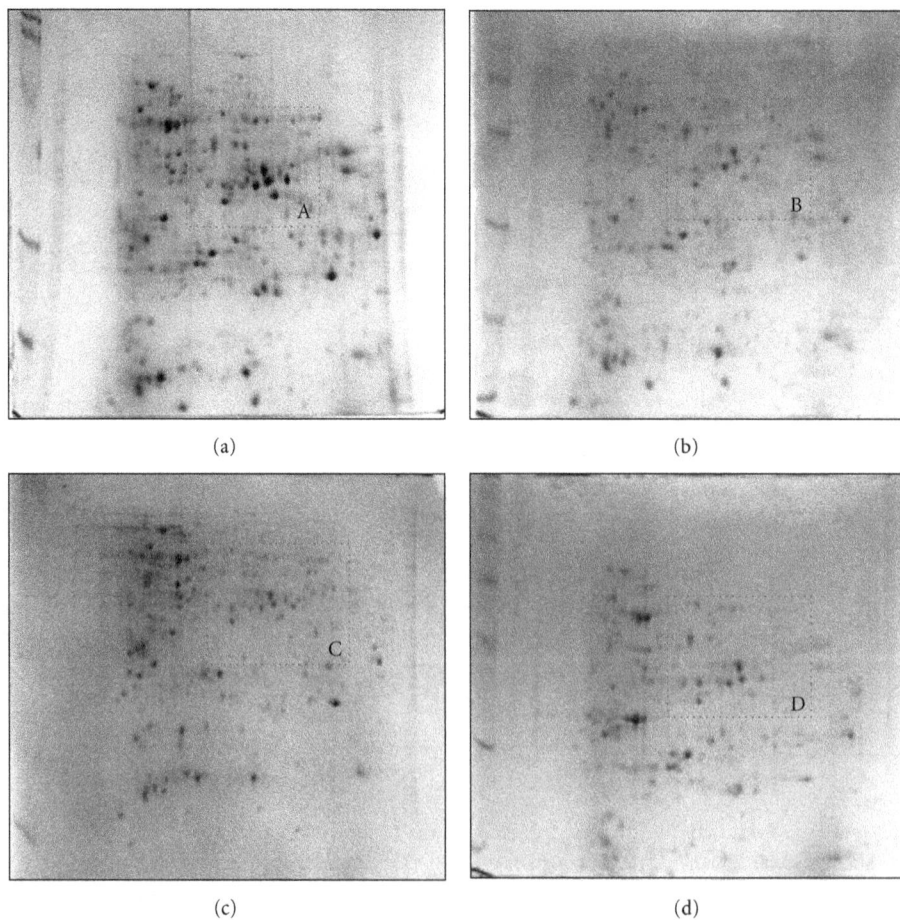

(a)

(b)

(c)

(d)

FIGURE 3: 2DE profiles of chickpea root proteins of JG 62. Profile of proteins isolated using PSWS (a), PSWOS (b), PEWS (c), and PEWOS (d) extraction protocols. Inset A, B, C, D represents a close-up view of an area showing spot resolution: in PSWS (a), PSWOS (b), PEWS (c), and PEWOS (d), respectively.

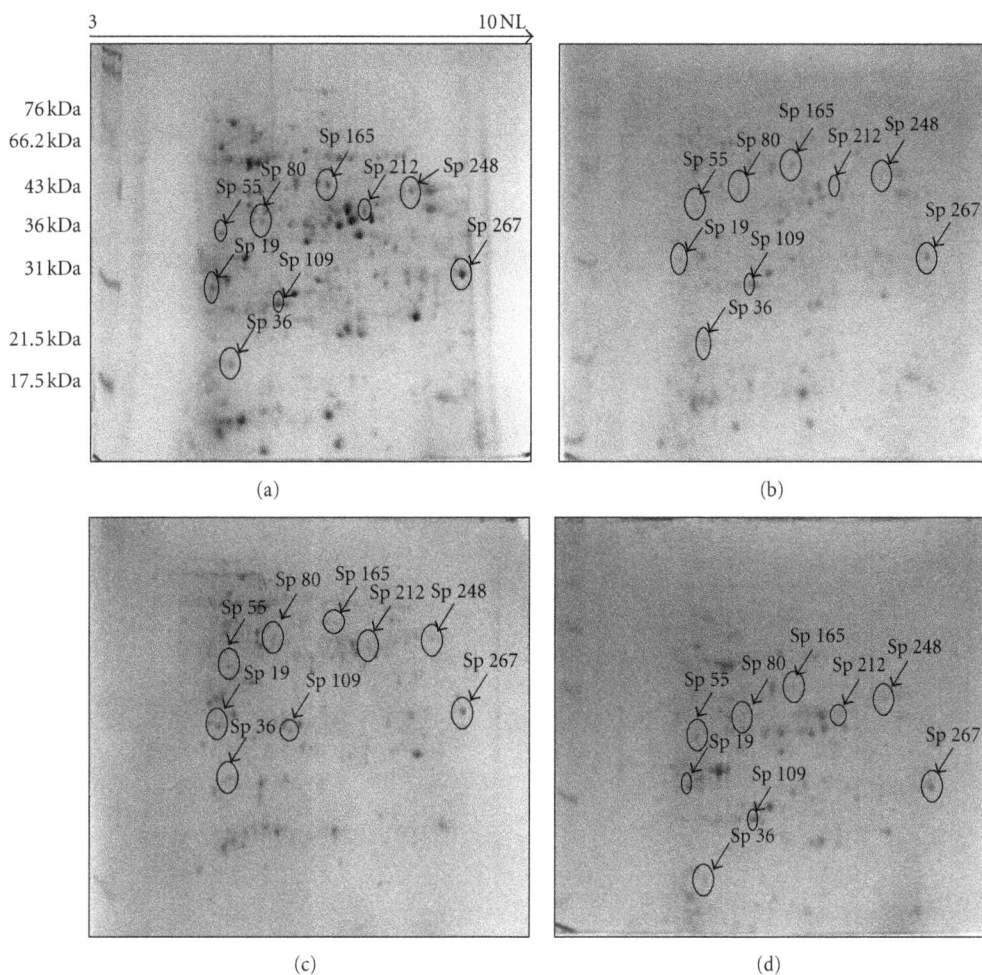

FIGURE 4: 2DE profiles with marked spots selected for MALDI-TOF MS and MS/MS. (a) 2DE profile using PSWS, (b) 2DE profile using PSWOS, (c) 2DE profile using PEWS, and (d) 2DE profile using PEWOS.

## 4. Discussion

Secondary metabolites are known to play important role in structural composition and defense of plants. These metabolites accumulate in various soluble forms in vacuoles and cause severe interference in protein extraction as well as separation in 2DE gels [18, 19]. Chickpea roots are rich in phenolic compounds like tannic acid, gallic acid, o-coumaric acid, chlorogenic acid, cinnamic acid; flavanoids, isoflavanoids like daidzein, genistein, as well as tannins, lignins, and carbohydrates [20, 21]. These compounds form hydrogen bonds with proteins. Besides they also form irreversible complexes with proteins by oxidation and covalent condensation which leads to charge heterogeneity resulting in streaking of gels [22]. Carbohydrates block gel pores causing precipitation and prolonged focusing time, which also results in loss of protein spots and streaks in the gels [15]. Although the amount of these secondary metabolites is comparatively low in etiolated tissues like roots, but low protein content and limiting tissue amounts demand for a competent protein extraction method. In our study TCA-acetone method and phenol-based method using two different extraction buffers

(SDS buffer and extraction buffer without SDS) with and without sonication were evaluated. Comparison was done on the basis of protein yield, spot focusing, resolution, number of resolved spots, and also intensities of the spot and their downstream analysis using high throughput technology (MALDI/MS) of the optimized method.

Quantitative comparison of protein extracts revealed that phenol-based methods gave higher protein yield as compared to TCA-acetone method. The major reason for low protein yield in TCA-acetone method which constrained it for further downstream processing could probably be attributed to the insolubility of protein pellet in IEF buffer as compared to phenol-based methods [23]. Moreover TCA-acetone protocol is known to be effective with tissues from young plants and was found not to be the best choice for more complex tissues [5, 10, 15].

In case of phenol extraction, the proteins were first homogenized in two different extraction buffers; both the buffers contained sucrose which was added to create phase inversion. These buffers formed the aqueous lower phase containing carbohydrates, nucleic acid, insoluble cell debris, while the upper phenol phase contained cytosolic and

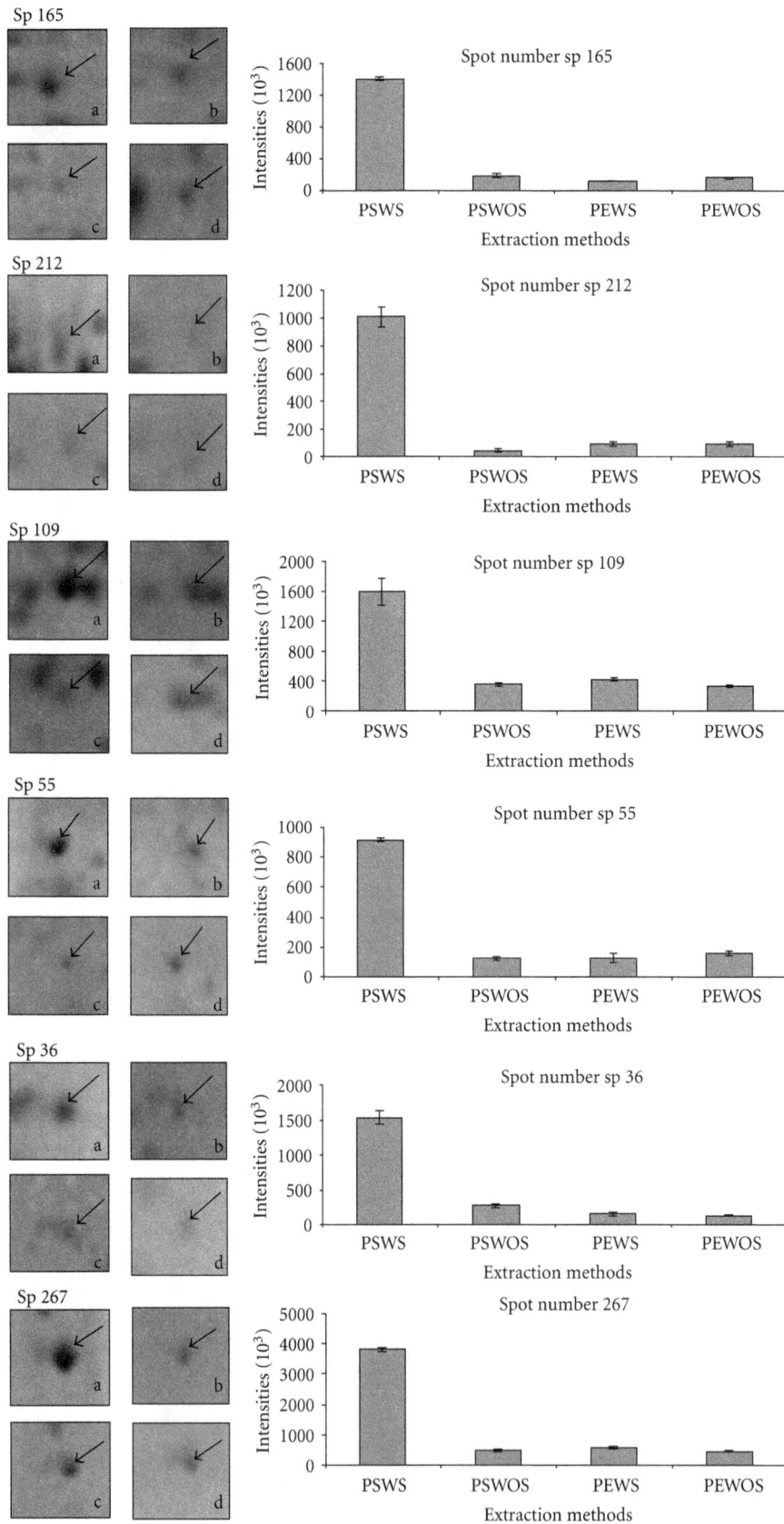

FIGURE 5: Continued.

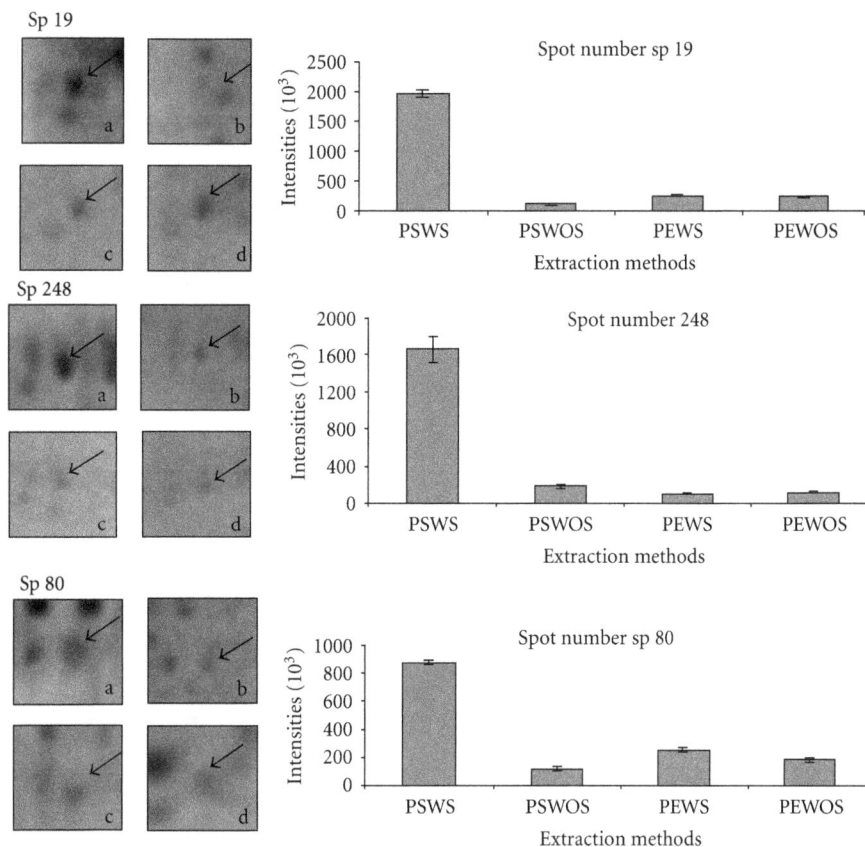

FIGURE 5: 2DE gel profiles showing individual spots and their relative intensities in graphical form using PSWS, PSWOS, PEWS, and PEWOS protein extraction protocols. (a), (b), (c), (d) represent the spot obtained by PSWS, PSWOS, PEWS, and PEWOS, respectively.

membrane proteins, lipids, and pigment [15]. SDS buffer contained about 30% sucrose which helped in better phase separation as compared to extraction buffer (24%). The high pH buffers inhibit common activity of the proteases [24] and cause ionization of phenolic compounds, thus preventing them from forming hydrogen bonding with the protein [22]. It also neutralizes the acids that are released by disrupted vacuoles. PMSF and $\beta$-mercaptoethanol which were used in both buffers in the present study were reported to irreversibly inhibit serine protease action and act as a reducing agent which prevents protein oxidation, respectively. KCl and EDTA were used in case of extraction buffer without SDS (PEWS and PEWOS). KCl facilitates the extraction of proteins by its salting in effect and EDTA inhibits metalloprotease and polyphenoloxidase by chelating metal ions [15]. Although the salting in effect or chelation of metal ions could not improve the protein yield as compared to SDS buffer with sonication, SDS is known to act as an excellent solubilizing agent, which allows the recovery of membrane-bound proteins [10]. The solubilization of protein was found to increase with sonication as evident from the increase in protein yield and spot resolution after sonication in PSWS compared to PSWOS. Sonication results in better disruption of cell membrane and release of intracellular proteins and thus provides explanation for SDS to have efficiently solubilized the protein in PSWS method. In contrary, in case of extraction buffer, sonication could not improve protein yield

or resolution, presumably due to the interference with constituents of buffer (KCl or EDTA) or due to lack of better solubilizing agent like SDS and/or both.

The phenol used in this method was buffered to pH 8.0 to ensure that nucleic acids are partitioned to the buffer phase and not to phenol-rich phase [25], and thus proteins in phenol phase were purified and concentrated simultaneously by subsequent methanol ammonium acetate precipitation. Phenol acts as one of the strongest dissociaters known to decrease molecular interaction between proteins and other materials [15]. It can minimize protein degradation resulting from endogenous proteolytic activity [26]. Phenol extraction method though with high clean-up capacity has a little tendency to dissolve polysaccharides and nucleic acids.

We found that in PSWS method the spots obtained were well resolved and showed high intensity (Figures 3 and 5) as compared to PSWOS, PEWS, and PEWOS. About 25% unique spots were obtained in PSWS and the rest 75% spots though existed in PSWOS, PEWS, and PEWOS, however, resolved with variable clarity. Streaking was absent in all the gels. We could see that the difference in number of spots between PSWS and PSWOS was more as compared to PEWS and PEWOS, which confirmed that the effectivity of SDS increased in presence of sonication. However in the latter case (PEWS, PEWOS) sonication did not have much influence.

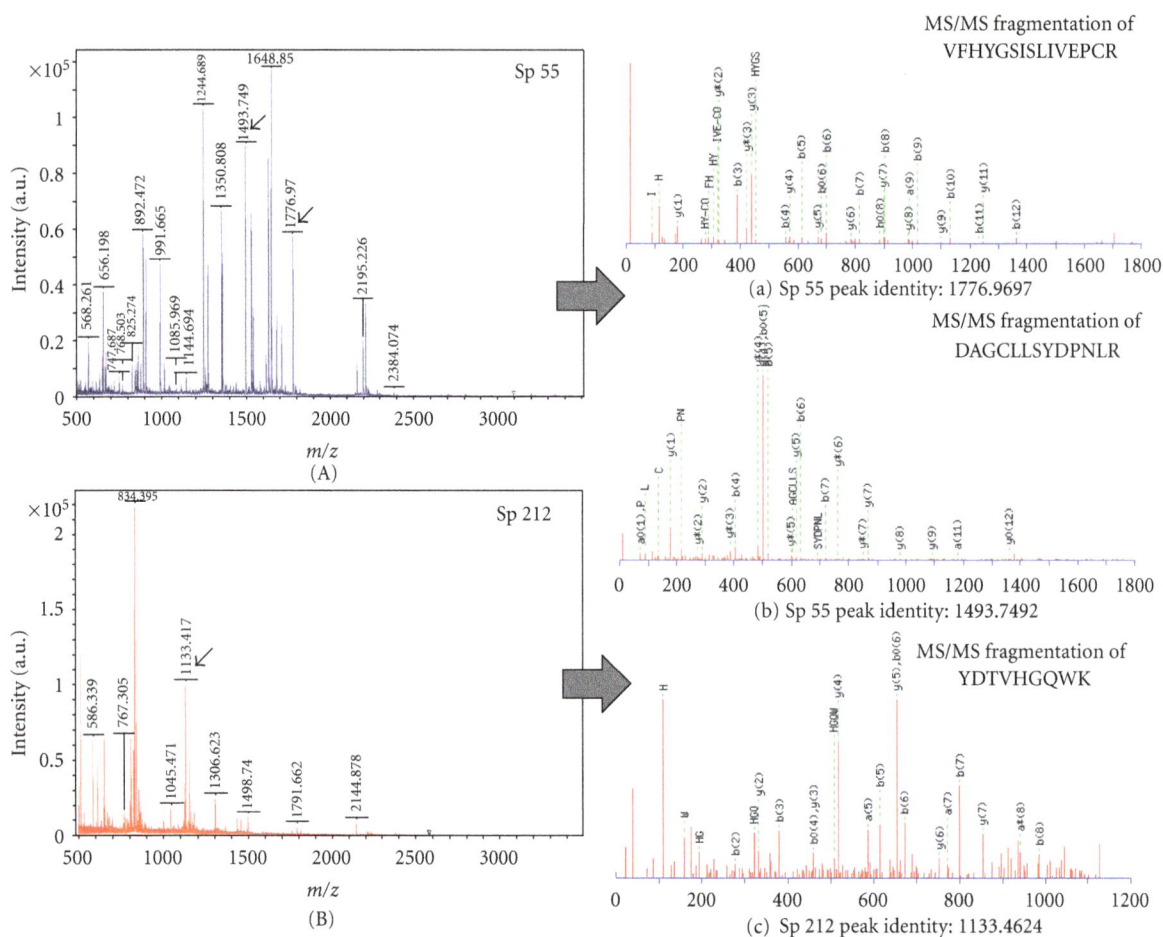

Figure 6: Spectral profiles obtained by MALDI-TOF MS and MS/MS. (A) MALDI spectra of sp 55 and (a), (b), show MS/MS spectra of two selected peaks of sp 55 (1776.9697 and 1493.7492). (B) MALDI spectra of sp 212 and (c) shows MS/MS spectra of the selected peak of sp 212 (1133.4624).

Improvisation of the extraction buffer was also made by adding 2% SDS, which resulted in precipitation. Interference between constituents of the extraction buffer and SDS was assumed to be the cause of such precipitation. However further experimentation needs to be performed for confirmation of such predictions.

All protein spots selected for MALDI-TOF/MS and MS/MS from PSWS resulted in successful identification. High intense spot like sp 55, (fructokinase-like protein) and less intense spot like sp 212, (glyceraldehyde 3-phosphate dehydrogenase) both resulted in high quality spectra with low background noise (Figure 6). These results further indicated the compatibility of PSWS method with both MS and MS/MS and its reliability for downstream processing.

## 5. Conclusion

The present study emphasizes PSWS as the optimized phenol-based method for chickpea root protein extraction. This method successfully isolated high quality protein suitable for downstream processing. Hence, the data obtained projects this protocol as an effective and efficient one that could be applied for other recalcitrant leguminous root tissues as well. Nevertheless, it should be kept in mind that one generalized protein extraction protocol applicable for global protein profiling of variable tissues irrespective of their origins though theoretically conceivable, but fails to meet practical feasibility.

## Acknowledgments

M. Chatterjee is thankful to NMTILI, Council of Scientific and Industrial Research for financial support. S. Gupta is thankful to Department of Biotechnology, Government of India for financial assistance. A. Bhar is also thankful to Council of Scientific and Industrial Research for financial support. Besides, all the authors are thankful to Bose Institute for infrastructure. Special thanks are offered to Rajesh Vashisth (Bruker Daltonics) for providing technical help for conducting mass spectrometry. The help provided by the Central Instrumentation Facility, Bose Institute on proteomic services is duly acknowledged. The authors would also like to thank International Crops Research Institute for the Semi Arid Tropics (ICRISAT), Patancheru, Andhra Pradesh, for seeds. Finally the authors thank Mr. Arup Kumar Dey for providing backup support.

Optimization of an Efficient Protein Extraction Protocol Compatible with Two-Dimensional Electrophoresis and Mass
Spectrometry from Recalcitrant Phenolic Rich Roots of Chickpea (Cicer arietinum L.)

201

# References

[1] J. K. C. Rose, S. Bashir, J. J. Giovannoni, M. M. Jahn, and R. S. Saravanan, "Tackling the plant proteome: practical approaches, hurdles and experimental tools," *The Plant Journal*, vol. 39, no. 5, pp. 715–733, 2004.

[2] P. Gegenheimer, "Preparation of extracts from plants," *Methods in Enzymology*, vol. 182, pp. 174–193, 1990.

[3] A. Tsugita and M. Kamo, "2-D electrophoresis of plant proteins," *Methods in Molecular Biology*, vol. 112, pp. 95–97, 1999.

[4] W. Wang, F. Tai, and S. Chen, "Optimizing protein extraction from plant tissues for enhanced proteomics analysis," *Journal of Separation Science*, vol. 31, no. 11, pp. 2032–2039, 2008.

[5] R. S. Saravanan and J. K. C. Rose, "A critical evaluation of sample extraction techniques for enhanced proteomic analysis of recalcitrant plant tissues," *Proteomics*, vol. 4, no. 9, pp. 2522–2532, 2004.

[6] C. Damerval, D. D. Vienne, M. Zivy, and H. Thiellement, "Technical improvements in two-dimensional electrophoresis increase the level of genetic variation detected in wheat-seedling proteins," *Electrophoresis*, vol. 7, no. 1, pp. 52–54, 1986.

[7] W. J. Hurkman and C. K. Tanaka, "Solubilization of plant membrane proteins for analysis by two-dimensional gel electrophoresis," *Plant Physiology*, vol. 81, no. 3, pp. 802–806, 1986.

[8] Y. Meyer, J. Grosset, Y. Chartier, and J. C. Cleyet-Marel, "Preparation by two-dimensional electrophoresis of proteins for antibody production: antibodies against proteins whose synthesis is reduced by auxin in tobacco mesophyll protoplasts," *Electrophoresis*, vol. 9, no. 11, pp. 704–712, 1988.

[9] K. V. Mijnsbrugge, H. Meyermans, M. Van Montagu, G. Bauw, and W. Boerjan, "Wood formation in poplar: identification, characterization, and seasonal variation of xylem proteins," *Planta*, vol. 210, no. 4, pp. 589–598, 2000.

[10] W. Wang, M. Scali, R. Vignani et al., "Protein extraction for two-dimensional electrophoresis from olive leaf, a plant tissue containing high levels of interfering compounds," *Electrophoresis*, vol. 24, no. 14, pp. 2369–2375, 2003.

[11] T. Isaacson, C. M. B. Damasceno, R. S. Saravanan et al., "Sample extraction techniques for enhanced proteomic analysis of plant tissues," *Nature Protocols*, vol. 1, no. 2, pp. 769–774, 2006.

[12] T. J. Raharjo, I. Widjaja, S. Roytrakul, and R. Verpoorte, "Comparative proteomics of *Cannabis sativa* plant tissues," *Journal of Biomolecular Techniques*, vol. 15, no. 2, pp. 97–106, 2004.

[13] A. M. Schuster and E. Davies, "Ribonucleic acid and protein metabolism in pea epicotyls: II. Response to wounding in aged tissue," *Plant Physiology*, vol. 73, no. 3, pp. 817–821, 1983.

[14] S. Gupta, D. Chakraborti, A. Sengupta, D. Basu, and S. Das, "Primary metabolism of chickpea is the initial target of wound inducing early sensed *Fusarium oxysporum* f. sp. *ciceri* race I," *PLoS ONE*, vol. 5, no. 2, Article ID e9030, 2010.

[15] S. C. Carpentier, E. Witters, K. Laukens, P. Deckers, R. Swennen, and B. Panis, "Preparation of protein extracts from recalcitrant plant tissues: an evaluation of different methods for two-dimensional gel electrophoresis analysis," *Proteomics*, vol. 5, no. 10, pp. 2497–2507, 2005.

[16] M. M. Bradford, "A rapid and sensitive method for the quantitation of microgram quantities of protein utilizing the principle of protein-dye binding," *Analytical Biochemistry*, vol. 72, no. 1-2, pp. 248–254, 1976.

[17] A. Shevchenko, H. Tomas, J. Havliš, J. V. Olsen, and M. Mann, "In-gel digestion for mass spectrometric characterization of proteins and proteomes," *Nature Protocols*, vol. 1, no. 6, pp. 2856–2860, 2007.

[18] F. Granier, "Extraction of plant proteins for two-dimensional electrophoresis," *Electrophoresis*, vol. 9, no. 11, pp. 712–718, 1988.

[19] C. M. Vâlcu and K. Schlink, "Reduction of proteins during sample preparation and two-dimensional gel electrophoresis of woody plant samples," *Proteomics*, vol. 6, no. 5, pp. 1599–1605, 2006.

[20] S. Maurya, U. P. Singh, D. P. Singh, K. P. Singh, and J. S. Srivastava, "Secondary metabolites of chickpea (*Cicer arietinum*) and their role in pathogenesis after infection by *Sclerotium rolfsii*," *Journal of Plant Diseases and Protection*, vol. 112, no. 2, pp. 118–123, 2005.

[21] M. Chérif, A. Arfaoui, and A. Rhaiem, "Phenolic compounds and their role in bio-control and resistance of chickpea to fungal pathogenic attacks," *Tunisian Journal of Plant Protection*, vol. 2, no. 1, pp. 7–21, 2007.

[22] W. D. Loomis and J. Battaile, "Plant phenolic compounds and the isolation of plant enzymes," *Phytochemistry*, vol. 5, no. 3, pp. 423–438, 1966.

[23] S. X. Chen and A. C. Harmon, "Advances in plant proteomics," *Proteomics*, vol. 6, no. 20, pp. 5504–5516, 2006.

[24] D. F. Hochstrasser, M. G. Harrington, A. C. Hochstrasser, M. J. Miller, and C. R. Merril, "Methods for increasing the resolution of two-dimensional protein electrophoresis," *Analytical Biochemistry*, vol. 173, no. 2, pp. 424–435, 1988.

[25] A. Pusztai, "Interactions of proteins with other polyelectrolytes in a two-phase system containing phenol and aqueous buffers at various pH values," *Biochemical Journal*, vol. 99, no. 1, pp. 93–101, 1966.

[26] A. Schuster and E. Davies, "Ribonucleic acid and protein metabolism in pea epicotyls II. Response to wounding in aged tissue," *Plant Physiology*, vol. 73, no. 3, pp. 817–821, 1983.

# Permissions

The contributors of this book come from diverse backgrounds, making this book a truly international effort. This book will bring forth new frontiers with its revolutionizing research information and detailed analysis of the nascent developments around the world.

We would like to thank all the contributing authors for lending their expertise to make the book truly unique. They have played a crucial role in the development of this book. Without their invaluable contributions this book wouldn't have been possible. They have made vital efforts to compile up to date information on the varied aspects of this subject to make this book a valuable addition to the collection of many professionals and students.

This book was conceptualized with the vision of imparting up-to-date information and advanced data in this field. To ensure the same, a matchless editorial board was set up. Every individual on the board went through rigorous rounds of assessment to prove their worth. After which they invested a large part of their time researching and compiling the most relevant data for our readers. Conferences and sessions were held from time to time between the editorial board and the contributing authors to present the data in the most comprehensible form. The editorial team has worked tirelessly to provide valuable and valid information to help people across the globe.

Every chapter published in this book has been scrutinized by our experts. Their significance has been extensively debated. The topics covered herein carry significant findings which will fuel the growth of the discipline. They may even be implemented as practical applications or may be referred to as a beginning point for another development. Chapters in this book were first published by Hindawi Publishing Corporation; hereby published with permission under the Creative Commons Attribution License or equivalent.

The editorial board has been involved in producing this book since its inception. They have spent rigorous hours researching and exploring the diverse topics which have resulted in the successful publishing of this book. They have passed on their knowledge of decades through this book. To expedite this challenging task, the publisher supported the team at every step. A small team of assistant editors was also appointed to further simplify the editing procedure and attain best results for the readers.

Our editorial team has been hand-picked from every corner of the world. Their multi-ethnicity adds dynamic inputs to the discussions which result in innovative outcomes. These outcomes are then further discussed with the researchers and contributors who give their valuable feedback and opinion regarding the same. The feedback is then collaborated with the researches and they are edited in a comprehensive manner to aid the understanding of the subject.

Apart from the editorial board, the designing team has also invested a significant amount of their time in understanding the subject and creating the most relevant covers. They scrutinized every image to scout for the most suitable representation of the subject and create an appropriate cover for the book.

The publishing team has been involved in this book since its early stages. They were actively engaged in every process, be it collecting the data, connecting with the contributors or procuring relevant information. The team has been an ardent support to the editorial, designing and production team. Their endless efforts to recruit the best for this project, has resulted in the accomplishment of this book. They are a veteran in the field of academics and their pool of knowledge is as vast as their experience in printing. Their expertise and guidance has proved useful at every step. Their uncompromising quality standards have made this book an exceptional effort. Their encouragement from time to time has been an inspiration for everyone.

The publisher and the editorial board hope that this book will prove to be a valuable piece of knowledge for researchers, students, practitioners and scholars across the globe.

# List of Contributors

**Rovshan G. Sadygov**
Sealy Center for Molecular Medicine, Department of Biochemistry and Molecular Biology, The University of Texas Medical Branch, Galveston, TX 77555, USA

**Shintaro Kikkawa and Osamu Yokosuka**
Department of Medicine and Clinical Oncology, Graduate School of Medicine, Chiba University, 1-8-1 Inohana, Chuo-ku, Chiba, Chiba City 260-8670, Japan

**Kazuyuki Sogawa and Mamoru Satoh**
Clinical Proteomics Center, Chiba University Hospital, 1-8-1 Inohana, Chuo-ku, Chiba, Chiba City 260-8670, Japan

**Fumio Nomura**
Clinical Proteomics Center, Chiba University Hospital, 1-8-1 Inohana, Chuo-ku, Chiba, Chiba City 260-8670, Japan
Department of Molecular Diagnosis, Graduate School of Medicine, Chiba University, 1-8-1 Inohana, Chuo-ku, Chiba, Chiba City 260-8670, Japan

**Hiroshi Umemura and Kazuyuki Matsushita**
Department of Molecular Diagnosis, Graduate School of Medicine, Chiba University, 1-8-1 Inohana, Chuo-ku, Chiba, Chiba City 260-8670, Japan

**Yoshio Kodera**
Clinical Proteomics Center, Chiba University Hospital, 1-8-1 Inohana, Chuo-ku, Chiba, Chiba City 260-8670, Japan
Department of Physics, School of Science, Kitasato University, 1-15-1 Kitasato, Minami-ku, Kanagawa, Sagamihara City 228-8555, Japan

**Takeshi Tomonaga**
Clinical Proteomics Center, Chiba University Hospital, 1-8-1 Inohana, Chuo-ku, Chiba, Chiba City 260-8670, Japan
Laboratory of Proteome Research, National Institute of Biomedical Innovation, 7-6-8 Saito Asagi, Osaka, Ibaraki City 567-0085, Japan

**Masaru Miyazaki**
Department of General Surgery, Graduate School of Medicine, Chiba University, 1-8-1 Inohana, Chuo-ku, Chiba, Chiba City 260-8670, Japan

**Roslyn N. Brown, Jea H. Park, Brooke L. Deatherage, Boyd L. Champion, Richard D. Smith and Joshua N. Adkins**
Biological Sciences Division, Pacific Northwest National Laboratory, 902 Battelle Boulevard, Richland, WA 99352, USA

**James A. Sanford**
Biomedical Sciences Graduate Program, University of California San Diego, 9500 Gilman Dive, La Jolla, CA 92063, USA

**Fred Heffron**
Department of Molecular Microbiology and Immunology, Oregon Health and Science University, 3181 SW Sam Jackson Park Road, Portland, OR 97239, USA

**Alessandra Tessitore, Agata Gaggiano, Germana Cicciarelli, Daniela Verzella, Daria Capece, Mariafausta Fischietti, Francesca Zazzeroni and Edoardo Alesse**
Department of Biotechnological and Applied Clinical Sciences, University of L'Aquila, Via Vetoio Coppito 2, 67100 L'Aquila, Italy

**Karen de Morais-Zani and Anita Mitico Tanaka-Azevedo**
Laboratorio de Herpetologia, Instituto Butantan, Avenida Vital Brazil 1500, 05503-900 Sao Paulo, SP, Brazil
Programa de Pos-Graduac ao Interunidades em Biotecnologia, Universidade de Sao Paulo, Avenida Professor Lineu Prestes 2415, 05508-900 Sao Paulo, SP, Brazil

**Kathleen Fernandes Grego**
Laboratorio de Herpetologia, Instituto Butantan, Avenida Vital Brazil 1500, 05503-900 Sao Paulo, SP, Brazil

**Aparecida Sadae Tanaka**
Departamento de Bioqu´ımica, Universidade Federal de Sao Paulo, Rua Tres de Maio 100, 04044-020 Sao Paulo, SP, Brazil

**Yunki Yau**
Bioanalytical Mass Spectrometry Facility, Mark Wainwright Analytical Centre, The University of New South Wales, Sydney, NSW2052, Australia
Department of Gastroenterology and Liver Services, Concord Repatriation General Hospital, Sydney, NSW2139, Australia

**Valerie C. Wasinger and Ming Zeng**
Bioanalytical Mass Spectrometry Facility, Mark Wainwright Analytical Centre, The University of New South Wales, Sydney, NSW2052, Australia

**John E. Hale**
Hale Biochemical Consulting, 6341 Wyatt Lane, Klamath Falls, OR 97601, USA

**Chandra Kirana, Hongjun Shi, Kylie Hood, Mark Hayes and Richard Stubbs**
Wakefield Biomedical Research Unit, University of Otago, Wellington 6242, New Zealand

**Emma Laing**
School of Medicine and Health Sciences, University of Otago, Wellington 6242, New Zealand

**Rose Miller and Peter Bethwaite**
Department of Pathology and Molecular Medicine, University of Otago, Wellington 6242, New Zealand

**John Keating**
Capital & Coast District Health Board, Wellington Hospital, Wellington 6021, New Zealand

**T. William Jordan**
Centre for Biodiscovery, School of Biological Sciences, Victoria University of Wellington, Wellington 6012, New Zealand

**Jochen M. Schwenk**
NMI Natural and Medical Sciences Institute at the University of Tubingen, Markwiesenstraße 55, 72770 Reutlingen, Germany
Science for Life Laboratory Stockholm, School of Biotechnology, KTH Royal Institute of Technology, P.O. Box 1031, 171 21 Solna, Sweden

**Oliver Poetz and Thomas O. Joos**
NMI Natural and Medical Sciences Institute at the University of Tubingen, Markwiesenstraße 55, 72770 Reutlingen, Germany

**Robert Zeillinger**
Molecular Oncology Group, Department of Obstetrics and Gynecology, Medical University of Vienna, Wahringer Gurtel 18-20, 5Q, 1090 Vienna, Austria

**Si Wu, Rui Zhao, Nikola ToliT and Ljiljana Paša-ToliT**
Environmental Molecular Science Laboratory, Pacific Northwest National Laboratory, P.O. Box 999/MS K8-98, Richland, WA 99352, USA

**Roslyn N. Brown**
Center for Bioproducts and Bioenergy, Washington State University, Richland, WA, USA

**Samuel H. Payne, Anil Shukla, Matthew E. Monroe, Ronald J. Moore and Mary S. Lipton**
Biological Sciences Division, Pacific Northwest National Laboratory, Richland, WA, USA

**Da Meng**
Computational Sciences and Mathematics Division, Pacific Northwest National Laboratory, Richland, WA, USA

**Li Cao**
Department of Neurobiology, 720Westview Drive SW, Atlanta, GA, USA

**Martijn W. H. Pinkse and Eda Bener-Aksam**
Department of Biotechnology, Netherlands Proteomics Centre, Delft University of Technology, Julianalaan 67, 2628BC Delft, The Netherlands

**Emanuel Weber**
Department of Biotechnology, Netherlands Proteomics Centre, Delft University of Technology, Julianalaan 67, 2628BC Delft, The Netherlands
Institute of Sensor and Actuator Systems, Vienna University of Technology, Gusshausstrasse 27-29/E366, 1040 Vienna, Austria

**Michael J. Vellekoop**
Institute of Sensor and Actuator Systems, Vienna University of Technology, Gusshausstrasse 27-29/E366, 1040 Vienna, Austria

**Peter D. E. M. Verhaert**
Biomedical Research Institute (BJOMED), Hasselt University, Agoralaan building C, 3590 Diepenbeek, Belgium

**Zhongping Liao**
Greenebaum Cancer Center, University of Maryland School of Medicine, Baltimore, MD 21201, USA

**Stefani N. Thomas**
Department of Pharmacology and Molecular Sciences, Johns Hopkins University School of Medicine, Baltimore, MD 21205, USA

**Yunhu Wan**
Department of Epidemiology and Public Health, University of Maryland School of Medicine, Baltimore, MD 21201, USA

**H. Helen Lin and David K. Ann**
Department of Molecular Pharmacology, Beckman Research Institute, City of Hope Medical Center, Duarte, CA 91010, USA

**Austin J. Yang**
Greenebaum Cancer Center, University of Maryland School of Medicine, Baltimore, MD 21201, USA
Department of Anatomy and Neurobiology, University of Maryland School of Medicine, Baltimore, MD 21201, USA

**Hercules Moura, Adrian R. Woolfitt, Yulanda M. Williamson, Thomas A. Blake, Maria I. Solano and John R. Barr**
Division of Laboratory Sciences, National Center for Environmental Health, Centers for Disease Control and Prevention (CDC), MS F-50, 4770 Buford Hwy NE, Atlanta, GA 30341, USA

**Rebecca R. Terilli**
Division of Laboratory Sciences, National Center for Environmental Health, Centers for Disease Control and Prevention (CDC), MS F-50, 4770 Buford Hwy NE, Atlanta, GA 30341, USA
Association of Public Health Laboratories, Silver Spring, MD 20910, and Oak Ridge Institute for Scientific Education, Oak Ridge, TN 37380, USA

**Glauber Wagner**
Division of Laboratory Sciences, National Center for Environmental Health, Centers for Disease Control and Prevention (CDC), MS F-50, 4770 Buford Hwy NE, Atlanta, GA 30341, USA
Universidade do Oeste de Santa Catarina, 89600 Joacaba, SC, Brazil

**Jonas Bergquist**
Analytical Chemistry, Department of Chemistry, Biomedical Center and SciLife Lab, Uppsala University, P.O. Box 599, 751 24 Uppsala, Sweden

**Gokhan Baykut, Matthias Witt and Franz-Josef Mayer**
Bruker Daltonik GmbH, 28359 Bremen, Germany

**Maria Bergquist**
Analytical Chemistry, Department of Chemistry, Biomedical Center and SciLife Lab, Uppsala University, P.O. Box 599, 751 24 Uppsala, Sweden
Department of Medical Sciences, Hedenstierna Laboratory, Uppsala University, 75185 Uppsala, Sweden

**Doan Baykut**
Institute of Biophysics, University of Frankfurt, 60438 Frankfurt/M, Germany

**Lisandra E. de Castro Bras, Kristine Y. De Leon, Yonggang Ma, Qiuxia Dai and Merry L. Lindsey**
San Antonio Cardiovascular Proteomics Center, The University of Texas Health Science Center at San Antonio, San Antonio, TX 78245, USA
Division of Geriatrics, Gerontology & Palliative Medicine, Department of Medicine, UTHSCSA, San Antonio, TX 78245, USA

**Kevin Hakala and Susan T. Weintraub**
San Antonio Cardiovascular Proteomics Center, The University of Texas Health Science Center at San Antonio, San Antonio, TX 78245, USA
Department of Biochemistry, UTHSCSA, San Antonio, TX 78245, USA

**Francesco Giorgianni and Sarka Beranova-Giorgianni**
Department of Pharmaceutical Sciences, The University of Tennessee Health Science Center, Memphis, TN 38163, USA

**Valentina Mileo and Silvia Catinella**
Corporate Preclinical R&D, Analytics and Early Formulations Department, Chiesi Farmaceutici S.p.A., 43122 Parma, Italy

**Dominic M. Desiderio**
Department of Neurology, The University of Tennessee Health Science Center, Memphis, 38163 TN, USA
Charles B. Stout Neuroscience Mass Spectrometry Laboratory, The University of Tennessee Health Science Center, Memphis, 38163 TN, USA

**Amedeo Cappione III, Janet Smith, Masaharu Mabuchi and Timothy Nadler**
EMD Millipore, Merck KGaA, 17 Cherry Hill Drive, Danvers, MA 01923, USA

**Kiyoshi Yanagisawa**
Institute for Advanced Research, Nagoya University, Furo-cho, Chikusa-ku, Nagoya 464-8601, Japan
Division of Molecular Carcinogenesis, Center for Neurological Diseases and Cancer, Nagoya University Graduate School of Medicine, Nagoya 466-8550, Japan

**Chinatsu Arima, Takashi Takahashi and Shuta Tomida**
Division of Molecular Carcinogenesis, Center for Neurological Diseases and Cancer, Nagoya University Graduate School of Medicine, Nagoya 466-8550, Japan

**Keitaro Matsuo and Kazuo Tajima**
Division of Epidemiology and Prevention, Aichi Cancer Center, Nagoya 464-8681, Japan

**Yoko Kuroyanagi, Toshiyuki Takeuchi and Miyoko Kusumegi**
Division of Research and Development, Oncomics Co., Ltd., Nagoya 464-0858, Japan

**Yukihiro Yokoyama, Takeo Kawahara and Masato Nagino**
5Division of Surgical Oncology, Department of Surgery, Nagoya University Hospital, Nagoya 466-8550, Japan

**Hidemi Goto and Shigeru B. H. Ko**
Department of Gastroenterology, Nagoya University Hospital, Nagoya 466-8550, Japan

**Kenji Yamao and Nobumasa Mizuno**
Department of Gastroenterology, Aichi Cancer Center, Nagoya 464-8681, Japan

**Brian McDonagh, C. Alicia Padilla and Jose Antonio Barcena**
Department of Biochemistry and Molecular Biology, University of Cordoba and IMIBIC, 14071 Cordoba, Spain

**Pablo Martınez-Acedo and Jesus Vazquez**
Cardiovascular Proteomics Laboratory, National Center for Cardiovascular Research, 28026 Madrid, Spain

**David Sheehan**
Department of Biochemistry, University College Cork, Cork, Ireland

**Amita R. Oka, Matthew P. Kuruc, Ketan M. Gujarathi and Swapan Roy**
ProFACT Proteomics Inc., 1 Deer Park Drive, Suite M, Monmouth Junction, NJ 08852, USA

**Moniya Chatterjee, Sumanti Gupta, Anirban Bhar and Sampa Das**
Division of Plant Biology, Bose Institute, Centenary Campus, P 1/12, CIT Scheme VII-M, Kankurgachi, West Bengal, Kolkata 700054, India